Representation Theory and Higher Algebraic K-Theory

A Program of Monographs, Textbooks, and Lecture Notes

E. Hansen and G. W. Walster, Global Optimization Using Interval Analysis, Second Edition, Revised and Expanded (2004)

M. M. Rao, Measure Theory and Integration, Second Edition, Revised and Expanded (2004)

W. J. Wickless, A First Graduate Course in Abstract Algebra (2004)

R. P. Agarwal, M. Bohner, and W-T Li, Nonoscillation and Oscillation Theory for Functional Differential Equations (2004)

J. Galambos and I. Simonelli, Products of Random Variables: Applications to Problems of Physics and to Arithmetical Functions (2004)

Walter Ferrer and Alvaro Rittatore, Actions and Invariants of Algebraic Groups (2005)

Christof Eck, Jiri Jarusek, and Miroslav Krbec, Unilateral Contact Problems: Variational Methods and Existence Theorems (2005)

M. M. Rao, Conditional Measures and Applications, Second Edition (2005)

A. B. Kharazishvili, Strange Functions in Real Analysis, Second Edition (2006)

Vincenzo Ancona and Bernard Gaveau, Differential Forms on Singular Varieties: De Rham and Hodge Theory Simplified (2005)

Santiago Alves Tavares, Generation of Multivariate Hermite Interpolating Polynomials (2005)

Sergio Macías, Topics on Continua (2005)

Mircea Sofonea, Weimin Han, and Meir Shillor, Analysis and Approximation of Contact Problems with Adhesion or Damage (2006)

Marwan Moubachir and Jean-Paul Zolésio, Moving Shape Analysis and Control: Applications to Fluid Structure Interactions (2006)

Alfred Geroldinger and Franz Halter-Koch, Non-Unique Factorizations: Algebraic, Combinatorial and Analytic Theory (2006)

Kevin J. Hastings, Introduction to the Mathematics of Operations Research *Mathematica*®, Second Edition (2006)

Robert Carlson, A Concrete Introduction to Real Analysis (2006)

John Dauns and Yiqiang Zhou, Classes of Modules (2006)

N. K. Govil, H. N. Mhaskar, Ram N. Mohapatra, Zuhair Nashed, and J. Szabados, Frontiers in Interpolation and Approximation (2006)

Luca Lorenzi and Marcello Bertoldi, Analytical Methods for Markov Semigroups (2006)

M. A. Al-Gwaiz and S. A. Elsanousi, Elements of Real Analysis (2006)

R. Sivaramakrishnan, Certain Number-Theoretic Episodes in Algebra (2006)

Aderemi Kuku, Representation Theory and Higher Algebraic K-Theory (2006)

Representation Theory and Higher Algebraic K-Theory

Aderemi Kuku

International Centre for Theoretical Physics
Trieste, Italy

CRC Press
Taylor & Francis Group
Boca Raton London New York

CRC Press is an imprint of the
Taylor & Francis Group, an **informa** business

A CHAPMAN & HALL BOOK

CRC Press
Taylor & Francis Group
6000 Broken Sound Parkway NW, Suite 300
Boca Raton, FL 33487-2742

First issued in paperback 2019

ISBN-13: 978-1-58488-603-7 (hbk)
ISBN-13: 978-0-367-39030-3 (pbk)

**Visit the Taylor & Francis Web site at
http://www.taylorandfrancis.com**

**and the CRC Press Web site at
http://www.crcpress.com**

Dedicated to Funke

Contents

Introduction

A representation of a discrete group G in the category $\mathcal{P}(F)$ of finite dimensional vector spaces over a field F could be defined as a pair $(V, \rho : G \to Aut(V))$ where $V \in \mathcal{P}(F)$ and ρ is a group homomorphism from G to the group $Aut(V)$ of bijective linear operators on V. This definition makes sense if we replace $\mathcal{P}(F)$ by more general linear structures like $\mathcal{P}(R)$, the category of finitely generated projective modules over any ring R with identity.

More generally, one could define a representation of G in an arbitrary category $\mathcal{C}$ as a pair $(X, \rho : G \to Aut(X))$ where $X \in ob(\mathcal{C})$ and ρ is a group homomorphism from G to the group of $\mathcal{C}$-automorphisms of X. The representations of G in $\mathcal{C}$ also form a category $\mathcal{C}_G$ which can be identified with the category $[G/G, \mathcal{C}]$ of covariant functors from the translation category $\underline{G/G}$ of the G-set $\overline{G/G}$ (where G/G is the final object in the category of G-sets (see 1.1). The foregoing considerations also apply if G is a topological group and $\mathcal{C}$ is a topological category, i.e. a category whose objects X and $Hom_{\mathcal{C}}(X, Y)$ are endowed with a topology such that the morphisms are continuous. Here, we have an additional requirement that $\rho : G \to Aut(X)$ be continuous. For example, G could be a Lie group and $\mathcal{C}$ the category of Hilbert spaces over $\mathbb{C}$, in which case we have unitary representations of G.

It is the aim of this book to explore connections between $\mathcal{C}_G$ and higher algebraic K-theory of $\mathcal{C}$ for suitable categories (e.g. exact, symmetric monoidal and Waldhausen categories) when G could be a finite, discrete, profinite or compact Lie group.

When $\mathcal{C} = \mathcal{P}(\mathbb{C})$, ($\mathbb{C}$ the field of complex numbers) and G is a finite or compact Lie group, the Grothendieck group $K_0(\mathcal{C}_G)$ can be identified with the group of generalized characters of G and thus provides the initial contact between representation theory and K-theory. If F is an arbitrary field, G a finite group, $\mathcal{P}(F)_G$ can be identified with the category $\mathcal{M}(FG)$ of finitely generated FG-modules and so, $K_0(\mathcal{P}(F)_G) \cong K_0(\mathcal{M}(FG)) \cong G_0(FG)$ yields K-theory of the group algebra FG, thus providing initial contact between K-theory of $\mathcal{P}(F)_G$ and K-theory of group algebras (see 1.2). This situation extends to higher dimensional K-theoretic groups i.e. for all $n \geq 0$ we have $K_n(\mathcal{P}(F)_G) \cong K_n(\mathcal{M}(FG)) \cong G_n(FG)$ (see 5.2).

More generally, if R is any commutative ring with identity and G is a finite group, then the category $\mathcal{P}(R)_G$ can be identified with the category $\mathcal{P}_R(RG)$ of RG-lattices (i.e. RG-modules that are finitely generated and projective over R) and so, for all $n \geq 0$, $K_n(\mathcal{P}(R)_G)$ can be identified with $K_n(\mathcal{P}_R(RG))$ which, when R is regular, coincides with $K_n(\mathcal{M}(RG))$ usually denoted by $G_n(RG)$ (see $(5.2)^B$).

When R is the ring of integers in a number field or p-adic field F or more generally R a Dedekind domain with quotient field F or more generally still R a regular ring, the notion of a groupring $RG(G$ finite) generalizes to the notion of R-orders Λ in a semi-simple F-algebra Σ when char(F) does not divide the order of G and so, studying K-theory of the category $\mathcal{P}_R(\Lambda)$ of Λ-

lattices automatically yields results on the computations of K-theory of the category $\mathcal{P}_R(RG)$ of RG-lattices and so, K-theory of orders is appropriately classified as belonging to Integral representation theory.

Now the classical K-theory (K_0, K_1, K_2, K_{-n}) of orders and grouprings (especially K_0 and K_1) have been well studied via classical methods and documented in several books [20, 39, 159, 168, 211, 213] and so, we only carefully review the classical situation in Part I of this book (chapters 1-4), with clear definitions, examples, statements of important results (mostly without proofs) and refer the reader to one of the books or other literature for proofs. We include, in particular, classical results which have higher dimensional versions for which we supply proofs once and for all in the context of higher K-theory. Needless to say that some results proved for higher K-theory with no classical analogues invariably apply to the classical cases also. For example, there was no classical result that $K_2(\Lambda), G_2(\Lambda)$, are finite for arbitrary orders Λ in semi-simple algebras over number fields, but we prove in this book that $K_{2n}(\Lambda), G_{2n}(\Lambda)$, are finite for all $n \geq 1$, thus making this result also available for $K_2(\Lambda)$.

Some of the impetus for the growth of Algebraic K-theory from the beginning had to do with the fact that the classical K-group of grouprings housed interesting topological/geometric invariants, e.g.

(1) Class groups of orders and grouprings (which also constitute natural generalizations to number theoretic class groups of integers in number fields) also house Swan-Wall invariants (see $(2.3)^C$ and [214, 216]) etc.

(2) Computations of the groups $G_0(RG)$, R Noetherian, G Abelian is connected with the calculations of the group 'SSF' (see $(2.4)^B$ or [19]) which houses obstructions constructed by Shub and Francs in their study of Morse-Smale diffeomorphisms (see [19]).

(3) Whitehead groups of integral grouprings house Whitehead torsion which is also useful in the classification of manifolds (see [153, 195]).

(4) If G is a finite group and $Orb(G)$ the orbit category of G (an 'EI' category (see 7.6)). X a G-CW-complex with round structure (see [137]), then the equivariant Riedemester torsion takes values in $Wh(Q\ orb(G))$ where $Wh(Q\ orb(G))$ is the quotient of $K_1(QOrb(G))$ by subgroups of "trivial units" see [137].

(5) K_2 of integral grouprings helps in the understanding of the pseudo-isotopy of manifolds (see [80]).

(6) The negative K-theory of grouprings can also be interpreted in terms of bounded h-corbordisms (see $(4.5)^E$ or [138]).

It is also noteworthy that several far-reaching generalizations of classical concepts have been done via higher K-theory. For example, the K-theoretic definition of higher dimensional class groups $C\ell_n(\Lambda)(n \geq 0)$ of orders Λ generalize to higher dimensions the notion of class group $C\ell(\Lambda)$ of orders and grouprings which in turn generalizes the number-theoretic notion of class groups of Dedekind domains and integers R in number fields (see 7.4). Note that $C\ell_1(\Lambda)$ for $\Lambda = RG$ is intimately connected with Whitehead torsion (see 7.4 or [159]) and as already observed $C\ell(\Lambda) = C\ell_0(\Lambda)$ houses some topological/geometric invariants (see $(2.3)^C$).

Moreover, the profinite higher K-theory for exact categories discussed in chapter 8 is a cohomology theory which generalizes classical profinite topological K-theory (see [199]) as well as K-theory analogues of classical continuous cohomology of schemes rooted in Arithmetic algebraic geometry.

Part II (chapters 5 to 8) is devoted to a systematic exposition of higher algebraic K-theory of orders and grouprings. Again, because the basic higher K-theoretic constructions have already appeared with proofs in several books (e.g. [25, 88, 198]), the presentation in chapters 5 and 6 is restricted to a review of important results (with examples) relevant to our context. Topics reviewed in chapter 5 include the 'plus' construction as well as higher K-theory of exact, symmetric monoidal and Waldhausen categories. We try as much as possible to emphasize the utility value of the usually abstract topological constructions.

In chapter 7, we prove quite a number of results on higher K-theory of orders and grouprings. In $(7.1)^A$ we set the stage for arbitrary orders by first proving several finiteness results for higher K-theory of maximal orders in semi-simple algebras over p-adic fields and number fields as well as higher K-theory of associated division and semi-simple algebras.

In $(7.1)^B$, we prove among other results that if R is the ring of integers in a number field F, Λ as R-order in a semi-simple F-algebra Σ, then for all $n \geq 0$, $K_n(\Lambda), G_n(\Lambda)$ are finitely generated Abelian groups, $SK_n(\Lambda), SG_n(\Lambda), SK_n(\hat{\Lambda}_p)$ and $SG_n(\hat{\Lambda}_p)$ are finite groups (see [108, 110, 112, 113]) and $SG_n(RG)$ are trivial (see [131]) where G is a finite group. In 7.2 we prove that rank $K_n(\Lambda)$ = rank $G_n(\Lambda)$ = rank $K_n(\Gamma)$ if Γ is a maximal R-order containing Λ, see [115]. We consequently prove that for all $n \geq 1$, $K_{2n}(\Lambda), G_{2n}(\Lambda)$ are actually finite groups. Hence for any finite group G, $K_{2n}(RG), G_{2n}(RG)$ are finite (see $(7.2)^B$ or [121]).

Next, we obtain in 7.3 a decomposition (for G Abelian) $G_n(RG) \cong \oplus G_n(R < C >)$ for all $n \geq 0$ where R is a Noetherian ring, and C ranges over all cyclic quotients of G and $R < C >= R\zeta_{|C|}(\frac{1}{|C|}), \zeta_{|C|}$ being a primitive $|C|^{\text{th}}$ root of unity (see [232]). (This decomposition is a higher dimensional version of that of $G_0(RG)$ (see $(2.4)^A$.) The decomposition of $G_n(RG)$ is extended to some non-Abelian groups e.g. dihedral, quaternion and nilpotent groups (see [231, 233]). We conclude 7.3 with a discussion of a conjecture due to Hambleton, Taylor and Williams on the decomposition for $G_n(RG)$, G any finite groups (see [76]), and the counter-example provided for this conjecture by D. Webb and D. Yao (see [235].

Next, in 7.4 we define and study higher-dimensional class groups $Cl_n(\Lambda)$ of R-orders Λ which generalize the classical notion of class groups $Cl(\Lambda)(= Cl_0(\Lambda))$ of orders. We prove that $\forall\, n \geq 0$, $Cl_n(\Lambda)$ is a finite group and identify p-torsion in $Cl_{2n-1}(\Lambda)$ for arbitrary orders Λ (see [102]) while we identify p-torsion for all $Cl_{2n}(\Lambda)$ when Λ is an Eichler or hereditary order (see [74, 75]).

In 7.5, we study higher K-theory of grouprings of virtually infinite cyclic groups V in the two cases when $V = G \rtimes_\alpha T$, the semi-direct product of a finite group G (of order r, say) and an infinite cyclic group $T =< t >$ with respect to the automorphism $\alpha : G \to G$ $g \to tgt^{-1}$ and when $V = G_0 *_H G_1$ where the groups $G_i = 0, 1$ and H are finite and $[G_i : H] = 2$. These groups V are conjectured by Farrell and Jones (see [54]) to constitute building blocks for the understanding of K-theory of grouprings of an arbitrary discrete group G - hence their importance. We prove that when $V = G \rtimes_\alpha T$, then for all $n \geq 0$, $G_n(RV)$ is a finitely generated Abelian group and that $NK_n(RV)$ is r-torsion. For $V = G_0 *_H G_1$ we prove that the nil groups of V are $|H|$-torsion (see [123]).

The next section of chapter 7 is devoted to the study of higher K and G-theory of modules over 'EI'-categories. Modules over 'EI'-categories constitute natural generalizations for the notion of modules over grouprings and K-theory of such modules are known to house topological and geometric invariants and are also replete with applications in the theory of transformation groups (see [137]). Here, we obtain several finiteness and other results which are extension of results earlier obtained for higher K-theory of grouprings of finite groups.

In 7.7 we obtain several finiteness results on the higher K-theory of the category of representations of a finite group G in the category of $\mathcal{P}(\Gamma)$ where Γ is a maximal order in central division algebra over number fields and p-adic fields. These results translate into computations of $G_n(\Gamma G)$ as well as lead to showing via topological and representation theoretic techniques that a non-commutative analogue of a fundamental result of R.G. Swan at the zero-dimensional level does not hold (see [110]).

In chapter 8, we define and study profinite higher K and G-theory of exact categories, orders and grouprings. This theory is an extraordinary cohomology theory inspired by continuous cohomology theory in algebraic topology and arithmetic algebraic geometry. The theory yields several ℓ-completeness theorems for profinite K and G-theory of orders and grouprings as well as yields some interesting computations of higher K-theory of p-adic orders otherwise inaccessible. For example we use this theory to show that if Λ is a p-adic order in a p-adic semi-simple algebra Σ, then for all $n \geq 1$, $K_n(\Lambda)_\ell, G_n(\Lambda)_\ell, K_n(\Sigma)_\ell$ are finite groups provided ℓ is a prime $\neq p$. We also define and study continuous K-theory of p-adic orders and obtain a relationship between profinite and continuous K-theory of such orders (see [117]).

Now if $\underline{S}$ is the translation category of any G-set S, and $\mathcal{C}$ is a small category, then the category $[\underline{S}, \mathcal{C}]$ of covariant functors from $\underline{S}$ to $\mathcal{C}$ is also called the category of G-equivariant $\mathcal{C}$-bundles on S because if $\mathcal{C} = \mathcal{P}(\mathbb{C})$, then $[\underline{S}, \mathcal{P}(\mathbb{C})]$ is just the category of G-equivariant $\mathbb{C}$-bundles on the discrete G-space S so that $K_0[\underline{S}, \mathcal{P}(\mathbb{C})] = K_0^G(S, \mathcal{P}(\mathbb{C}))$ is the zero-dimensional G-equivariant K-theory of S. Note that if S is a G-space, then the translation category $\underline{S}$ of S as well as the category $[\underline{S}, \mathcal{C}]$ are defined similarly. Indeed, if S is a compact G-space then $K_0^G(S, \mathcal{P}(\mathbb{C}))$ is exactly the Atiyah-Segal equivariant K-theory of S (see [184]).

One of the goals of this book is to exploit representation theoretic techniques (especially induction theory) to define and study equivariant higher algebraic K-theory and their relative generalizations for finite, profinite and compact Lie group actions, as well as equivariant homology theories for discrete group actions in the context of category theory and homological algebra with the aim of providing new insights into classical results as well as open avenues for further applications. We devote Part III (chapters 9 - 14) of this book to this endeavour.

Induction theory has always aimed at computing various invariants of a given group G in terms of corresponding invariants of certain classes of subgroups of G. For example if G is a finite group, it is well know by Artin induction theorem that two G-representations in $\mathcal{P}(\mathbb{C})$ are equivalent if their restrictions to cyclic subgroups of G are isomorphic. In other words, given the exact category $\mathcal{P}(\mathbb{C})$, and a finite group G, we have found a collection $D(\mathcal{P}(\mathbb{C}), G)$ of subgroups (in this case cyclic subgroups) of G such that two G-representations in $\mathcal{P}(\mathbb{C})$ are equivalent iff their restrictions to subgroups in $D(\mathcal{P}(\mathbb{C}), G)$ are equivalent. One could then ask the following general question: Given a category $\mathcal{A}$ and a group G, does there exist a collection $D(\mathcal{A}, G)$ of proper subgroups of G such that two G-representations in $\mathcal{A}$ are equivalent if their restrictions to subgroups in $D(\mathcal{A}, G)$ are equivalent?

As we shall see in this book, Algebraic K-theory is used copiously to answer these questions. For example, if G is a finite group, T any G-set, $\mathcal{C}$ an exact category, we construct in 10.2 for all $n \geq 0$, equivariant higher K-functors.

$$K_n^G(-, \mathcal{C}, T), \ K_n^G(-, \mathcal{C}, T), K_n^G(-, \mathcal{C})$$

as Mackey functors from the category $G\mathcal{S}et$ of G-sets to the category $\mathbb{Z}$-$\mathcal{M}od$ of Abelian groups (i.e. functors satisfying certain functorial properties, in particular, categorical version of Mackey subgroup theorem in representation theory) in such a way that for any subgroup H of G we identify $K_n^G(G/H, \mathcal{M}(R))$ with $K_n(\mathcal{M}(RH)) := G_n(RH), K_n^G(G/H, \mathcal{P}(R))$ with $K_n(\mathcal{P}_R(RH)) := G_n(R, H)$ and $P_n^G(G/H, \mathcal{P}(R), G/e)$ with $K_n(RH)$ for all $n \geq 0$ (see [52, 53]). Analogous constructions are done for profinite group actions (chapter 11) and compact Lie group actions (chapter 12), finite group actions in the context of Waldhausen categories, chapter 13, as well as e-quivariant homology theories for the actions of discrete groups (see chapter 14).

For such Mackey functors M, one can always find a canonical smallest class U_M of subgroups of G such that the values of M on any G-set can be computed from their restrictions to the full subcategory of G-sets of the form G/H with $H \in U_M$. The computability of the values of M from its restriction to G-sets of the form $G/H, H \in U_M$ for finite and profinite groups is expressed in terms of vanishing theorems for a certain cohomology theory associated with M (U_M) - a cohomology theory which generalizes group cohomology. In 9.2, we discuss the cohomology theory (Amitsur cohomology) of Mackey functors, defined on an arbitrary category with finite coproducts, finite pullbacks and final objects in 9.1 and then specialize as the needs arise for the cases of interest-category of G-sets for G finite (in chapter 9 and chapter 10), G profinite (chapter 11) - yielding vanishing theorems for the cohomology of the K-functors as well as cohomology of profinite groups (11.2) (see [109]).

The equivariant K-theory discussed in this book yields various computations of higher K-theory of grouprings. For example apart from the result that higher K-theory of RG (G finite or compact Lie group) can be computed by restricting to hyper elementary subgroups of G (see 10.4 and 12.3.3) (see [108, 116]), we also show that if R is a field k of characteristic p and G a finite or profinite group, then the Cartan map $K_n(kG) \rightarrow G_n(kG)$ induces an isomorphism $\mathbb{Z}(\frac{1}{p}) \otimes K_n(kG) \cong \mathbb{Z}(\frac{1}{p}) \otimes G_n(kG)$ leading to the result that for all $n \geq 1$, $K_{2n}(kG)$ is a p-group for finite groups G. We also have an interesting result that if R is the ring of integers in a number field, G a finite group then the Waldhausen K-groups of the category $(Ch_b(\mathcal{M}(RG), \omega)$ of bounded complexes of finitely generated RG-modules with stable quasi-isomorphisms as weak equivalences are finite Abelian groups.

The last chapter (chapter 14) which is devoted to Equivariant homology theories, also aims at computations of higher algebraic K-groups for grouprings of discrete groups via induction techniques also using Mackey functors. In fact, an important criteria for a G-homology theory $\mathcal{H}_n^G : GSet \rightarrow \mathbb{Z}\text{-}\mathcal{M}od$ is that it is isomorphic to some Mackey functor: $GSet \rightarrow \mathbb{Z}\text{-}\mathcal{M}od$. The chapter is focussed on a unified treatment of Farrell and Baum-Connes isomorphism conjectures through Davis-Lück assembly maps (see 14.2 or [40]) as well as some specific induction results due to W. Lück, A. Bartels and H. Reich (see 14.3 or [14]). One other justification for including Baum-Connes conjecture in this unified treatment is that it is well known by now that Algebraic K-theory and Topological K-theory of stable C^*-algebras do coincide (see [205]). We review the state of knowledge of both conjectures (see 14.3, 14.4) and in the case of Baum-Connes conjecture also discuss its various formulations including the most recent in terms of quantum group actions.

Time, space and the heavy stable homotopy theoretic machinery involved (see [147]) (for which we could not prepare the reader) has prevented us from including a G-spectrum formulation of the equivariant K-theory developed in chapters 10, 11, 12, 13. In [192, 193, 194], K. Shimakawa provided, (for G a finite group) a G-spectrum formulation of part of the (absolute) equivariant theory discussed in 10.1. It will be nice to have a G-spectrum formulation of the relative theory discussed in 10.2 as well as a G-spectrum formulation for the equivariant theory discussed in 11.1 and 12.2 for G profinite and G compact Lie group. However, P. May informs me that equivariant infinite loop space theory itself is only well understood for finite groups. He thinks that profinite groups may be within reach but compact Lie groups are a complete mystery since no progress has been made towards a recognition principle in that case. Hence, there is currently no idea about how to go from the type of equivariant Algebraic K-theory categories defined in this book to a G-spectrum when G is a compact Lie group.

Appendix A contains some known computations while Appendix B consists of some open problems.

The need for this book

1) So far, there is no book on higher Algebraic K-theory of orders and grouprings. The results presented in the book are only available in scattered form in journals and other scientific literature, and there is a need for a coordinated presentation of these ideas in book form.

2) Computations of higher K-theory even of commutative rings (e.g. $\mathbb{Z}$) have been notoriously difficult and up till now the higher K-theory of $\mathbb{Z}$ is yet to be fully understood. Orders and grouprings are usually non-commutative rings that also involve non-commutative arithmetic and computations of higher K-theory of such rings are even more difficult, since methods of etále cohomology etc. do not work. So it is desirable to collect together in book form methods that have been known to work for computations of higher K-theory of such non-commutative rings as orders and grouprings.

3) This is the first book to expose the characterization of all higher algebraic K-theory as Mackey functors leading to equivariant higher algebraic K-theory and their relative generalization, also making computations of higher K-theory of grouprings more accessible. The translation of the abstract topological constructions into representation theoretic language of Mackey functors has simplified the theory some what and it is desirable to have these techniques in book form.

4) Interestingly, obtaining results on higher K-theory of orders Λ (and hence grouprings) for all $n \geq 0$ have made these results available for the first time for some classical K-groups. For instance, it was not known classically that if R is the ring of integers in a number field F, and Λ any R-order in a semi-simple F-algebra, then $K_2(\Lambda), G_2(\Lambda)$ (or even $SK_s(\Lambda), SG_2(\Lambda)$) are finite groups. Having these results for $K_{2n}(\Lambda), G_{2n}(\Lambda)$ and hence $SK_{2n}(\Lambda), SG_{2n}(\Lambda)$ for all $n \geq 1$ makes these results available now for $n = 1$.

5) Also computations of higher K-theory of orders which automatically yield results on higher K-theory of $RG(G$ finite) also extends to results on higher K-theory of some infinite groups e.g. computations of higher K-groups of virtually infinite cyclic groups that are fundamental to the subject.

Who can use this book?

It is expected that readers would already have some working knowledge of algebra in a broad sense including category theory and homological algebra, as well as working knowledge of basic algebraic topology, representation theory, algebraic number theory, some algebraic geometry and operator algebras. Nevertheless, we have tried to make the book as self-contained as possible by defining the most essential ideas.

As such, the book will be useful for graduate students who have completed at least one year of graduate study, professional mathematicians and researchers of diverse backgrounds who want to learn about this subject as well as specialists in other aspects of K-theory who want to learn about this approach to the subject. Topologists will find the book very useful in updating their knowledge of K-theory of orders and grouprings for possible applications and representation theorists will find this innovative approach to and applications of their subject very enlightening and refreshing while number theorists and arithmetic algebraic geometers who want to know more about non-commutative arithmetics will find the book very useful.

Acknowledgments

It is my great pleasure to thank all those who have helped in one way or the other to make the writing of this book a reality.

First I like to thank Hyman Bass for his encouragement and inspiration over the years. I also like to thank Andreas Dress also for his inspiration as well as his reading and offering helpful comments when an earlier version of chapter 9 was being written. I feel very grateful to Joshua Leslie who introduced me to K-theory. I thank Bruce Magurn for reading some parts of the book and offering useful comments.

I thank all my collaborators over the years - A. Dress, D.N. Diep, D. Goswami, X. Guo, M. Mahdavi-Hezavehi, H. Qin and G. Tang. I thank the various institutions where I held permanent or visiting positions while conducting the various researches reported in the book - University of Ibadan, Nigeria; International Centre for Theoretical Physics (ICTP), Trieste, Italy; Columbia University, New York; University of Chicago; Cornell University, Ithaca, New York; University of Bielefeld; Max Planck Institute, Bonn; Institute for Advanced Study (IAS), Princeton, New Jersey; Mathematical Sciences Research Institute (MSRI), Berkeley, CA; Queen's University, Kingston, Canada; The Ohio State University; Howard University, Washington DC and Miami

University, Oxford, Ohio. My special thanks go to IAS, Princeton, MSRI, Berkely, ICTP (Trieste) and Miami University for their hospitality during the writing stages of this book.

A substantial part of this book was written while I was employed by the Clay Mathematical Research Institute and I thank the Institute most heartily.

It is my great pleasure to thank most heartily Dilys Grilli of ICTP, Trieste, for selfless help in organizing the typing of the manuscript as well as doing the formatting and general preparation of the book for publication. I am also very grateful to Linda Farriell of Miami University, Oxford, Ohio, who helped with the typing of the exercises and open problems.

I feel grateful also to my wife, Funke, and children, Dolapo, Kemi, Yemisi and Solape, who over the years endured frequent absences from home while a lot of the work reported in this book was being done.

Finally, I thank the Editors - Jim McGovern, David Grubbs, Kevin Sequeira and Fred Coppersmith, as well as Jessica Vakili, the project coordinator at Taylor and Francis for their cooperation.

Notes on Notations

Notes on Notation

- $mor_{\mathcal{C}}(A, B), Hom_{\mathcal{C}}(A, B) :=$ set of $\mathcal{C}$-morphisms from A to B ($\mathcal{C}$ a category)

- $\bar{A} = K(A) =$ Gröthendieck group associated to a semi-group A

- $VB_F(X) =$ category of finite dimensional vector bundles on X ($F = \mathbb{R}$ or $\mathbb{C}$)

- $X(G) =$ set of cyclic quotients of a finite group G

- $\mathcal{A}[z] =$ polynomial extension of an additive category $\mathcal{A}$

- $\mathcal{A}[z, z^{-1}]$, Laurent polynomal extension of $\mathcal{A}$

- $\mathcal{O}_{r_{\mathcal{F}}}(G) = \{G/H | H \in \mathcal{F}\}$, $\mathcal{F}$ a family of subgroups of G

- $A_{R,\mathcal{F}} =$ assembly map

- $\mathcal{H}_G(S, B) := \{G\text{-maps } f : S \to B\}$, S a G-set, B a $\mathbb{Z}G$-module

- $B(S) =$ set of all set-theoretic maps $S \to B$. Note that

- $\mathcal{H}_G(S, B)$ is the subgroup of G-invariant elements in $B(S)$

- $\mathcal{A}(G) =$ category of homogeneous G-spaces (G a compact Lie group)

- $\Omega(\mathcal{B}) :=$ Burnside ring of a based category $\mathcal{B}$

- $\Omega(G) :=$ Burnside ring of a group G

- $\|\mathcal{B}\| :=$ Artin index of a based category $\mathcal{B}$
  $:=$ exponent of $\bar{\Omega}(\mathcal{B})/\Omega(\mathcal{B})$

- $M_m^n :=$ n-dimensional mod-m Moore space

- $H_n(X, \underline{E}) = \underline{E}_n(X) :=$ homology of a space X with coefficient in a spectrum $\underline{E}$

- $H^n(X, \underline{E}) = \underline{E}^n(X) :=$ cohomology of a space X with coefficients in a spectrum $\underline{E}$

- $\mathcal{P}(A) :=$ category of finitely generated projective A-modules (A a ring with identity)

- $\mathcal{M}(A) :=$ category of finitely generated A-modules

- $\mathcal{M}'(A) :=$ category of finitely presented A-modules

- $A - \mathcal{M}\text{od} :=$ category of left A-modules

- $\mathcal{P}(X) :=$ category of locally free sheaves of O_X-modules (X a scheme)

- $\mathcal{M}(X) :=$ category of coherent sheaves of O_X-modules (X a Noetherian scheme)

- $\mathcal{P}_R(A) :=$ category of A-modules finitely generated and projective as R-modules (A an R-algebra)

- $\underline{S} :=$ translation category of a G-set S (see 1.1.3) (G a group)
  $GSet :=$ category of G-sets
  $[\underline{S}, \mathcal{C}] :=$ category of covariant functors $\underline{S} \to \mathcal{C}$ ($\mathcal{C}$ any category)

- For any Abelian group G, and a rational prime ℓ
  G_ℓ or $G(\ell) :=$ ℓ-primary subgroup of G
  $G[\ell^s] = \{g \in G | \ell^s g = 0\}$
  Note. $G(\ell) = UG[\ell^s] = \varinjlim\ G[\ell^s]$

- If $\mathcal{G} : GSet \to \mathbb{Z}\text{-}\mathcal{M}\text{od}$ is a Green functor and A a commutative ring with identity then $\mathcal{G}^A := A \otimes_{\mathbb{Z}} \mathcal{G} : GSet \to A\text{-}\mathcal{M}\text{od}$ is a Green functor given by $(A \otimes_{\mathbb{Z}} \mathcal{G})(S) = A \otimes \mathcal{G}(S)$ If $\mathcal{D}_\mathcal{G}$ is a defect basis for $\mathcal{G}$ (see 9.6.1) write $\mathcal{D}_\mathcal{G}^A$ for the defect basis of $\mathcal{G}^A$ If $\mathcal{P}$ is a set of primes and $A = \mathbb{Z}_\mathcal{P} = \mathbb{Z}[\frac{1}{q}|q \notin \mathcal{P})$ write $\mathcal{D}_\mathcal{G}^\mathcal{P}$ for $\mathcal{D}_\mathcal{G}^A$

- $Ch_b(\mathcal{C}) =$ category of bounded chain complexes in an exact category $\mathcal{C}$

- Let C be a cyclic group of order t
  $\mathbb{Z}(C) = \mathbb{Z}[\zeta]$ where ζ is a primitive t^{th} root of 1
  $\mathbb{Z} < C >= \mathbb{Z}(C)(\frac{1}{t}) = \mathbb{Z}[\zeta_t, \frac{1}{t}]$
  For any ring R, $R(C) = R \otimes \mathbb{Z}(C)$, $R < C >= R \otimes \mathbb{Z} < C >$

- For any finite group G, p a rational prime
  $G(p)$ or $S_p(G) :=$ Sylow p-subgroup of G
  For G Abelian, $G(p') = \bigoplus\limits_{\substack{q\ prime \\ q \neq p}} G(q)$. So $G = G(p) \times G(p')$
  If $\mathcal{P}$ is a set of primes, $G(\mathcal{P}) = \bigoplus\limits_{p \in \mathcal{P}} G(p)$ i.e. $\mathcal{P}$-torsion part of G.

- If R is the ring of integers in a number field F, Λ an R-order in a semi-simple F-algebra, then
  $P(\Lambda) :=$ finite set of prime ideals $\underline{p}$ of R for which $\hat{\Lambda}_{\underline{p}}$ is not maximal
  $\check{P}(\Lambda) :=$ set of rational primes lying below the prime ideals in $P(\Lambda)$

- For a category $\mathcal{C}$, $\mathbb{P}(\mathcal{C}) :=$ idempotent completion of $\mathcal{C}$

- For a discrete group G

  $All :=$ all subgroups of G

  $Fin :=$ all finite subgroups of G

  $VCy :=$ all virtually cyclic subgroups of G

  $Triv :=$ trivial family consisting of only one element i.e. the identity element of G

  $FCy :=$ all finite cyclic subgroups of G

- $\mathcal{M}od^R_{\mathcal{F}}(G) :=$ category of contravariant functors

$$Or_{\mathcal{F}}(G) \longrightarrow R - \mathcal{M}od$$

  $G - \mathcal{M}od^R_{\mathcal{F}} :=$ category of covariant functors

$$Or_{\mathcal{F}}(G) \longrightarrow R - \mathcal{M}od$$

- $E_{\mathcal{F}}(G) :=$ classifying space for a family $\mathcal{F}$ of subgroups of a discrete group G

  $:=$ universal G-space with stabilizers in $\mathcal{F}$

  $\underline{E}G = E_{Fin}(G) =$ universal space for proper actions of G

Part I

Review of Classical Algebraic K-Theory and Representation Theory

Chapter 1

Category of representations and constructions of Grothendieck groups and rings

1.1 Category of representations and G-equivariant categories

1.1.1 Let G be a discrete group, V a vector space over a field F. A representation of G on V is a group homomorphism $\rho : G \to \mathrm{Aut}(V)$ where $\mathrm{Aut}(V)$ is the group of invertible linear operators on V. Call V a representation space of ρ.

An action of G on V is a map $\rho : G \times V \to V$ $\rho(g \, . \, v) := gv$ such that $ev = v, (gh)v = g(h(v))$. Note that an action $\rho : G \times V \to V$ gives rise to a representation $\rho_V : G \to \mathrm{Aut}(V)$ where $\rho_V(g) := \rho_g : v \to \rho(g,v)$, and conversely, any representation $\rho_V : G \to \mathrm{Aut}(V)$ defines an action $\rho : G \times V \to V : (g,v) \to \rho_g(v)$.

Two representations ρ, ρ' with representation spaces V, V' are said to be equivalent if there exists an F-isomorphism β of V onto V' such that

$$\rho'(g) = \beta\rho(g).$$

The dimension of V over F is called the degree of ρ.

Remarks 1.1.1   (i) For the applications, one restricts V to finite-dimensional vector spaces. We shall be interested in representations on V ranging from such classical spaces as vector spaces over complex numbers to more general linear structures like finitely generated projective modules over such rings as Dedekind domains, integers in number fields, and p-adic fields, etc.

   (ii) When G, V have topologies, we have an additional requirement that ρ be continuous.

   (iii) More generally, G could act on a finite set S, i.e., we have a permutation $s \to gs$ of S satisfying the identities $1s = s, g(hs) = (gh)s$ for $g, h \in G, s \in S$. Let V be the vector space having a basis $(e_s)_{s \in S}$ indexed by

$s \in S$. So, for $g \in G$, let ρ_g be the linear map $V \to V$ sending e_x to e_{gx}. Then $\rho : G \to \mathrm{Aut}(V)$ becomes a linear representation of G called permutation representation associated to S.

(iv) Let A be a finite-dimensional algebra over a field F, and V a finite-dimensional vector space over F. A representation of A on V is an algebra homomorphism $\overline{\rho} : A \to \mathrm{Hom}_F(V,V) = \mathrm{End}_F(V)$, i.e., a mapping $\overline{\rho}$, which satisfies:

$$\overline{\rho}(a+b) = \overline{\rho}(a) + \overline{\rho}(b) , \qquad \overline{\rho}(ab) = \overline{\rho}(a)\overline{\rho}(b)$$
$$\overline{\rho}(\alpha a) = \alpha\overline{\rho}(a) , \qquad \overline{\rho}(e) = 1 , a, b \in A, \alpha \in F ,$$

where e is the identity element of A.

Now, if $\rho : G \to \mathrm{Aut}(V)$ is a representation of G with representation space V, then there is a unique way to extend ρ to a representation $\overline{\rho}$ of FG with representation space V, i.e., $\overline{\rho}\left(\Sigma a_g \rho(g)\right) = \Sigma a_g \rho(g)$. Conversely, every representation of FG, when restricted to G, yields a representation of G. Hence there is a one-one correspondence between F-representations of G with representation space V and FG-modules.

Definition 1.1.1 means that we have a representation of G in the category $\mathcal{P}(F)$ of finite-dimensional vector spaces over F. This definition could be generalized to any category as follows.

1.1.2 Let $\mathcal{C}$ be a category and G a group. A G-object in $\mathcal{C}$ (or a representation of G in $\mathcal{C}$) is a pair $(X, \rho), X \in ob\mathcal{C}, \rho : G \to \mathrm{Aut}(X)$ a group homomorphism. We shall write ρ_g for $\rho(g)$.

The G-objects in $\mathcal{C}$ form a category $\mathcal{C}_G$ where for $(X, \rho), (X', \rho') \in ob\mathcal{C}_G$, $mor_{\mathcal{C}_G}((X, \rho), (X', \rho'))$ is the set of all $\mathcal{C}$-morphisms $\varphi : X \to X'$ such that for each $g \in G$, the diagram

$$
\begin{array}{ccc}
X & \xrightarrow{\rho_g} & X \\
\downarrow{\varphi} & & \downarrow{\varphi} \\
X' & \xrightarrow{\rho'_g} & X'
\end{array}
\quad \text{commutes}
$$

Examples 1.1.1   (i) When $\mathcal{C} = FSet$, the category of finite sets, $\mathcal{C}_G = GSet$, the category of finite G-sets.

(ii) When F is a field, and $\mathcal{C} = \mathcal{P}(F)$, then $\mathcal{P}(F)_G$ is the category $\mathcal{M}(FG)$ or finitely generated FG-modules.

(iii) When R is a commutative ring with identity and $\mathcal{C} = \mathcal{M}(R)$ the category of finitely generated R-modules, then $\mathcal{C}_G = \mathcal{M}(RG)$, the category of finitely generated RG-modules.

(iv) If R is a commutative ring with identity, and $\mathcal{C} = \mathcal{P}(R)$, then $\mathcal{P}(R)_G = \mathcal{P}_R(RG)$, the category of RG-lattices, i.e., RG-modules that are finitely generated and projective over R.

Note that for a field F, every $M \in \mathcal{M}(FG)$ is an FG-lattice and so $\mathcal{M}(FG) = \mathcal{P}_F(FG)$.

1.1.3 Let S be a *GSet* (G a discrete group). We can associate to S a category $\underline{S}$ as follows:

$$ob\ \underline{S} = \text{elements of } S; \qquad \text{mor}_{\underline{S}}(s,t) = \{(g,s)|g \in G, gs = t\}.$$

Composition of morphisms is defined by $(h,t) \circ (g,s) = (hg,s)$, and the identity morphism $s \to s$ is (e,s) where e is the identity of G. $\underline{S}$ is called the translation category of S.

- For any category $\mathcal{C}$, let $[\underline{S},\mathcal{C}]$ be a category of (covariant) functors $\zeta : \underline{S} \to \mathcal{C}$, which associates to an element $s \in S$ a $\mathcal{C}$-object ζ_s and to a morphism (g,s) a $\mathcal{C}$-map $\zeta_{(g,s)} : \zeta_s \to \zeta_{gs}$, $s \in S$ $\zeta_{(e,s)} = id_{\zeta_s}$ and $\zeta_{(g,hs)} \circ \zeta_{(h,s)} = \zeta_{(gh,s)}$ for all $g,h \in G, s \in S$. Call such a functor a G-equivariant $\mathcal{C}$-bundle on S.

The motivation for this terminology is that if $\mathcal{C}$ is the category of finite-dimensional vector spaces over the field $\mathbb{C}$ of complex numbers, then ζ is indeed easily identified with a G-equivariant $\mathbb{C}$-vector bundle over the finite discrete G-sets S.

1.1.4 Note that the category $\underline{S}$ defined above is a groupoid, i.e., a category in which every morphism is an isomorphism. More generally, for any small groupoid $\mathcal{G}$, and any small category $\mathcal{C}$, we shall write $[\mathcal{G},\mathcal{C}]$ for the category of covariant functors $\mathcal{G} \to \mathcal{C}$ and $[\mathcal{G},\mathcal{C}]'$ for the category of contravariant functors $\mathcal{G} \to \mathcal{C}$. We shall extend the ideas of this section from $[\underline{S},\mathcal{C}]$ to $[\mathcal{G},\mathcal{C}]$ for suitable $\mathcal{C}$ in chapter 14 when we study equivariant homology theories vis-a-vis induction techniques.

1.1.5 Examples and some properties of $[\underline{S},\mathcal{C}]$

(i) For any category $\mathcal{C}$, there exists an equivalence of categories $[\overline{G/G},\mathcal{C}] \to \mathcal{C}_{\mathcal{G}}$ given by $\zeta \mapsto (\zeta_*, \rho : G \to \text{Aut}(\zeta_*); g \to \zeta_{(g,*)})$ where $\zeta_{g,*} \in \text{Aut}(\zeta_*)$, since $\zeta^{-1}_{(g,*)} = \zeta_{(g^{-1},*)}$.

Hence if G is a finite group, we have

- $[\overline{G/G},\mathcal{M}(R)] \simeq \mathcal{M}(R)_G \simeq \mathcal{M}(RG)$, if R is a commutative ring with identity
- $[\overline{G/G},\mathcal{P}(R)] \simeq \mathcal{P}_R(RG)$.

(ii) (a) Let $\mathcal{C}$ be a category and X a fixed $\mathcal{C}$-object. Define a new category $\mathcal{C}/X$ (resp. $X/\mathcal{C}$) called the category of $\mathcal{C}$-objects over X (resp. under X) as follows:

The objects of $\mathcal{C}/X$ (resp. $X/\mathcal{C}$) are pairs $(A, \varphi : A \to X)$ (resp. $(B, \delta : X \to B)$) where A (resp. B) runs through the objects of $\mathcal{C}$ and φ through $\mathrm{mor}_\mathcal{C}(A, X)$ (resp. δ through $\mathcal{C}(X, B)$). If (A, φ), $(A', \varphi') \in \mathcal{C}/X$ (resp. $(B, \delta), (B', \delta') \in X/\mathcal{C}$), then $\mathcal{C}/X((A, \varphi), (A', \varphi')) = \{\psi \in \mathcal{C}(A, A') | \varphi = \varphi'\psi\}$ (resp. $X/\mathcal{C}((B, \delta), (B', \delta')) = \{\rho \in \mathcal{C}(B, B') | \delta' = \rho\gamma\}$), i.e., a morphism from (A, φ) to (A', φ') (resp. (B, δ) to (B', δ')) is a commutative triangle

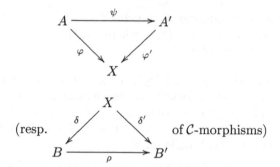

(resp. of $\mathcal{C}$-morphisms)

(b) If in (a) $\mathcal{C} = GSet, S \in GSet$, we have $GSet/S$ (resp. $S/GSet$). Note that if $S \in GSet$, the category $\underline{S}$ can be realized as a full subcategory of $GSet/S$ whose objects are all maps $G/e \to S$ where $s \in S$ is identified with $f_s : G/e \to S$ $g \mapsto gs$) and (g, s) is identified with

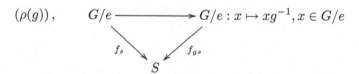

$(\rho(g))$, $G/e \longrightarrow G/e : x \mapsto xg^{-1}, x \in G/e$

(c) If $\mathcal{C} = FSet$ in (i), then we have an equivalence of categories $[S, FSet] \simeq GSet/S$ defined as follows.

For $\zeta \in [S, FSet]$, the set $|\zeta| = \{(s, x) | s \in S, x \in \zeta_s\}$ is a G-set w.r.t. $G \times |\zeta| \to |\zeta| : (g, (s, x)) \mapsto (gs, \zeta_{(g,s)}x)$ and $|\zeta| \to S :$ $(s, x) \mapsto s$ is a G-map (note $|\zeta|$ could be described as the disjoint union of fibres of ζ.

Conversely, if $\varphi : S' \to S$ is a G-map over S, then $\varphi : S' \to S$ gives rise to a $\mathcal{C}$-bundle ζ over S with fibres $\zeta_s = \varphi^{-1}(s)$ and maps $\zeta_{(g,s)} : \zeta_s \to \zeta_{gs} : x \mapsto gx$. It is easily checked that a $\mathcal{C}$-bundle morphism $\mu : \zeta \to \zeta'$ between two $\mathcal{C}$-bundles corresponds to a G-map between the corresponding G-sets over S and vice versa and that this way we get, indeed, an equivalence of categories.

(iii) (a) Let $\underline{G/H}$ (resp. $\underline{H/H}$) be the category associated with the G-set $G/\overline{H}$ (resp. H-set $\overline{H/H}$). Then the functor $\underline{H/H} \to \underline{G/H}$ given by $*_H \to \overline{H} \in \underline{G/H}$ and $(u, *_H) \to (u, \overline{H})$ for $u \in \overline{H}$ is an equivalence of categories.

Proof. The proof follows from the fact that if $\mathcal{C}_1$ is a full subcategory of a category $\mathcal{C}_2$ and if for any $\mathcal{C}_2$-object X there exists $X' \in \mathcal{C}_1$ such that $X' \cong X$, then $\mathcal{C}_1 \to \mathcal{C}_2$ is an equivalence of categories (see [141]). Now, since $\underline{H/H}$ is full in $\underline{G/H}$ and any object in $\underline{G/H}$ is isomorphic to $*_H = \overline{H} \in \underline{G/H}$, we may apply this fact to $\overline{\mathcal{C}_1} = \underline{H/H}$ and $\mathcal{C}_2 = \underline{G/H}$.

(b) The equivalence $\underline{H/H} \to \underline{G/H}$ of categories in (a) defines an equivalence of categories $\overline{[\underline{G/H}, \mathcal{C}]} \to [\underline{H/H}, \mathcal{C}]$ for any category $\mathcal{C}$. Hence $\mathcal{C}_H$, which has been shown to be isomorphic to $[\underline{H/H}, \mathcal{C}]$ already, is equivalent to $[\underline{G/H}, \mathcal{C}]$.

Remark. Note that it follows from (ii) and (iii)(b) that we now have an equivalence of categories between $GSet/(G/H)$ and $HSet \simeq HSet/(H/H)$ defined by associating to any G-set S over G/H the pre-image of $*_H$ in G/H considered as an H-set.

(c) Let G be a finite group, $H \leq G$, T a H-set, and $G \underset{H}{\times} T$ the induced G-set defined as a set of H-orbits $\overline{(g,t)} \subseteq G \times T$ with respect to the H-action $h(g,t) = (gh^{-1}, ht), h \in H, g \in G, t \in T$. Then the functor $\underline{T} \to \underline{G \underset{H}{\times} T}$ given by $t \mapsto (e, t), (g, t) \mapsto (g, \overline{(e, t)})$ is an equivalence of categories, and so, for any category $\mathcal{C}$, $[\underline{G \underset{H}{\times} T}, \mathcal{C}] \to [\underline{T}, \mathcal{C}]$ is an equivalence of categories. Note that $h \in G$ acts on $\overline{(g,t)} \in G \underset{H}{\times} T$ by $h(g,t) = (hg, t)$.

Proof. Again, with $\mathcal{C}_1 = \underline{T}$, $\mathcal{C}_2 = \underline{G \underset{H}{\times} T}$, the embedding $\mathcal{C}_1 \to \mathcal{C}_2$ defined above makes $\mathcal{C}_1$ a full subcategory of $\mathcal{C}_2$ such that any object in $\mathcal{C}_2$ is isomorphic to some object in (the image of) $\mathcal{C}_1$. (Details are left to the reader as an exercise.)

(iv) If $\varphi : H \to G$ is a group homomorphism, then we have a functor $GSet \to HSet$ given by $S \mapsto S|_H$. Now, we can associate to any $\zeta \in [\underline{S}, \mathcal{C}]$ the H-equivariant $\mathcal{C}$-bundle $\zeta|_H$ over $S|_H$, which has the same fibres as ζ with the H-action defined by restricting the G-action to H via φ. We thus get a functor $[\underline{S}, \mathcal{C}] \to [\underline{S|_H}, \mathcal{C}]$. Note that this functor can also be derived from the canonical functor $\underline{S|_H} \to \underline{S}$ given by $s \mapsto s, (g, s) \mapsto (\varphi(g), s)$.

(v) If $\mathcal{C}$ is any category and S_1, S_2 are two G-sets, then

$$[\underline{S_1 \dot\cup S_2}, \mathcal{C}] \simeq [\underline{S_1}, \mathcal{C}] \times [\underline{S_2}, \mathcal{C}].$$

(vi) If $\varphi : S \to T$ is a G-map, then we have a functor $\underline{\varphi} : \underline{S} \to \underline{T}$ given by $s \mapsto \varphi(s), (g,s) \mapsto (g, \varphi(s))$ and hence a functor $\varphi^* : [\underline{T}, \mathcal{C}] \to [\underline{S}, \mathcal{C}]$ given by $\zeta \mapsto \zeta\underline{\varphi} = \varphi^*(\zeta)$, the bundle ζ restricted to S via φ.

Now, if $\mathcal{C}$ has finite sums, then φ also induces a functor $\varphi_* : [\underline{S}, \mathcal{C}] \to [\underline{T}, \mathcal{C}]$ defined as follows: if $\zeta \in [\underline{S}, \mathcal{C}]$, define $\varphi_*(\zeta) = \zeta_*$ where $\zeta_{*t} = \coprod_{s \in \varphi^{-1}(t)} \zeta_s$ and

$$\zeta_{*(g,t)} = \coprod_{s \in \varphi^{-1}(t)} \zeta_{(g,s)} : \zeta_{*t} = \coprod_{s \in \varphi^{-1}(t)} \zeta_s \to \coprod_{s \in \varphi^{-1}(t)} \zeta_{gs} = (\zeta_*)_{gt}$$

and for a morphism $\mu : \zeta \to \eta$ in $[\underline{S}, \mathcal{C}]$ define $\varphi_*(\mu) = \mu_* : \zeta_* \mapsto \eta_*$ in $[\underline{T}, \mathcal{C}]$ by $\mu_*(t) = \coprod_{s \in \varphi^{-1}(t)} \mu(s) : \zeta_{*t} \mapsto \eta_{*t}$.

Similarly, if $\mathcal{C}$ has finite products, then φ induces a functor $\check{\varphi}_* : [\underline{S}, \mathcal{C}] \to [\underline{T}, \mathcal{C}]$ where the fibres of $\check{\varphi}_*(\zeta)$ are defined by $\check{\varphi}_*(\zeta)_t = \prod_{s \in \varphi^{-1}(t)} \zeta_s (t \in T)$ and the G-action is defined accordingly.

(vii) In (vi) above, we saw that if $\mathcal{C}$ is a category with finite sums (resp. products), G a finite group, and $\varphi : S \to T$ a G-map, then we have functors $\varphi^* : \mathcal{C}^{\underline{T}} =: [\underline{T}, \mathcal{C}] \to \mathcal{C}^{\underline{S}} =: [\underline{S}, \mathcal{C}]$ and φ_* (resp. $\check{\varphi}_*$):$[\underline{S}, \mathcal{C}] \to [\underline{T}, \mathcal{C}]$.

We now realize that φ_* ($\check{\varphi}_*$) is the left (right) adjoint of φ^*, i.e., that $\mathcal{C}^{\underline{S}}(\zeta, \varphi^*(\eta)) \cong \mathcal{C}^{\underline{T}}(\varphi_*(\zeta), \eta)$ (resp. $\mathcal{C}^{\underline{S}}(\varphi^*(\eta), \zeta) \cong \mathcal{C}^{\underline{T}}(\eta, \check{\varphi}_*(\zeta))$).

This isomorphism is given by associating to each $\mu : \zeta \to \varphi^*\eta$ (resp. $\mu : \varphi^*(\eta) \to \zeta$) (i.e., to any family of maps $\mu(s) : \zeta_s \to \eta_{\varphi(s)}$ (resp. $\mu(s) : \eta_{\varphi(s)} \to \zeta_s$) compatible with the G-action) the morphism

$$\mu' : \varphi_*(\zeta) \to \eta \qquad (\mu' : \eta \to \check{\varphi}_*(\zeta))$$

given by

$$\mu'(t) = \coprod_{s \in \varphi^{-1}(t)} \mu(s) : \coprod_{s \in \varphi^{-1}(t)} \zeta_s \to \eta_t \quad /$$

$$\left(\text{resp.} \mu'(t) = \prod_{s \in \varphi^{-1}(t)} \mu(s) : \eta_t \to \prod_{s \in \varphi^{-1}(t)} \zeta_s \right).$$

1.2   Grothendieck group associated with a semi-group

1.2.1 Let $(A, +)$ be an Abelian semi-group. Define a relation '$\sim$'on $A \times A$ by $(a,b) \sim (c,d)$ if there exists $u \in A$ such that $a + d + u = b + c + u$. One can easily check that '$\sim$' is an equivalence relation. Let $\overline{A}$ denote the set of equivalence classes of '$\sim$', and write $[a, b]$ for the class of (a, b) under '$\sim$'. We define addition ($\dotplus$) on $\overline{A}$ by $[a, b] \dotplus [c, d] = [a + c, b + d]$. Then $(\overline{A}, \dotplus)$ is an Abelian group in which the identity element is $[a, a]$ and the inverse of $[a, b]$ is $[b, a]$.

Moreover, there is a well-defined additive map $f : A \rightarrow \overline{A} : a \rightarrow [a + a, a]$ which is, in general, neither injective nor surjective. However, f is injective iff A is a cancellation semi-group, i.e., iff $a + c = b + c$ implies that $a = b$ for all $a, b \in A$ (see [95]).

1.2.2 It can be easily checked that $\overline{A}$ possesses the following universal property with respect to the map $f : A \rightarrow \overline{A}$. Given any additive map $h : A \rightarrow B$ from A to an Abelian group B, then there exists a unique map $g : \overline{A} \rightarrow B$ such that $h = gf$.

Definition 1.2.1 $\overline{A}$ *is usually called the Grothendieck group of A or the group completion of A and denoted by* $K(A)$.

Remarks 1.2.1 (i) The construction of $K(A) = \overline{A}$ above can be shown to be equivalent to the following:

Let $(F(A), \dotplus)$ be the free Abelian group freely generated by the element of A, and $R(A)$ the subgroup of $F(A)$ generated by all elements of the form $a \dotplus b - (a + b)$, $a, b \in A$. Then $K(A) \simeq F(A)/R(A)$.

(ii) If A, B, C are Abelian semi-groups together with bi-additive map $f : A \times B \rightarrow C$, then f extends to a unique bi-additive map $\overline{f} : \overline{A} \times \overline{B} \rightarrow \overline{C}$ of the associated Grothendieck groups. If A is a semi-ring, i.e., an additive Abelian group together with a bi-additive multiplication $A \times A \rightarrow A \, (a, b) \rightarrow ab$, then the multiplication extends uniquely to a multiplication $\overline{A} \times \overline{A} \rightarrow \overline{A}$, which makes $\overline{A}$ into a ring (commutative if A is commutative) with identity $1 = [1 + 1, 1]$ in $\overline{A}$ if $1 \in A$.

(iii) If B is a semi-module over a semi-ring A, i.e., if B is an Abelian semi-group together with a bi-additive map $A \times B \rightarrow B : (a, b) \rightarrow a \cdot b$ satisfying $a'(ab) = (a'a)b$ for $a, a' \in A$, $b \in B$, then the associated Grothendieck group $\overline{B}$ is an $\overline{A}$-module.

(iv) If $A = \{1, 2, 3, \ldots\}$, $\overline{A} = K(A) = \mathbb{Z}$. Hence the construction in 1.2.1 is just a generalization of the standard procedure of constructing integers from the natural numbers.

(v) A sub-semi-group A of an Abelian semi-group B is said to be cofinal in B if for any $b \in B$, there exists $b' \in B$ such that $b + b' \in A$. It can be easily checked that $K(A)$ is a subgroup of $K(B)$ if A is cofinal in B.

1.2.3 K_0 of a ring. For any ring A with identity, let $\mathcal{P}(\Lambda)$ be the category of finitely generated projective Λ-modules. Then the isomorphism classes $I\mathcal{P}(\Lambda)$

of objects of $\mathcal{P}(\Lambda)$ form an Abelian semi-group under direct sum '$\oplus$'. We write $K_0(\Lambda)$ for $K(I\mathcal{P}(\Lambda))$ and call $K_0(\Lambda)$ the Grothendieck group of Λ. For any $P \in \mathcal{P}(\Lambda)$, we write (P) for the isomorphism class of P (i.e., an element of $I\mathcal{P}(\Lambda)$) and $[P]$ for the class of (P) in $K_0(\Lambda)$.

If Λ is commutative, then $I\mathcal{P}(\Lambda)$ is a semi-ring with tensor product $\otimes_\Lambda$ as multiplication, which distributes over '$\oplus$'. Hence $K_0(\Lambda)$ is a ring by remarks 1.2.1(ii).

Remarks 1.2.2   (i) $K_0 : \mathcal{R}ings \to \mathcal{A} : \Lambda \to K_0(\Lambda)$ is a functor — since any ring homomorphism $f : \Lambda \to \Lambda'$ induces a semi-group homomorphism $I\mathcal{P}(\Lambda) \to I\mathcal{P}(\Lambda') : P \to P \otimes \Lambda'$ and hence a group homomorphism $K_0(\Lambda) \to K_0(\Lambda')$.

(ii) K_0 is also a functor: $\mathcal{C}\mathcal{R}ings \to \mathcal{C}\mathcal{R}ings$.

(iii) $[P] = [Q]$ in $K_0(\Lambda)$ iff P is stably isomorphic to Q in $\mathcal{P}(\Lambda)$, i.e., iff $P \oplus \Lambda^n \simeq Q \oplus \Lambda^n$ for some integer n. In particular $[P] = \Lambda^n]$ for some n iff P is stably free (see [17, 20]).

Examples 1.2.1   (i) If Λ is a field or a division ring or a local ring or a principal ideal domain, then $K_0(\Lambda) \simeq \mathbb{Z}$.

Note. The proof in each case is based on the fact that any finitely generated Λ-module is free and Λ satisfies the invariant basis property (i.e., $\Lambda^r \simeq \Lambda^s \Rightarrow r = s$). So, $I\mathcal{P}(\Lambda) \simeq \{1, 2, 3, \ldots\}$, and so, $K_0(\Lambda) \simeq \mathbb{Z}$ by remarks 1.2.1(iv) (see [17] or [181]).

(ii) Any element of $K_0(\Lambda)$ can be written as $[P] - r[\Lambda]$ for some integer $r > 0$, $P \in \mathcal{P}(\Lambda)$, or as $s[\Lambda] - [Q]$ for some $s > 0$, $Q \in \mathcal{P}(\Lambda)$ (see [20, 211]). If we write $\tilde{K}_0(\Lambda)$ for the quotient of $K_0(\Lambda)$ by the subgroup generated by $[\Lambda]$, then every element of $\tilde{K}_0(\Lambda)$ can be written as $[P]$ for some $P \in \mathcal{P}[\Lambda]$ (see [20] or [224]).

(iii) If $\Lambda \simeq \Lambda_1 \times \Lambda_2$ is a direct product of two rings Λ_1, Λ_2, then $K_0(\Lambda) \simeq K_0(\Lambda_1) \times K_0(\Lambda_2)$ (see [17] for a proof).

(iv) Let G be a semi-simple connected affine algebraic group over an algebraically closed field. Let A be the coordinate ring of G. Then $K_0(A) \simeq \mathbb{Z}$.

Remarks. See [188] for a proof of this result, which says that all algebraic vector bundles on G are stably trivial. The result is due to A. Grothendieck.

(v) $K_0(k[x_0, x_1, \ldots, x_n]) \simeq \mathbb{Z}$. This result is due to J.P. Serre (see [188]).

Remarks 1.2.3 Before providing more examples of Grothendieck group constructions, we present in the next section 1.3 a generalization of 1.2.1 in the context of K_0 of symmetric monoidal categories.

1.3 K_0 of symmetric monoidal categories

Definition 1.3.1 *A symmetric monoidal category is a category* $\mathcal{C}$ *equipped with a functor* $\perp: \mathcal{C} \times \mathcal{C} \to \mathcal{C}$ *and a distinguished object "0" such that* $\perp$ *is "coherently associative and commutative" in the sense of Maclane (see [139]), that is,*

(i) $A \perp 0 \simeq A \simeq 0 \perp A$.

(ii) $A \perp (B \perp C) \simeq (A \perp B) \perp C$.

(iii) $A \perp B \simeq B \perp A$ *for all* $A, B, C \in \mathcal{C}$.

Moreover, the following diagrams commute.

(i)
$$(A \perp (0 \perp B)) \xrightarrow{\ \sim\ } (A \perp 0) \perp B$$
$$\downarrow{\scriptstyle \wr} \qquad\qquad\qquad \downarrow{\scriptstyle \wr}$$
$$A \perp B \xrightarrow{\ \sim\ } B \perp A$$

(ii)
$$A \perp 0 \xrightarrow{\ \sim\ } 0 \perp A$$
$$\searrow{\scriptstyle \sim} \qquad \swarrow{\scriptstyle \sim}$$
$$A$$

(iii)
$$A \perp (B \perp (C \perp D)) \xrightarrow{\ \sim\ } (A \perp B) \perp (C \perp D)$$
$$\downarrow{\scriptstyle \wr} \qquad\qquad\qquad\qquad \downarrow{\scriptstyle \wr}$$
$$A \perp ((B \perp (C \perp D)) \qquad ((A \perp B) \perp C) \perp D$$
$$\downarrow{\scriptstyle \wr} \qquad \nearrow{\scriptstyle \sim}$$
$$(A \perp (B \perp C)) \perp D$$

Let $I\mathcal{C}$ be the set of isomorphism classes of objects of $\mathcal{C}$. Clearly, if $\mathcal{C}$ is small, then $(I\mathcal{C}, \perp)$ is an Abelian semi-group (in fact a monoid), and we write $K_0^{\perp}(\mathcal{C})$ for $K(I\mathcal{C}, \perp)$ or simply $K_0(\mathcal{C})$ when the context is clear.

In other words, $K_0^{\perp}(\mathcal{C}) = F(\mathcal{C})/R(\mathcal{C})$ where $F(\mathcal{C})$ is the free Abelian group on the isomorphism classes $(\mathcal{C})$ of $\mathcal{C}$-objects, and $R(\mathcal{C})$ the subgroup of $F(\mathcal{C})$ generated by $(C' \perp C'') - (C') - (C'')$ for all C', C'' in ob$(\mathcal{C})$.

Remarks 1.3.1 (i) $K_0^\perp(\mathcal{C})$ satisfies a universal property as in 1.2.2.

(ii) If $\mathcal{C}$ has another composition 'o' that is associative and distributive with respect to $\perp$, then $K_0^\perp(\mathcal{C})$ can be given a ring structure through 'o' as multiplication and we shall sometimes denote this ring by $K_0^\perp(\mathcal{C}, \perp, o)$ or $K_0(\mathcal{C}, \perp, o)$ or just $K_0(\mathcal{C})$ if the context is clear.

Examples 1.3.1 (i) If Λ is any ring with identity, then $(\mathcal{P}(\Lambda), \oplus)$ is a symmetric monoidal category (s.m.c.) and $K_0^\oplus(\Lambda) = K_0(\Lambda)$ as in 1.2.5. If Λ is commutative, then $K_0^\oplus(\Lambda)$ is a ring where $(\mathcal{P}(\Lambda), \oplus)$ has the further composition '$\otimes$'.

(ii) Let $F\mathcal{S}et$ be the category of finite sets, $\dot\cup$ the disjoint union. Then $(F\mathcal{S}et, \dot\cup)$ is a symmetric monoidal category and $K_0^{\dot\cup}(F\mathcal{S}et) \simeq \mathbb{Z}$.

(iii) Let G be a finite group, $\mathcal{C}$ any small category. Let $\mathcal{C}_G$ be the category of G-objects in $\mathcal{C}$, or equivalently, the category of G-representations in $\mathcal{C}$, i.e., objects of $\mathcal{C}_G$ are pairs $(X, U : G \to \mathrm{Aut}(X))$ where $X \in \mathrm{ob}(\mathcal{C})$ and U is a group homomorphism from G to the group of $\mathcal{C}$-automorphism of X. If $(\mathcal{C}, \perp)$ is a symmetric monoidal category, so is $(\mathcal{C}_G, \dot\perp)$ where for

$$(X, U : G \to \mathrm{Aut}(X)), \qquad (X', U' : G \to \mathrm{Aut}(X'))$$

in $\mathcal{C}_G$, we define

$$(X, U)\dot\perp(X', U') := (X \perp X', U \perp U' : G \to \mathrm{Aut}(X \perp X')),$$

where $U \perp U'$ is defined by the composition

$$G \overset{U \perp U'}{\longrightarrow} \mathrm{Aut}(X) \times \mathrm{Aut}(X') \to \mathrm{Aut}(X \perp X').$$

So we obtain the Grothendieck group $K_0^\perp(\mathcal{C}_G)$.

If $\mathcal{C}$ possesses a further associative composition 'o' such that $\mathcal{C}$ is distributive with respect to $\perp$ and 'o', then so is $\mathcal{C}_G$, and hence $K_0^\perp(\mathcal{C}_G)$ is a ring.

Examples 1.3.2 (a) If $\mathcal{C} = \mathcal{P}(R)$, $\perp = \oplus$, 'o' $= \otimes_R$ where R is a commutative ring with identity, then $\mathcal{P}(R)_G$ is the category of RG-lattices (see [39]), and $K_0(\mathcal{P}(R)_G)$ is a ring usually denoted by $G_0(R, G)$. Observe that when $R = \mathbb{C}$, $G_0(\mathbb{C}, G)$ is the usual representation ring of G denoted in the literature by $R(G)$.

(b) If $\mathcal{C} = F\mathcal{S}ets$, '$\perp$' $=$ disjoint union, 'o' $=$ Cartesian product. Then $K_0(\mathcal{C}_G)$ is the Burnside ring of G usually denoted by $\Omega(G)$. See 9.3 and 9.4 for a detailed discussion of Burnside rings.

(iv) Let G be a finite group, S a G-set. As discussed in 1.1.3, we can associate with S a category **S** as follows: $\text{ob}(\mathbf{S}) = \{s|s \in S\}$. For $s, t \in S$, $\text{Hom}_{\mathbf{S}}(s, t) = \{(g, s)|s \in G, gs = t\}$ where the composition is defined for $t = gs$ by $(h, t) \cdot (g, s) = (hg, s)$, and the identity morphism $s \to s$ is given by (e, s) where e is the identity element of G. Now let $(\mathcal{C}, \perp)$ be a symmetric monoidal category and let $[\mathbf{S}, \mathcal{C}]$ be the category of covariant functors $\zeta : \mathbf{S} \to \mathcal{C}$. The $([\mathbf{S}, \mathcal{C}], \perp)$ is also a symmetric monoidal category where $(\zeta \perp \eta)_{(g,s)} : \zeta_s \perp \eta_s \to \zeta_{gs} \perp \eta_{gs}$. We write $K_0^G(S, \mathcal{C})$ for the Grothendieck group of $[\mathbf{S}, \mathcal{C}]$.

If $(\mathcal{C}, \perp)$ possesses an additional composition 'o' that is associative and distributive with respect to '$\perp$', then $K_0(\mathbf{S}, \mathcal{C})$ can be given a ring structure (see [111]). As we shall see in chapter 9, for any symmetric monoidal category $(\mathcal{C}, \perp)$, $K_0^G(-, \mathcal{C}) : GSet \to \mathcal{A}b$ is a 'Mackey functor' (see example 9.1.1(iv)), and when $\mathcal{C}$ possesses an additional composition 'o' discussed before, then $K_0^G(-, \mathcal{C}) : GSet \to \mathcal{A}b$ is a 'Green functor' (see example 9.1.1(iv)).

(v) Suppose that G, H are finite groups, and $\theta : H \to G$ a group homomorphism. By restricting the action of G on a G-set S to H via θ, one defines a functor $\hat{\theta} : GSet \to HSet$, which can easily be checked to commute with finite sums, products, and pullbacks (and more generally, with limit and colimits). Moreover, by restricting the action of G on G-equivariant $\mathcal{C}$-bundle ζ over S to H through θ, we have a natural transformation of functors from $K_0^G(-, \mathcal{C}) : GSet \to \mathbb{Z}\text{-}\mathcal{M}od$ and $K_0^H \circ \theta : GSet \to HSet \to \mathbb{Z}\text{-}\mathcal{M}od$. In particular, if $H \le G$, T an H-set, we have a homomorphism $K_0^G(\underset{H}{G \times T}, \mathcal{C}) \to K_0^H(T, \mathcal{C})$ where the second map is induced by $T \to \underset{H}{G \times T} : t \to (e, t)$. We now observe that

$$K_0^G \left(\underset{H}{G \times T}, \mathcal{C} \right) \to K_0^H(T, \mathcal{C}) \qquad (\text{I})$$

is an isomorphism since by 1.1.5 iii(c) $[\underline{\underset{H}{G \times T}, \mathcal{C}}] \to [\underline{T}, \mathcal{C}]$ is an equivalence of categories.

Note that if $T = H/K$ for some subgroup $K \le H$, we have $\underset{H}{G \times T} = G/K$, and the above isomorphism (I) is the map $K_0^G(G/K, \mathcal{C}) \to K_0(H/K, \mathcal{C})$ defined by restricting a $\mathcal{C}$-bundle over G/K to H/K and the action of G to H at the same time, i.e., the map $K_0^G(G/K, \mathcal{C}) \to K_0(H/K, \mathcal{C})$ is defined by the functor $\underline{H/K \to G/K}$.

(vi) Let X be a compact topological space and for $= \mathbb{R}$ or $\mathbb{C}$, let $\mathbf{VB}_F(X)$ be the (symmetric monoidal) category of (finite-dimensional) vector bundles on X. Then $I\!\mathbf{VB}_F(X)$ is an Abelian monoid under Whitney sum '$\oplus$'. It is usual to write $KO(X)$ for $K_0^\oplus(\mathbf{VB}_{\mathbb{R}}(X))$ and $KU(X)$ for $K_0^\oplus(\mathbf{VB}_{\mathbb{C}}(X))$. Note that if X, Y are homotopy equivalent, then $KO(X) = KO(Y)$ and $KU(X) = KU(Y)$. Moreover, if X is contractible, we have $KO(X) = KU(X) = \mathbb{Z}$ (see [10] or [95]).

Let X be a compact space, $\mathbb{C}(X)$ the ring of $\mathbb{C}$-valued functions on X. By a theorem of R.G. Swan [208, 214], there exists an equivalence of categories $\Gamma : VB_{\mathbb{C}}(X) \to \mathcal{P}(\mathbb{C}(X))$ taking a vector bundle $E \xrightarrow{p} X$ to $\Gamma(E)$, where $\Gamma(E) = \{\text{sections } s : X \to E | ps = 1\}$. This equivalence induces a group isomorphism $KU(X) \simeq K_0(\mathbb{C}(X))$ (I).

The isomorphism (I) provides the basic initial connection between algebraic K-theory (right-hand side of (I)) and topological K-theory (left-hand side of (I)) since the K-theory of $\mathcal{P}(\Lambda)$ for an arbitrary ring Λ could be studied instead of the K-theory of $\mathcal{P}(\mathbb{C}(X))$.

Now, $\mathcal{C}(X)$ is a commutative C^*-algebra, and the Gelfand–Naimark theorem [35] says that any commutative C^* algebra Λ has the form $\Lambda = \mathbb{C}(X)$ for some locally compact space X. Indeed, for any commutative C^*-algebra Λ, we could take X as the spectrum of Λ, i.e., the set of all nonzero homomorphisms from Λ to $\mathbb{C}$ with the topology of pointwise convergence. Noncommutative geometry is concerned with the study of noncommutative C^*-algebras associated with "noncommutative" s-paces and K-theory (algebraic and topological) of such C^*-algebras has been extensively studied and connected to some (co)homology theories (see, e.g., Hochschild and cyclic (co)homology theories) of such algebras through Chern characters (see, e.g., [35, 43, 44, 136]).

(vii) Let G be a group acting continuously on a topological space X. The category $VB_G(X)$ of complex G-vector bundles on X is symmetric monoidal under Whitney sum '$\oplus$', and we write $K_G^0(X)$ for the Grothendieck group $K_0(VB_G(X))$. If X is a point, $VB_G(X)$ is the category of representations of G in $\mathcal{P}(\mathbb{C})$ and $K_G^0(X) = R(G)$, the representation ring of G.

If G acts trivially on X, then $K_G^0(X) \simeq KU(X) \otimes_{\mathbb{Z}} R(G)$ (see [184, 186]).

(viii) Let R be a commutative ring with identity. Then $\mathrm{Pic}(R)$, the category of finitely generated projective R-modules of rank one (or equivalently the category of algebraic line bundles L over R) is a symmetric monoidal category, and $K_0^{\otimes}(\mathrm{Pic}(R)) = \mathrm{Pic}(R)$, the Picard group of R.

(ix) The category $\mathrm{Pic}(X)$ of line bundles on a locally ringed space is a symmetric monoidal category under '$\otimes$', and $K_0^{\otimes}(\mathrm{Pic}(X)) := \mathrm{Pic}(X)$ is called the Picard group of X. Observe that when $X = \mathrm{Spec}(R)$, we recover $\mathrm{Pic}(R)$ in (viii). It is well known that $\mathrm{Pic}(X) \simeq H^1(X, O_X^*)$ (see [79] or [198]).

The significance of discussing ringed spaces and schemes in this book lies in the fact that results on affine schemes are results on commutative rings and hence apply to commutative orders and grouprings.

(x) Let R be a commutative ring with identity. An R-algebra Λ is called an Azumaya algebra if there exists another R-algebra Λ' such that $\Lambda \otimes_R \Lambda' \simeq M_n(R)$ for some positive integer n. Let $\text{Az}(R)$ be the category of Azumaya algebras. Then $(\text{Az}(R), \otimes_R)$ is a symmetric monoidal category. Moreover, the category $FP(R)$ of faithfully projective R-modules is symmetric monoidal with respect to $\perp = \otimes_R$ if the morphisms in $FP(R)$ are restricted to isomorphisms. There is a monoidal functor $FP(R) \to \text{Az}(R) : P \to \text{End}(P)$ inducing a group homomorphism $K_0(FP(R)) \xrightarrow{\varphi} K_0(\text{Az}(R))$. The cokernel of φ is called the Brauer group of R and is denoted by $\text{Br}(R)$. Hence $\text{Br}(R)$ is the Abelian group generated by isomorphism classes of central simple F-algebras with relations $[\Lambda \otimes \Lambda'] = [\Lambda] + [\Lambda']$ and $[M_n(F)] = 0$ (see [181]).

(xi) Let A be an involutive Banach algebra and $\text{Witt}(A)$ the group generated by isomorphism classes $[Q]$ of invertible Hermitian forms Q on $P \in \mathcal{P}(A)$ with relations $[Q_1 \oplus Q_2] = [Q_1] + [Q_2]$ and $[Q] + [-Q] = 0$. Define a map $\varphi : K_0(A) \to \text{Witt}(A)$ by $[P] \mapsto$ class of (P, Q) with Q positive. If A is a C^*-algebra with 1, then there exists on any $P \in \mathcal{P}(A)$ an invertible form Q satisfying $Q(x,x) \geq 0$ for all $x \in P$ and in this case $\varphi : K_0(A) \to \text{Witt}(A)$ is an isomorphism. However, φ is not an isomorphism in general for arbitrary involutive Banach algebras (see [35]).

(xii) Let F be a field and $\text{Sym}B(F)$ the category of symmetric inner product spaces (V, β) – V a finite-dimensional vector space over F and $\beta : V \otimes V \to F$ a symmetric bilinear form. Then $(\text{Sym}B(F), \perp)$ is a symmetric monoidal category where $(V, \beta) \perp (V', \beta')$ is the orthogonal sum of (V, β) and (V', β') is defined as the vector space $V \oplus V'$ together with a bilinear form $\beta^* : (V \oplus V', V \oplus V') \to F$ given by $\beta^*(v \oplus v', v_1 \oplus v_1') = \beta(v, v_1) + \beta'(v', v_1')$.

If we define composition $(V, \beta) \odot (V', \beta')$ as the tensor product $V \otimes V'$ together with a bilinear form $\beta^*(v \otimes v', v_1 \otimes v_1') = \beta(v, v_1)\beta'(v', v_1')$, then $K_0(\text{Sym}B(F), \perp, \odot)$ is a commutative ring with identity.

The Witt ring $W(F)$ is defined as the quotient of $K_0(\text{Sym}B(F))$ by the subgroup $\{n, H\}$ generated by the hyperbolic plane $H = \left(F^2, \left(\begin{smallmatrix} 0 & 1 \\ 1 & 0 \end{smallmatrix}\right)\right)$.

For more details about $W(F)$ see [182].

(xiii) Let A be a ring not necessarily unital (i.e., not necessarily with identity). The ring A_+ obtained by adjoining a unit to A is defined as follows. As an Abelian group, $A_+ = A \oplus \mathbb{Z}$ with multiplication defined by $(a, r)(b, s) = (ab + rb + sa, rs)$ where $a, b \in A$, $r, s \in \mathbb{Z}$. Here the unit of A_+ is $(0, 1)$.

If A already has unit e, say, there is a unital isomorphism $\varphi : A_+ \to A \times \mathbb{Z}$ given by $\varphi(a, r) = (a + re, r)$.

If A is not-unital, there is a split exact sequence $0 \to A \to A_+ \to \mathbb{Z} \to 0$.

Define. $K_0(A) := \mathrm{Ker}(K_0(A_+) \to K_0(\mathbb{Z})) \simeq \mathbb{Z}$.

(xiv) For any Λ with identity, let $M_n(\Lambda)$ be the set of $n \times n$ matrices over Λ, and write $M(\Lambda) = \bigcup_{n=1}^{\infty} M_n(\Lambda)$. Also $GL_n(\Lambda)$ be the group of invertible $n \times n$ matrices over Λ and write $GL(\Lambda) = \bigcup_{n=1}^{\infty} GL_n(\Lambda)$. For $P \in \mathcal{P}(\Lambda)$ there exists $Q \in \mathcal{P}(\Lambda)$ such that $P \oplus Q \simeq \Lambda^n$ for some n. So, we can identify with each $P \in \mathcal{P}(\Lambda)$ an idempotent matrix $p \in M_n(\Lambda)$ (i.e., $p : \Lambda^n \to \Lambda^n$), which is the identity on P and '0' on Q.

Note that if p, q are idempotent matrices in $M(\Lambda)$, say $p \in M_r(\Lambda)$, $q \in M_s(\Lambda)$, corresponding to $P, Q \in \mathcal{P}(\Lambda)$, then $P \simeq Q$ iff it is possible to enlarge the sizes of p, q (by possibly adding zeros in the lower right-hand corners) such that p, q have the same size ($t \times t$, say) and are conjugate under the action of $GL_t(\Lambda)$ (see [181]).

Let $\mathrm{Idem}(\Lambda)$ be the set of idempotent matrices in $M(\Lambda)$. It follows from the last paragraph that $GL(\Lambda)$ acts by conjugation on $\mathrm{Idem}(\Lambda)$, and so, we can identify the semi-group $I\mathcal{P}(\Lambda)$ with the semi-group of conjugation orbits $(\mathrm{Idem}(\Lambda))^{\check{}}$ of the action of $GL(\Lambda)$ on $\mathrm{Idem}(\Lambda)$ where the semi-group operation is induced by $(p,q) \to \begin{pmatrix} p & 0 \\ 0 & q \end{pmatrix}$. $K_0(\Lambda)$ is the Grothendieck group of this semi-group $(\mathrm{Idem}(\Lambda))^{\check{}}$.

1.4 K_0 of exact categories – definitions and examples

Definition 1.4.1 *An exact category is an additive category $\mathcal{C}$ embeddable as a full subcategory of an Abelian category $\mathcal{A}$ such that $\mathcal{C}$ is equipped with a class $\mathcal{E}$ of short exact sequences $0 \to M' \to M \to M'' \to 0$ (I) satisfying*

(i) *$\mathcal{E}$ is a class of sequences (I) in $\mathcal{C}$ that are exact in $\mathcal{A}$.*

(ii) *$\mathcal{C}$ is closed under extensions in $\mathcal{A}$, i.e., if (I) is an exact sequence in $\mathcal{A}$ and $M', M'' \in \mathcal{C}$, then $M \in \mathcal{C}$.*

Definition 1.4.2 *For a small exact category $\mathcal{C}$, define the Grothendieck group $K_0(\mathcal{C})$ of $\mathcal{C}$ as the Abelian group generated by isomorphism classes (C) of $\mathcal{C}$-objects subject to the relation $(C') + (C'') = (C)$ whenever $0 \to C' \to C \to C'' \to 0$ is an exact sequence in $\mathcal{C}$.*

Remarks 1.4.1 (i) $K_0(\mathcal{C}) \simeq \mathcal{F}/\mathcal{R}$ where $\mathcal{F}$ is the free Abelian group on the isomorphism classes (C) of $\mathcal{C}$-objects and $\mathcal{R}$ the subgroup of $\mathcal{F}$ generated by all $(C) - (C') - (C'')$ for each exact sequence $0 \to C' \to C \to C'' \to 0$ in $\mathcal{C}$. Denote by $[C]$ the class of (C) in $K_0(\mathcal{C}) = \mathcal{F}/\mathcal{R}$.

(ii) The construction satisfies the following property: If $\chi : \mathcal{C} \to A$ is a map from $\mathcal{C}$ to an Abelian group A given that $\chi(C)$ depends only on the isomorphism class of $\mathcal{C}$ and $\chi(C) = \chi(C') + \chi(C'')$ for any exact sequence $0 \to C' \to C \to C'' \to 0$, then there exists a unique $\chi' : K_0(\mathcal{C}) \to A$ such that $\chi(C) = \chi'([C])$ for any $\mathcal{C}$-object C.

(iii) Let $F : \mathcal{C} \to \mathcal{D}$ be an exact functor between two exact categories $\mathcal{C}, \mathcal{D}$ (i.e., F is additive and takes short exact sequences in $\mathcal{C}$ to such sequences in $\mathcal{D}$). Then F induces a group homomorphism $K_0(\mathcal{C}) \to K_0(\mathcal{D})$.

(iv) Note that an Abelian category $\mathcal{A}$ is also an exact category and the definition of $K_0(\mathcal{A})$ is the same as in definition 1.4.2.

Examples 1.4.1   (i) Any additive category is an exact category as well as a symmetric monoidal category under '$\oplus$', and $K_0(\mathcal{C})$ is a quotient of the group $K_0^{\oplus}(\mathcal{C})$ defined in 1.3.1.

If every short exact sequence in $\mathcal{C}$ splits, then $K_0(\mathcal{C}) = K_0^{\oplus}(\mathcal{C})$. For example, $K_0(\Lambda) = K_0(\mathcal{P}(\Lambda)) = K_0^{\oplus}(\mathcal{P}(\Lambda))$ for any ring Λ with identity.

(ii) Let Λ be a (left) Noetherian ring. Then the category $\mathcal{M}(\Lambda)$ of finitely generated (left)-Λ-modules is an exact category and we denote $K_0(\mathcal{M}(\Lambda))$ by $G_0(\Lambda)$. The inclusion functor $\mathcal{P}(\Lambda) \to \mathcal{M}(\Lambda)$ induces a map $K_0(\Lambda) \to G_0(\Lambda)$ called the Cartan map. For example, $\Lambda = RG$ (R a Dedekind domain, G a finite group) yields a Cartan map $K_0(RG) \to G_0(RG)$.

If Λ is left Artinian, then $G_0(\Lambda)$ is free Abelian on $[S_1], \ldots, [S_r]$ where the $[S_i]$ are distinct classes of simple Λ-modules, while $K_0(\Lambda)$ is free Abelian on $[I_1], \ldots, [I_l]$ and the $[I_i]$ are distinct classes of indecomposable projective Λ-modules (see [39]). So, the map $K_0(\Lambda) \to G_0(\Lambda)$ gives a matrix a_{ij} where a_{ij} = the number of times S_j occurs in a composition series for I_i. This matrix is known as the Cartan matrix.

If Λ is left regular (i.e., every finitely generated left Λ-module has finite resolution by finitely generated projective left Λ-modules), then it is well known that the Cartan map is an isomorphism (see [215]).

(iii) Let R be a Dedekind domain with quotient field F. An important example of (ii) above is when Λ is an R-order in a semi-simple F-algebra Σ. Recall (see [39, 174]) that Λ is a subring of Σ such that R is contained in the centre of Λ, Λ is a finitely generated R-module, and $F \otimes_R \Lambda = \Sigma$. For example, if G is any finite group, then the group-ring RG is an R-order in the group algebra FG.

Recall also that a maximal R-order Γ in Σ is an order that is not contained in any other R-order. Note that Γ is regular (see [38, 39]). So, as in (ii) above, we have Cartan maps $K_0(\Lambda) \to G_0(\Lambda)$ and when Λ, is a maximal order, we have $K_0(\Lambda) \simeq G_0(\Lambda)$.

(iv) Let R be a commutative ring with identity, Λ an R-algebra. Let $\mathcal{P}_R(\Lambda)$ be the category of left Λ-lattices, i.e., Λ-modules that are finitely generated and projective as R-modules. Then $\mathcal{P}_R(\Lambda)$ is an exact category and we write $G_0(R, \Lambda)$ for $K_0(\mathcal{P}_R(\Lambda))$. If $\Lambda = RG$, G, a finite group, we write $\mathcal{P}_R(G)$ for $\mathcal{P}_R(RG)$ and also write $G_0(R, G)$ for $G_0(R, RG)$. If $M, N \in \mathcal{P}_R(\Lambda)$, then, so is $(M \otimes_R N)$, and hence the multiplication given in $G_0(R, G)$ by $[M][N] = (M \otimes_R N)$ makes $G_0(R, G)$ a commutative ring with identity.

(v) If R is a commutative regular ring and Λ is an R-algebra that is finitely generated and projective as an R-module (e.g., $\Lambda = RG$, G a finite group or R is a Dedekind domain with quotient field F, and Λ is an R-order in a semi-simple F-algebra), then $G_0(R, \Lambda) \simeq G_0(\Lambda)$.

Sketch of proof. Define a map $\varphi : G_0(R, \Lambda) \to G_0(\Lambda)$ by $\varphi[M] = [M]$. Then φ is a well-defined homomorphism. Now for $M \in \mathcal{M}(\Lambda)$, there exists an exact sequence $0 \to L \to P_{n-1} \xrightarrow{\varphi_{n-1}} P_{n-2} \to \cdots P_0 \to M \to 0$ where $P_i \in \mathcal{P}(\Lambda)$, $L \in \mathcal{M}(\Lambda)$. Now, since $\Lambda \in \mathcal{P}(R)$, each $P_i \in \mathcal{P}(R)$ and hence $L \in \mathcal{P}(R)$. So $L \in \mathcal{P}_R(\Lambda)$. Now define $\delta[M] = [P_0] - [P_1] + \cdots + (-1)^{n-1}[P_{n-1}] + (-1)^n[L] \in G_0(R, \Lambda)$. One easily checks that $\delta f = 1 = f\delta$.

(vi) Let G be a finite group, S a G-set, $\underline{S}$ the category associated to S (see 1.1.3), $\mathcal{C}$ an exact category, and $[\underline{S}, \mathcal{C}]$ the category of covariant functors $\zeta : \underline{S} \to \mathcal{C}$. We write ζ_s for $\zeta(s)$, $s \in S$. Then, $[\underline{S}, \mathcal{C}]$ is an exact category where the sequence $0 \to \zeta' \to \zeta \to \zeta'' \to 0$ in $[\underline{S}, \mathcal{C}]$ is defined to be exact if $0 \to \zeta'_s \to \zeta_s \to \zeta''_s \to 0$ is exact in $\mathcal{C}$ for all $s \in S$. Denote by $K_0^G(S, \mathcal{C})$ the K_0 of $[\underline{S}, \mathcal{C}]$. Then $K_0^G(-, \mathcal{C}) : G\mathcal{S}et \to Ab$ is a functor that can be seen to be a 'Mackey' functor. We shall prove this fact for $K_n^G(-, \mathcal{C}), n \geq 0$ in chapter 10 (see theorem 10.1.2).

As seen earlier in 1.1.5, if $\underline{S} = G/G$, the $[G/G, \mathcal{C}] \simeq \mathcal{C}_G$ in the notation of 1.1.2. Also, constructions analogous to the one above will be done for G, a profinite group, in chapter 11, and compact Lie groups in chapter 12.

Now if R is a commutative Noetherian ring with identity, we have $[G/G, \mathcal{P}(R)] \simeq \mathcal{P}(R)_G \simeq \mathcal{P}_R(RG)$, and so, $K_0^G((G/G, \mathcal{P}(R)) \simeq G_0(R, G) \simeq G_0(RG)$. This provides an initial connection between K-theory of representations of G in $\mathcal{P}(R)$ and K-theory of the group ring RG. As observed in (iv) above $G_0(R, G)$ is also a ring.

In particular, when $R = \mathbb{C}$, $\mathcal{P}(\mathbb{C}) = \mathcal{M}(\mathbb{C})$, and $K_0(\mathcal{P}(\mathbb{C})_G \simeq G_0(\mathbb{C}, G) = G_0(\mathbb{C}G) =$ the Abelian group of characters, $\chi : G \to \mathbb{C}$ (see [39]), as already observed in this chapter.

If the exact category $\mathcal{C}$ has a pairing $\mathcal{C} \times \mathcal{C} \to \mathcal{C}$, which is naturally associative and commutative, and there exists $E \in \mathcal{C}$ such that $\langle E, M \rangle =$

$\langle M, E \rangle = M$ for all $M \in \mathcal{C}$, then $K_0^G(-, \mathcal{C})$ is a Green functor (see [52]), and we shall see also in 10.1.6 that for all $n \geq 0$, $K_n^G(-, \mathcal{C})$ is a module over $K_0^G(-, \mathcal{C})$.

(vii) Let k be a field of characteristic p, G a finite group. We write $a(G)$ for $K_0(\mathcal{M}(kG))$. Let H be a subgroup of G.

A sequence $0 \to M' \to M \to M'' \to 0$ (I) of modules in $\mathcal{M}(kG)$ is said to be H-split if upon restriction to H, (I) is a split exact sequence. Write $a(G, H)$ for K_0 of the exact category $\mathcal{M}(kG)$ with respect to the collection of H-split exact sequences. For $X, Y \in \mathcal{M}(kG)$, $X \otimes_R Y \in \mathcal{M}(kG)$, with $g(x \otimes y) = gx \otimes gy$, $g \in G$. If $0 \to M' \to M \to M'' \to 0$ is an H-split exact sequence in $\mathcal{M}(kG)$, so is $0 \to M' \otimes X \to M \otimes X \to M'' \otimes X \to 0$. If we put $[M][X] = [M \otimes_R X]$, then $a(G, H)$ is a commutative ring with identity element $[1_G]$. Call $a(G, H)$ the relative Grothendieck ring with respect to H. This ring has been well studied (see [128, 129, 130]).

For example, if $H = 1$, then $a(G, 1)$ is $\mathbb{Z}$-free on $[F_1], \ldots, [F_s]$ where $\{[F_i]\}$ is a finite set of non-isomorphic irreducible G-modules. Also $a(G, 1) \simeq \sum_{i=1}^{s} \mathbb{Z} \varphi_i$ where $\{\varphi_i\}$ are the irreducible Brauer characters of G relative to k. Also $a(G, 1)$ contains no non-zero nilpotent element. If $H = G$, $a(G, G)$ is a free $\mathbb{Z}$-module spanned by the indecomposable modules, and $a(G, G)$ is called the representation ring of kG.

(viii) Let $H \leq G$. A module $N \in \mathcal{M}(kG)$ is (G, H)-projective if every exact sequence $0 \to M' \to M \to N \to 0$ of kG-modules that is H-split is also G-split. Note that N is (G, H)-projective iff N is a direct summand of some induced module $V^G := kG \otimes_{kH} V$ where $V \in \mathcal{M}(kH)$.

- Let $\mathcal{P}_H :=$ category of all (G, H)-projective modules $P \in \mathcal{M}(kG)$, $\mathcal{E}_H :=$ collection of H-split (and hence G-split) sequences in $\mathcal{P}_H$. Let $\underline{p}(G, H)$ be the K_0 of the exact category $\mathcal{P}_H$ with respect to $\mathcal{E}_H$.
  Then, $\underline{p}(G, G) = a(G)$, $\underline{p}(G, H)$ is an ideal of $a(G)$, and $\underline{p}(G, H)$ is $\mathbb{Z}$-free on the indecomposable projective kG-modules. If $i(G, H)$ is the additive subgroup of $a(G)$ generated by all $[M] - [M'] - [M'']$ where $0 \to M' \to M \to M'' \to 0$ ranges over all H-split exact sequences of kG-modules, then $i(G, H)$ is an ideal of $a(G)$ and $a(G, H) \simeq a(G)/i(G, H)$.

Also we have the Cartan map $\underline{p}(G, H) \to a(G, H)$ defined by $\underline{p}(G, H) \hookrightarrow a(G) \to a(G)/i(G, H) = a(G, H)$.

(ix) We have the following generalization of (vii) and (viii) above (see [49]). Let G be a finite group, S a G-set, k a field of characteristic p. Then

we have an exact functor: $F_S : \mathcal{M}(kG) \to \mathcal{M}(kG)$ defined by $M \to$
$M[S] := \left\{ \sum_{s \in S} m_s s \mid m_s \in M \right\}$, the generalized permutation module.

- Note that $M[S]$ is a kG-module and $M[S] \simeq M \otimes_k k[S]$.

- An exact sequence of modules in $\mathcal{M}(kG)$ is said to be S-split if
  $0 \to M'[S] \to M[S] \to M''[S] \to 0$ is split exact.

- A kG-module P is S-projective if any S-split sequence $0 \to M' \to$
  $M \to P \to 0$ of kG-modules splits.

- Note that M is S-projective, iff M is a direct summand of $M[S]$
  (see [49] and that if $P \in \mathcal{M}(kG)$ is S-projective, and $M \in \mathcal{M}(kG)$,
  then $M \otimes_k P$ is S-projective.

- E.g., If $H \leq G$, $S = G/H$, then a sequence $0 \to M' \to M \to$
  $M'' \to 0$ in $\mathcal{M}(kG)$ is S-split iff it is H-split, and $M \in \mathcal{M}(kG)$ is
  (G, H)projective iff it is (G/H)-projective.

- Let $\underline{p}_S(G)$ be the additive subgroup of $a(G)$ generated by isomorphism classes of S-projective modules, and $i_S(G)$ the additive subgroup of $a(G)$ generated by Euler characteristics $[M'] - [M] + [M'']$
  of S-split exact sequences $0 \to M' \to M \to M'' \to 0$. Then $p_S(G)$,
  $i_S(G)$ are ideals of $a(G)$, and $\underline{p}_S(G) \cdot i_S(G) = 0$ (see [49]).

- If $S^G \neq \emptyset$, then $\underline{p}_S(G) = a(G)$, $i_S(G) = 0$.

- Define relative Grothendieck ring of kG-modules with respect to S
  by $a_S(G) := a(G)/i_S(G)$.

- We also have Cartan map $\underline{p}_S(G) \hookrightarrow a(G) \twoheadrightarrow a_S(G)$.

1.4.1

(i) We noted in 1.1.4 that we shall need to discuss in chapter 14 K-theory
    of the functor category $[\mathcal{G}, \mathcal{C}]$ where $\mathcal{G}$ is a groupoid. We also note that,
    if $\mathcal{C}$ is an exact category, then $[\mathcal{G}, \mathcal{C}]$, $[\mathcal{G}, \mathcal{C}]'$ are also exact categories. We
    shall write $Sw(\mathcal{G})$ for the 'Swan group' $K_0([\mathcal{G}, \mathcal{M}(\mathbb{Z})]')$ and $Sw^f(\mathcal{G}) :=$
    $K_0([\mathcal{G}, \mathcal{F}(\mathbb{Z})]')$ where $\mathcal{F}(\mathbb{Z})$ is the category of finitely generated free $\mathbb{Z}$-
    module. Note that the forgetful functor yields a bijection $Sw^f(\mathcal{G}) \simeq$
    $Sw(\mathcal{G})$ (see [14, 209]).

    For $M, N \in [\mathcal{G}, \mathcal{M}(\mathbb{Z})]'$, $M \otimes_{\mathbb{Z}} N \in \mathcal{G}, \mathcal{M}(\mathbb{Z})]'$ and $\otimes_{\mathbb{Z}}$ induce a parcing
    $Sw^f(\mathcal{G}) \otimes Sw^f(\mathcal{G}) \to Sw^f(\mathcal{G})$.

    Hence $Sw^f(\mathcal{G})$ is a commutative ring with the class of constant con-
    travariant functor $M : \mathcal{G} \to \mathcal{M}(\mathbb{Z})$ with constant value $\mathbb{Z}$ as a unit.

(ii) Let $\mathcal{G}, \mathcal{G}'$ be groupoids and $F : \mathcal{G} \to \mathcal{G}'$ a functor. Restriction with
     F yields an exact functor $[\mathcal{G}', \mathcal{F}(\mathbb{Z})]' \to [\mathcal{G}, \mathcal{F}(\mathbb{Z})]'$ and $[\mathcal{G}', \mathcal{M}(\mathbb{Z})]' \to$

$[\mathcal{G}, \mathcal{M}(\mathbb{Z})]'$ and hence a ring homomorphism $F^* : Sw^f(\mathcal{G}') \to Sw^f(\mathcal{G}')$ and $F^* : Sw(\mathcal{G}') \to Sw(\mathcal{G})$.

Similarly, induction with F yields $[\mathcal{G}, \mathcal{F}(\mathbb{Z})]' \to [\mathcal{G}', \mathcal{F}(\mathbb{Z})]'$ and $[\mathcal{G}, \mathcal{M}(\mathbb{Z})]' \to [\mathcal{G}, \mathcal{M}(\mathbb{Z})]'$, also ring homomorphisms $F_* : Sw^f(\mathcal{G}) \to Sw^f(\mathcal{G}')$ and $F_* : Sw(\mathcal{G}) \to Sw(\mathcal{G}')$.

(iii) In the notation of (ii), F is said to be admissible if for every object $c \in \mathcal{G}$, the group homomorphism in $\mathrm{aut}_{\mathcal{G}}(c) \to \mathrm{aut}_{\mathcal{G}}(F(c))$ induced by F is injective and its image has finite index, and also if the map $\pi_0(F) : \pi_0(\mathcal{G}) \to \pi_0(\mathcal{G}')$ has the property that the pre-image of any element in $\pi_0(\mathcal{G}')$ is finite.

E.g., If G is a discrete group and $f : S \to T$ a map of finite G-sets, then $\underline{f} : \underline{S} \to \underline{T}$ is admissible.

(iv) If $E, F : \mathcal{G} \to \mathcal{G}'$ are functors that are naturally equivalent, then

$$E^* = F^* : Sw^f(\mathcal{G}') \to Sw^f(\mathcal{G})$$
$$E^* = F^* : Sw(\mathcal{G}') \to Sw(\mathcal{G})$$

Note that E is admissible iff F is admissible, in which case, $E_* = F_* : Sw^f(\mathcal{G}) \to Sw^f(\mathcal{G}')$ and $E_* = F_* : Sw(\mathcal{G}) \to Sw(\mathcal{G}')$.

Exercises

1.1 Let H be a subgroup of a finite group G, $S^H = \{s \in S | gs = s \text{ for all } g \in H\}$

(i) Show that the map $\mathrm{Hom}_{G\mathcal{S}et}(G/H, S) \to S : \phi \to \phi(eH)$ induces a bijection $\mathrm{Hom}_{G\mathcal{S}et}(G/H, S) \simeq S^H$.

(ii) If S, T are G-sets, show that $(S \times T)^H = S^H \times T^H$ and $(S \dot{\cup} T)^H \simeq S^H \dot{\cup} T^H$.

1.2 If H, H' are subgroups of G, show that $G/H \leq G/H'$ iff there exists $g \in G$, such that $g^{-1}Hg \leq H'$.

1.3 Let S_1, S_2 be G-sets. Show that $[\underline{S_1 \cup S_2}, \mathcal{C}] \simeq [\underline{S_1}, \mathcal{C}] \times [\underline{S_2}, \mathcal{C}]$ for any category $\mathcal{C}$.

1.4 Let S, T be G-sets, and X a simple G-set (i.e., any G-subset of S is either empty or X). Show that there exists a bijective map

$$\mathrm{Hom}_{G\mathcal{S}et}(X, S) \dot{\cup} \mathrm{Hom}_{G\mathcal{S}et}(X, T) \simeq \mathrm{Hom}_{G\mathcal{S}et}(X, S \dot{\cup} T).$$

1.5 Let S, T, Y be G-sets. Show that there exists a natural isomorphism

$$\text{Hom}_{GSet}(T \times S, Y \simeq \text{Hom}_{GSet}(S, Y^T)$$

where Y^T is the set of all set-theoretic maps $T \to Y$ made into a G-set by defining

$$gf : T \to Y : t \to gf(g^{-1}t) \text{ for all } g \in G, f \in Y^T, t \in T.$$

1.6 Let Λ be a ring with identity and $P, Q \in \mathcal{P}(\Lambda)$. Show that

(a) $[P] = [Q]$ in $K_0(A)$ iff P is stably isomorphic to Q in $(\mathcal{P}(\Lambda))$, i.e., if $P \oplus \Lambda^n \simeq Q \oplus \Lambda^n$ for some integer n.

(b) Any element of $K_0(\Lambda)$ can be written as $[P] - r[\Lambda]$ for some integer r.

1.7 Let R be a commutative ring with identity $(\mathcal{P}(R), \oplus, \otimes)$, a symmetric monoidal category (see 1.3 in the text). In the notation of 1.1 in the text, verify that $K_0(\mathcal{P}(R)_G)$ is a ring with identity.

1.8 Let R_1, R_2 be two rings with identity. Show that $K_0(R_1 \times R_2) \simeq K_0(R) \times K_0(R_2)$.

1.9 Let A be a ring and $M, N \in \mathcal{M}(A)$. Show that $[M] = [N]$ in $G_0(A)$ iff there exists a pair of short exact sequences in $\mathcal{M}(A)$ of the form

$$0 \to L_1 \to L_2 \oplus M \to L_3 \to 0; \quad 0 \to L_1 \to L_2 \oplus N \to L_3 \to 0.$$

1.10 Let Λ be a (left) Artinian ring. Show that

(a) $G_0(\Lambda)$ is a free Abelian group on $[S_1], \cdots, [S_r]$, say where the $[S_i]$ are distinct classes of simple Λ-modules.

(b) $K_0(\Lambda)$ is a free Abelian group on $[I_1], [I_2], \cdots, [I_t]$, say, where the $[I_j]$ are distinct classes of indecomposable projective Λ-modules.

1.11 Let A be a finite-dimensional algebra over a field F, and L a field extension of F. Show that the maps $K_0(A) \to K_0(A \otimes_F L)$ and $G_0(A) \to G_0(A \otimes_F L)$ induced by $A \to A \otimes_F L$ are monomorphisms.

Chapter 2

Some fundamental results on K_0 of exact and Abelian categories – with applications to orders and grouprings

In the following section we discuss some of the results whose higher-dimensional analogues will be given when higher K-groups are treated in chapter 6.

2.1 Some fundamental results on K_0 of exact and Abelian categories

$(2.1)^A$ Devissage theorem and example

Definition 2.1.1 *Let $\mathcal{C}_0 \subset \mathcal{C}$ be exact categories. The inclusion functor $\mathcal{C}_0 \to \mathcal{C}$ is exact and hence induces a homomorphism $K_0(\mathcal{C}_0) \to K_0(\mathcal{C})$. A $\mathcal{C}_0$-filtration of an object A in $\mathcal{C}$ is a finite sequence of the form: $0 = A_0 \subset A_1 \subset \cdots \subset A_n = A$ where each $A_i/A_{i-1} \in \mathcal{C}_0$.*

Lemma 2.1.1 *If $0 \neq A_0 \subset A_1 \subset \cdots \subset A_n = A$ is a $\mathcal{C}_0$-filtration, then $[A] = \Sigma[A_i/A_{i-1}]$, $1 \leq i \leq n$, in $K_0(\mathcal{C})$.*

Theorem 2.1.1 (Devissage theorem) *Let $\mathcal{C}_0 \subset \mathcal{C}$ be exact categories such that $\mathcal{C}_0$ is Abelian. If every $A \in \mathcal{C}$ has $\mathcal{C}_0$-filtration, then $K_0(\mathcal{C}_0) \to K_0(\mathcal{C})$ is an isomorphism.*

PROOF   Since $\mathcal{C}_0$ is Abelian, any refinement of a $\mathcal{C}_0$-filtration is also $\mathcal{C}_0$-filtration. So, by the Zassenhaus lemma, any two finite filtrations have equivalent refinements, that is, refinements such that the successive factors of the first refinement are, up to a permutation of the order of their occurrences, isomorphic to those of the second.

So, if $0 \subset A_0 \subset A_1 \subset \cdots \subset A_n = A$ is any $\mathcal{C}_0$-filtration of A in $\mathcal{C}$, then

$$J(A) = \Sigma[A_i/A_{i-1}] \qquad (1 \le i \le n)$$

is well defined since $J(A)$ is unaltered by replacing the given filtration with a refinement.

Now let $0 \to A' \xrightarrow{\alpha} A \xrightarrow{\beta} A'' \to 0$ be an exact sequence in $\mathcal{C}$. Obtain a filtration for A by $0 = A_0 \subset A_1 \subset \cdots \subset A_n = A'$ for A' and $\beta^{-1}(A^0) \subset \beta^{-1}(A^1) \subset \cdots \subset \beta^{-1}(A'')$ if $A^0 \subset A^1 \subset \cdots \subset A''$ is a $\mathcal{C}_0$-filtration of A''. Then $0 = A_0 \subset A_1 \subset \cdots \subset A_k \subset \beta^{-1}(A^0) \subset \beta^{-1}(A^1) \subset \cdots \subset \beta^{-1}(A^n)$ is a filtration of A.

So $J(A) = J(A') + J(A'')$. Hence J induces a homomorphism $K_0(\mathcal{C}) \to K_0(\mathcal{C}_0)$. We also have a homomorphism $i : K_0(\mathcal{C}_0) \to K_0(\mathcal{C})$ induced by the inclusion functor $i : \mathcal{C}_0 \to \mathcal{C}$. Moreover, $i \circ J = 1_{K_0(\mathcal{C})}$ and $J \circ i = 1_{K_0(\mathcal{C})}$. Hence $K_0(\mathcal{C}) \simeq K_0(\mathcal{C})$. $\square$

Corollary 2.1.1 Let **a** be a nilpotent two-sided ideal of a Noetherian ring R. Then $G_0(R/\mathbf{a}) \simeq G_0(R)$.

PROOF If $M \in \mathcal{M}(R)$, then $M \supset \mathbf{a}M \supset \cdots \supset \mathbf{a}^k M = 0$ is an $\mathcal{M}(R/\mathbf{a})$ filtration of $\mathcal{M}$. Result follows from 2.1.1. $\square$

Example 2.1.1   (i) Let R be an Artinian ring with maximal ideal **m** such that $\mathbf{m}^r = 0$ for some r. Let $k = R/\mathbf{m}$ (e.g., $R = Z/p^r, k = F_p$).

In 2.1.1, put $\mathcal{C}_0 =$ category of finite-dimensional k-vector spaces, and $\mathcal{C}$ the category of finitely generated R-modules. Then, we have a filtration

$$0 = \mathbf{m}^r M \subset \mathbf{m}^{r-1}M \subset \cdots \subset \mathbf{m}M \subset M \text{ of } M,$$

where $M \in \mathrm{ob}\mathcal{C}$. Hence by 2.1.1, $K_0(\mathcal{C}_0) \simeq K_0(\mathcal{C})$.

$(2.1)^B$ Resolution theorem and examples

2.1.1   Resolution theorem [20, 165]

Let $\mathcal{A}_0 \subset \mathcal{A}$ be an inclusion of exact categories. Suppose that every object of $\mathcal{A}$ has a finite resolution by objects of $\mathcal{A}_0$ and that if $0 \to M' \to M \to M'' \to 0$ is an exact sequence in $\mathcal{A}$, then $M \in \mathcal{A}_0$ implies that $M', M'' \in \mathcal{A}_0$. Then $K_0(\mathcal{A}_0) \simeq K_0(\mathcal{A})$.

Examples 2.1.1   (i) Let R be a regular ring. Then, for any $M \in \mathrm{ob}\mathcal{M}(R)$, there exists $P_i \in \mathcal{P}(R)$, $i = 0, 1, \ldots, n$, such that the sequence $0 \to P_n \to P_{n-1} \to \cdots \to M \to 0$ is exact. Put $\mathcal{A}_0 = \mathcal{P}(R)$, $\mathcal{A} = \mathcal{M}(R)$ in 2.1.6. Then we have $K_0(R) \simeq G_0(R)$.

(ii) Let $\mathcal{H}(R)$ be the category of all R-modules having finite homological dimension, i.e., having a finite resolution by finitely generated projective R-modules and $\mathcal{H}_n(R)$ the subcategory of modules having resolutions of length $\leq n$. Then by the resolution theorem 2.1.1 applied to $\mathcal{P}(R) \subseteq \mathcal{H}(R)$, we have $K_0(R) \simeq K_0\mathcal{H}(R) \simeq K_0\mathcal{H}_n(R)$ for all $n \geq 1$ (see [20] or [213]).

(iii) Let $\mathcal{C}$ be an exact category and $\mathrm{Nil}(\mathcal{C})$ the category whose objects are pairs (M, ν) where $M \in \mathcal{C}$ and ν is a nilpotent endomorphism of M, i.e., $\nu \in \mathrm{End}_{\mathcal{C}}(M)$. Let $\mathcal{C}_0 \subset \mathcal{C}$ be an exact subcategory $\mathcal{C}$ such that every object of $\mathcal{C}$ has a finite $\mathcal{C}_0$-resolution. Then every object of $\mathrm{Nil}(\mathcal{C})$ has a finite $\mathrm{Nil}(\mathcal{C}_0)$-resolution and so, by 2.1.1, $K_0(\mathrm{Nil}(\mathcal{C}_0)) \simeq K_0(\mathrm{Nil}(\mathcal{C}))$.

(iv) In the notation of (iii), we have two functors $Z : \mathcal{C} \rightarrow \mathrm{Nil}(\mathcal{C}) : Z(M) = (M, 0)$ (where '0' denotes zero endomorphism) and $F : \mathrm{Nil}(\mathcal{C}) \rightarrow \mathcal{C} :$ $F(M, \nu) = M$ satisfying $FZ = 1_{\mathcal{C}}$ and hence a split exact sequence $0 \rightarrow K_0(\mathcal{C}) \overset{Z}{\rightarrow} K_0(\mathrm{Nil}(\mathcal{C})) \rightarrow \mathrm{Nil}_0(\mathcal{C}) \rightarrow 0$, which defines $\mathrm{Nil}_0(\mathcal{C})$ as cokernel of Z.

If Λ is a ring, and $\mathcal{H}(\Lambda)$ is the category defined in (ii) above, then we denote $\mathrm{Nil}_0(\mathcal{P}(\Lambda))$ by $\mathrm{Nil}_0(\Lambda)$. If S is a central multiplicative system in Λ, $\mathcal{H}_S(\Lambda)$ the category of S-torsion objects of $\mathcal{H}(\Lambda)$, and $\mathcal{M}_S(\Lambda)$ the category of finitely generated S-torsion Λ-modules, one can show that if $S = T_+ = \{t^i\}$, a free Abelian monoid on the generator t, then there exists isomorphisms of categories $\mathcal{M}_{T_+}(\Lambda[t]) \simeq \mathrm{Nil}(\mathcal{M}(\Lambda))$ and $\mathcal{H}_{T_+}(\Lambda[t]) \simeq \mathrm{Nil}(\mathcal{H}(\Lambda))$ and an isomorphism of groups: $K_0(\mathcal{H}_{T_+}(\Lambda[t])) \simeq K_0(\Lambda) \oplus \mathrm{Nil}_0(\Lambda)$. Hence $K_0\mathrm{Nil}(\mathcal{H}(\Lambda)) \simeq K_0(\Lambda) \oplus \mathrm{Nil}_0(\Lambda)$. See [20, 215] for further information.

(v) The fundamental theorem for K_0 says that

$$K_0\left(\Lambda\left[t, t^{-1}\right]\right) \simeq K_0(\Lambda) \oplus K_{-1}(\Lambda) \oplus NK_0(\Lambda) \oplus NK_0(\Lambda)$$

where $NK_0(\Lambda) := \mathrm{Ker}(K_0(\Lambda[t]) \overset{\tau_+}{\rightarrow} K_0(\Lambda)$ where τ_+ is induced by augmentation $t = 1$, and K_{-1} is the negative K-functor $K_{-1} : \mathbf{Rings} \rightarrow$ Abelian groups defined by H. Bass in [20]. See 4.4 for more details. For generalization of this fundamental theorem to higher K-theory, see chapter 6.

$(2.1)^C$ K_0 and localization in Abelian categories plus examples

We close this section with a discussion leading to a localization short exact sequence and then give copious examples to illustrate the use of the sequence.

2.1.1 A full subcategory $\mathcal{B}$ of an Abelian category $\mathcal{A}$ is called a Serre subcategory if whenever $0 \rightarrow M' \rightarrow M \rightarrow M'' \rightarrow 0$ is an exact sequence in $\mathcal{A}$, then $M \in \mathcal{B}$ if and only if $M', M'' \in \mathcal{B}$. We now construct a quotient Abelian category $\mathcal{A}/\mathcal{B}$ whose objects are just objects of $\mathcal{A}$.

$\text{Hom}_{\mathcal{A}/\mathcal{B}}(M, N)$ is defined as follows: If $M' \subset M, N' \subset N$ are subobjects such that $M/M' \in \text{ob}(\mathcal{B}), N' \in \text{ob}(\mathcal{B})$, then there exists a natural homomorphism $\text{Hom}_A(M, N) \to \text{Hom}_A(M', N/N')$. As M', N' range over such pairs of objects, the groups $\text{Hom}_A(M', N/N')$ form a direct system of Abelian groups and we define

$$\mathcal{A}/\mathcal{B}(M, N) = \varinjlim_{(M', N')} \mathcal{A}((M', N/N')).$$

The quotient functor $T : \mathcal{A} \to \mathcal{A}/\mathcal{B}$ defined by $M \to T(M)$ is such that

(i) $T : \mathcal{A} \to \mathcal{A}/\mathcal{B}$ is an additive functor.

(ii) If $\nu \in \text{Hom}_A(M, N)$, then $T(\nu)$ is null if and only if $\text{Im}(\nu)) \in \text{ob}(\mathcal{B})$. Also $T(\nu)$ is an epimorphism if and only if coker $\mu \in \text{ob}(\mathcal{B})$ and it is a monomorphism iff $\text{Ker}(\mu) \in \text{ob}(\mathcal{B})$. Hence $T(\nu)$ is an isomorphism if and only if μ is a $\mathcal{B}$-isomorphism.

Remarks 2.1.1 Note that $\mathcal{A}/\mathcal{B}$ satisfies the following universal property: If $T' : \mathcal{A} \to \mathcal{D}$ is an exact functor such that $T'(M) \simeq 0$ for all $M \in \mathcal{B}$, then there exists a unique exact functor $U : A/\mathcal{B} \to \mathcal{D}$ such that $T' = U \circ T$.

Theorem 2.1.2 [20, 81, 215] *Let $\mathcal{B}$ be a Serre subcategory of an Abelian category A. Then there exists an exact sequence*

$$K_0(\mathcal{B}) \to K_0(\mathcal{A}) \to K_0(\mathcal{A}/\mathcal{B}) \to 0.$$

Examples 2.1.2 (i) Let Λ be a Noetherian ring, $S \subset \Lambda$ a central multiplicative subset of $\Lambda, \mathcal{M}_S(\Lambda)$ the category of finitely generated S-torsion Λ-modules. Then $\mathcal{M}(\Lambda)/\mathcal{M}_S(\Lambda) \simeq \mathcal{M}(\Lambda_S)$ (see [81, 215]), and so, the exact sequence in 2.1.2 becomes

$$K_0(\mathcal{M}_S(\Lambda)) \to G_0(\Lambda) \to G_0(\Lambda_S) \to 0.$$

(ii) If Λ in (i) is a Dedekind domain R with quotient field F, and $S = R - 0$, then $K_0(\mathcal{M}_S(R)) \simeq \oplus_{\mathbf{m}} G_0(R/\mathbf{m}) = \oplus_{\mathbf{m}} K_0(R/\mathbf{m})$ where $\mathbf{m}$ runs through the maximal ideals of R. Now, since $K_0(R/\mathbf{m}) \simeq \mathbb{Z}$ and $K_0(R) \simeq \mathbb{Z} \oplus Cl(R)$, the sequence (I) yields the exactness of

$$\bigoplus \mathbb{Z} \to \mathbb{Z} \oplus Cl(R) \to \mathbb{Z} \to 0.$$

(iii) Let Λ be a Noetherian ring, $S = \{s^i\}$ for some $s \in S$. Then $K_0(\mathcal{M}_S(R)) \simeq G_0(R/sR)$ (by Devissage), yielding the exact sequence

$$G_0(\Lambda/s\Lambda) \to G_0(\Lambda) \to G_0\left(\Lambda\left(\frac{1}{s}\right)\right) \to 0.$$

(iv) Let R be the ring of integers in a p-adic field F, Γ a maximal R-order in a semi-simple F-algebra Σ, and $S = R - 0$, then $K_0(\mathcal{M}_S(\Gamma)) \simeq G_0(\Gamma/\pi\Gamma) \simeq K_0(\Gamma/\mathrm{rad}\Gamma)$ where πR is the unique maximal ideal of R.

(v) If R is the ring of integers in a p-adic field F, Λ an R-order in a simple-field F-algebra Σ, let $S = R - 0$. Then $K_0(\mathcal{M}_S(\Lambda)) \simeq \oplus G_0(\Lambda/p\Lambda)$ where p runs through all the prime ideals of R.

(vi) Let Λ be a (left) Noetherian ring, $\Lambda[t]$ the polynomial ring in the variable t, $\Lambda\left[t, t^{-1}\right]$ the Laurent polynomial ring. Then $\Lambda\left[t, t^{-1}\right] = \Lambda[t]_S$ where $S = [t^i]$. Now, the map $\epsilon : \Lambda[t] \rightarrow \Lambda$, $t \rightarrow 0$ induces an inclusion $\mathcal{M}(\Lambda) \subset \mathcal{M}(\Lambda[t])$ and the canonical map $i : \Lambda[t] \rightarrow \Lambda[t]_S = \Lambda\left[t, t^{-1}\right]$, $t \rightarrow t/1$, yields an exact functor $\mathcal{M}(\Lambda[t]) \rightarrow \mathcal{M}(\Lambda\left[t, t^{-1}\right]$. So, from theorem 2.1.2, we have the localization sequence

$$G_0(\Lambda) \overset{\epsilon_*}{\rightarrow} G_0\left(\Lambda[t]\right) \rightarrow G_0(\Lambda\left[t, t^{-1}\right]) \rightarrow 0. \tag{II}$$

Now $\epsilon_* = 0$ since for any Λ the exact sequence of $\Lambda[t]$-modules $0 \rightarrow N[t] \overset{t}{\rightarrow} N[t] \rightarrow N \rightarrow 0$ yields

$$\epsilon_*[N] = [N[t]] - [N[t]] = 0.$$

So, $G_0(\Lambda[t]) \simeq G_0\left(\Lambda\left[t, t^{-1}\right]\right)$ from (II) above. This proves the first part of fundamental theorem for G_0 of rings 2.1.3.

Theorem 2.1.3 (Fundamental theorem for G_0 of rings) *If Λ is a left Noetherian ring, then the inclusion $\Lambda \overset{i}{\hookrightarrow} \Lambda[t] \overset{j}{\hookrightarrow} \Lambda\left[t, t^{-1}\right]$ induces isomorphisms*

$$G_0(\Lambda) \cong G_0(\Lambda[t]) \cong G_0\left(\Lambda\left[t, t^{-1}\right]\right).$$

PROOF   See [20, 215] for the proof of the second part.   ▯

Remarks 2.1.2   (i) There is a generalization of theorem 2.1.3 due to Grothendieck as follows: let R be a commutative Noetherian ring, Λ a finite R-algebra, T a free Abelian group or monoid with finite basis. Then $G_0(\Lambda) \rightarrow G_0(\Lambda[T])$ is an isomorphism (see [20]).

(ii) If Λ is a (left) Noetherian regular ring, so are $\Lambda[t]$ and $\Lambda\left[t, t^{-1}\right]$. Since $K_0(R) \cong G_0(R)$ for any Noetherian regular ring R, we have from theorem 2.1.3 that $K_0(\Lambda) \simeq K_0(\Lambda[t]) \simeq K_0\Lambda\left[t, t^{-1}\right]$. Furthermore, if T is a free Abelian group or monoid with a finite basis, then $K_0(A) \rightarrow K_0(\Lambda[t])$ is an isomorphism (see [20]).

2.2 Some finiteness results on K_0 and G_0 of orders and grouprings

In this section we call attention to some finiteness results on K_0 and G_0 of orders that will have higher-dimensional analogues in 7.1. The results here are due to H. Bass (see [20]).

Definition 2.2.1 *An integral domain R with quotient field F is called a D-edekind domain if it satisfies any of the following equivalent conditions:*

(i) *Every ideal in R is projective (i.e., R is hereditary).*

(ii) *Every nonzero ideal $\mathbf{a}$ of R is invertible (that is $\mathbf{aa}^{-1} = R$ where $\mathbf{a}^{-1} = \{x \in F | x\mathbf{a} \subset R\}$.*

(iii) *R is Noetherian, integrally closed, and every nonzero prime ideal is maximal.*

(iv) *R is Noetherian and $R_{\mathbf{m}}$ is a discrete valuation ring for all maximal ideals $\mathbf{m}$ of R.*

(v) *Every nonzero ideal is uniquely a product of prime ideals.*

Example 2.2.1 $\mathbb{Z}, F[x]$, are Dedekind domains. So is the ring of integers in a number field.

Definition 2.2.2 *Let R be a Dedekind domain with quotient field F. An R-order Λ in a finite-dimensional semi-simple F-algebra Σ is a subring of Σ such that*

(i) *R is contained in the centre of Λ.*

(ii) *Λ is finitely generated R-module.*

(iii) *$F \otimes_R \Lambda = \Sigma$.*

Example For a finite group G, the groupring RG is an R-order in FG when $\mathrm{char}(F)$ does not divide $|G|$.

Definition 2.2.3 *Let R, F, Σ be as in definition 2.2.2. A maximal R-order Γ in Σ is an order that is not properly contained in any other R-order in Σ.*

Example

(i) R is maximal R-order in F.

(ii) $M_n(R)$ is a maximal R-order in $M_n(F)$.

Remarks 2.2.1 Let R, F, Σ be as in definition 2.2.2. Then

(i) Any R-order Λ is contained in at least one maximal R-order in Σ (see [38, 171]).

(ii) Every semi-simple F-algebra Σ contains at least one maximal order. However, if Σ is commutative, then Σ contains a unique maximal order, namely, the integral closure of R in Σ (see [171]).

(iii) If Λ is an R-order in Σ, then $\Lambda_{\mathbf{p}}$ is an $R_{\mathbf{p}}$-order in Σ for any prime $=$ maximal ideal $\mathbf{p}$ of R. Moreover, $\Lambda = \bigcap_{\mathbf{p}} \Lambda_{\mathbf{p}}$ (intersection within Σ) (see [180]).

(iv) For $n > 0$, there exists only a finite number of isomorphism classes of $M \in M(\Lambda)$ that are torsion free of rank $\leq n$ as R-modules. See [174].

Theorem 2.2.1 [20] *Let R be the ring of integers in a number field F, Λ an R-order in a semi-simple algebra Σ. Then $K_0(\Lambda), G_0(\Lambda)$ are finitely generated Abelian groups.*

Definition 2.2.4 *Let R, F, Λ, Σ be as in theorem 2.2.1. Λ is said to satisfy Cartan condition if for each prime ideal $\mathbf{p}$ in R the Cartan homomorphism $K_0(\Lambda/\mathbf{p}\Lambda) \to G_0(\Lambda/\mathbf{p}\Lambda)$ is a monomorphism, i.e., the Cartan matrix of $\Lambda/\mathbf{p}\Lambda$ has a non-zero determinant (see 1.4.1 (ii)).*

2.2.1 Let R, F, Λ, Σ be as in theorem 2.2.1. We shall write $SK_0(\Lambda) :=$ kernel of the canonical map $K_0(\Lambda) \to K_0(\Sigma)$ and $SG_0(\Lambda) := \mathrm{Ker}(G_0(\Lambda) \to G_0(\Sigma)$.

Theorem 2.2.2 [20] *Let R, F, Λ, Σ be as in 2.2.1. Suppose that Λ satisfies the Cartan condition. Then $SK_0(\Lambda)$ and $SG_0(\Lambda)$ are finite. So also is the kernel of the Cartan map $K_0(\Lambda) \to G_0(\Lambda)$. Moreover, $rank\,(G_0(\Lambda)) = $ number of simple factors of Λ.*

2.3 Class groups of Dedekind domains, orders, and grouprings plus some applications

For a Dedekind domain R with quotient field F, the notion of class groups of R-orders Λ is a natural generalization of the notion of class groups of rings of integers in number fields as well as class groups of grouprings RG where G is a finite group. The class groups of grouprings, apart from their intimate connections with representation theory and number theory, also house some topological invariants where G is usually the fundamental group of some spaces.

$(2.3)^A$ Class groups of Dedekind domains

Definition 2.3.1 *A fractional ideal of Dedekind domain R (with quotient field F) is an R-submodule $\mathbf{a}$ of F such that $s\mathbf{a} \subset R$ for some $s \neq 0$ in F. Then $\mathbf{a}^{-1} = \{x \in F | x\mathbf{a} \subset R\}$ is also a fractional ideal. Say that $\mathbf{a}$ is invertible if $\mathbf{a}\mathbf{a}^{-1} = R$. The invertible fractional ideals form a group, which we denote by I_R. Also each element $u \in F^*$ determines a principal fractional ideal Ru. Let P_R be the subgroup of I_R consisting of all principal fractional ideals. The (ideal) class group of R is defined as I_R/P_R and denoted by $Cl(R)$.*

It is well known that if R is the ring of integers in a number field, then $Cl(R)$ is finite (see [39]).

Definition 2.3.2 *Let R be a Dedekind domain with quotient field F. An R-lattice is a finitely generated torsion free R-module. Note that any R-lattice M is embeddable in a finite-dimensional F-vector space V such that $F \otimes_R M = V$. Moreover, every R-lattice M is R-projective (since R is hereditary and M can be written as a direct sum of ideals) (see 2.3.1 below — Steinitz's theorem). For $P \in \mathcal{P}(R)$ define the R-rank of P as the dimension of the vector space $F \otimes_R P$ and denote this number by $rk(P)$.*

Theorem 2.3.1 [39], Steinitz theorem *Let R be a Dedekind domain. Then*

(i) *If $M \in \mathcal{P}(R)$, then $M = \mathbf{a}_1 \oplus \mathbf{a}_2 \oplus \cdots \oplus \mathbf{a}_n$ where n is the R-rank of M and each $\mathbf{a}_i$ is an ideal of R.*

(ii) *Two direct sums $\mathbf{a}_1 \oplus \mathbf{a}_2 \oplus \cdots \oplus \mathbf{a}_n$ and $\mathbf{b}_1 \oplus \mathbf{b}_2 \oplus \cdots \oplus \mathbf{b}_m$ of nonzero ideals of R are R-isomorphic if and only if $n = m$ and the ideal class of $\mathbf{a}_1\mathbf{a}_2 \cdots \mathbf{a}_n = $ ideal class of $\mathbf{b}_1\mathbf{b}_2 \cdots \mathbf{b}_n$.*

Definition 2.3.3 *The ideal class associated to M as in theorem 2.3.1 is called the Steinitz class and is denoted by $st(M)$.*

Theorem 2.3.2 *Let R be a Dedekind domain. Then*

$$K_0(R) \simeq \mathbb{Z} \oplus C\ell R.$$

Sketch of proof. Define a map

$$Q = (\mathrm{rk, \ st}) : K_0(R) \to \mathbb{Z} \times C\ell R$$

by

$$(\mathrm{rk, \ st})[P] = (\mathrm{rk}P, \mathrm{st}(P)) \ ,$$

where $\mathrm{rk}P$ is the R-rank of P definition 2.3.2 and $\mathrm{st}(P)$ is the Steinitz class of P. We have $\mathrm{rk}(P \oplus P^1) = \mathrm{rk}(P) + \mathrm{rk}(P^1)$ and $\mathrm{st}(P \oplus P^1) = \mathrm{st}(P) \oplus \mathrm{st}(P^1)$. So, φ is a homomorphism that can easily be checked to be an isomorphism, the inverse being given by $\eta : \mathbb{Z} \times C\ell R \to K_0(R), (n, \mathbf{a}) \to n[R] + [\mathbf{a}]$.

Remarks 2.3.1 (i) It follows easily from Steinitz's theorem that $\text{Pic}(R) \simeq C\ell(R)$ for any Dedekind domain R.

(ii) Let R be a commutative ring with identity, $\text{Spec}(R)$ the set of prime ideals of R. For $P \in \mathcal{P}(R)$ define $r_P : \text{Spec}(R) \to \mathbb{Z}$ by $r_P(\mathbf{p}) = \text{rank}$ of $P_\mathbf{p}$ over $R_\mathbf{p} = $ dimension of $P_\mathbf{p}/\mathbf{p}_\mathbf{p}P_\mathbf{p}$. Then r_P is continuous where $\mathbb{Z}$ is given the discrete topology (see [20], [215]). Let $H_0 :=$ group of continuous functions $\text{Spec}(R) \to \mathbb{Z}$. Then we have a homomorphism $r : K_0(R) \to H_0(R) : r([P]) = r_P$ (see [20]). One can show that if R is a one-dimensional commutative Noetherian ring, then $(\text{rk, det}) : K_0(R) \to H_0(R) \oplus \text{Pic}(R)$ is an isomorphism – a generalization of theorem 2.3.2 that we recover by seeing that for Dedekind domains R, $H_0(R) \simeq \mathbb{Z}$. Note that $\det : K_0(R) \to \text{Pic}(R)$ is defined by $\det(P) = \Lambda^n P$ if the R-rank of P is n. (See [20]).

(iii) Since a Dedekind domain is a regular ring, $K_0(R) \simeq G_0(R)$.

$(2.3)^B$ Class groups of orders and grouprings

Definition 2.3.4 *Let R, F, Σ, Λ be as in definition 2.2.2. A left Λ-lattice is a left Λ-module that is also an R-lattice (i.e., finitely generated and projective as an R-module).*

A Λ-ideal in Σ is a left Λ-lattice $M \subset \Sigma$ such that $FM \subset \Sigma$.

Two left Λ-lattices M, N are said to be in the same genus if $M_\mathbf{p} \simeq N_\mathbf{p}$ for each prime ideal $\mathbf{p}$ of R. A left Λ-ideal is said to be locally free if $M_\mathbf{p} \simeq \Lambda_\mathbf{p}$ for all $\mathbf{p} \in \text{Spec}(R)$. We write $M \vee N$ if M and N are in the same genus.

Definition 2.3.5 *Let R, F, Σ be as in 2.2.2, Λ an R-order in Σ. Let*

$$S(\Lambda) = \left\{ \mathbf{p} \in Spec(R) | \, \widehat{\Lambda}_\mathbf{p} \text{ is not a maximal } \widehat{R}_\mathbf{p}\text{-order in } \widehat{\Sigma} \right\} .$$

Then $S(\Lambda)$ is a finite set and $S(\Lambda) = \emptyset$ iff Λ is a maximal R-order. Note that the genus of a Λ-lattice M is determined by the isomorphism classes of modules $\{M_\mathbf{p} | \mathbf{p} \in S(\Lambda)\}$ (see [39], [213]).

Theorem 2.3.3 [39] *Let L, M, N be lattices in the same genus. Then $M \oplus N \simeq L \oplus L'$ for some lattice L' in the same genus. Hence, if M, M' are locally free Λ-ideals in Σ, then $M \oplus M' = \Lambda \oplus M''$ for some locally free ideal M''.*

Definition 2.3.6 *Let R, F, Σ be as in definition 2.2.2. The idèle group of Σ, denoted $J(\Sigma)$, is defined by $J(\Sigma) := \{(\alpha_\mathbf{p} \in \prod(\widehat{\Sigma}_\mathbf{p})^* | \alpha_\mathbf{p} \in \widehat{\Lambda}_\mathbf{p}^*$ almost everywhere$\}$. For $\alpha = (\alpha_\mathbf{p} \in J(\Sigma)$, define*

$$\Lambda\alpha := \Sigma \cap \left\{ \underset{\mathbf{p}}{\cap} \widehat{\Lambda}_\mathbf{p}\alpha_\mathbf{p} \right\} = \underset{\mathbf{p}}{\cap} \left\{ \Sigma \cap \widehat{\Lambda}_\mathbf{p}\alpha_\mathbf{p} \right\} .$$

The group of principal idèles, denoted $u(\Sigma)$ is defined by

$$u(\Sigma) = \{\alpha = (\alpha_{\mathbf{p}})|\alpha_{\mathbf{p}} = x \in \Sigma^* \text{ for all } \mathbf{p} \in Spec(R)\}.$$

The group of unit idèles is defined by

$$U(\Lambda) = \prod_{\mathbf{p}} (\Lambda_{\mathbf{p}})^* \subseteq J(\Sigma).$$

Remarks

(i) $J(\Sigma)$ is independent of the choice of the R-order Λ in Σ since if Λ' is another R-order, then $\Lambda_{\mathbf{p}} = \Lambda'_{\mathbf{p}}$ almost everywhere.

(ii) $\Lambda\alpha$ is isomorphic to a left ideal of Λ, and $\Lambda\alpha$ is in the same genus as Λ. Call $\Lambda\alpha$ a locally free (rank 1) Λ-lattice or a locally free fractional Λ-ideal in Σ. Note that any $M \in g(\Lambda)$ can be written in the form $M = \Lambda\alpha$ for some $\alpha \in J(\Sigma)$ (see [39]).

(iii) If $\Sigma = F$ and $\Lambda = R$, we also have $J(F), u(F)$ and $U(R)$ as defined above.

(iv) For $\alpha, \beta \in J(\Sigma)$, $\Lambda\alpha \oplus \Lambda\beta \cong \Lambda \oplus \Lambda\alpha\beta$ (see [39]).

Definition 2.3.7 *Let F, Σ, R, Λ be as in definition 2.2.2. Two left Λ-modules M, N are said to be stably isomorphic if $M \oplus \Lambda^{(k)} \simeq N \oplus \Lambda^{(k)}$ for some integer k. If F is a number field, then $M \oplus \Lambda^{(k)} \simeq N \oplus \Lambda^{(k)}$ iff $M \oplus \Lambda \simeq N \oplus \Lambda$. We write $[M]$ for the stable isomorphism class of M.*

Theorem 2.3.4 [39, 213] *The stable isomorphism classes of locally free ideals form an Abelian group $C\ell(\Lambda)$ called the locally free class group of Λ where addition is given by $[M] + [M'] = [M'']$ whenever $M \oplus M' \simeq \Lambda \oplus M''$. The zero element is (Λ) and the inverses exist since $(\Lambda\alpha) \oplus (\Lambda\alpha^{-1}) \simeq \Lambda \oplus \Lambda$ for any $\alpha \in J(\Sigma)$.*

Theorem 2.3.5 [39, 213] *Let R, F, Λ, Σ be as in definition 2.2.2. If F is an algebraic number field, then $C\ell(\Lambda)$ is a finite group.*

Sketch of proof. If L is a left Λ-lattice, then there exists only a finite number of isomorphism classes of the left Λ-lattices M such that $FM \simeq FL$ as Σ-modules. In particular, there exists only a finite number of isomorphism classes of left Λ ideals in Σ (see [39, 173]).

Remarks 2.3.2 Let R, F, Λ, Σ be as in 2.2.2.

(i) If $\Lambda = R$, then $C\ell(\Lambda)$ is the ideal class group of R.

(ii) If Γ is a maximal R-order in Σ, then every left Γ-ideal in Σ is locally free. So, $C\ell(\Gamma)$ is the group of stable isomorphism classes of all left Γ-ideals in Σ.

(iii) Define a map $J(\Sigma) \to C\ell(\Lambda)$ $\alpha \to [\Lambda\alpha]$. Then one can show that this map is surjective and that the kernel is $J_0(\Sigma)\Sigma^*U(\Lambda)$ where $J_0(\Sigma)$ is the kernel of the reduced norm acting on $J(\Sigma)$. So $J(\Sigma)/(J_0(\Sigma)\Sigma^*U(\Lambda)) \simeq C\ell(\Lambda)$ (see [39, 213]).

(iv) If G is a finite group such that no proper divisor of $|G|$ is a unit in R, then $C\ell(RG) \simeq SK_0(RG)$. Hence $C\ell(\mathbb{Z}G) \simeq SK_0(\mathbb{Z}G)$ for every finite group G (see [39, 211]).

For computations of $C\ell(RG)$ for various R and G see [39, 170, 173, 175].

Remarks 2.3.3 In 7.4, we shall define and obtain results on higher dimensional class groups $C\ell(\Lambda)$ of orders Λ for all $n \geq 0$ in such a way that $C\ell_0(\Lambda) = C\ell(\Lambda)$. Indeed one can show that

$$C\ell(\Lambda) \simeq \operatorname{Ker}\left(SK_0(\Lambda) \to \bigoplus_{\underline{p},} SK_0(\widehat{\Lambda}_{\underline{p}})\right)$$

where $\underline{p}$ ranges over prime ideals of R. See [39].

As we shall see in 7.4, $C\ell_n(\Lambda) := \operatorname{Ker}\left(SK_n(\Lambda) \to \bigoplus_{\underline{p},} SK_n(\widehat{\Lambda}_{\underline{p}})\right)$ for all $n \geq 0$.

$(2.3)^C$ Applications – Wall finiteness obstruction

2.3.1 An application – the Wall finiteness obstruction theorem

Let R be a ring. A bounded chain complex $C = (C_*, d)$ of R-modules is said to be of finite type if all the C_j's are finitely generated. The Euler characteristic of $C = (C_*, d)$ is given by $\chi(C) = \sum_{i=-\infty}^{\infty}(-1)^i[C_i]$, and we write $\overline{\chi}(C)$ for the image of $\chi(C)$ in $\widetilde{K}_0(R)$.

The initial motivation for Wall's finiteness obstruction theorem stated on the following page was the desire to find out when a connected space has the homotopy type of a CW-complex. If X is homotopically equivalent to a CW-complex, the singular chain complex $S_*(X)$ with local coefficient is said to be finitely dominated if it is chain homotopic to a complex of finite type. Let $R = \mathbb{Z}\pi_1(X)$, the integral group-ring of the fundamental group of X. Wall's finite obstruction theorem implies that a finitely dominated complex has a finiteness obstruction in $\widetilde{K}_0(R)$ and is chain homotopic to a complex of finite type of free R-modules if and only if the finiteness obstruction vanishes. More precisely, we have the following, where we observe that for $R = \mathbb{Z}\pi_1(X)$, $\widetilde{K}_0(R) = Cl(R)$.

Theorem 2.3.6 [225] *Let $C = (C_*, d)$ be a chain complex of projective R-modules that is homotopic to a chain complex of finite type of projective R-modules. Then $C = (C_*, d)$ is chain homotopic to a chain complex of finite type of free R-modules if and only if $\overline{\chi}(C) = 0$ in $\widetilde{K}_0(R)$.*

2.3.1 Note that the end invariant of Siebermann [195] interprets Wall's finiteness obstruction $[M] \in \overline{K}_0(\mathbb{Z}(\pi_1(M^n)))$ of an open n-dimensional manifold M^n with one tame end as an obstruction to closing the end assuming that $n \geq 6$ and that $G = \pi_1(M^n)$ is also a fundamental group of the end. In fact, the following conditions on M^n are equivalent

(i) $[M^n] = 0 \in \widetilde{K}_0(\mathbb{Z}G)$.

(ii) M^n is homotopy equivalent to a finite CW-complete.

(iii) M^n is homeomorphic to the interior of a closed n-dimensional manifold M.

(iv) The cellular chain complex $C(\widetilde{M^n})$ of the inversal cover $\widetilde{M^n}$ of M is chain equivalent to a finitely generated free $\mathbb{Z}G$-module chain complex.

2.4 Decomposition of $G_0(RG)$ (G Abelian group) and extensions to some non-Abelian groups

The aim of the section is to present a decomposition of $G_0(RG)$ (R a commutative ring, G an Abelian group) due to H. Lenstra [133] and extensions of these results to dihedral and quaternion groups due to D. Webb [230].

Because we are going to present a higher-dimensional version of these decompositions in 7.3, we develop notations here for later use and prove the results on $G_n(RG)$, $n \geq 0$ once and for all in 7.3.

The calculation of $G_0(RG)$, G Abelian, is connected with the calculation of the group "SSF" which houses obstructions constructed by Shub and Franks in their study of Morse - Smale diffeomorphisms. See [19].

$(2.4)^A$ Decomposition of $G_0(RG)$, G Abelian

2.4.1 Let n be a positive integer, $\mathbb{Z}(\frac{1}{n})$ the subring of $\mathbb{Q}$ generated by $\frac{1}{n}$. If M is an Abelian group, we put $M(\frac{1}{n}) = M \otimes_{\mathbb{Z}} (\frac{1}{n})$. Then if R is a ring, $R(\frac{1}{n}) \simeq R[x]/(nx - 1)R[x]$ (see [133]).

Denote $1 \otimes \frac{1}{n} \in R\left(\frac{1}{n}\right)$ by $\frac{1}{n}$. Note that $M(\frac{1}{n})$ is an $R\left[\frac{1}{n}\right]$-module and that the functor R-$\mathcal{M}od \to R\left(\frac{1}{n}\right) - \mathcal{M}od$ given by $M \to M\left(\frac{1}{n}\right)$ is exact.

2.4.2 Let R be a (left) Noetherian ring and C a finite cyclic group of order n generated by, say, t. If f is a ring homomorphism $\mathbb{Z}C \xrightarrow{f} \mathbb{C}$ which is injective when restricted to C, then $\mathrm{Ker} f$ is generated by $\Phi_n(t)$, the nth cyclotonic polynomial. Then the ideal $\Phi_n(t)\mathbb{Z}C$ is independent of the choice of generator t of C.

Define $R(C) := RC/\Phi_n(t)RC$ for any ring R. Then $\mathbb{Z}(C)$ is a domain isomorphic to $\mathbb{Z}[\zeta_n]$ where ζ_n is the primitive n^{th} root of 1. We identify $Q(\zeta_n)$, the field of fractions of $\mathbb{Z}[\xi_n]$, with $Q(C)$.

Now write $R\langle C\rangle := R(C)[x]/(nx-1) \simeq R(C)(\frac{1}{n}) \simeq R\left[\zeta_n, \frac{1}{n}\right]$.

Remarks 2.4.1   (i) As an R-module, $R(C)$ is free on $\varphi(n)$ generators.

(ii) If R is left Noetherian, so is $R(C)$.

(iii) If $C' \leq C$, then there exists a natural inclusion $R(C') \subseteq R(C)$.

Theorem 2.4.1 [133] *Let A be a left Noetherian ring. Then $A\left(\frac{1}{n}\right)$ is a left Noetherian ring and $G_0\left(A(\frac{1}{n})\right) \simeq G_0(A)/H$ where H is the subgroup of $G_0(A)$ generated by all symbols $[M]$ where M ranges over $M \in \mathcal{M}(A)$ such that $nM = 0$.*

If C is a finite cyclic group of order n, then $G_0(R\langle C\rangle) \simeq G_0(R(C))/H$ where H is the subgroup of $G_0(R(C))$ generated by symbols $M \in (R(C))$ such that $nM = 0$.

2.4.3 Let G be a finite Abelian group, $X(G)$ the set of cyclic quotient of G. If $C \in X(G)$, R a ring, then there is a natural surjection $RG \twoheadrightarrow RC \twoheadrightarrow R(C)$, and so, one can identify $R(C)$-modules with RG-modules annihilated by $\mathrm{Ker}(RG \to R(C)) := k_C$, i.e., RG-modules M such that $k_C \cdot M = 0$.

For $C \in X(G)$, let $\underline{m}_C := \mathrm{Ker}(QG \to Q(C))$. Since $Q(C)$ is a field, $(\simeq Q(\zeta_{|C|}))$, $\underline{m}_C$ is a maximal ideal of QG, and for any $g \in G$, $g - 1 \in \underline{m}_C$ iff $g \in \mathrm{Ker}(G \twoheadrightarrow C)$. So, $\underline{m}_C \neq \underline{m}_{C'}$, for $C \neq C'$.

Hence $QG \simeq \prod_{C \in X(G)} Q(C)$ (see [133, 230]).

Theorem 2.4.2 [133] *Let R be a left Noetherian ring with identity, G a finite Abelian group.*

Then $G_0(RG) \simeq \bigoplus_{C \in X(G)} G_0(R\langle C\rangle)$.

Remarks 2.4.2 Recall from 2.3.2 that if R is a Dedekind domain (a regular ring), then $K_0(R) \simeq G_0(R) \simeq \mathbb{Z} \oplus C\ell(R)$. Also, if C is a cyclic group of order n, then C has exactly one cyclic factor group of order d for every divisor d of n. Hence theorem 2.4.2 yields the following result.

Theorem 2.4.3 [133] *If C is a finite cyclic group of order n, then*
$$G_0(\mathbb{Z}C) \simeq \bigoplus_{d|n} \mathbb{Z} \oplus C\ell\left(\mathbb{Z}\left[\zeta_d, \tfrac{1}{d}\right]\right).$$

2.4.4 Let R be a Noetherian domain with field of fractions K, and G a finite group. A character of G is a group homomorphism from G to the multiplicative group of $\overline{K}$. Two characters θ, θ' of G are called conjugate over K if $\theta = \sigma \circ \theta'$ for some K-automorphism σ of $\overline{K}$. If θ is a character of G, then $G/\mathrm{Ker}(\theta) \in X(G)$ and has order not divisible by $char(K)$. Conversely, if $C \in X(G)$ has order n with n not divisible by $char(K)$, then the set of K-conjugacy of characters θ for which $= G/\mathrm{Ker}(\theta)$ is in bijective correspondence with the set of monic irreducible factors of Φ_n in the polynomial ring $K[X]$. For a character θ, let $R\langle\theta\rangle$ be the subring of $\overline{K}$ generated by R, the image of θ and the inverse of $\theta[G]$. Note that the exponent of $G = \ell \cdot c \cdot m$ of orders of the elements of G.

We now record the following result due to H. Lenstra.

Theorem 2.4.4 [133]

(i) *Let R be a Noetherian domain and G a finite Abelian group. Suppose that for every n dividing the exponent of G but not divisible by $char(K)$, at least one of the irreducible factors of Φ_n in $K[X]$ has coefficients in $R\left[\tfrac{1}{n}\right]$ and leading coefficients 1, then $G_0(RG) \simeq \bigoplus_{\theta \in Y} G_0\left(R\langle\theta\rangle\right)$ where Y is a set of representatives for the K-conjugacy classes of characters of G.*

(ii) *If R is a Dedekind domain, θ, Y as in (i), then for each $\theta \in Y$, the ring $R\langle\theta\rangle$ is a Dedekind ring and $G_0(RG) \cong \bigoplus_{\theta \in Y} \mathbb{Z} \oplus C\ell\left(R\langle\theta\rangle\right).$*

$(2.4)^B$ Connections to the group "SSF"

2.4.5 As an application of the foregoing idea, we next discuss the group "SSF". As we shall see below in 2.4.6, the above computations of $G_0(RG)$, G Abelian is connected with the calculations of "SSF", which houses obstructions constructed by Shub and Francs in their study of Morse - Smale diffeomorphism (see [19]).

2.4.6 Let S denote the category of pairs (H, u), $H \in \mathcal{M}(\mathbb{Z})$, $u \in \mathrm{Aut}(H)$ such that $u^n - id_H$ is nilpotent for some n.

The morphism $\mathrm{Hom}_S\left((H, u), (H', u')\right)$ from (H, u) to (H', u') consists of group homomorphisms $f : H \to H'$ such that $f \circ u = u' \circ f$. Call (H, u) a permutation module if H has a $\mathbb{Z}$-basis permutated by u. Let P be the subgroup of $G_0(S)$ generated by all permutation modules.

Definition 2.4.1

$$SSF \cong G_0(\mathcal{S})/P$$

Theorem 2.4.5 [19]

$$SSF \cong \bigoplus_{n \geq 0} C\ell \left(\mathbb{Z} \left[\zeta_n, \frac{1}{n} \right] \right).$$

$(2.4)^C$ Extensions to some Non-Abelian groups (Dihedral and Quaternion groups)

The results in this subsection are due to D. Webb (see [230]). First, we record the following information on G-rings, twisted groupring, and crossed-product ring, which will be useful even in other contexts. For proof and other information see [230].

Definition 2.4.2 *Let R be a ring, G a group acting on R by ring automorphism. Call R a G-ring.*

The twisted groupring $R\#G$ is defined as the R-module $R \otimes \mathbb{Z}G$ with elements $a \otimes g$ ($a \in R, g \in G$) denoted $a\#g$ and multiplication defined by $(a\#g) \cdot (a'\#g') = ag(a')\#gg'$. If G acts trivially on R, then $R\#G = RG$.

If R is a G-ring and M an R-module upon which G acts $\mathbb{Z}$-linearly, then the action is R-semilinear if $g(a \cdot m) = g(a)g(m)$ for all $g \in G$, $a \in R$, $m \in M$. An $R\#G$-module is simply an R-module with semi-linear G-action.

Remarks 2.4.3   (i) If $R \to S$ is a ring homomorphism, and $G \to S^*$ is a group homomorphism, and if the resulting G-action on S is R-semilinear, then there exists an induced ring map $R\#G \to S$.

(ii) If R_1, R_2 are G-rings, and if $R_1 \to R_2$ a G-equivariant ring map, then there exists an induced ring map

$$R_1\#G \to R_2\#G. \tag{I}$$

(iii) Let R be a G-ring, $\underline{a} \subset R$ a G-stable two-sided ideal so that $R/\underline{a}$ is a G-ring. Then $\underline{a}\#G$ is a two-sided ideal of $R\#G$ and the natural map $R\#G \to R/\underline{a}\#G$ induces an isomorphism

$$\frac{R\#G}{\underline{a}\#G} \cong R/\underline{a}\#G. \tag{II}$$

(iv) Let S be a commutative ring on which G acts: $R = S^G$, the G-invariant subring. Define

$$\alpha : S\#G \to \operatorname{End}_R(S)$$
$$\beta : S \otimes_R S \to \operatorname{Hom}_{\operatorname{Set}}(G, S)$$

as follows:

- The regular representation $S \to \operatorname{End}_R(S)$ and the action $G \to \operatorname{Aut}_R(S)$ induces by (ii) a map $\alpha : S\#G \to \operatorname{End}_R(S)$ of R-algebras.
- The map β is given by $\beta(x \otimes y)(g) = xg(y)$ (β is an S-algebra map).

(v) Let S be a commutative ring on which a finite group G acts faithfully. Let $R = S^G$. Suppose that $S \in \mathcal{P}(R)$. Then in (iv) on the previous page, α is an isomorphism iff β is an isomorphism.

Hence, if K is a field on which a finite group acts faithfully, and $k = K^G$, then the map in (iv) is an isomorphism $K\#G \xrightarrow{\sim} \operatorname{End}_R(K)$.

(vi) Let R be a Dedekind domain with field of fraction F, K a finite Galors extension of F with group G, $\overline{R}$ the integral closure of R in K, $\underline{p}$ a prime ideal of R. Then the map

$$\alpha : \frac{\overline{R}}{\underline{p}\overline{R}}\#G \to \operatorname{End}_{R/\underline{p}}\left(\overline{R}/\underline{p}\overline{R}\right)$$

is an isomorphism iff $\underline{p}$ does not ramify in $\overline{R}$.

If $\overline{R}/R$ is unramified, then $\alpha : \overline{R}\#G \xrightarrow{\sim} \operatorname{End}_R(\overline{R})$.

2.4.7   (i) Let R be a commutative G-ring, $c : G \times G \to R^*$ a normalized 2-cocycle with values in R^* (see [230]). Then the crossed-product ring $R\#_cG$ is the R-module $R \otimes \mathbb{Z}G$ with multiplication given by $(a\#g) \cdot (a'\#g') = ag(a')c(g, g')\#gg'$. Then $R\#G$ is an associative ring with identity $1\#1$. Note that if $c \equiv 1$, then we obtain the twisted groupring.

(ii) If R, S are commutative G-rings and $G \times G \to R^*$ normalized cocycle, and $\alpha : R \to S$ a G-equivariant ring map, then the composite $G \times G \xrightarrow{c} R^* \to S^*$ is a 2-cocycle and there exists induced ring homomorphism $R\#_cG \to S\#_cG$.

If $\underline{a} = \operatorname{Ker}\alpha$ is nilpotent, then $\underline{a}\#_cG = \operatorname{Ker}(\alpha\#G)$.

(iii) Let $\underline{a}$ be a G-stable ideal of R, and $\pi : R \to R/\underline{a}$ a map of G-rings. Let $\underline{a}\#_cG$ be the R-submodule generated by $a\#g$, $a \in \underline{a}$, $g \in G$. Then $\underline{a}\#_cG$ is a two-sided ideal of $R\#_cG$ generated by $a\#g$ and there exists a ring isomorphism

$$\frac{R\#_cG}{\underline{a}\#_cG} \xrightarrow{\sim} (R/\underline{a})\#_cG$$

(iv) Let R be a Dedekind ring, F its field of fractions and K a finite Galois extension of F with Galois group $G = Gal(K/F)$, $\overline{R}$ the integral closure of R in K, $c : G \times G \to \overline{R}^*$ a normalized 2-cocycle. If $\overline{R}$ is unramified over R, then $(\overline{R}\#_R\overline{R} \simeq M_{|G|}(\overline{R})$.

If $\overline{R}$ is unramified over R, then $\overline{R}\#_cG$ is a maximal R-order in $K\#_cG$.

(v) Let $1 \to G \to H \to G_1 \to 1$ be a group extension with associated normalized 2-cocycle $c : G_1 \times G_1 \to G$, R any G-ring. Then $RH \simeq RG\#_cG_1$.

Theorem 2.4.6 [230] *Let $H = G \rtimes G_1$ be the semi-direct product of G and G_1, where G is a finite Abelian group, G_1 any finite group such that the action of G_1 on G stabilizes every cocyclic subgroup of G.*

Then $G_0(\mathbb{Z}H) \cong \underset{C \in X(G),}{\oplus} G_0 (\mathbb{Z} < C > \#G_1)$.

Remarks 2.4.4 (see [230]) If $H = G \rtimes G_1$, as in 2.4.8, then $QH \simeq \underset{C \in X(G)}{\prod} Q(C)\#G_1$ and $\mathbb{Z} \longmapsto \underset{C \in X(G)}{\prod} \mathbb{Z}(C)\#G_1$ is a subdirect embedding.

Theorem 2.4.7 [230] *Let H be a dihedral extension of a finite group G, i.e., H is a split extension $G \rtimes G_1$, of G by 2-element group G_1, whose non-trivial element acts on G by $x \to x^{-1}$. Then* $G_0(\mathbb{Z}H) \cong \underset{\substack{C \in X(G) \\ |C|=2}}{\oplus} (\mathbb{Z} \oplus$

$\mathbb{Z}) \underset{\substack{C \in X(G) \\ |C|=2}}{\oplus} \left(\mathbb{Z} \oplus C\ell \left(\mathbb{Z}(C)_+ \left(\frac{1}{|C|} \right) \right) \right)$ *where $\mathbb{Z}(C)_+$ is the ring of integers of* $Q(C)_+ = Q(C)^{G_1}$, *the maximal real subfield of $Q(C) \cong \mathbb{Q}(\zeta_{|C|})$.*

Remarks 2.4.5 It follows from theorem 2.4.7 that if D_{2n} is a dihedral group of order $2n$, then $G_0(\mathbb{Z}D_{2n}) \simeq \mathbb{Z}^\epsilon \oplus \underset{d|n}{\oplus} \left(\mathbb{Z} \oplus C\ell \left(\mathbb{Z} \left[\zeta_d, \frac{1}{d} \right]_+ \right) \right)$ where

$$\epsilon = \begin{cases} 2 \text{ if } n \text{ is odd, } \zeta_d \text{ a primitive } d^{\text{th}} \text{ root of } 1 \\ 4 \text{ if } n \text{ is even} \end{cases}$$

and where $\mathbb{Z} \left[\zeta_d, \frac{1}{d} \right]_+$ is the ring of integers in $Q(\zeta_d)_+ := Q(\zeta_d + \zeta_d^{-1})$, the maximal subfield of the cyclotomic field $Q(\zeta_d)$.

Definition 2.4.3 *Let R a Dedekind ring with field of fractions F, and A a central simple F-algebra. Then the ray-class group $C\ell_A(R)$ of R relative to A is the quotient of the group $I(R)$ of fractional ideals of R (see definition 2.3.1) by the subgroup $P_A(R)$ consisting of principal ideals generated by reduced norms on A (see 3.2.1). So, we have an exact sequence*

$$A^* \xrightarrow{Nrd} I(R) \longrightarrow C\ell_A(R) \longrightarrow 0.$$

Remarks 2.4.6   (i) In the notation of definition 2.4.3, if Γ is a maximal R-order in A, then $G_0(\Gamma) \cong G_0(A) \oplus C\ell_A(R)$.

(ii) With notations as in definition 2.4.3, the kernel of the surjection β : $C\ell_A(R) \twoheadrightarrow C\ell(R)$ sending the ray-class of a fractional ideal to its ideal class is an elementary Abelian 2-group.

(iii) If $A = M_n(F)$, $C\ell_A(R) = C\ell(R)$.

(iv) Let Q_{4n} be the generalized quaternion group $Q_{4n} = \langle x, y | x^n = y^2, y^4 = 1, yxy^{-1} = x^{-1} \rangle$. $G = \{1, y\}$, the 2-element group that acts on $Q(\zeta_n)$, the non-trivial element acting as complex conjugation.
Let $c : G \times G \to Q(\zeta_n)^*$ be the normalized 2-cocycle given by $c(\gamma, \gamma) = -1$. Let $H_n = Q(\zeta_n)\#_c G$. If $n > 2$, $G = Gal\,(Q(\zeta_n)/Q(\zeta_n)_+)$. So $H_n = (Q(\zeta_n)/Q(\zeta_n)_+)$. If $n = 1$ or 2, then G acts trivially and $H_n = Q(i)$ since $1\#y$ has square -1. If n is odd, $Q(\zeta_n) = Q(\zeta_{2n})$ since $-\zeta_n$ is a primitive $2n$-th root of unity.

Theorem 2.4.8 [230] *Let G be the quaternion group Q_{4n}, $n = 2^s n'$, n' odd. Then*

$$G_0(\mathbb{Z}G) \cong$$

$$\mathbb{Z}^\epsilon \bigoplus_{d|n'} \left(\mathbb{Z}^{s+2} \bigoplus_{i=0}^{s} C\ell\left(\mathbb{Z}\left[\zeta_{2^i d}, \frac{1}{2^i d} \right] \right) \bigoplus C\ell_{H_{2^{s+1}d}}\left(\mathbb{Z}\left[\zeta_{2^{s+1}d}, \frac{1}{2^{s+1}d} \right] \right) \right)$$

where

$$\epsilon = \begin{cases} 2 \text{ if } s > 0 \\ 1 \text{ if } s = 0 \end{cases}$$

Exercises

2.1 Let K be a field of characteristic p, G a finite group; H is a subgroup of G. Let $a(G, H)$ be the relative Gröthendieck ring with respect to H (see example 1.4.1 (vii)). Show that

(a) $a(G, 1)$ is the $\mathbb{Z}$-free Abelian group on $[F_1], \ldots, [F_t]$ where $\{F_i\}$ are a finite set of non-isomorphic irreducible G-modules.

(b) $a(G, 1)$ contains no non-zero idempotents.

2.2 Let G be a finite group, S a G-set, k a field of characteristic p, $M[S]$ the generalized permutation module (see example 1.4.1 (ix)).
Show that

(a) $M[S])$ is a kG-module and $M[S] \simeq M \underset{k}{\otimes} k[S]$.

(b) Let $\underline{p}_S(G)$ be the additive subgroup of $G_0(kG)$ generated by isomorphism classes of S-projective modules and $i_S(G)$ the additive subgroup of $G_0(kG)$ generated by Euler characteristics $[M'] - [M] + (M'')$ of S-split sequences $0 \to M' \to M \to M'' \to 0$. Show that $\underline{p}_S(G) i_S(G) = 0$.

2.3 Let G be an Abelian p-group. Show that $G_0(\mathbb{Z}G) \simeq G_0(\Gamma)$ where Γ is a maximal order in $Q G$ containing $\mathbb{Z}G$.

2.4 Let R be a semi-local Dedekind domain with quotient field F. Show that there exists an isomorphism $G_0(R, G) \cong G_0(FG)$.

2.5 Let R be a Dedekind domain whose quotient field F is a global field, Γ a maximal order in a semi-simple F-algebra A, $C\ell_A(R)$ the ray class group as defined in 2.4.3. Show that $G_0(\Gamma) \simeq G_0(A) \oplus Cl_A(R)$.

2.6 Let (K, R, k) be a p-modular system, i.e., R is a discrete valuation ring with quotient field K, $k = R/\underline{p}$ where $\underline{p}$ is the unique maximal ideal of R. Let G be a finite group. Show that

(i) $K_0(kG) = \Sigma i_{H*}(K_0(kH))$ where $i_H : K_0(kH) \to K_0(kG)$ is induced by the inclusion. $kH \to kG$ as H runs through all K-elementary subgroups of G.

(ii) The cokernel of the Cartan map $K_0(kG) \to G_0(kG)$ is a p-group.

2.7 Let A be a left Noetherian ring, n a positive integer. Show that $G_0(A(\frac{1}{n}) \simeq G_0(A)/H$ where H is a subgroup of $G_0(A)$ generated by all symbols $[M]$ where M ranges over $M \in \mathcal{M}(A)$ such that $nM = 0$.

2.8 Let A be a ring, T_+ the free Abelian monoid on generator t. In the notation of example 2.1.1 (ii) and (iv), show that there are isomorphisms of categories $M_{T_+}(A[t]) \simeq [Nil(\mathcal{M}(A)$ and $\mathcal{H}_{T_+}(A[t]) \simeq Nil(\mathcal{H}(A))$ and an isomorphism of groups

$$K_0(\mathcal{H}_{T_+}(A[t])) \simeq K_0(A) \oplus K_0 Nil_0(A).$$

2.9 Let G be a finite group, R a G-ring, and $\underline{a}$ a G-stable two-sided ideal of R so that $R/\underline{a}$ is a G-ring. Show that $\underline{a}\#G$ is a two-sided ideal of $R\#G$ and that the natural map $R\#G \to (R/\underline{a})\#G$ induces an isomorphism $(R\#G)/\underline{a}\#G \simeq (R/\underline{a})\#G$.

2.10 Let R be a complete Noetherian local ring, Λ an R-algebra finitely generated and projective as R-module; $\underline{a}$ a two-sided ideal of Λ contained

in rad Λ. Show that there is an isomorphism $K_0(\Lambda) \cong K_0(\Lambda/\underline{a})$ given by $[P] \rightarrow [P/\underline{a}P]$ where $P \in \mathcal{P}(\Lambda)$.
If $\underline{a} = rad(\Lambda)$, show that $K_0(\Lambda/\underline{a}) \simeq G_0(\Lambda/\underline{a})$.

2.11 Let R be Dedekind domain and G a finite group such that no proper divisor of $|G|$ is a unit in R. Show that $Cl(RG) \simeq SK_0(RG)$.

2.12 Let R be a Dedekind domain and Λ an R-order in a semi-simple F-algebra. Show that $Cl(\Lambda) \simeq Ker(SK_0(\Lambda)) \rightarrow \underset{\underline{p}}{\oplus} SK_0(\hat{\Lambda}_{\underline{p}})$ where $\underline{p}$ ranges over the prime ideals of R.

2.13 Let R be a Dedekind domain and Λ, Λ' R-orders in a semi-simple algebra Σ such that $\Lambda \subset \Lambda'$. Show that this inclusion induces a surjection of $C\ell(\Lambda)$ onto $C\ell(\Lambda^1)$.

2.14 Let R be the ring of integers in a number field F, Λ an R-order in a semi-simple F-algebra Σ. Suppose that Λ satisfies the Cartan condition (see definition 2.2.4). Show that $SK_0(\Lambda)$, $SG_0(\Lambda)$ are finite groups.

Chapter 3

K_1, K_2 of orders and grouprings

3.1 Definitions and basic properties

$(3.1)^A$ K_1 of a ring

3.1.1 Let R be a ring with identity, $GL_n(R)$ the group of invertible $n \times n$ matrices over R. Note that $GL_n(R) \subset GL_{n+1}(R) : A \to \left(\begin{smallmatrix} A & 0 \\ 0 & 1 \end{smallmatrix}\right)$. Put $GL(R) = \varinjlim GL_n(R) = \bigcup_{n=1}^{\infty} GL(R)$. Let $E_n(R)$ be the subgroup of $GL_n(R)$ generated by elementary matrices $e_{ij}(a)$ where $e_{ij}(a)$ is the $n \times n$ matrix with 1's along the diagonal, a in the (i,j)-position with $i \neq j$ and zeros elsewhere. Put $E(R) = \varinjlim E_n(R)$.

Note The $e_{ij}(a)$ satisfy the following.

(i) $e_{ij}(a)e_{ij}(b) = e_{ij}(a+b)$ for all $a, b \in R$.

(ii) $[e_{ij}(a), e_{jk}(b)] = e_{ik}(ab)$ for all $i \neq k, a, b \in R$.

(iii) $[e_{ij}(a), e_{kl}(b)] = 1$ for all $i \neq \ell, j \neq k$.

Lemma 3.1.1 *If* $A \in GL_n(R)$, *then* $\left(\begin{smallmatrix} A & 0 \\ 0 & A^{-1} \end{smallmatrix}\right) \in E_{2n}(R)$.

PROOF First observe that for any $C \in M_n(R)$, $\left(\begin{smallmatrix} I_n & 0 \\ C & I_n \end{smallmatrix}\right)$ and $\left(\begin{smallmatrix} I_n & C \\ 0 & I_n \end{smallmatrix}\right)$ are in $E_{2n}(R)$, where I_n is the identity $n \times n$ matrix. Hence

$$\begin{pmatrix} A & 0 \\ 0 & A^{-1} \end{pmatrix} = \begin{pmatrix} I_n & A \\ 0 & I_n \end{pmatrix} \begin{pmatrix} I_n & 0 \\ -A^{-1} & I_n \end{pmatrix} \begin{pmatrix} 0 & -I_n \\ I_n & 0 \end{pmatrix} \begin{pmatrix} I_n & 0 \\ -A & I_n \end{pmatrix} \in E_{2n}(R)$$

since

$$\begin{pmatrix} 0 & -I_n \\ I_n & 0 \end{pmatrix} = \begin{pmatrix} I_n & -I_n \\ 0 & I_n \end{pmatrix} \begin{pmatrix} I_n & 0 \\ I_n & I_n \end{pmatrix} \begin{pmatrix} I_n & -I_n \\ 0 & I_n \end{pmatrix}.$$

$\square$

Theorem 3.1.1 (Whitehead lemma)

(i) $E(R) = [E(R), E(R)]$, *i.e.*, $E(R)$ *is perfect.*

(ii) $E(R) = [GL(R), GL(R)]$.

Sketch of proof

(i) It follows from property (ii) of elementary matrices that $[E(R), E(R)] \subset E(R)$. Also, $E_n(R)$ is generated by elements of the form $e_{ij}(a) = [e_{ij}(a), e_{kj}(1)]$, and so, $E(R) \subset [E(R), E(R)]$. So, $E(R) = [E(R), E(R)]$.

(ii) For $A, B \in GL_n(R)$,

$$\begin{pmatrix} ABA^{-1}B^{-1} & 0 \\ 0 & I_n \end{pmatrix} = \begin{pmatrix} AB & 0 \\ 0 & (AB)^{-1} \end{pmatrix} \begin{pmatrix} A^{-1} & 0 \\ 0 & A \end{pmatrix} \begin{pmatrix} B^{-1} & 0 \\ 0 & B \end{pmatrix} \in E_{2n}(R).$$

Hence $[GL(R), GL(R)] \subset E(R)$ by 3.1.2.
Also, from (i) above, $E(R) \subset [E(R), E(R)] \subset [GL(R), GL(R)]$. Hence $E(R) = [GL(R), GL(R)]$.

Definition 3.1.1

$$K_1(R) := GL(R)/E(R) = GL(R)/[GL(R), GL(R)]$$
$$= H_1(GL(R), \mathbb{Z}).$$

Remarks 3.1.1 (i) For an exact category $\mathcal{C}$, the Quillen definition of $K_n(\mathcal{C})$, $n \geq 0$ coincides with the above definition of $K_1(R)$ when $\mathcal{C} = \mathcal{P}(R)$ (see [62, 165]). We shall discuss the Quillen construction in chapter 5.

(ii) The above definition 3.1.1 is functorial, i.e., any ring homomorphism $R \to R'$ induces an Abelian group homomorphism $K_1(R) \to K_1(R')$.

(iii) $K_1(R) = K_1(M_n(R))$ for any positive integer n and any ring R.

(iv) $K_1(R)$, as defined above, coincides with $K_{\det}(\mathcal{P}(R))$, a quotient of the additive group generated by all isomorphism classes $[P, \mu]$, $P \in \mathcal{P}(R)$, $\mu \in \text{Aut}(P)$ (see [20, 39]). (Also see exercise 3.1.)

3.1.2 If R is commutative, the determinant map $\det : GL_n(R) \to R^*$ commutes with $GL_n(R) \to GL_{n+1}(R)$ and hence defines a map $\det : GL(R) \to R^*$, which is surjective since given $a \in R^*$, there exists $A = \begin{pmatrix} a & 0 \\ 0 & 1 \end{pmatrix}$ such that $\det A = a$. Now, $\det$ induces a map $\det : GL(R)/[GL(R), GL(R)] \to R^*$, i.e., $\det : K_1(R) \to R^*$. Moreover, $\alpha(a) = \begin{pmatrix} a & 0 \\ 0 & 1 \end{pmatrix}$ for all $a \in R^*$ defines a map $\alpha : R^* \to K_1(R)$ and $\det \alpha = 1_R$. Hence $K_1(R) \simeq R^* \oplus SK_1(R)$ where $SK_1(R) := \text{Ker}(\det : K_1(R) \to R^*)$. Note that $SK_1(R) = SL(R)/E(R)$ where $SL(R) = \lim_{n \to \infty} SL_n(R)$ and $SL_n(R) = \{A \in GL_n(R) | \det A = 1\}$. Hence $SK_1(R) = 0$ if and only if $K_1(R) \simeq R^*$.

Examples 3.1.1 (i) If F is a field, then $K_1(F) \simeq F^*$, $K_1(F[x]) \simeq F^*$.

(ii) If R is a Euclidean domain (for example, $\mathbb{Z}$, $\mathbb{Z}[i] = \{a + ib;\ a, b \in \mathbb{Z}\}$, polynomial ring $F[x]$, F a field), then $SK_1(R) = 0$, i.e., $K_1(R) \simeq R^*$ (see [155, 181]).

(iii) If R is the ring of integers in a number field F, then $SK_1(R) = 0$ (see [21, 181]).

$(3.1)^B$ K_1 of local rings and skew fields

Theorem 3.1.2 [41, 180] *Let R be a noncommutative local ring. Then there exists a homomorphism* det $: GL_n(R) \to R^*/[R^*, R^*]$ *for each positive integer n such that*

(i) $E_n(R) \subset Ker(\text{det})$

(ii)

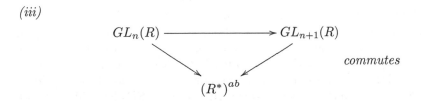

$$\text{det}\begin{pmatrix} \alpha_1 & & & & \\ & \alpha_2 & & 0 & \\ & & \ddots & & \\ & 0 & & \ddots & \\ & & & & \alpha_n \end{pmatrix} = \overline{\alpha}_1\overline{\alpha}_2\cdots\overline{\alpha}_n \qquad \text{where } \alpha_i \in R^*$$

for all i and $\alpha \to \overline{\alpha}$ is the natural map

$$R^* \to (R^*)^{ab} = R^*/[R^*, R^*] \ .$$

(iii)

$$GL_n(R) \longrightarrow GL_{n+1}(R)$$
$$\searrow \qquad \swarrow \qquad \qquad commutes$$
$$(R^*)^{ab}$$

Note. The homomorphism 'det' above is usually called Dieudonné determinant because it was J. Dieudonné who first introduced the ideas in 3.1 for skew fields (see [20]).

Theorem 3.1.3 [181] *Let R be a noncommutative local ring. Then the natural map $GL_1(R) = R^* \hookrightarrow GL(R)$ induces a surjection $R^*/[R^*, R^*] \to K_1(R)$ whose kernel is the subgroup generated by the images of all elements $(1 - xy)/(1 - yx)^{-1} \in R^*$ where x or y is the unique maximal ideal $\mathbf{m}$ of R.*

Theorem 3.1.4 [181] *If R is a skew field, $K_1(R) \cong R^*/[R^*, R^*]$.*

$(3.1)^C$ Mennicke symbols

3.1.3 Let R be a commutative ring with identity $a, b \in R$. Choose $c, d \in R$ such that $ad - bc = 1$, i.e., such that $\left(\begin{smallmatrix} a & b \\ c & d \end{smallmatrix} \right) \in SL_2(R)$. Let $[a, b]$ = class of $\left(\begin{smallmatrix} a & b \\ c & d \end{smallmatrix} \right)$ in $SK_1(R)$.
Then

(i) $[a, b]$ is well defined.

(ii) $[a, b] = [b, a]$ if $a \in R^*$.

(iii) $[a_1 a_2, b] = [a_1, b][a_2, b]$ if $a_1 a_2 R + bR = R$.

(iv) $[a, b] = [a + rb, b]$ for all $r \in R$.

We have the following result:

Theorem 3.1.5 [20] *If R is a commutative ring of Krull dimension ≤ 1, then the Menicke symbols generate $SK_1(R)$.*

Remarks See [20, 21, 142] for further details on Mennicke symbols.

$(3.1)^D$ Stability for K_1

3.1.4 Stability results are very useful for reducing computations of $K_1(R)$ to computations of matrices over R of manageable size.

Let A be any ring with identity. An integer n is said to satisfy stable range condition (SR_n) for $GL(A)$ if whenever $r > n$, and $(a_1, a_2, \ldots, a_r)$ is a unimodular row, then there exists $b_1, b_2, \ldots, b_{r-1} \in A$ such that $(a_1 + a_r b_1, a_2 + a_r b_2, \ldots, a_{r-1} + a_r b_{r-1})$ is unimodular. Note that $(a_1, a_2, \ldots, a_r) \in A^r$ unimodular says that $(a_1, a_2, \ldots, a_r)$ generates the unit ideal, i.e., $\sum A a_i = A$ (see [20]). For example, any semi-local ring satisfies SR_2 (see [20] or [215]).

Theorem 3.1.6 [20, 215, 221] *If SR_n is satisfied, then*

(i) *$GL_m(A)/E_m(A) \to GL(A)/E(A)$ is onto for all $m \geq n$, and injective for all $m > n$.*

(ii) *$E_m(A) \lhd GL_m(A)$ if $m \geq n + 1$.*

(iii) *$GL_m(A)/E_m(A)$ is Abelian for $m > n$.*

For further information on K_1-stability, see [20, 21, 215, 221, 142].

3.2 K_1, SK_1 of orders and grouprings; Whitehead torsion

$(3.2)^A$ K_1, SK_1 of orders and grouprings

Let R be the ring of integers in a number field F, Λ an R-order in a semi-simple F-algebra Σ. First we have the following result (see [20]).

Theorem 3.2.1 $K_1(\Lambda)$ *is a finitely generated Abelian group.*

PROOF The proof relies on the fact that $GL_n(\Lambda)$ is finitely generated and also that $GL_2(\Lambda) \to K_1(\Lambda)$ is surjective (see [20]). □

Remarks 3.2.1 Let R be a Dedekind domain with quotient field F, Λ an R-order in a semi-simple F-algebra Σ. The inclusion $\Lambda \hookrightarrow \Sigma$ induces a map $K_1(\Lambda) \to K_1(\Sigma)$. Putting $SK_1(\Lambda) = \operatorname{Ker}(K_1(\Lambda) \to K_i(\Sigma))$, it means that understanding $K_1(\Lambda)$ reduces to understanding $K_1(\Sigma)$ and $SK_1(\Lambda)$. Since Σ is semi-simple, $\Sigma = \oplus \Sigma_i$ where $\Sigma_i = M_{n_i}(D_i)$, D_i is a skew field. So $K_1(\Sigma) = \oplus K_1(D_i)$.

One way of studying $K_1(\Lambda)$ and $SK_1(\Lambda)$, $K_1(\Sigma)$ is via reduced norms. We consider the case where R is the ring of integers in a number field or p-adic field F.

Let R be the ring of integers in a number field or p-adic field F. Then there exists a finite extension E of F such that $E \otimes \Sigma$ is a direct sum of full matrix algebras over E, i.e., E is a splitting field of Σ. If $a \in \Sigma$, the element $1 \otimes a \in E \otimes \Sigma$ may be represented by a direct sum of matrices, and the reduced norm of a, written $\operatorname{nr}(a)$, is defined as the direct product of their determinants. We then have $\operatorname{nr}: GL(\Sigma) \to C^*$ where $C = $ centre of Σ (if $\Sigma = \oplus_{i=1}^m \Sigma_i$ and $C = \oplus_{i=1}^m C_i$ we could compute $\operatorname{nr}(a)$ componentwise via $GL(\Sigma_i) \to C_i^*$). Since C^* is Abelian, we have $\operatorname{nr}: K_1(\Sigma) \to C^*$. Composing this with $K_1(\Lambda) \to K_1(\Sigma)$ we have a reduced norm $\operatorname{nr}: K_1(\Lambda) \to K_1(\Sigma) \to C^*$.

From the discussion below, it will be clear that an alternative definition of $SK_1(\Lambda)$ is $\{x \in K_1(\Lambda)|\operatorname{nr}(x) = 1\}$.

Theorem 3.2.2 *Let R be the ring of integers in a number field F, Λ an R-order in a semi-simple F-algebra Σ. In the notation of remarks 3.2.1, let U_i be the group of all nonzero elements $a \in C_i$ such that $\beta(a) > 0$ for each embedding $\beta: C_i \to \mathbb{R}$ at which $\mathbb{R} \otimes_{C_i} \Sigma_i$ is not a full matrix algebra over $\mathbb{R}$. Then*

(i) The reduced norm map yields an isomorphism $\operatorname{nr}: K_1(\Sigma) \cong \prod_{i=1}^m U_i$.

(ii) $\operatorname{nr}: K_1(\Lambda) \subset \prod_{i=1}^m (U_i \cap R_i^*)$ where R_i is the ring of integers in C_i.

PROOF See [39, 213]. ⧠

Remarks 3.2.2 (i) If Γ is a maximal R-order in Σ, then we have equality in (ii) of theorem 3.2.2, i.e., nr $(K_1(\Gamma)) = \prod_{i=1}^{m}(U_i \cap R_i^*)$. (See [39, 213]). Hence rank$K_1(\Gamma) = \text{rank}\prod_{i=1}^{m}(U_i \cap R_i^*)$.

(ii) For all $n \geq 1$, $K_n(\Lambda)$ finitely generated and $SK_n(\Lambda)$ is finite (see 7.1).

Theorem 3.2.3 *Let R be the ring of integers in a number field F, and Λ any R-order in a semi-simple F-algebra Σ. Then $SK_1(\Lambda)$ is a finite group.*

PROOF See [20]. The proof involves showing that $SK_1(\Lambda)$ is torsion and observing that $SK_1(\Lambda)$ is also finitely generated as a subgroup of $K_1(\Lambda)$ (see theorem 3.2.1). ⧠

The next results are local version of theorem 3.2.3.

Theorem 3.2.4 *Let R be the ring of integers in a p-adic field F, Γ a maximal R-order in a semi-simple F-algebra Σ. In the notation of 3.2.1, we have*

(i) nr $: K_1(\Sigma) \simeq C^*$

(ii) nr $: K_1(\Gamma) \cong S^*$ where $S = \oplus R_i$ and R_i is the ring of integers in C_i.

Theorem 3.2.5 *(i) Let F be a p-adic field (i.e., any finite extension of $\widehat{Q}_p$), R the ring of integers of F, Λ any R-order in a semi-simple F-algebra Σ. Then $SK_1(\Lambda)$ is finite.*

(ii) Let R be the ring of integers in a p-adic field F, $\mathbf{m}$ the maximal ideal of R, $q = |R/\mathbf{m}|$. Suppose that Γ is a maximal order in a central division algebra over F. Then $SK_1(\Gamma)$ is a cyclic group of order $(q^n - 1)/q - 1$. $SK_1(\Gamma) = 0$ iff $D = F$.

Remarks 3.2.3 (i) For the proof of theorem 3.2.5, see [104, 105, 99].

(ii) If in theorem 3.2.2 and remark 3.2.3 $R = Z$, $F = Q$, G a finite group, we have that rank of $K_1(ZG) = s - t$ where s is the number of real representations of G, and t is the number of rational representations of G. (See [20]).

(iii) Computation of $SK_1(ZG)$ for various groups has attracted extensive attention because of its applicability in topology. For details of such computations, see [142, 143, 159].

(iv) That for all $n \geq 1$, $SK_n(\mathbb{Z}G)$, $SK_n(\widehat{\mathbb{Z}_p}(G))$are finite groups are proved in 7.1.

(v) It also is known that if Γ is a maximal order in a p-adic semi-simple F-algebra Σ, then $SK_{2n-1}(\Gamma) = 0$ for all $n \geq 1$ iff Σ is unramified over its centre (see [107]). These generalizations will be discussed in 7.1.

$(3.2)^B$ Applications – Whitehead torsion and s-cobordism theorem

3.2.1 J.H.C. Whitehead (see [244]) observed that if X is a topological space with fundamental group G, and $R = \mathbb{Z}G$, then the elementary row and column transformations of matrices over R have some natural topological meaning. To enable him to study homotopy between spaces, he introduced the group $\mathrm{Wh}(G) = K_1(\mathbb{Z}G)/\omega(\pm G)$ where ω is the map $G \to GL_1(\mathbb{Z}G) \to GL(\mathbb{Z}G) \to K_1(\mathbb{Z}G)$, such that if $f : X \to Y$ is a homotopy equivalence, then there exists an invariant $\tau(f)$ in $\mathrm{Wh}(G)$ such that $\tau(f) = 0$ iff f is induced by elementary deformations transforming X to Y. The invariant $\tau(f)$ is known as Whitehead torsion (see [153]).

Now, it follows from theorem 3.2.1 that $\mathrm{Wh}(G)$ is finitely generated when G is a finite group. Moreover, it is also well known that $\mathrm{Tor}(K_1(\mathbb{Z}G)) = (\pm 1) \times G^{ab} \times SK_1(\mathbb{Z}G)$ where $SK_1(\mathbb{Z}G) = \mathrm{Ker}\,(K_1(\mathbb{Z}G) \to K_1(QG))$ (see [159]). So, $\mathrm{rank}K_1(\mathbb{Z}G) = \mathrm{rank}\mathrm{Wh}(G)$, and it is well known that $SK_1(\mathbb{Z}G)$ is the full torsion subgroup of $\mathrm{Wh}(G)$ (see [159]). So, computations of $\mathrm{Tor}\,(K_1(\mathbb{Z}G))$ reduce essentially to computations of $SK_1(\mathbb{Z}G)$. The last two decades have witnessed extensive research on computations of $SK_1(\mathbb{Z}G)$ for various groups G (see [159]). More generally, if R is the ring of integers in a number field or a p-adic field F, there has been extensive effort in understanding the groups $SK_n(RG) = \mathrm{Ker}\,(K_n(RG) \to K_n(FG))$ for all $n \geq 1$ (see [112, 113, 115]). More generally still, if Λ is an R-order in a semi-simple F-algebra Σ (i.e., Λ is a subring of Σ, finitely generated as an R-module, and $\Lambda \otimes_R F = \Sigma$), there has been extensive effort to compute $SK_n(\Lambda) = \mathrm{Ker}\,(K_n(\Lambda) \to K_n(\Sigma))$ (see [112, 113, 115]), the results of which apply to $\Lambda = RG$. We shall discuss these computations further in the forthcoming chapter on higher K-theory (see chapter 7).

Note also that Whitehead torsion is useful in the classifications of manifolds (see [153] or [159]). Indeed , one important application of Whitehead torsion is the "s-cobordism theorem", which is the main tool for classifying manifolds of dimension ≥ 5.

3.2.2 Let M^n be a connected compact n-manifold of dimension ≥ 5; an h-cobordism on M^n is a connected manifold W^{n+1} with exactly two boundary components, one of which is M^n and the other another manifold M'^n such that W has deformation retraction on both M^n and M'^n.

3.2.3 "s-cobordism theorem" [153, 138] In the notation of 3.2.2, let $\mathcal{F}$ be a family of "h-cobordisms" built on the n-manifold M^n with fundamental group $G = \pi_1(M^n)$. Then there is a map $\tau : \mathcal{F} \to Wh(G)$ called "Whitehead torsion" and τ induces a natural one-to-one correspondence from $\mathcal{F}/\backsim$ to $Wh(G)$ where $\backsim$ is the equivalence relation induced by homeomorphisms $W \to W'$, which are the identity on M. If W is the "trivial" "h-cobordism" $W = M \times [0,1]$, then $\tau(W) = 1$.

Corollary 3.2.1 Let M^n be a connected compact n-manifold of dimension ≥ 5 with fundamental group G. If $Wh(G) = 1$, then every h-cobordism built on M is homeomorphic relative to M to a product $M \times [0,1]$. In particular, the other boundary component M' is homeomorphic to M.

Remarks 3.2.4   (i) S. Smale observed that 3.2.1 can be applied to prove Poincare conjecture in dimension ≥ 6, i.e., any manifold M^n homotopy equivalent to S^n is also homeomorphic to S^n (see [181] for a proof).

(ii) Theorem 3.2.3 and corollary 3.2.1 (i) apply to the three major categories of manifolds: (1) topological manifolds and continuous maps, (2) PL manifolds and PL maps, and (3) smooth manifolds and C^∞ maps.

3.3   The functor K_2

In this section, we provide a brief review of the functor K_2 due to J. Milnor (see [155]). Classical results on K_2 of orders and grouprings are rather scanty, but we shall obtain in chapters 7 and 8 higher K-theoretic results that also hold for K_2. We review here some results on K_2 of rings, fields, and division algebras that are also relevant to computations of K-theory of orders and group-rings.

$(3.3)^A$ K_2 of rings and fields

3.3.1 Let A be a ring. The Steinberg group of order n $(n \geq 1)$ over A, denoted $\mathrm{St}_n(A)$, is the group generated by $x_{ij}(a)$, $i \neq j$, $1 \leq i, j \leq n$, $a \in A$, with realations

(i) $x_{ij}(a)x_{ij}(b) = x_{ij}(a+b)$.

(ii) $[x_{ij}(a), x_{kl}(b)] = 1$, $j \neq k$, $i \neq l$.

(iii) $[x_{ij}(a), x_{jk}(b)] = x_{ik}(ab)$, i, j, k distinct.

(iv) $[x_{ij}(a), x_k(b)] = x_{ij}(-ba)$, $j \neq k$.

Since the generators $e_{ij}(a)$ of $E_n(A)$ satisfy relations (i) - (iv) above, we have a unique surjective homomorphism $\varphi_n : \text{St}_n(A) \to E_n(A)$ given by $\varphi_n(x_{ij}(a)) = e_{ij}(a)$. Moreover, the relations for $\text{St}_{n+1}(A)$ include those of $\text{St}_n(A)$, and so, there are maps $\text{St}_n(A) \to \text{St}_{n+1}(A)$. Let $\text{St}(A) = \varinjlim_n \text{St}(A)$, $E(A) = \varinjlim_n E_n(A)$. Then we have a canonical map $\varphi : \text{St}(A) \to E(A)$.

Definition 3.3.1 *Define $K_2^M(A)$ as the kernel of the map $\varphi : \text{St}(A) \to E(A)$.*

Theorem 3.3.1 [155] *$K_2^M(A)$ is Abelian and is the centre of $\text{St}(A)$. So, $\text{St}(A)$ is a central extension of $E(A)$.*

PROOF   See [155].   ⧠

Definition 3.3.2 *An exact sequence of groups of the form $1 \to A \to E \xrightarrow{\varphi} G \to 1$ is called a central extension of G by A if A is central in E. Write this extension as (E, φ). A central extension (E, φ) of G by A is said to be universal if for any other central extension (E', φ') of G, there is a unique morphism $(E, \varphi) \to (E', \varphi')$.*

Theorem 3.3.2 *$\text{St}(A)$ is the universal central extension of $E(A)$. Hence there exists a natural isomorphism $K_2^M(A) \simeq H_2(E(A), Z)$.*

PROOF   The last statement follows from the fact that for a perfect group G (in this case $E(A)$), the kernel of the universal central extension (E, φ) (in this case $(\text{St}(A), \varphi)$) is naturally isomorphic to $H_2(G, Z)$ (in this case $H_2(E(A), \mathbb{Z})$).   ⧠

For the proof of the first part see [155].

Definition 3.3.3 *Let A be a commutative ring $u \in A^*$*

$$w_{ij}(u) := x_{ij}(u)x_{ji}(-u^{-1})x_{ij}.$$

Define $h_{ij}(u) := w_{ij}(u)w_{ij}(-1)$. For $u, v \in A^*$, one can easily check that $\varphi\left([h_{12}(u), h_{13}(v)]\right) = 1$.

So, $[h_{12}(u), h_{13}(v)] \in K_2(A)$. It can be shown that $[h_{12}(u), h_{13}(v)]$ is independent of the indices $1, 2, 3$. We write $\{u, v\}$ for $[h_{12}(u), h_{13}(v)]$ and call this the Steinberg symbol.

Theorem 3.3.3 *Let A be a commutative ring. The Steinberg symbol $\{,\} : A^* \times A^* \to K_2(A)$ is skew symmetric and bilinear, i.e.,*

$$\{u, v\} = \{v, u\}^{-1} \qquad and \qquad \{u_1 u_2, v\} = \{u_1, v\}\{u_2, v\}.$$

PROOF   See [155].   ☐

Theorem 3.3.4 (Matsumoto) *If F is a field, then $K_2^M(F)$ is generated by* $\{u,v\}$, $u,v \in F^*$, *with relations*

   (i) $\{uu',v\} = \{u,v\}\{u',v\}$.

  (ii) $\{u,vv'\} = \{u,v\}\{u,v'\}$.

 (iii) $\{u,1-u\} = 1$.

That is, $K_2^M(F)$ is the quotient of $F^* \otimes_\mathbb{Z} F^*$ by the subgroup generated by the elements $x \otimes (1-x)$, $x \in F^*$.

Examples 3.3.1 Writing K_2 for K_2^M:

   (i) $K_2(\mathbb{Z})$ is cyclic of order 2. See [155].

  (ii) $K_2(\mathbb{Z}(i)) = 1$, so is $K_2(\mathbb{Z}\sqrt{-7})$. See [155].

 (iii) $K_2(\mathbb{F}_q) = 1$ if F_q is a finite field with q elements. See [155].

 (iv) If F is a field, $K_2(F[t]) \simeq K_2(F)$. See [155].

More generally, $K_2(R[t]) \simeq K_2(R)$ if R is a commutative regular ring.

Remarks 3.3.1   (i) There is a definition by J. Milnor of higher K-theory of fields $K_n^M(F)$, $n \geq 1$, which coincides with $K_2^M(F)$ above for $n = 2$. More precisely,

$$K_n^M(F) := \underbrace{F^* \otimes F^* \otimes \cdots F^*}_{n \text{ times}} / \{a_1 \otimes \cdots \otimes a_n | a_i + a_j = 1 \text{ for some } i \neq j, a_i \in$$

i.e., $K_n^M(F)$ is the quotient of $F^* \otimes F^* \otimes \cdots F^*$ (n times) by the subgroup generated by all $a_1 \otimes a_2 \otimes \cdots \otimes a_n$, $a_i \in F$ such that $a_i + a_j = 1$. $\oplus_{n \geq 0}^\infty K_n^M(F)$ is a ring (see [154]).

(ii) The higher K-groups defined by D. Quillen (see chapter 5), namely $K_n(\mathcal{C})$, $\mathcal{C}$ an exact category $n \geq 0$ and $K_n(A) = \pi_n(BGL(A)^+)$, $n \geq 1$, coincides with $K_2^M(A)$ above when $n = 2$ and $\mathcal{C} = \mathcal{P}(A)$.

Remarks 3.3.2 Let E/F be a finite field extension of degree n. There are interesting questions about when K_2E is generated by symbols $\{a,b\}$, $a \in F^*, b \in E^*$. Now, it is well known that if $n \geq 3$, K_2E is generated by symbols $\{a,b\}$, $a \in F^*, b \in E^*$ (see [149, 169]). However, if $n > 3$, then in general $K_2(E)$ is not generated by symbols $\{a,b\}$.

If F is a global field, then X. Guo, A. Kuku, and H. Qin (G/K/Q) provided a counterexample in [73], namely, for $F = \mathbb{Q}$, $E = Q(\sqrt{-1}, \sqrt{-3})$, K_2E cannot be generated by the Steinberg symbols $\{a,b\}$, $a \in F^*$, $b \in E^*$. In fact, they proved the following general result:

Theorem 3.3.5 [73] *Let F be a global field. Then*

(i) *For any integer $n > 3$, there is an extension field E over F of degree n such that K_2E is not generated by the Steinberg symbols $\{a, b\}$, $a \in F^*$, $b \in E^*$.*

(ii) *For any extension field E over F of degree n, the subgroup H of K_2E generated by Steinberg symbols $\{a, b\}$ with $a \in F^*$, $b \in E^*$ is of finite index.*

PROOF See [73]. ⬚

3.3.2 Our next aim is to be able to state the reciprocity uniqueness theorem of C. Moore [157].

So, let F be a number field and v either a discrete valuation of F or an archimedean absolute value. Then the completion F_v is either a local field, or the field of real numbers, or the field of complex numbers. In the local and real cases the group $\mu(F_v)$, consisting of all roots of unity in the completed field, is finite cyclic. But in the complex case the group $\mu(F_v) \cong \mu(C)$ is not a finite group.

Let $m(v)$ denote the order of the finite cyclic group $\mu(F_v)$. The $m(v)$-th *power norm residue symbol* $(x, y)_v$ is a continuous Steinberg symbol on $\mu(F_v)$. For the definition of this symbol, see [155].

Since Moore has shown that any continuous Steinberg symbol on the complex field is identically equal to 1, we shall exclude the complex case so that F_v is either a local field or the real field.

Now suppose that x and y are non-zero elements of the field $F \subset F_v$. In order to put all of the various symbol $(x, y)_v$ into one group, we apply the surjection

$$\mu(F_v) \to \mu(F),$$

which carries each root of unity ξ to the power $\xi^{m(v)/m}$. The m-th *power reciprocity theorem* then asserts that

$$\prod_v (x, y)_v^{m(v)/m} = 1.$$

See [155].

We now state the reciprocity uniqueness theorem of C. Moore.

Theorem 3.3.6 [157] *The sequence $K_2F \to \oplus_v \mu(F_v) \to \mu(F) \to 1$ is exact, where the first homomorphism carries each generator $\{x, y\}$ of K_2F to the element whose v-th component is the $m(v)$th power residue symbol $(x, y)_v$, and the second homomorphism carries each element $\{\xi\}$ to the product*

$\prod_v \xi_v^{m(v)/m}$. *Here v ranges over all discrete valuations, and all archimedean absolute values with completion $F_v \cong \mathbb{R}$.*

Here is one consequence of theorem 3.3.6. We shall briefly discuss a generalization of theorem 3.3.6 to division algebras in the next subsection.

Corollary 3.3.1 Let F be a number field with ring of integers R, $d : K_2F \to \bigoplus_{\underline{p}} R/\underline{p}$ the homomorphism that takes each generator $\{x,y\}$ of K_2F to the element in the direct sum whose pth coordinate is the tame symbol $d_{\underline{p}}(x,y)$ (see [155]). Then d is surjective.

PROOF   See [155]. ⬜

We now close this subsection with a discussion of connections of K_2 with Brauer group of fields and Galois cohomology.

3.3.3 Let F be a field and $B_r(F)$ the Brauer group of F, i.e., the group of stable isomorphism classes of central simple F-algebras with multiplication given by tensor product of algebras. See [149].

A central simple F-algebra A is said to be split by an extension E of F if $E \otimes A$ is E-isomorphic to $M_r(E)$ for some positive integer r. It is well known (see [149]) that such E can be taken as some finite Galois extension of F. Let $B_r(F,E)$ be the group of stable isomorphism classes of E-split central simple F-algebras. Then $B_r(F) := B_r(F,F_s)$ where F_s is the separable closure of F.

Theorem [149] *Let E be a Galois extension of a field F, $G = Gal(E/F)$. Then there exists an isomorphism $H^2(G,E^*) \cong B_r(F;E)$. In particular, $B_r(F) \cong H^2(G,F_s^*)$ where $G = Gal(F_s/F) = \varinjlim Gal(E_i/F)$, where E_i runs through the finite Galois extensions of F.*

Now, for any $m > 0$, let μ_m be the group of m-th roots of 1, $G = Gal(F_s/F)$. We have the Kummer sequence of G-modules

$$0 \to \mu_m \to F_s^* \to F_s^* \to 0$$

from which we obtain an exact sequence of Galois cohomology groups

$$F^* \xrightarrow{m} F^* \to H^1(F,\mu_m) \to H^1(F,F_s^*) \to \cdots,$$

where $H^1(F,F_s^*) = 0$ by Hilbert theorem 90. So, we obtain isomorphism $\chi_m : F^*/mF^* \cong F^* \otimes \mathbb{Z}/m \to H^1(F,\mu_m)$.

Now, the composite

$$F^* \otimes_{\mathbb{Z}} F^* \to (F^* \otimes_{\mathbb{Z}} F^*) \otimes \mathbb{Z}/m \to H^1(F,\mu_m) \otimes H^1(F,\mu_m)$$
$$\to H^2\left(F,\mu_m^{\otimes^2}\right)$$

is given by $a \otimes b \to \chi_m(a) \cup \chi_m(b)$ (where $\cup$ is cup product), which can be shown to be a Steinberg symbol inducing a homomorphism $g_{2,m} : K_2(F) \otimes \mathbb{Z}/m\mathbb{Z} \to H^2\left(F, \mu_m^{\otimes^2}\right)$.

We then have the following result due to A.S. Merkurjev and A.A. Suslin (see [150]).

Theorem 3.3.7 [150] *Let F be a field, m an integer > 0 such that the characteristic of F is prime to m. Then the map*

$$g_{2,m} : K_2(F)/mK_2(F) \to H^2\left(F, \mu_m^{\otimes^2}\right)$$

is an isomorphism where $H^2\left(F, \mu_m^{\otimes^2}\right)$ can be identified with the m-torsion subgroup of $B_r(F)$.

Remarks 3.3.3 By generalizing the process outlined in 8.2.3 above, we obtain a map

$$g_{n,m} : K_n^M(F)/mK_n^M(F) \to H^n\left(F, \mu_m^{\otimes}\right).$$

It is a conjecture of Bloch - Kato that $g_{n,m}$ is an isomorphism for all F, m, n. So, theorem 3.3.7 is the $g_{2,m}$ case of Bloch - Kato conjecture when m is prime to the characteristic of F. Furthermore, A. Merkurjev proved in [149] that theorem 3.3.7 holds without any restriction on F with respect to m.

It is also a conjecture of Milnor that $g_{n,2}$ is an isomorphism. In 1996, V. Voevodsky proved that $g_{n,2^r}$ is an isomorphism for any r.

$(3.3)^B$ K_2 of division algebras and maximal orders

In this subsection, we discuss several results on K_2 of division algebras, which are generalizations of those for fields. First, we have the following generalization of Moore reciprocity uniqueness theorem 3.3.6 due to A. Bak and U. Rehmann (see [11]).

Theorem 3.3.8 [11] *Let D be a central division algebra over a global field E. Then the cokernel of the map $K_2(D) \to \underset{p}{\oplus} K_2(\widehat{D}_p)$ is isomorphic to $\mu(E)$ or $\mu(E)/(\pm 1)$ except in the case that $\widehat{D}_v$ is split at some real place v of E. In this case, the cokernel is trivial.*

Remarks 3.3.4 (i) The ambiguity (± 1) in theorem 3.3.8 was removed in some cases by R. Oliver. See [159], theorem 4.1.3.

(ii) A. Merkurjev and A. Suslin remarked in [150] that the ambiguity can be completely removed if the degree of D is square free, and in [102] M. Kolster and R. Laubenbacher provided a proof for this situation.

(iii) We next discuss the question of when $K_2(D)$ is generated by $\{a, b\}$, $a \in F^*$, $b \in D^*$ when D is a central division algebra over F and state the following result 3.3.9 (see [73]) due to G/K/Q. It is well known that $K_2(D)$ is generated by $\{a, b\}$, $a, b \in D^*$.

Theorem 3.3.9 [73] *Let D be a division algebra over a field F. Then K_2D is generated by the Steinberg symbols $\{a, d\}$ with $a \in F^*$ and $d \in D^*$ in the following cases:*

(i) *F is a number field and the index of D is square free.*

(ii) *F is a nonarchimedean local field and the index of D is square free.*

(iii) *F is a nonarchimedean local field and the characteristic of the residue field is prime to the index of D.*

PROOF   See [73].   ▯

The next result says that in the number field case, every element of $K_2(D)$ is a symbol when the index of D is square free.

Theorem 3.3.10 [73] *Let F be a number field and D a central division algebra over F with square-free index. Then $K_2D = \{\{a, b\} | a \in F^*, b \in D^*\}$, i.e., every element of $K_2(D)$ is a symbol of the form $\{a, b\}$ for $a \in F^*, b \in D^*$.*

PROOF   See [73].   ▯

The next result, also due to G/K/Q, computes K_2 of a maximal order in a central division algebra D over a number field when the index of D is square free.

Theorem 3.3.11 [73] *Let F be a number field and R the ring of integers in F. Let D be a finite-dimensional central division F-algebra with square-free index, and Λ a maximal R-order of D. Then*

$$K_2\Lambda \simeq K_2^+ R = \ker \left(K_2 R \to \prod_{\text{real ramified}} \{\pm 1\} \right).$$

PROOF   See [73].   ▯

$(3.3)^C$ K_2 and pseudo-isotropy

Let $R = \mathbb{Z}G$, G a group. For $u = R^*$, put $w_{ij}(u) := x_{ij}(u)x_{ji}(-u^{-1})x_{ij}(u)$. Let W_G be the subgroup of $St(R)$ generated by all $w_{ij}(g), g \in G$. Define $Wh_2(G) = K_2(R)/(K_2(R) \cap W_G)$.

Now let M be a smooth n-dimensional compact connected manifold without boundary. Two diffeomorphisms h_0, h_1 of M are said to be isotropic if they lie in the same path component of the diffeomorphism group. h_0, h_1 are said to be pseudo-isotropic if there is a diffeomorphism of the cylinder $M \times [0,1]$ restricted to h_0 on $M \times (0)$ and to h_1 on $M \times (1)$. Let $P(M)$ be the pseudo-isotropy space of M, i.e., the group of diffeomorphism h of $M \times [0,1]$ restricting to the identity on $M \times (0)$. Computation of $\pi_0(P(M^n))$ helps to understand the differences between isotopies, and we have the following result due to A. Hatcher and J. Wagoner.

Theorem 3.3.12 [80] *Let M be an n-dimension ($n \geq 5$) smooth compact manifold with boundary. Then there exists a surjective map*

$$\pi_0(P(M)) \to Wh_2(\pi_1(X)),$$

where $x_1(X)$ is the fundamental group of X.

Exercises

3.1 $K_{det}(\mathcal{C})$

Let A be a ring with identity and $\mathcal{C}$ a category of A-modules (e.g., $\mathcal{P}(A)$). Define a category $\overline{\mathcal{C}}$ as follows: $ob\ \overline{\mathcal{C}} = \{(P, \mu)|P \in \mathcal{C}, \mu \in Aut(P)\}$. A morphism $f : (P, \mu) \to (Q, \nu)$ is defined as an element $f \in Hom_A(\mathcal{P}, Q)$ such that $\nu f = f\mu$. A sequence $0 \to (P', \mu) \to (P, \lambda) \to (P'', \nu) \to 0(I)$ is said to be exact if $0 \to P' \xrightarrow{f} P \xrightarrow{g} P'' \to 0$ is exact in $\mathcal{C}$. Let G be the free Abelian group generated by isomorphism classes of objects of $\overline{\mathcal{C}}$. Suppose that G_0 is the subgroup of G generated by all expressions $(P, \lambda) - (P', \mu) - (P'', \nu)$, one for each short exact sequence (I), together with all expressions $(P, \lambda'\lambda) - (P, \lambda) - (P, \lambda')$ for all $P \in \mathcal{C}, \lambda, \lambda', \in Aut(P)$.

Define $K_{det}(\mathcal{C}) = G/G_0$.

(i) Show that $K_{det}(\mathcal{P}(A)) \simeq K_1(A)$ where $K_1(A) := GL(A)/E(A)$.

(ii) Now define $G_1(A) = K_{det}(\mathcal{M}(A))$. If R is a commutative ring with identity and Λ is an R-algebra finitely generated and projective as R-modules, define $G_1(R, \Lambda) := K_{det}(\mathcal{P}_R(\Lambda))$. Show that if R is regular, then $G_1(R, \Lambda) \simeq G_1(\Lambda)$.

3.2 Relative K_1

Let F be a p-adic field or number field with ring of integers R, Λ any R-order in a semi-simple F-algebra Σ, $\underline{a}$ a two-sided ideal of Λ such that $F \underset{R}{\otimes} \underline{a} = \Sigma$. Let $GL_n(\Lambda, \underline{a})$ be the kernel of the map $GL_n(\Lambda) \to GL_n(\Lambda/\underline{a})$ induced by the canonical map $\Lambda \to \Lambda/\underline{a}$. Write $GL(\Lambda, \underline{a})$ for $\varinjlim_n GL_n(\Lambda, \underline{a})$. Let $E_n(\Lambda, \underline{a})$ be the normal subgroup of $E_n(\Lambda)$ generated by $\{e_{ij}(a) | a \in \underline{a}, 1 \leq i, j \leq n\}$. Put $E(\Lambda, \underline{a}) = \varinjlim E_n(\Lambda, \underline{a})$ and $K_1(\Lambda, \underline{a}) = GL(\Lambda, \underline{a})/E(\Lambda, \underline{a})$.

Show that

(i) $K_1(\Lambda, \underline{a})$ is an Abelian group equal to the center of $GL(\Lambda)/E(\Lambda, \underline{a})$.

(ii) $SK_1(\Lambda, \underline{a}) := Ker(K_1(\Lambda, \underline{a}) \to K_1(\Sigma))$ is a finite group.

3.3 Let F be a number field and D a finite dimensional division algebra over F. For an Abelian group A, let $div(A)$ be the subgroup of divisible elements in A and $WK_2(D)$ the wild kernel of D. (See definition 7.1.1.)

Show that

(i) $div(K_2(D)) \subseteq WK_2(D)$ and that when the index of D is square free.

(ii) $divK_2(D) \simeq divK_2F$.

(iii) $WK_2(D) \simeq WK_2(F)$ and $|WK_2(D)/div(K_2(D))| \leq 2$.

3.4 Let R be a commutative ring with identity, G a finite group. Show that the torsion subgroup of $K_1(RG) \simeq$ (torsion subgroup of R^*) $\times G^{ab} \times SK_1(RG)$.

3.5 Let F be a p-adic field, R the ring of integers of F, Γ a maximal order in a central division algebra D over F, $\underline{m} = rad\ \Gamma$. Show that $SK_1(\Gamma, \underline{m}) = 0$.

3.6 Let F be a number field and D a central division algebra over F with square-free index. Show that every element of $K_2(D)$ is a symbol of the form $\{a, b\}$ where $a \in F^*, b \in D^*$.

3.7 Let R be a Dedekind domain and Λ an R-algebra finitely generated as an R-module. Show that Λ has stable range 2 and that the map $GL_2(\Lambda) \to K_1(\Lambda)$ is surjective.

3.8 Let R be a complete discrete valuation ring with quotient field F and finite residue class field. Let Γ be a maximal R-order in a central simple F-algebra. Show that

$$nr(K_1(\Gamma)) = nr(\Gamma^*) = R^*.$$

3.9 Let R be the ring of integers in a p-adic field, G a finite group such that G has an Abelian Sylow-p-subgroup. Show that $SK_1(RG) = 0$.

3.10 Let G be a finite Abelian group, R a Dedekind domain. Show that $Cl_1(RG) = SK_1(RG)$.

3.11 Show that $Cl_1(\mathbb{Z}G) = 0$ if G is a dihedral or generalized quaternion 2-group.

3.12 Let F be a finite field of characteristic p, G a finite group whose Sylow-p-subgroup is a cyclic direct factor. Show that $K_2(FG) = 0$. Hence show that for a finite Abelian group G, $K_2(FG) = 0$ if and only if the Sylow-p-subgroup of G is cyclic.

3.13

(a) Let C_n be a cyclic group of order n, R the ring of integers in a number field F. Show that $SK_1(RC_n) = 0$.

(b) Let G be a finite group. Show that $SK_1(\mathbb{Z}G)(p) = 0$ if the Sylow-p-subgroup of G is isomorphic to C_{p^n} or $C_{p^n} \times C_p$ (any n).

3.14 Let G be a finite group. Show that $SK_1(\hat{\mathbb{Z}}_p G) = 0$ if the Sylow-p-subgroup of G is Abelian.

3.15 Let F be a number field and D a central division algebra over F with square-free index. Show that every element of $K_2(D)$ is a symbol of the form $\{a, b\}$ where $a \in F^*, b \in D^*$.

Chapter 4

Some exact sequences; negative K-theory

In this chapter, we discuss $K_1 - K_0$ exact sequences and define negative K-theory (K_{-n}) for rings, additive categories, etc., yielding the extensions of the $K_1 - K_0$ sequences to the right and consequent computations of negative K-groups of orders and grouprings. We also discuss the Farrell - Jones conjecture for lower K-theory.

4.1 Mayer - Vietoris sequences

4.1.1 Let

$$
\begin{array}{ccc}
A & \xrightarrow{f_1} & A_1 \\
\downarrow{f_2} & & \downarrow{g_1} \\
A_2 & \xrightarrow{g_2} & A'
\end{array}
\qquad\qquad (I)
$$

be a commutative square of ring homomorphisms satisfying

(i) $A = A_1 \times_A A_2 = \{(a_1, a_2) \in A_1 \times A_2 | g_1(a_1) = g_2(a_2)\}$, i.e., given $a_1 \in A_1$, $a_2 \in A_2$ such that $g_1 a_1 = g_2 a_2$, then there exists one and only one element $a \in A$ such that $f_1(a) = a_1$, $f_2(a) = a_2$.

(ii) At least one of the two homomorphisms g_1, g_2 is surjective. The square (I) is then called a Cartesian square of rings.

Theorem 4.1.1 [155] *Given a Cartesian square of rings as in 4.1.1, there exists an exact sequence* $K_1(A) \xrightarrow{\alpha_1} K_1(A_1) \oplus K_1(A_2) \xrightarrow{\beta_1} K_1(A') \xrightarrow{\delta} K_0(A) \xrightarrow{\alpha_0} K_0(A_1) \oplus K_0(A_2) \xrightarrow{\beta_0} K_0(A')$.

Note. Call this sequence the Mayer - Vietoris sequence associated to the Cartesian square (I). For details of the proof of theorem 4.1.1, see [155]. Theorem 4.1.1 is due to J. Mihor.

Sketch of proof. The maps $\alpha_i, \beta_i (i = 0, 1)$ are defined as follows: for $x \in K_i(A), \alpha_i(x) = (f_{1*}, f_{2*})$, and for $(y, z) \in K_i(A_1) \oplus K_i(A_2)$ $i = 0, 1$, $\beta_i(y, z) = g_{1*}y \to g_{2*}z$. The boundary map $\delta : K_1(A') \to K_0(A)$ is defined as follows: represent $x \in K_1(A')$ by a matrix $\gamma = (a_{ij}$ in $GL_n(A')$. This matrix determines an automorphism $\gamma : A'^n \to A'^n$. Let $\gamma(z_j) = a_{ij}z_j$ where $\{z_j\}$ is a standard basis for A'^n. Let $P(\gamma)$ be the subgroup of $A_1^n \times A_2^n$ consisting of $\{(x, y) | \gamma g_1^n(x) = g_2^n(y)\}$ where $g_1^n : A_1^n \to A'^n$, $g_2^n : A_2^n \to A'^n$ are induced by g_1, g_2, respectively. We need the following

Lemma 4.1.1 *(i) If there exists $(b_{ij}) \in GL_n(A_2)$, which maps to $\gamma(a_{ij})$, then $P(\gamma) \simeq A^n$.*

(ii) If g_2 is surjective, then $P(\gamma)$ is a finitely generated projective A-module.

For the proof of lemma 4.1.1 see [155].
Conclusion of definition of δ: now define

$$\delta[\gamma] = [P\gamma)] - [A^n] \in K_0(A)$$

and verify exactness of the sequence in theorem 4.1.1 as an exercise.

Corollary 4.1.1 If A is a ring and $\mathbf{a}_1$ and $\mathbf{a}_2$ ideals of A such that $\mathbf{a}_1 \cap \mathbf{a}_2 = 0$, then there exists an exact sequence

$$K_1(A) \to K_1(A/\mathbf{a}_1) \oplus K_1(A/\mathbf{a}_2) \to K_1(A/(\mathbf{a}_1 + \mathbf{a}_2)$$
$$\xrightarrow{\delta} K_0(A) \to K_0(A/\mathbf{a}_1 \oplus K_0(A/\mathbf{a}_2) \to K_0(A/(\mathbf{a}_1 \oplus \mathbf{a}_2)).$$

PROOF Follows by applying theorem 4.1.1 to the Cartesian square:

$$
\begin{array}{ccc}
A & \xrightarrow{f_1} & A/\mathbf{a}_1 \\
\downarrow{\scriptstyle f_2} & & \downarrow{\scriptstyle g_1} \\
A/\mathbf{a}_2 & \xrightarrow[g_2]{} & A/(\mathbf{a}_1 \oplus \mathbf{a}_2)
\end{array}
\qquad \text{(II)}
$$

□

Example 4.1.1 Let G be a finite group of order n, and $A = ZG$. Let $\mathbf{a}_1$ be the principal ideal of A generated by $b = \sum_{g \in G} g$, $\mathbf{a}_2$ the augmentation ideal $= \{\sum r_g g | \sum r_g = 0\}$. Then $\mathbf{a}_1 \cap \mathbf{a}_2 = 0$. So, $A_2 = A/\mathbf{a}_2 \simeq \mathbb{Z}$, $A' = A/(\mathbf{a}_1 \oplus \mathbf{a}_2) \simeq \mathbb{Z}/n\mathbb{Z}$ from the Cartesian squares (I) and (II) above.

Now, suppose that $|G| = p$, a prime. Let $G = \langle x \rangle$. Put $t = f_1(x)$. Then, A_1 has the form $\mathbb{Z}[t]$ with a single relation $\sum_{i=1}^{p-1} t^i = 0$. So, A_1 may be identified with $Z[\xi]$ where ξ is a primitive p-th root of unity.
We now have the following:

Theorem 4.1.2 *If $|G| = p$, then $f_1 : K_0(ZG) \cong K_0(Z[\xi])$ is an isomorphism. Hence $K_0(ZG) \simeq Z \oplus C\ell(\mathbb{Z}[\xi])$.*

PROOF From theorem 4.1.1, we have an exact sequence

$$K_1(Z[\xi]) \oplus K_1(Z) \to K_1(\mathbb{Z}/p\mathbb{Z}) \xrightarrow{\delta} K_0(\mathbb{Z}G)$$
$$\to K_0(Z[\xi]) \oplus K_0(Z) \to K_0(Z/pZ) \,.$$

⬚

Now, since $g_{2*} : K_0(\mathbb{Z}) \simeq K_0(Z/pZ)$ is an isomorphism, the result will follow once we show that $\delta = 0$. To show that $\delta = 0$, it suffices to show that $K_1(Z[\xi]) \to K_1(Z/pZ)$ is onto. Let r be a positive integer prime to p. Put $u = 1 + \xi + \cdots + \xi^{r-1} \in Z[\xi]$. Let $\xi^r = \eta$, $\eta^s = \xi$, for some $s > 0$. Then $v = 1 + \eta + \cdots + \eta^{s-1} \in Z[\xi]$. In $Q(\xi)$, we have

$$v = (\eta^s - 1)/(\eta - 1) = (\xi - 1)(\xi^r - 1) = 1/u \,.$$

So, $u \in (Z[\xi])^*$, i.e., given $r \in (Z/pZ)^* \simeq K_1(Z/pZ)$, there exists $u \in (Z[\xi])^*$ such that $g_{1*}(u) = r$. That $K_0(\mathbb{Z}G) \simeq Z \oplus C\ell(Z[\xi])$ follows from 2.3.5.

Remarks 4.1.1 (i) The Mayer - Vietoris sequence in theorem 4.1.1 can be extended to the right to negative K-groups defined by H. Bass in [20]. More precisely, there exists functors K_{-n}, $n \geq 1$ from rings to Abelian groups such that the sequence

$$\cdots K_0(A') \to K_{-1}(A) \to K_{-1}(A_1) \oplus K_{-1}(A_2) \to K_{-1}(A') \to \cdots$$

is exact. This sequence will be derived in 4.4.

(ii) The Mayer - Vietoris sequence in theorem 4.1.1 can be extended beyond K_2 under special circumstances that will be discussed in chapter 6 (see 6.4). In general, it cannot be extended beyond K_2 (see [212]).

4.2   Localization sequences

4.2.1 Let S be a central multiplicative system in a ring A, $\mathcal{H}_S(A)$ the category of finitely generated S-torsion A-modules of finite projective dimension. Note that an A-module M is S-torsion if there exists $s \in S$ such that $sM = 0$, and that an A-module has finite projective dimension if there exists a finite $\mathcal{P}(A)$-resolution, i.e., there exists an exact sequence (I) $0 \to P_n \to P_{n-1} \to \cdots P_0 \to M \to 0$ where $P_i \in \mathcal{P}(A)$. Then we have the following theorem.

Theorem 4.2.1 *With notation as in 4.2.1, there exist natural homomorphisms δ, ϵ such that the following sequence is exact:*

$$K_1(A) \to K_1(A_S) \xrightarrow{\delta} K_0\left(\mathcal{H}_S(A)\right) \xrightarrow{\epsilon} K_0(A) \to K_0(A_S)\,,$$

where A_S is the ring of fractions of A with respect to S.

PROOF   We shall not prove exactness in detail but indicate how the maps δ and ϵ are defined, leaving details of proof of exactness at each point as an exercise.

Let $M \in \mathcal{H}_S(A)$ have a finite $\mathcal{P}(A)$-resolution as in 4.2.1 above. Define $\epsilon([M]) = \Sigma(-1)^i[P_i] \in K_0(A)$. We define δ as follows: if $\alpha \in GL_m(A_S)$, let $s \in S$ be a common denominator for all entries of α such that $\beta = s\alpha$ has entries in A. We claim that $A^n/\beta A^n$ and $A^n/sA^n \in \mathcal{H}_S(A)$. That they have finite $\mathcal{P}(A)$-resolutions follow from the exact sequences

$$0 \to A^n \xrightarrow{\beta} A^n \to A^n/\beta A^n \to 0 \qquad \text{and}$$
$$0 \to A^n \xrightarrow{s} A^n \to A^n/sA^n \to 0\,.$$

To see that $A/\beta A^n$ is S-torsion, let $t \in S$ be such that $\alpha^{-1}t = \gamma$ has entries in A. Then $\gamma A^n \subset A^n$ implies that $tA^n \subset \alpha A^n$ and hence that $stA^n \subset s\alpha A^n = \beta A^n$. Then $st \in S$ annihilates $A^n/\beta A^n$.

We now define

$$\delta[\alpha] = [A^n/\beta A^n] - [A^n/sA^n]\,.$$

So

$$\begin{aligned}
\epsilon\delta[\alpha] &= \epsilon[A^n/\beta A^n] - \epsilon[A^n/sA^n] \\
&= [A^n] - [A^n]) - ([A^n] - A^n]) = 0\,.
\end{aligned}$$

$\square$

Remarks 4.2.1   (i) Putting $A = \Lambda[t]$ and $S = \left\{t^i\right\}_{i \geq 0}$ in theorem 4.2.1, we obtain an exact sequence

$$\begin{aligned}
K_1\left(\Lambda[t]\right) &\to K_1\left(\Lambda\left[t, t^{-1}\right]\right) \xrightarrow{\partial} K_0\left(H_{\{t^i\}}\left(A[t]\right)\right) \\
&\to K_0\left(\Lambda[t]\right) \to K_0\left(\Lambda\left[t, t^{-1}\right]\right)\,,
\end{aligned}$$

which is an important ingredient in the proof of the following result called the fundamental theorem for K_1 (see [20]).

(ii) Fundamental theorem for K_1:

$$K_1\left(\Lambda\left[t, t^{-1}\right]\right) \simeq K_1(\Lambda) \oplus K_0(\Lambda) \oplus NK_1(\Lambda) \oplus NK_1(\Lambda)\,,$$

where $NK_1(\Lambda) = \operatorname{Ker}\left(K_1(\Lambda[t]) \xrightarrow{\tau} K_1(\Lambda)\right)$ and τ is induced by the augmentation $\Lambda[t] \to \Lambda$ $(t = 1)$.

(iii) In chapter 6 on higher K-theory, we shall discuss the extension of the localization sequence in theorem 4.2.1 to the left for all $n \geq 1$, as well as some further generalizations of the sequence.

4.3 Exact sequence associated to an ideal of a ring

4.3.1 Let A be a ring, $\mathbf{a}$ any ideal of A. The canonical map $f : A \to A/\mathbf{a}$ induces $f_* : K_i(A) \to K_i(A/\mathbf{a})$, $i = 0, 1$. We write $\overline{A}$ for $A/\mathbf{a}$ and for $M \in \mathcal{P}(A)$ we put $\overline{M} = M/\mathbf{a}M \simeq \overline{A} \otimes_A M$. Let $K_0(A, \mathbf{a})$ be the Abelian group generated by expressions of the form $[M, f, N]$, $M, N \in \mathcal{P}(A)$, where $f : \overline{A} \otimes_A M \simeq \overline{A} \otimes_A N$ with relations defined as follows:

For $L, M, N \in \mathcal{P}(A)$ and $\overline{A}$-isomorphism $f : \overline{L} \simeq \overline{M}$, $g : \overline{M} \simeq \overline{N}$, we have

$$[L, gf, N] = [L, d, M] + [M, g, N].$$

(ii) Given exact sequences

$$0 \to M_1 \to M_2 \to M_3 \to 0; \qquad 0 \to N_1 \to N_2 \to N_3 \to 0,$$

where $M_i, N_i, N_i \in \mathcal{P}(A)$, and given $\overline{A}$-isomorphisms $f_1 : \overline{M}_i \simeq \overline{N}_i$ ($i = 1, 2, 3$) that commute with the maps associated with the given sequences, we have $[M_2, f_2, N_2] = [M_1, f_1, N_1] + [M_3, f_3, N_3]$.

Theorem 4.3.1 [155] *There exists an exact sequence*

$$K_1(A) \to K_1(\overline{A}) \xrightarrow{\delta} K_0(A, \mathbf{a}) \xrightarrow{\eta} K_0(A) \to K_0(\overline{A}).$$

Remarks 4.3.1 (i) We shall not prove the above result in detail but indicate how the maps δ, η are defined, leaving the rest as an exercise. It is clear how the maps $K_1(A) \to K_i(\overline{A})$, $i = 0, 1$ are defined. The map δ assigns to each $f \in GL_n(\overline{A})$ the triple $[A^n, f, A^n] \in K_0(A, \mathbf{a})$, while the map η takes $[M, f, N]$ onto $[M] - [N]$ for $M, N \in \mathcal{P}(A)$ such that $f : \overline{M} \simeq \overline{N}$.

(ii) The exact sequence in theorem 4.3.1 could be extended to K_2 and beyond with appropriate definitions of $K_i(A, \mathbf{a})$, $i \geq 1$. We shall discuss this in the context of higher K-theory in chapter 6. Also the sequence can be extended to the right. (see 4.4.3).

4.4　Negative K-theory K_{-n}, n positive integer

In [20], H. Bass defined the functors K_{-n} using the notion of LF, NF functors where F is "contracted". We shall briefly present these ideas, which lead to the extension of the Mayer - Vietoris sequence 4.1.1 to the right.

$(4.4)^A$ LF, NF functors and the functors K_{-n}

Definition 4.4.1 *An oriented cycle is an infinite cyclic group T with a designated generated t, together with submonoid T_+ generated by t and submonoid T_- generated by t^{-1}. We shall denote such a cycle by $(T, T_\pm)$.*

Let $F : (Rings) \to \mathbb{Z}\text{-}\mathcal{M}od$ be a functor. Define $NF, LF : (Rings) \to \mathbb{Z}\text{-}\mathcal{M}od$ as follows: $NF(A) = N_{T_+}F(A) := \text{Ker}\,(F(A[T_+]) \to F(A))$ where ϵ_+ is induced by the augmentation. Hence we have $F(A[T_+]) \simeq F(A) \oplus N_{T_+}F(A)$, which is functorial in A.

The inclusions $\tau_\pm : A[T_\pm] \subset A[T]$ induce a homomorphism $F(A(T_+)) \oplus F(A[T_-]) \overset{\tau=(\tau_+, \tau_-)}{\longrightarrow} F(A(T))$, and we now define $LF(A) = L_T F(A)$ as cokernel of τ.

4.4.1 Let $\text{Seq}F(A) = \text{Seq}_T F(A)$ denote the complex

$$0 \to F(A) \overset{e}{\to} F(A[T_+]) \oplus F(A[T_-]) \overset{\tau}{\to} F(A(T)) \to L_T F(A) \to 0 \qquad \text{(I)}$$

where $e(x) = (x, -x)$. Note that $\text{Seq}F(A)$ is exact iff $\tau_\pm$ are both monomorphisms and $F(A) = \text{Im}(\tau_+) \cap \text{Im}(\tau_-)$.
In this case, we could regard $\tau_\pm$ as inclusions and have $\text{Im}(\tau) = F(A[T_+]) + F(A[T_-]) = F(A) \oplus N_{T_+}F(A) \oplus N_{T_-}F(A)$.

Definition 4.4.2 *Let $(T, T_\pm)$ be an oriented cycle, $F : Rings \to \mathbb{Z}\text{-}\mathcal{M}od$ a functor. A contraction of F is a natural homomorphism $h = h_{T,A} : L_T F(A) \to F(A[T])$, which is right inverse for the canonical projection $p : F(A(T)) \to L_T F(A)$.*

The pair (F, h) is called a contracted functor if further $\text{Seq}_T F(A)$ is acyclic for all A.
Note that if (F, h) is a contracted functor, we have a decomposition

$$F(A[T]) = F(A) \oplus N_{T_+}F(A) \oplus N_{T_-}F(A) \oplus Im(h_{T,A}) \qquad \text{(II)}$$

where $Im(h_{t,A})$ is isomorphic to $L_t F(A)$.
We also have the notion of morphism of contracted functors $(F, h), (F', h')$ (see [20], p.661).

Theorem 4.4.1 [20] *Let* $\varphi : (F, h) \to (F_1, h_1)$ *be a morphism of contracted functors. Then there are* h' *and* h'_1 *induced by* h *and* h_1, *respectively, and* $(Ker(\varphi), h_1)$ *and* $(Coker(\varphi), h'_1)$ *are contracted functors.*

In particular, there are contracted functors (NF, Nh) *and* (LF, Lh). *Moreover, there is a natural isomorphism* $(LNF, LNh) \simeq (NLF, NLh)$ *of contracted functors.*

Theorem 4.4.2 [20] "Fundamental theorem" *Let* (T, T_+) *be an oriented cycle,* A *a ring,* $\partial_+ : Nil_0(A) \to N_{T_+} K_1(A)$ *be the homomorphism defined by* $\partial_+[P, v] = [P(t), \partial_+(v)]$ *where* $\partial_+(v) : \partial_1(v)^{-1} \partial_1(tv)$ *and* $\partial_1(v) = 1 - v$.

Define $h = h_{T,A} : K_0(A) \to K_1(A[T])$ *by* $h[P] = \left[P[T], t^1_{P[T]} \right]$, $P \in \mathcal{P}(A)$. Then

(i) $\partial_+ : Nil_0 \to NK_1$ *is an isomorphism of functors.*

(ii) *The homomorphism* h *induces on passing to the quotient an isomorphism* $K_0 \to L_T K_1$. *Using this isomorphism to identify* K_0 *with* LK_1, *we now have that* (K_1, h) *is a contracted functor.*

Definition 4.4.3 *Since* $K_0 = LK_1$ *by theorem 4.4.2, we now define* K_{-1} *as* LK_0 *and* K_{-n} *as* $L^n K_0$.

Remarks 4.4.1 (i) Theorem 4.4.2 and definition 4.4.3 above yield the fundamental theorem for K_0 and K_1, stated, respectively, in 2.1.1(v) and 4.2.1(ii).

(ii) Our next aim is to briefly indicate how to use the foregoing ideas on LF and NF functors to extend the Mayer - Vietoris sequence in theorem 4.1.1 to the right.

$(4.4)^B$ Mayer - Vietoris sequence

Definition 4.4.4 *Let* $\mathcal{C}art$ *be the category whose objects are Cartesian squares*

$$ C: \quad \begin{array}{ccc} A & \xrightarrow{f_2} & A_2 \\ \downarrow{f_1} & & \downarrow{g_2} \\ A_1 & \xrightarrow{g_1} & A' \end{array} \quad\quad (I) $$

in Rings such that g_1 *or* g_2 *is surjective. If* $F : Rings \to \mathbb{Z}\text{-}\mathcal{M}od$ *is a functor we associate to* F *and* C, *the sequence*

$$ F(A) \xrightarrow{(f_1, -f_2)} F(A_1) \oplus F(A_2) \xrightarrow{\binom{g_1}{g_2}} F(A'). \quad\quad (II) $$

A Mayer - Vietoris pair is a triple (F_1, F_0, δ) where $F_1, F_0 : (Rings) \to \mathbb{Z}$-$\mathcal{M}od$ are functors where δ associates to each $C \in \mathcal{C}art$ as above a homo-morphism $\delta_C : F_1(A') \to F_0(A)$, which is natural in C and is such that the sequence

$$F_1(A) \to F_1(A_1) \oplus F_1(A_2) \to F_1(A') \xrightarrow{\delta} F_0(A) \to$$

$$\to F_0(A_1) \oplus F_0(A_2) \to F_0(A') \qquad\qquad (III)$$

is exact.

Theorem 4.4.3 [20] Let $((F_1, h_1), (F_0, h_0), \delta)$ be a Mayer - Vietoris pair of contracted functors and let J denote either N or L. Then $((JF_1, Jh_1), (JF_0, Jh_0), J\delta)$ is also a Mayer - Vietoris pair of contracted functors.

Corollary 4.4.1 [20] Let (F, h) be a contracted functor. Assume that there is a δ such that $((F, h), (LF, Lh), \delta)$ is a Mayer - Vietoris pair. Then for the Cartesian square

$$C = \begin{array}{ccc} A & \xrightarrow{f_2} & A_2 \\ \downarrow{\scriptstyle f_1} & & \downarrow{\scriptstyle g_2} \\ A_1 & \xrightarrow[g_1]{} & A' \end{array}$$

there is a long exact Mayer - Vietoris sequence

$$F(A) \cdots \to L^{n-1}F(A') \to L^n F(A) \to L^n F(A_1) \oplus L^n F(A_2)$$
$$\to L^n F(A) \to L^{n+1} F(A).$$

Moreover, $((N^i F, N^i h), (LN^i F, LN^i h), \delta)$ is also an $M - V$ pair, and so, there is a corresponding long Mayer - Vietoris sequence for the functors $(L^n N^i F)$ $n \geq 0$ for each $i \geq 0$.

Remarks 4.4.2 Since by theorem 4.1.1, (K_1, K_0, δ) is an $M - V$ pair, and by theorem 4.4.2 we can identify LK_1 with K_0, we can apply 4.4.1 and deduce 4.4.4 below. For some applications see [103].

Theorem 4.4.4 Let C be as in 4.4.1. Then for each $i \geq 0$, there is a long exact $M - V$ sequence

$$F(A) \to \cdots \to L^{n-1}F(A') \to L^n F(A) \to L^n F(A_1) \oplus L^n F(A_2) \to L^n F(A')$$
$$\to L^{n+1}F(A)$$

where $F = N^i K_1$.

In case $i = 0$, the sequence above becomes

$$K_1(A) \to \cdots \to K_0(A') \to K_{-1}(A) \to K_1(A) \oplus K_{-1}(A_2)$$
$$\to K_{-1}(A') \to K_{-2} \to \cdots$$

where we unite $K_n(A) = L^n K_0(A) = L^{n+1} K_1(A)$.

$(4.4)^C$ Exact sequence associated to an ideal

Remarks 4.4.3 Next we record the following result theorem 4.4.5, which says that the sequence 4.3.1 can also be extended to the right. Recall Example 1.3.1(xii) that if R is a non-unital ring, we can define $K_0(R) := \text{Ker}(K_0(R_+) \to K_0(\mathbb{Z}) \simeq \mathbb{Z})$.
As such $LK_0(R) := \text{Ker}(LK_0(R_+) \to LK_0(\mathbb{Z}))$ is defined since $LK_0(R_+)$ and $LK_0(\mathbb{Z})$ are defined. So we have the following result.

Theorem 4.4.5 *Let A be a ring with identity and $\underline{a}$ a two-sided ideal in A viewed as a non-unital ring. Then the exact sequence*

$$\to K_0(A) \xrightarrow{\pi_*} K_0(A/\mathbf{a}) \xrightarrow{\partial} K_{-1}(\mathbf{a}) \xrightarrow{i_*} K_{-1}(A) \xrightarrow{\pi_*} K_{-1}(A/\mathbf{a}) \xrightarrow{\partial} K_{-2}(\mathbf{a}).$$

PROOF See [181], p.155. ⬜

Example 4.4.1   (i) If $A = \mathbb{Z}$, $\mathbf{a} = \langle m \rangle$, $m \in \mathbb{Z}$, $m > 0$, $K_{-n}(\mathbb{Z}) = 0$, and $K_{-n}(\mathbb{Z}/m) \simeq K_{-n-1}(\langle m \rangle)$ for all $n > 1$. (See [181]).

(ii) If G is a cyclic group of prime order p, one can show that $K_{-n}(\mathbb{Z}G) = 0$ for $n > 1$ by showing that if $\mathbf{a} = \text{Ker}(\mathbb{Z}G \to \mathbb{Z}[\xi])$ where $\xi = e^{2\pi i/p}$, then $K_{-n}(\mathbf{a}) = 0$ for all $n > 1$.

$(4.4)^D$ Localization sequence

In this subsection, we shall indicate how the localization sequence 4.2.1 can be extended to the right. The results in this subsection are due to D.W. Carter (see [32]).

4.4.2 Let R be a commutative ring and S a multiplicative set of elements of R. We denote by $\mathcal{A}(R, S)$ the category whose objects are R-algebras on which multiplication by any s in S is injective and whose morphisms are exactly those R-algebra homomorphisms $\Lambda \to \Gamma$ for which Γ is flat as a right Λ-module. We shall be interested in various functors from $\mathcal{A}(R, S)$ to Abelian groups such as Grothendieck groups K_0 and G_0, the Whitehead groups K_1 and G_1, and three related functors $K_0 H_S, K_i S^{-1}(i = 0, 1)$, which we define as follows:

$$K_i S^{-1}(\Lambda) := K_i \left(S^{-l}\Lambda \right), \qquad i = 0,1$$
$$K_0 H_S(\Lambda) := K_0 \left(\mathcal{H}_S\Lambda \right).$$

Note that any functor from rings to Abelian groups may be viewed as functor from $\mathcal{A}(R,S)$.

Definition 4.4.5 *Let F be a functor from $\mathcal{A}(R,S)$ to Abelian groups. Then we define another such functor, denoted LF, by*

$$LF(\Lambda) = Coker\left(F(\Lambda[t]) \oplus F(\Lambda[t^{-1}]) \to F(\Lambda[t,t^{-1}]) \right)$$

where the indicated map is induced by the obvious inclusion (localizations).

Just as before (see 4.4.1, 4.4.2) we have the notion of F being contracted, and we can already make several observations.

Remarks 4.4.4   (i) $K_0, K_o S^{-1}$; $K_1, K_1 S^{-1}$ are contracted functors, and $LK_1 \simeq K_0$; $LK_1 S^{-1} \cong K_0 S^{-1}$.

Hence $K_{-i} = L^i K_0$, $K_{-i} S^{-i} = L^i K_0 S^{-i}$ $(i > 0)$.

(ii) If F, G are contracted functors, we also have the notion of morphism of contracted functors F, G, i.e., a natural transformation $\alpha : F \to G$, which respect the natural splittings, i.e., for all Λ, the square

$$
\begin{array}{ccc}
LF(\Lambda) & \longrightarrow & F\left(\Lambda\left[t,t^{-1}\right] \right) \\
\downarrow{\scriptstyle L_\alpha} & & \downarrow{\scriptstyle \alpha} \\
LG(\Lambda) & \longrightarrow & G\left(\Lambda\left[t,t^{-1}\right] \right)
\end{array}
$$

commutes.

If $\alpha : F \to G$ is a morphism of contracted functors $\mathcal{A}(R,S) \to \mathbb{Z}\text{-}\mathcal{M}od$, then $Ker(\alpha)$, $Coker(\alpha)$ are also contracted functors. In particular, LF is a contracted functor $\mathcal{A}(R,S) \to \mathbb{Z}\text{-}\mathcal{M}od$.

(iii) Let $0 \to F \to G \to H \to 0$ be a short exact sequence of functors $\mathcal{A}(R,S) \to \mathbb{Z}\text{-}\mathcal{M}od$. Suppose that F and H are contracted functors $\mathcal{A}(R,S) \to \mathbb{Z}\text{-}\mathcal{M}od$, then there exists a short exact sequence $0 \to LF \to LG \to LH \to 0$ where LF and LH are contracted functors $\mathcal{A}(R,S) \to \mathbb{Z}\text{-}\mathcal{M}od$, and G is acyclic.

Definition 4.4.6 $K_{-i}(\mathcal{H}_S(\Lambda)) := L^i K_0 \mathcal{H}_S(\Lambda).$

We now state the localization sequence for K_n $\forall n \in \mathbb{Z}$.

Theorem 4.4.6 [32] *Let R be a commutative ring and let S be a multiplicative set of elements of R. Let Λ be an R-algebra on which multiplication by any s in S is injective. Let $\mathcal{H}_S(\Lambda)$ denote the category of S-torsion left Λ-modules that admit a finite resolution by finitely generated projective left Λ-modules. Then the localization $\psi : \Lambda \to S^{-1}\Lambda$ gives rise to a long exact sequence*

$$K_{n+1}(\Lambda) \xrightarrow{\psi_*} K_{n+1}\left(S^{-1}\Lambda\right) \to K_n\left(\mathcal{H}_S(\Lambda)\right) \to K_n(\Lambda) \xrightarrow{\psi_*} K_n\left(S^{-1}\Lambda\right)$$

(for all integers n).

PROOF For the proof of the lower part of the sequence connecting $K_0, K_{-1}, K_{-2}, \ldots, K_{-i}$ see [32]. The part for higher K-theory is due to D. Quillen and will be discussed in chapter 6.
The $K_1 - K_0$ localization sequence is due to H. Bass (see [20]). Although a general localization sequence for higher K-theory was announced by D. Quillen, the proof in the context of theorem 4.4.6 was supplied by S. Gersten in [60]. ⊓

We now state other results of D. Carter [32], connected with orders and grouprings.

If R is a Dedekind domain with field of fractions F, and Λ is an R-algebra on which multiplication by any non-zero element of R is injective, we shall write $SK_n(\Lambda) := \mathrm{Ker}(K_n(\Lambda) \to K_n(\Sigma))$ which $\Sigma = \Lambda \otimes_R F$; $\overline{K_n(\Lambda)} = K_n(\Sigma)/\mathrm{Im}(K_n(\Lambda))$.

Theorem 4.4.7 [32] *Let R be the ring of integers in a number field F, $\Lambda = RG$, G a finite group of order s, $\Sigma = FG$. Then there exists an exact sequence*

$$0 \to K_0(\mathbb{Z}) \to \bigoplus_{\underline{p}|sR} K_0(\widehat{\Lambda}_{\underline{p}}) \oplus K_0(\Sigma) \to \bigoplus_{\underline{p}|sR} K_0(\widehat{\Sigma}_{\underline{p}}) \to K_{-1}(\Lambda).$$

Moreover, $K_{-n}(\Lambda) = 0$ for all $n > 1$.

PROOF See [32]. ⊓

Corollary 4.4.2 In the notation of 4.4.7, let $f, f_{\underline{p}}, r_{\underline{p}}$, respectively, be the number of isomorphism classes of irreducible F, $\widehat{F}_{\underline{p}}$, and $R/\underline{p}$ representations of G. Then $K_{-1}(\Lambda)$ is a finitely generated Abelian group and $\mathrm{rank} K_{-1}(\Lambda) = f + \sum_{\underline{p}/sR} f_{\underline{p}} - r_{\underline{p}})$.

PROOF See [32]. ⊓

$(\mathbf{4.4})^E$ $K_{-n}(A) := K_0(S^n A)$

4.4.3 Remarks/Definition There is another definition of negative K-groups $K_{-n}A$ of a ring A, due to M. Karoubi. The cone of A, CA, is defined as the ring of matrices $(a_{ij})\, 1 \leq i,j < \infty$, $a_{ij} \in A$ such that (1) the set $\{a_{ij} | 1 \leq i,j < \infty\} \subset A$ is finite; (2) The number of nonzero entries in each row and in each column is finite.

Then $M_\infty(A) := \varinjlim M_n(A)$ is a two-sided ideal of CA. Define the suspension of A, SA as the quotient $CA/M_\infty(A)$. By iteration, we obtain $S^n A$-called n^{th} suspension of A.

Definition 4.4.7 *The negative K-groups $K_{-n}(A)$ is defined as $K_0(S^n A)$.*

One can show that this definition of $K_{-n}(A)$ coincides with the earlier one by H. Bass.

$(\mathbf{4.4})^F$ $K_{-n}(\mathcal{A})$, $\mathcal{A}$ an additive category

Definition. The functor of Waldhausen's negative nil groups depends on the definition of negative K-groups for an additive category $\mathcal{A}$. This in turn depends on the definitions of polynomial extension $\mathcal{A}[z]$ and finite Laurent extension $\mathcal{A}\left[z, z^{-1}\right]$ of $\mathcal{A}$. We now briefly introduce these ideas.

Define a metric on $\mathbb{Z}^n$ by $d(J,K) := \max\{|j_i - k_i| \mid 1 \leq i \leq n\} \geq 0$ where $J = (j_1, j_2, \ldots, j_n)$, $K = (k_1, k_2, \ldots, k_n) \in \mathbb{Z}^n$. Let $\mathcal{C}_{\mathbb{Z}^n}(\mathcal{A}) := C_n(\mathcal{A})$ be the $\mathbb{Z}^n$-graded category of a filtered additive category $\mathcal{A}$ (see [168]). We shall write $\mathbb{P}_n(\mathcal{A})$ for $\mathbb{P}_0(C_n(\mathcal{A}))$ $(n \leq 1)$ the idempotent completion of the additive category $C_n(A)$. (Recall that if $\mathcal{B}$ is an additive category, the idempotent completion $\mathcal{P}_0(\mathcal{B})$ of $\mathcal{B}$ is defined as follows: $\text{ob}\mathbb{P}_0(\mathcal{B})$ are pairs (a,p) where $p : a \to a$ is a morphism such that $p^2 = p$. A morphism $f : (a,p) \to (b,q)$ is a morphism $f : a \to b$ such that $qfp = f$).

Note that $C_{n+1}(\mathcal{A}) = C_1(C_n(\mathcal{A}))$ and we have the identification $K_1(C_{n+1}(\mathcal{A}) = K_0(\mathbb{P}(\mathcal{A}))$ $n \geq 0$ (see [168], 6.2).
Now define

$$K_{-n}(\mathcal{A}) := K_1(C_{n+1}(\mathcal{A})$$
$$= K_0(\mathbb{P}_n(\mathcal{A})) \ n \geq 1 \qquad \qquad \text{(II)}$$

Note If A is a ring, $K_{-n}(A) = K_{-n}(\mathcal{B}^f(A))$ where $\mathcal{B}^f(A)$ is the category of based finitely generated free A-modules.

4.4.4 Our next aim is to indicate how $K_{-m}(\mathcal{A})$ fits into split exact sequences

$$0 \to K_{1-n}(\mathbb{P}_0(\mathcal{A})) \to K_{1-n}(\mathbb{P}_0(\mathcal{A})[z]) \oplus (\mathbb{P}_0(\mathcal{A})[z^{-1}]))$$
$$\to K_{1-n}(\mathbb{P}_0(\mathcal{A})[z, z^{-1}]) \to K_{-n}(\mathcal{A}) \to 0$$

in the spirit of ideas in $(4.4)^A$.

Definition 4.4.8 *Define the polynomial extension $\mathcal{A}[z]$ of an additive category $\mathcal{A}$ as the additive category having, for each object L of $\mathcal{A}$, one object $L[z] = \sum_{k=0}^{\infty} z^k L$ and one morphism $f = \sum_{k=0}^{\infty} z^k f_k : L[z] \to L'[z]$ for each collection $\{f_k \in Hom_{\mathcal{A}}(L, L') \mid k \geq 0\}$ of morphisms in $\mathcal{A}$ with $\{k \geq 0 | f_k \neq 0\}$ finite.*

We can regard the category $\mathcal{A}[z]$ as a subcategory of $C_1(\mathcal{A})$ with objects M such that

$$M(j) = \begin{cases} z^j M(0) & \text{if } j \geq 0 \\ 0 & \text{if } j < 0 \end{cases}$$

One has the functor $j_+ : \mathcal{A}[z] \to \mathcal{A}[z, z^{-1}]$; $L[z] \to L[z, z^{-1}]$. One defines the polynomial extension $\mathcal{A}[z^{-1}]$ similarly with an inclusion $j_- : \mathcal{A}[z^{-1}] \to \mathcal{A}[z, z^{-1}]$, $L[z^{-1}] \to L[z, z^{-1}]$. The inclusion functors define a commutative square of additive functors:

$$
\begin{array}{ccc}
\mathcal{A} & \xrightarrow{\;\;j_-\;\;} & \mathcal{A}[z] \\
\downarrow{\scriptstyle i_+} & & \downarrow{\scriptstyle j_+} \\
\mathcal{A}[z^{-1}] & \longrightarrow & \mathcal{A}[z, z^{-1}]
\end{array}
$$

4.4.5 Now, given a functor

$$F : \{Additive\ categories\} \longrightarrow \mathbb{Z}\text{-}\mathcal{M}od,$$

define the functor

$$LF : \{Additive\ categories\} \longrightarrow \mathbb{Z}$$

$$\mathcal{A} \longrightarrow LF(\mathcal{A}) \tag{III}$$

by $LF(\mathcal{A}) = \operatorname{Coker}(j_+, j_-) : F(\mathcal{A}[z]) \oplus F(\mathcal{A}[z^1]) \to F(\mathcal{A}[z, z^{-1}])$.

Definition 4.4.9 *Let $\mathcal{A}$ be an additive category. The lower K-groups of an additive category $\mathcal{A}$ is defined by*

$$K_{-n}(\mathcal{A}) = L^n K_0(\mathbb{P}_0(\mathcal{A})) \; (n \geq 1) \tag{II}$$

Definition 4.4.10 *Following H. Bass, (see $(4.4)^A$), we now say that a functor $F : \{Additive\ categories\} \longrightarrow \mathbb{Z}\text{-}\mathcal{M}od$ is contracted if the chain complex*

$$0 \to F(\mathcal{A}) \stackrel{\binom{i_+}{i_-}}{\to} F(\mathcal{A}[z]) \oplus F(\mathcal{A}[z^{-1}]) \stackrel{(j_+, j_-)}{\to} F(\mathcal{A}[z, z^{-1}])$$
$$\stackrel{q}{\to} LF(\mathcal{A}) \to 0$$

has a natural chain contraction with q the natural projection.

Proposition 4.4.1 [168] The functors

$$L^n K_1 : \{\text{Additive categories}\} \longrightarrow \mathbb{Z}\text{-}\mathcal{M}od$$
$$\mathcal{A} \longrightarrow L^n K_1(\mathcal{A})$$

for $n \geq 1$ are contracted ($L^0 K_1 = K_1$), we have a natural identification

$$L^n K_1(\mathcal{A}) = K_{1-n}(\mathbb{P}_0(\mathcal{A})) = K_1(C_n(\mathcal{A})) = K_0(\mathbb{P}_{n-1}(\mathcal{A})).$$

Definition 4.4.11 *For $n \geq 1$, we define the n-fold Laurent polynomial extension of an additive category $\mathcal{A}$ as the additive category*

$$\mathcal{A}[\mathbb{Z}^n] = \mathcal{A}[\mathbb{Z}^{n-1}][z_n, z_n^{-1}]$$
$$= \mathcal{A}[z, z_1^{-1}, z_2, z_2^{-1}, \ldots z_n, z_n^{-1}]$$

where $\mathcal{A}[\mathbb{Z}] = \mathcal{A}[z_1, z_1^{-1}]$.

We could also view $\mathcal{A}[\mathbb{Z}^n]$ as the subcategory $C_n(\mathcal{A}) = C_{\mathbb{Z}^n}(\mathcal{A})$ with one object $M[\mathbb{Z}^n]$ for each object M in $\mathcal{A}$ graded by $M[\mathbb{Z}^n](j_1, j_2, \ldots, j_m) = z_1^{j_1} z_2^{j_2} \ldots z_m^{j_m} M$ with $\mathbb{Z}^n$-equivariant morphism.

Theorem 4.4.8 [168] *The torsion group of the n-fold Laurent polynomial extension of $\mathcal{A}$ is such that up to natural isomorphism*

$$K_1(\mathcal{A}[\mathbb{Z}^n]) = \sum_{i=0}^n \binom{n}{i} K_{1-i}(\mathbb{P}_0(\mathcal{A})) \oplus 2\left(\sum_{i=1}^n \binom{n}{i} \widetilde{Nil}_{1-i}(\mathcal{A})\right)$$

where $\widetilde{Nil}_*(\mathcal{A}) = Coker(K_*(\mathcal{A}) \to Nil_*(\mathcal{A}))$ and $Nil_*(\mathcal{A}) = K_* Nil_*(\mathcal{A})$.

Remarks 4.4.5 If A is a ring, and $\mathcal{A} = \mathbb{B}^f(A)$, we recover the LF functors defined on rings in $(4.4)^A$, and the fundamental theorem of Lower K-theory is that the functors $K_{1-n} : (Rings) \to \mathbb{Z}\text{-}Mod : A \to K_{1-n}(A)$ ($n \geq 0$) are contracted with natural identification $LK_{1-n}(A) = K_{-n}(A)$.

4.5 Lower K-theory of grouprings of virtually infinite cyclic groups

$(4.5)^A$ Farrell - Jones isomorphism conjecture

4.5.1 The aim of this section is to discuss some applications of the Mayer - Vietoris sequences obtained above for negative K-theory to the computation of lower K-theory of some infinite groups. The motivation for the K-theory (lower and higher) of virtually cyclic groups is the Farrell - Jones conjecture, which says essentially that the K-theory of group rings of any discrete group G can be computed from the K-theory of virtually cyclic subgroups of G. However, we briefly introduce this conjecture in 4.5.1 below with the observation that this conjecture will be fully discussed in chapter 14 together with Baum - Connes conjectures as induction results under the umbrella of equivariant homology theories.

4.5.2 Definition/Notations

(i) Let A be a ring with identity, α an automorphism of A, $T = \langle t \rangle$ an infinite cyclic group generated by t, $A_\alpha[T]$ the α-twisted Laurent series ring, i.e., $A_\alpha[t] = A[t]$ additively with multiplication defined by $\left(rt^i \right) \left(st^j \right) = r\alpha^{-1}(s)t^{i+j}$, $r, s \in A$.
Let $A_\alpha[t]$ be the subgroup of $A_\alpha[T]$ generated by A and t, i.e., $A_\alpha[t]$ is the α-twisted polynomial ring.

(ii) For any $n \in \mathbb{Z}$, we shall write $NK_n(A, \alpha) := \mathrm{Ker}\left(K_n(A_\alpha[t]) \overset{\epsilon_n}{\to} K_n(A) \right)$ where ϵ is induced by the augmentation $A_\alpha[t] \to A$. If α is the identity automorphism, $A_\alpha[t] = A[t]$, and we recover
$NK_n(A) := \mathrm{Ker}\left(K_n(A[t]) \to K_n(A) \right)$.

(iii) Note that in our discussion of classical K-theory, we shall be concerned with $n \leq 1$. Discussion of NK_n for higher K-theory will be done in 7.5.

Definition 4.5.1 *A discrete group V is called virtually cyclic if it contains a cyclic subgroup of finite index, i.e., if V is finite or virtually infinite cyclic.*

By [183] theorem 5.12, virtually infinite cyclic groups are of two types:

(1) $V = G \rtimes_\alpha T$ is a semi-direct product where G is a finite group, $\alpha \in \mathrm{Aut}(G)$, and the action of T is given by $\alpha(g) = tgt^{-1}$ for all $g \in G$.

(2) V is a non-trivial amalgam of finite groups and has the form $V = G_0 \underset{H}{*} G_1$ where $[G_0 : H] = 2 = [G_1, H]$.

We denote by VCy the family of virtually cyclic subgroups of G.

4.5.1   Farrell - Jones isomorphism conjecture

Let G be a discrete group and $\mathcal{F}$ a family of subgroups of G closed under conjugation and taking subgroups, e.g., VCy.
Let $Or_{\mathcal{F}}(G) := \{G/H \mid H \in \mathcal{F}\}$, R any ring with identity.
There exists a "Davis - Lück" functor

$$\mathbb{K}R : Or_{\mathcal{F}}(G) \to \mathcal{S}\text{pectra} \qquad (\text{see [40] or chapter 14})$$
$$G/H \to \mathbb{K}R(G/H) = \underline{K}(RH)$$

where $\underline{K}(RH)$ is the K-theory spectrum such that $\pi_n(\underline{K}(RH)) = K_n(RH)$ (see 5.2.2(ii)).

There exists a homology theory (see chapter 14 or [40])

$$\mathbb{H}_n(-, \mathbb{K}R) : (GCWcxes \to \mathbb{Z} - \mathcal{M}\text{od}$$
$$X \to H_n(X, \mathbb{K}R).$$

Let $E_{\mathcal{F}}(G)$ be a G-CW-complex which is a model for the classifying space of $\mathcal{F}$.

Note that $E_{\mathcal{F}}(G)^H$ is homotopic to the one point space (i.e., contractible) if $H \in \mathcal{F}$ and $E_{\mathcal{F}}(G)^H = \emptyset$ if $H \notin \mathcal{F}$ and $E_{\mathcal{F}}(G)$ is unique up to homotopy.

There exists assembly map

$$A_{R,\mathcal{F}} : \mathbb{H}_n\left(E_{\mathcal{F}}(G), \mathbb{K}R\right) \to K_n(RG) \text{ (see chapter 14)}.$$

The Farrell - Jones isomorphism conjecture says that $A_{R,VCy}$: $\mathbb{H}_n(E_{VCy}(G), \mathbb{K}R) \simeq K_n(RG)$ is an isomorphism for all $n \in \mathbb{Z}$.

Remarks 4.5.1   (i) The Farrell - Jones (F/J) conjecture has been verified for $n \leq 1$ by A. Bartels, T. Farrell, L. Jones, and H. Reich (see [15]) for any ring R where G is the fundamental group of a Riemann manifold with strictly negative sectional curvature.

(ii) The F/J conjecture makes it desirable to study K-theory (higher and lower) of virtually cyclic groups as possible building blocks for understanding K-theory of discrete groups.

Farrell and Jones studied lower K-theory of virtually infinite cyclic groups in [55], while A. Kuku and G. Tang studied higher K-theory of virtually infinite cyclic groups (see [125]). Their results are discussed in 7.5.

(iii) The assembly maps are natural in the ring R and G. Hence there exists a split cofibration of spectra over $O_{r_{\mathcal{F}}}(G)$:

$$\mathbb{K}R \to \mathbb{K}R[t] \to \mathbb{N}R \text{ where } \pi_n(\mathbb{N}R) = NK_n(RG).$$

Let $NA_{R,\mathcal{F}} : \mathbb{H}_n\left(E_{\mathcal{F}}(G), \mathbb{N}R\right) \to NK_n(RG)$ be the assembly map corresponding to the spectrum valued functor $\mathbb{N}R$.

Proposition 4.5.1 [14] Assume that the assembly map $A_{R,\mathcal{F}}$ is an isomorphism. Then $A_{R[t],\mathcal{F}}$ is an isomorphism iff $NA_{R,\mathcal{F}}$ is also an isomorphism.

Corollary 4.5.1 [14] Let R be the ring of integers in a number field F. Assume that $A_{R,\mathcal{F}}$, $A_{R[t],\mathcal{F}}$ are isomorphism for $n \leq 1$. Then $NK_n(RG) = 0$ for $n \leq -1$.

PROOF   See [14].   ▯

$(4.5)^B$ A preliminary result

4.5.3 The construction of Cartesian and co-Cartesian squares of rings leading to computations of K-theory (higher and lower) of V will make use of the following result proved in [123] by G. Tang and A. Kuku in this generality. This result, proved in 7.5.9, is a generalization of an earlier result of T. Farrell and L. Jones in [55] for $R = \mathbb{Z}$, $\Lambda = \mathbb{Z}G$.

Theorem 4.5.1 [123] *Let R be the ring of integers in a number field F, Λ any R-order in a semi-simple F-algebra Σ. If $\alpha : \Lambda \to \Lambda$ is an R-automorphism, then there exists an R-order $\Gamma \subset \Sigma$ such that*

(1) $\Lambda \subset \Sigma$.

(2) Γ is α-invariant.

(3) Γ is a (right) regular ring. In fact, Γ is a (right) hereditary ring.

Remarks 4.5.2 Since any R-order is a $\mathbb{Z}$-order, it follows that there exists an integer s such that $\underline{q} = s\Gamma$ is both an ideal of Λ and Γ, and that we have a Cartesian square $\begin{array}{ccc} \Lambda & \longrightarrow & \Gamma \\ \downarrow & & \downarrow \\ \Lambda/\underline{q} & \longrightarrow & \Lambda/\underline{q} \end{array}$ that yields a long exact sequence of lower

K-groups, i.e., K_n, $n \leq 1$. In particular, if $\Lambda = \mathbb{Z}G$, $s = |G|$ fits the situation.

$(4.5)^C$ Lower K-theory for $V = G \rtimes_\alpha T$

In this subsection we briefly review some results on lower K-theory of V.

Theorem 4.5.2 *Let R be the ring of integers in a number field F, $V = G \rtimes_\alpha T$.*
Then

(i) for all $n < -1$, $K_n(RV) = 0$.

(ii) The inclusion $RG \to RV$ induces an epimorphism $K_{-1}(RG) \to K_{-1}(RV)$. Hence $K_{-1}(RV)$ is a finitely generated Abelian group.

Remarks 4.5.3 The proof of 4.5.2 due to G. Tang and A. Kuku [123] is given in 7.5 (i) and (ii) and constitutes a generalization of a similar result of T. Farrell and L. Jones in [55] for $\mathbb{Z}V$.

Note that $RV = (RG)_\alpha[T]$, and so, if $|G| = s$ and we put $\underline{q} = s\Gamma$ in the notation of 4.5.10, we have a Cartesian square

$$
\begin{array}{ccc}
RV & \longrightarrow & \Gamma_\alpha[T] \\
\downarrow & & \downarrow \\
(RG/\underline{q})_\alpha[T] & \longrightarrow & (\Gamma/\underline{q})_\alpha[T]
\end{array}
\qquad\qquad \text{(I)}
$$

whose associated lower K-theory $M - V$ sequence leads to required results (see 7.5.12).

$(4.5)^D$ Lower K-theory for $V = G_0 \underset{H}{*} G_1$

4.5.4 Let $\mathcal{J}$ be a category defined as follows:

$$
\text{ob}\mathcal{J} = \{\underline{R} = (R, B_0, B_1) \mid R \text{ a ring}, B_i \ R\text{-bimodules}\}.
$$

A morphism $(R, B_0, B_1) \to (S, C_0, C_1)$ is a triple $(f, \varphi_0, \varphi_1)$ where $f : R \to S$ is a unit preserving ring homomorphism and $\varphi_i : B_i \otimes_R S \to C_i$ is an $R - S$-bimodule homomorphism for $i = 0, 1$.

The composite of $(R, B_0, B_1) \overset{(f, \varphi_0, \varphi_1)}{\longrightarrow} (S, C_0, C_1)$ and $(S, C_0, C_1) \overset{(g, \varphi_0, \varphi_1)}{\longrightarrow}$ (T, D_0, D_1) is $(R; B_0, B_1) \overset{(gf, \psi_0(\varphi_0 \otimes 1_T), \psi_1(\varphi_1 \otimes 1_T))}{\longrightarrow} (T, D_0, D_1)$.

If $\underline{R} = (R, B_0, B_1) \in \mathcal{J}$, $f : R \to S$ a ring homomorphism, then f induces a morphism in $\mathcal{J} : (f, \varphi_0, \varphi_1) : (R, B_0, B_1) \to (S, S \otimes_R B_0 \otimes_R S, S \otimes_R B_1 \otimes_R S)$ where for $i = 0, 1$, $\varphi_i : B_i \otimes_R S \to S \otimes_R B_i \otimes_R S$ is defined by $\varphi_i(b \otimes s) = 1 \otimes b \otimes s$. One checks easily that φ_0, φ_1 are $R - S$-bimodule homomorphism.

4.5.5 In the notations of 4.5.4, there exists a functor $\rho : \mathcal{J} \to \mathcal{R}ings$ defined by

$$
\rho(\underline{R}) = R_\rho = \begin{pmatrix} T_R\left(B_1 \underset{R}{\otimes} B_0\right), & B_1 \otimes_R T_R\left(B_0 \underset{R}{\otimes} B_1\right) \\ B_0 \underset{R}{\otimes} T_R\left(B_1 \underset{R}{\otimes} B_0\right) & T_R(B_0 \otimes B_1) \end{pmatrix}
\qquad \text{(II)}
$$

where $T_R(B_1 \otimes_R B_0)$ and $T_R(B_1 \otimes_R B_0)$ are tensor algebras, and multiplication in R_ρ is given by the matrix multiplication, and on each entry there exists augmentation map $\epsilon : R_\rho \to \left(\begin{smallmatrix} R & 0 \\ 0 & R \end{smallmatrix} \right)$.

Define Nilgroup $NK_n(R_\rho)$ by

$$
NK_n(R_\rho) := \text{Ker}\left(K_n(R_\rho) \overset{\epsilon_*}{\to} K_n \begin{pmatrix} R & 0 \\ 0 & R \end{pmatrix} \right)
\qquad \text{(III)}
$$

for all $n \in \mathbb{Z}$. At this point, we will focus on discussing $NK_n(R_\rho)$ for $n \le 1$. Results on higher Nil groups will be discussed in 7.5. Our next aim is to compare $NK_n(R_\rho)$ with Waldhausen nil groups. First, we briefly discuss one important context in which the triples $\underline{R} = (R, B_0, B_1)$ occur, i.e., in the study of K-theory long exact sequences associated to co-Cartesian or pushout squares of rings.

4.5.6 A pure inclusion $\tau : R \to A$ of rings is an inclusion $\tau : R \to A$ such that there is a splitting $A = \tau(R) \oplus B$ as R-bimodules.

Now let

$$
\begin{array}{ccc}
R & \xrightarrow{\tau_0} & A_0 \\
\downarrow{\scriptstyle \tau_1} & & \downarrow{\scriptstyle j_0} \\
A_1 & \xrightarrow{j_1} & S
\end{array}
\tag{I}
$$

be a co-Cartesian (pushout) diagram of rings where τ_0, τ_1 are pure inclusions. In this case, $A_i = \tau_i(R) \oplus B_i$ $(i = 0, 1)$ and S contains the tensor algebras $T_R(B_0 \otimes_R B_1)$ and $T_R(B_1 \otimes_R B_0)$. The structure of (I) determines an object $\underline{R} = (R, B_0, B_1)$ of $\mathcal{J}$.

Moreover, there is associated to such a co-Cartesian square (I) a long exact sequence (see [224]) for all $n \in \mathcal{Z}$

$$
\cdots K_n(A_0) \oplus K_n(A_1) \to K_n(S) \to K_{n-1}(R) \oplus \mathrm{Nil}^W_{n-1}(\underline{R}) \to
$$

where $\mathrm{Nil}^W_{n-1}(\underline{R})$ are the Waldhausen nil groups defined next (see 4.5.2).

Definition 4.5.2 *Let* $\underline{R} = (R, B_0, B_1)$ *as in 4.5.4. Define a category* $\mathcal{N}il^W(\underline{R})$ *as follows:*

$$
ob\mathcal{N}il^W(\underline{R}) = \left\{ (P, Q; p, q) \left|
\begin{array}{l}
P \in \mathcal{P}(R), Q \in \mathcal{P}(R) \\
p : P \to Q \otimes_R B_0 \\
q : Q \to P \otimes_R B_1
\end{array}
\right. \right\}
$$

where p, q *are R-homomorphisms such that there exist filtrations* $0 = P_0 \subset P_1 \subset \cdots \subset P_n = P$, $0 = Q_0 \subset Q_1 \subset \cdots \subset Q_n = P$ *such that* $p(P_{i+1} \subset Q_i \otimes_R B_0$, $q(Q_{i+1} \subset P_i \otimes_R B_1$.
So, the following compositions

$$
P \xrightarrow{p} Q \otimes_R B_0 \xrightarrow{q \times 1} P \otimes_R B_1 \otimes B_0 \to \cdots
$$
$$
Q \xrightarrow{q} P \otimes_R B_1 \xrightarrow{p \times 1} Q \otimes_R B_0 \otimes B_1 \to \cdots
$$

vanish after a finite number of steps.

There exists a functor

$$
F : \mathcal{N}il^W(\underline{R}) \to \mathcal{P}(R) \times \mathcal{P}(R)
$$
$$
(P, Q, p, q) \to (P, Q)
$$

Define $\mathrm{Nil}_n^W(\underline{R}) := \mathrm{Ker}(K_n(\mathrm{Nil}(\underline{R}) \xrightarrow{F_*} K_n(\mathcal{P}(R) \times \mathcal{P}(R)))$ for all $n \in \mathbb{Z}$ (see [224]).

The next result due to F. Connolly and M. Da Silva [36] gives a connection between $NK_n(R_\rho)$ and $\mathrm{Nil}_{n-1}^W(\underline{R})$.

Proposition 4.5.2 [36] There are natural isomorphisms $NK_n(R_\rho) \cong \mathrm{Nil}_{n-1}^W(\underline{R})$ for $n \leq 1$.

Remarks. It will be of interest to find out if 4.5.2 holds for $n > 1$.

Here are some vanishing results for Waldhausen Nils. The first result 4.5.3 is due to F. Waldhausen [224].

Proposition 4.5.3 [224] Let R be a regular ring. Then for any triple $\underline{R} = (R, B_0, B_1)$ we have $\mathrm{Nil}_n^W(\underline{R}) = 0$ for all $n \in \mathcal{Z}$.

The next results 4.5.4 is due to F. Connolly and S. Passadis (see [93]). Recall that a ring R is quasi-regular if it has a two-sided nilpotent ideal $\underline{a}$ such that $R/\underline{a}$ is regular.

Proposition 4.5.4 [93] Let R be a quasi-regular ring. Then for any triple $\underline{R} = (R, B_0, B_1)$ we have $\mathrm{Nil}_n^W(\underline{R}) = 0$ for all $n \leq -1$.

4.5.7 (a) Now let $V = G_0 \underset{H}{*} G_1$ and consider $G_i - H$ as the right coset of H in G different from H. Then the free $\mathbb{Z}$-module $\mathbb{Z}[G_i - H]$ with basis $G_i - H$ is a $\mathbb{Z}H$-bimodule that is isomorphic to $\mathbb{Z}H$ as a left $\mathbb{Z}H$-module, but the right action is twisted by an automorphism of $\mathbb{Z}H$ induced by an automorphism of H. Then we have a triple $(\underline{\mathbb{Z}H}, \mathbb{Z}[G_0 - H], \mathbb{Z}[G_i - H])$.

(b) For α, β automorphisms of a ring R, we now consider the triple $\underline{R} = (R; R^\alpha, R^\beta)$, which encodes the properties of $\underline{\mathbb{Z}H}$ in (a).

For any automorphism α, let R^α be an $R - R$ bimodule, which is, R as a left R-module but with right multiplication given by $a \cdot r = a\alpha(r)$.

The following result 4.5.3, due to A. Kuku and G. Tang [123], expresses $R_\rho = \rho(\underline{R})$ as a twisted polynomial ring, thus facilitating computations of $K_n(\underline{R})$, $NK_n(\underline{R})$ when $\underline{R} = (R, R^\alpha, R^\beta)$. This result is proved in 7.5.

Theorem 4.5.3 [123] *For the triple $\underline{R} = (R; R^\alpha, R^\beta)$, let $R_\rho := \rho(\underline{R})$, and let γ be a ring automorphism of $\left(\begin{smallmatrix} R & 0 \\ 0 & R \end{smallmatrix}\right)$ given by $\gamma : \left(\begin{smallmatrix} a & 0 \\ 0 & b \end{smallmatrix}\right) \to \left(\begin{smallmatrix} \beta(b) & 0 \\ 0 & \alpha(b) \end{smallmatrix}\right)$. Then there is a ring isomorphism $\mu : R_\rho \xrightarrow{\sim} \left(\begin{smallmatrix} R & 0 \\ 0 & R \end{smallmatrix}\right)_\gamma [x]$.*

Here are some consequences of 4.5.3, also due to Kuku/Tang (see [123] or 7.5 for proofs).

Theorem 4.5.4 [123]

(i) *Let R be a regular ring; then $NK_n(R; R^\alpha, R^\beta) = 0$ for all $n \in \mathbb{Z}$.*

(ii) *If R is quasi-regular, then $NK_n(R; R^\alpha, R^\beta) = 0$ for all $n \leq 0$.*

Note

Theorem 4.5.4 gives the NK_n version of propositions 4.5.4 and 4.5.5.

We also have the following consequence of theorems 4.5.3 and 4.5.4 due to Kuku/Tang [123].

Theorem 4.5.5 [123] *Let* $V = G_0 \underset{H}{*} G_1$, *and* $[G_0 : H] = 2 = [G_1 : H]$. *Then* $NK_n(\mathbb{Z}H, \mathbb{Z}[G_0 - H], \mathbb{Z}[G_1 - H]) = 0$ *for* $n \leq 0$.

Remarks 4.5.4   (i) Let a discrete group G be an amalgamated free product given by $G = G_0 \underset{H}{*} G_1$ where G_0, G_1 are discrete groups and H a finitely generated central subgroup of G_0 and G_1. In [94], D. Juan-Pineda and S. Prassidis proved that the negative Waldhausen Nil groups that appear in the computations of K-theory of $\mathbb{Z}G$ vanish, i.e., $NK_n(\mathbb{Z}H, \mathbb{Z}[G_0 - H], \mathbb{Z}[G_1 - H]) = 0$ for $n \leq -1$. They also proved that if $H = H' \times T^s$ where H' is a finite group and T an infinite cyclic group, then $NK_0(\mathbb{Z}H, \mathbb{Z}[G_0 - H], \mathbb{Z}[G_1 - H])$ is $|H'|$-torsion.

(ii) The vanishing conjecture says that if G is a discrete group, then $K_n(\mathbb{Z}G) = 0$ for $n \leq -2$. This conjecture has been proved for all subgroups of cocompact discrete subgroups of Lie groups (see [15, 55]). In [93] Pineda/Prassidis also show that if G_0, G_1 are groups for which the vanishing conjecture holds, and H is a finitely generated central subgroup of G_i ($i = 0,1$), then the group $G_0 \underset{H}{\times} G_1$ also satisfies the vanishing conjecture.

$(4.5)^E$ Some Applications

4.5.8 One could interpret the negative K-groups in terms of bounded h-cobordisms.

Let W be a manifold equipped with a surjective proper map $p_W : W \to \mathbb{R}^s$, i.e., W is parametrized over $\mathbb{R}^s$. Assume also that the fundamental group $\pi_1(W)$ is bounded. Let W_1, W^1 be two manifolds parametrized over $\mathbb{R}^s$. A map $f, W \to W^1$ is said to be bounded if $\{p^1 \circ f(x) - p(x) | x \in W\}$ is a bounded subset of $\mathbb{R}^s$.

A cobordism (W, M^-, f^-, M^+, f^+) over M^- is a bounded cobordism if W is parametrized over $\mathbb{R}^s$, and we have a decomposition of its boundary ∂W into two closed $(n-1)$-dimensional manifolds $\partial^- W$ and $\partial^+ W$, two closed $(n-1)$-dimensional manifolds M^- and M^+, and diffeomorphisms $f^- : M^- \to \partial^- W$ and $f^+ : M^+ \to \partial^+ W$ such that the parametrization for $M^{\pm}$ is given by $p_W \circ f^{\pm}$. If we assume that the inclusions $i^{\pm} : \partial^{\pm} W \to W$ are homotopy equivalences, then there exist deformations $r^{\pm} : W \times I \to W$ $(x,t) \to r_t^{\pm}(x)$ such that $r_0^{\pm} = id_W$ and $r_1^{\pm}(W) \subset \partial^{\pm} W$.

A bounded cobordism is called a bounded h-cobordism if the inclusion $i^{\pm}$ are homotopy equivalences and, in addition, deformations can be chosen such that $S^{\pm} = \{p_W \circ r_t^{\pm}(x) - p_W \circ r_1^{\pm}(x) | x \in W, t \in [0,1]\}$ are bounded subsets of $\mathbb{R}^s$. We now have the following theorem.

Theorem 4.5.6 [138] (Bounded h-cobordism theorem) *Suppose that M^- is parametrized over $\mathbb{R}^s$ and satisfies* $\dim M^- \geq 5$. *Let G be its fundamental group(oid). Then there is a bijective correspondence between equivalence classes of bounded h-cobordisms over M^- modulo bounded diffeomorphism relative to M^- and elements in $\kappa_{1-s}(G)$ where*

$$\kappa_{1-s}(G) = \begin{cases} Wh(G) & if \quad s = 0 \\ \widetilde{K}_0(\mathbb{Z}G) & if \quad s = 1 \\ K_{1-s}(\mathbb{Z}G) & if \quad s \geq 2 \end{cases}$$

4.5.9 Let M be a compact manifold and $p : M \times \mathbb{R}^s \to \mathbb{R}^s$ the natural projection. The space $P_b(M : \mathbb{R}^k)$ of bounded pseudo-isotopics is the space of all self-homeomorphisms $h : M \times \mathbb{R}^s \times I \to M \times \mathbb{R}^s \times I$ such that when restricted to $M \times \mathbb{R}^s$ the map is h bounded (i.e., the set $\{p \circ h(y) - p(y) | y \in M \times \mathbb{R}^s \times I\}$ is a bounded subset of $\mathbb{R}$. There is a stabilization map $P_b(M; \mathbb{R}^s) \to P_b(M \times I, \mathbb{R}^s)$ and a stable bounded pseudo-isotopy space $P_b(M; \mathbb{R}^s) = \underset{j}{codim} P_b(M \times I^j; \mathbb{R}^s)$, as well as a homotopy equivalence $P_b(M; \mathbb{R}^s) \to \Omega P_b(M; \mathbb{R}^{s+1})$. Hence the sequence of spaces, $P_b(M; \mathbb{R}^s)$ $s = 0, 1, \ldots$ is an Ω-spectrum $\mathbf{P}(M)$ (see [138]).

One could also define in an analogous way the differentiable bounded pseudo-isotopics $P_b^{diff}(M; \mathbb{R}^s)$ and an Ω-spectrum $\mathbf{P}^{diff}(M)$.

We now have the following.

Theorem 4.5.7 [138] Negative homotopy groups of pseudo-isotopies
Let $G = \pi_1(M)$. Assume that n_1 and s are such that $n + s \geq 0$. Then for $s \geq 1$, we have isomorphisms

$$\pi_{n+s}(P_b(M; \mathbb{R}^s)) = \begin{cases} Wh(G) & if \quad n = -1 \\ \widetilde{K}_0(\mathbb{Z}G) & if \quad n = -2 \\ K_{n+2}(\mathbb{Z}G) & if \quad n < -2 \end{cases}$$

The result above also holds for $P_b^{diff}(M; \mathbb{R}^s)$.

Exercises

4.1 Given the Cartesian square (I) in 4.1.1, verify that there exists an exact sequence

$$K_1(A) \xrightarrow{\alpha_1} K_1(A_1) \oplus K_1(A_2) \xrightarrow{\beta_1} K_1(A') \xrightarrow{\delta} K_0(A) \xrightarrow{\delta_0} K_0(A_1) \oplus K_0(A_2) \xrightarrow{\beta_0} K_0(A').$$

4.2 Given the Cartesian square (I) in 4.1.1 where A, A_1, A' are right Noetherian rings such that A_1, A_2, A' are finitely generated A-modules, show that the restriction homomorphisms $A_i \to A'(i = 0, 1)$ induce epimorphisms $G_i(A_i) \oplus G_i(A_2) \longrightarrow G_i(A')(i = 0, 1)$.

Hence show that if $B = \prod B_i$ is a product of rings and $A \subset B$ is a subring that projects onto each B_i, then the homomorphisms $G_i(B) = \prod_j G_i(B_j) \longrightarrow G_i(A)(i = 0, 1)$ are surjective.

4.3 Let A be a ring with identity, $\underline{a}$ a two-sided ideal of A. The double of A along $\underline{a}$ is the ring D defined by $D = \{(x, y) \in A \oplus A | x \equiv y \mod \underline{a}\}$.

(i) Show that $K_0(D) \simeq K_0(A, \underline{a})$ where $K_0(A, \underline{a})$ is defined as in 4.3.1.

(ii) Show that $K_1(D) \simeq K_1(A, \underline{a}) \oplus K_1(A)$ where $K_1(A, \underline{a})$ is as defined in exercise 3.1.

4.4 Let F be a functor $\mathcal{R}ings \to \mathbb{Z}\text{-}\mathcal{M}od$, (T, T_+) an oriented cycle (see definition 4.4.1), A a ring with identity, and $seq\, F(A)$ the complex defined in 4.4.1. Show that $Seq\, F(A)$ is exact iff τ_+, τ_- are both monomorphisms and $F(A) = Im(\tau_+) \cap Im(\tau_-)$.

4.5 Let A be a commutative ring and (T, T_+) an oriented cycle. Show that

$$K_0(A[T_+^n]) = (1 + N)^n K_0(A)$$

and

$$K_0(A[T^n]) = (1 + 2N + L)^n K_0(A)$$

4.6 Let G be a finite Abelian group. Show that

(a) $K_0(\mathbb{Z}[G \times T^n]) = (1 + 2N)^n K_0(\mathbb{Z}G) \oplus nLK_0(\mathbb{Z}G)$

(b) $K_1(\mathbb{Z}[G \times T^n]) = (1 + 2N)^n K_1(\mathbb{Z}G) \oplus nK_0(\mathbb{Z}G) \oplus \frac{n(n-1)}{2} LK_0(\mathbb{Z}G)$.

4.7 In the notation of 4.4.2, let $0 \to F \to G \to H \to 0$ be a short exact sequence of functors from $\mathcal{A}(R, S)$ to $\mathbb{Z}\text{-}\mathcal{M}od$. Suppose that F and H are contracted functors on the category of R-algebras. Show that there is a short exact sequence $0 \to LF \to LG \to LH \to 0$ (where LF and LH are contracted functors on the category of R-algebras) and G is acyclic. (Note that a functor $F : \mathcal{A}(R, S) \to \mathbb{Z}\text{-}\mathcal{M}od$ is acyclic if the complex $0 \to F(\Lambda) \to F(\Lambda[t]) \oplus F(\Lambda[t^{-1}]) \to F(\Lambda[t, t^{-1}]) \to LF(\Lambda) \to 0$ is exact for all objects Λ of $\mathcal{A}(R, S)$, and F is said to be contracted if also the projection $F(\Lambda[t, t^{-1}]) \to LF(\Lambda)$ has natural splitting.)

Part II

Higher Algebraic
K-Theory and Integral
Representations

Chapter 5

Higher Algebraic K-theory – definitions, constructions, and relevant examples

In this chapter, we review mostly without proofs some basic definitions and constructions for Higher K-theory. There are already several books in which proofs can be found. See, for example, [25, 88, 198]. However, we give copious examples to illustrate the definitions, constructions, and results.

5.1 The plus construction and higher K-theory of rings

$(5.1)^A$ The plus construction

This construction, which leads to a definition of higher K-groups for rings, is due to D. Quillen. We shall, in 5.2, identify this construction as a special case of a categorical construction also due to Quillen.

The definition of $K_n(A)$, A any ring with identity, will make use of the following results.

Theorem 5.1.1 [88, 198] *Let X be a connected CW-complex, N a perfect normal subgroup of $\pi_1(X)$. Then there exists a CW-complex X^+ (depending on N) and a map $i : X \to X^+$ such that*

(i) *$i_* : \pi_1(X) \to \pi_1(X^+)$ is the quotient map $\pi_1(X) \to \pi_1(X^+)/N$.*

(ii) *For any $\pi_1(X^+)/N$-module L, there is an ismorphism $i_* : H_*(X, i^*L) \to H_*(X^+, L)$ where i^*L is L considered as a $\pi_1(X)$-module.*

(iii) *The space X^+ is universal in the sense that if Y is any CW-complex and $f : X \to Y$ is a map such that $f_* : \pi_1(X) \to \pi_1(Y)$ satisfies $f_*(N) = 0$, then there exists a unique map $f^+ : X^+ \to Y$ such that $f^+i = f$.*

5.1.1 Some properties of the plus construction

Let X, Y be connected CW-complexes and N, N' perfect normal subgroups of $\pi_1(X)$ and $\pi_1(Y)$, respectively. Then

(a) The map $(X \times Y)^+ \to X^+ \times Y^+$ is a homotopy equivalence.

(b) If $f : X \to Y$ is a continuous map such that $\pi_1(f)(N) = N'$ and $\pi_1(X)$ is perfect with $\pi_1(X) = N$, then the amalgamated sum $Z = X^+ \cup_X Y$ is homotopy equivalent to Y.

(c) If $\widetilde{X}$ is the covering space of X that corresponds to the subgroup N, then $\widetilde{X}^+$ is up to homotopy the universal covering space of X^+.

(For the proof of the statements above see [88].)

Definition 5.1.1 *Let A be a ring, $X = BGL(A)$ in theorem 5.1.1. Then $\pi_1 BGL(A) = GL(A)$ contains $E(A)$ as a perfect normal subgroup. Hence, by theorem 5.1.1, there exists a space $BGL(A)^+$. Define $K_n(A) = \pi_n(BGL(A)^+)$ (see $(5.2)^A$ for a discussion of classifying spaces).*

Examples/Remarks 5.1.1 For $n = 0, 1, 2, K_n(A)$ as in definition 5.1.1 above can be identified respectively with classical $K_n(A)$.

(i) $\pi_1(BGL(A)^+) = GL(A)/E(A) = K_1(A)$.

(ii) Note that $BE(A)^+$ is the universal covering space of $BGL(A)^+$, and so, we have

$$\pi_2(BGL(A)^+) \approx \pi_2(BE(A)^+) \approx H_2(BE(A)^+) \cong H_2(BE(A))$$
$$\cong H_2(E(A)) \approx K_2(A).$$

(iii) $K_3(A) = H_3(St(A))$. For proof see [58].

(iv) If A is a finite ring, then $K_n(A)$ is finite — (see (7.1.12) or [112]).

(v) For a finite field $\mathbb{F}_q$, $K_{2n}(\mathbb{F}_q) = 0$, $K_{2n-1}(\mathbb{F}_q) = \mathbb{Z}/(q^n - 1)$ (see [162]). In later chapters, we shall come across many computations of $K_n(A)$ for various rings, especially for orders and grouprings.

5.1.2 Hurewitz maps

The Hurewitz maps are very valuable for computations.
For any ring A with identity, there exists Hurewitz maps:

(i) $h_n : K_n(A) = \pi_n(BGL(A)^+) \to H_n(BGL(A)^+, \mathbb{Z}) \approx H_n(GL(A), \mathbb{Z})$ for all $n \geq 1$.

(ii) $h_n : K_n(A) = \pi_n(BE(A)^+) \to H_n(BE(A)^+, \mathbb{Z}) \approx H_n(E(A), \mathbb{Z})$ for all $n \geq 2$.

(iii) $h_n : K_n(A) = \pi_n(BSt(A)^+) \to H_n(BSt(A)^+, \mathbb{Z}) \approx H_n(St(A), \mathbb{Z})$ for all $n \geq 3$.

Note that $BGL(A)^+$ is connected, $BE(A)^+$ is simply connected (i.e., one-connected), and $BSt(A)^+$ is 2-connected.

For a comprehensive discussion of Hurewitz maps see [7].

5.1.3 Products

Let R, R' be two rings; $R \otimes_{\mathbb{Z}} R' := R \otimes R'$. Then the tensor product of matrices induce $GL_n(R) \times GL_n(R') \to GL_{mn}(R \otimes R')$ and hence the map $\eta_{m,n}^{R,R'} : BGL(R)^+ \times BGL_n(R')^+ \to BGL_{m,n}(R \otimes R')^+$, which can be shown to yield a map $\gamma^{R,R'} : BGL(R)^+ \times BGL(R')^+ \to BGL(R \otimes R')$ (see [136]. Now $\gamma^{R,R'}$ is homotopic to the trivial map on $BGL(R)^+ \vee BGL(R')^+$ and hence induces a map $\hat{\gamma}^{R,R'} : BGL(R)^+ \wedge BGL(R')^+ \to BGL(R \otimes R')^+$. $\hat{\gamma}^{R,R'}$ is natural in R, R' bilinear, associative and commutative up to weak homotopy (see [136]).

Now define the product map

$$* : K_i(R) \times K_j(R') \to K_{i+j}(R \otimes R')$$

as follows. Let $\alpha : S^i \to BGL(R)^+$, $\beta : S^j \to BGL(R')^+$ be representatives of $x \in K_i(R)$, $y \in K_j(R')$, respectively. Then

$$x * y = \left[S^{i+j} \approx S^i \wedge S^j \overset{\alpha \cap \beta}{\to} BGL(R)^+ \wedge BGL(R')^+ \overset{\hat{\gamma}^{R,R'}}{\to} BGL(R \otimes R')^+ \right].$$

The product '*' is natural in R, R', bilinear and associative for all $i, j \geq 1$. Hence we have a product $* : K_i(R) \otimes K_j(R') \to K_{i+j}(R \otimes R')$.

Now put $R = R'$. If R is commutative, then the ring homomorphism $\nabla : R \otimes R \to R$, $\nabla(a \otimes b) = ab$ induces a ring structure on $K_*(R)$, i.e., $* : K_i(R) \otimes K_j(R) \overset{*}{\to} K_{i+j}(R \otimes R) \overset{\nabla_*}{\to} K_{i+j}(R)$. If R is a commutative ring, then for all $x \in K_i(R)$, $y \in K_j(R)$, $i, j \geq 1$ we have

$$x * y = (-1)^{ij} y * x.$$

Note that the above construction of products is due to J. Loday (see [136]).

5.1.4 K_n^M-Milnor K-theory

Let A be a commutative ring with identity and $T(A^*)$ the tensor algebra over $\mathbb{Z}$ where A^* is the Abelian group of invertible elements of A. For any $x \in A^* - \{1\}$, the elements $x \otimes (1-x)$ and $x \otimes (-x)$ generate a 2-sided ideal I of $T(A^*)$. The quotient $T(A^*)/I$ is a graded Abelian group whose component in degree $0, 1, 2$ are, respectively, $\mathbb{Z}$, A^* and $K_2^M(A)$ where $K_2^M(A)$ is the classical K_2-group (see 3.3).

5.1.5   Connections with Quillen K-theory

(i) As remarked above, $K_n^M(A) = K_n^Q(A)$ for $n \leq 2$.

(ii) First observe that there is a well-defined product $K_m^Q(A) \times K_n^Q(A) \to K_{m+n}^Q(A)$, due to J. Loday (see 5.1.3 on page 89). Now, there exists a map $\phi : K_n^M(A) \to K_n^Q(A)$ constructed as follows: we use the isomorphism $K_1(A) \simeq A^*$ to embed A^* in $K_1(A)$ and use the product in Quillen K-theory to define inductively a map $(A^*)^n \to K_1(A)^n \to K_n(A)$, which factors through the exterior power $\Lambda^n A^*$ over $\mathbb{Z}$, and hence through the Milnor K-groups $K_n^M(A)$, yielding the map $\phi : K_n^M(A) \to K_n(A)$.

If F is a field, we have the following more precise result due to A. Suslin.

Theorem 5.1.2 [201] *The kernel of $\phi : K_n^M(F) \to K_n(F)$ is annihilated by $(n-1)!$.*

5.1.6 K_n^{k-v}-K-theory of Karoubi and Villamayor

Let $R(\Delta^n) = R[t_0, t_1, \ldots, t_n]/(\sum t_i - 1) \simeq R[t_1, \ldots, t_n]$. Applying the functor GL to $R(\Delta^n)$ yields a simplicial group $GL(R(\Delta^*))$.

Definition 5.1.2 *Let R be a ring with identity. Define the Karoubi - Villamayor K-groups by $K_n^{k-v}(R) = \pi_{n-1}(GL(R[\Delta^*])) = \pi_n(BGL(R[\Delta^*]))$ for all $n \geq 1$. Note that $\pi_0(GL(R[\Delta^*]))$ is the quotient $GL(R)/uni(R)$ of $K_1(R)$ where $Uni(R)$ is the subgroup of $GL(R)$ generated by unipotent matrices, i.e., matrices of the form $1 + N$ for some nilpotent matrix N.*

Theorem 5.1.3 [58]

(i) *For $p \geq 1$, $q \geq 0$, there is a spectral sequence $E_{pq}^1 = K_p(R[\Delta^q]) \implies K_{p+q}^{k-v}(R)$.*

(ii) *If R is regular, then the spectral sequence in (i) above degenerates and $K_n(R) = K_n^{k-v}(R)$ for all $n \geq 1$.*

Definition 5.1.3 *A functor $F : \mathcal{R}ings \to \mathbb{Z}\text{-}Mod$ (Chain complexes, etc.) is said to be homotopy invariant if for any ring R, the natural map $R \to R[t]$ induces an isomorphism $F(R) \approx F(R[t])$. Note that if F is homotopy invariant, then the simplicial object $F(R[\Delta^*])$ is constant.*

Theorem 5.1.4 *The functors $K_n^{k-v} : \mathcal{R}ings \to \mathbb{Z}\text{-}Mod$ is homotopy invariant, i.e., $K_n^{k-v}(R) \cong K_n^{k-v}(R[t])$ for all $n \geq 1$.*

Note.   In view of 5.1.13, K_n^{k-v} is also denoted K_n^h (h for homotopy) since it is a homotopy functor.

5.2 Classifying spaces and higher K-theory of exact categories – Constructions and examples

$(5.2.)^A$ Simplicial objects and classifying spaces

Definition 5.2.1 *Let Δ be a category defined as follows: $ob(\Delta) = $ ordered sets $\underline{n} = \{0 < 1 < \cdots < n\}$. The set $Hom_\Delta(\underline{m}, \underline{n})$ of morphisms from $\underline{m}$ to $\underline{n}$ consists of maps $f : \underline{m} \to \underline{n}$ such that $f(i) \le f(j)$ for $i < j$.*

Let $\mathcal{A}$ be any category. A simplicial object in $\mathcal{A}$ is a contravariant functor $X : \Delta \to \mathcal{A}$ where we write X_n for $X(\underline{n})$. Thus, a simplicial set (resp. group; resp. ring; resp. space, etc.) is a simplicial object in the category of sets (resp. group; resp. ring; resp. space, etc). A co-simplicial object is covariant functor $X : \Delta \to \mathcal{A}$.

Equivalently, one could define a simplicial object in a category $\mathcal{A}$ as a set of objects $X_n (n \ge 0)$ in $\mathcal{A}$ and a set of morphisms $\delta_i : X_n \to X_{n-1} (0 \le i \le n)$ called face maps as well as a set of morphisms $s_j : X_n \to X_{n+1} (0 \le j \le n)$ called degeneracy maps satisfying "simplicial identities" (see [238], p. 256). We shall denote the category of simplicial sets by $\mathcal{S}$ sets.

Definition 5.2.2 *The geometric n-simplex is the topological space*

$$\hat{\Delta}^n = \Big\{ (x_0, x_1, \ldots, x_n) \in \mathbb{R}^{n+1} \mid 0 \le x_i \le 1 \, \forall i \quad and \quad \sum x_i = 1 \Big\}.$$

The functor $\hat{\Delta} : \Delta \to $ spaces given by $\underline{n} \to \hat{\Delta}^n$ is a co-simplicial space.

Definition 5.2.3 *Let X_* be a simplicial set. The geometric realization of X_* written $|X_*|$ is defined by $|X_*| := X \times_\Delta \hat{\Delta} = \cup_{n \ge 0}(X_n \times \hat{\Delta}_n)/ \approx$ where the equivalence relation '$\approx$' is generated by $(x, \phi_*(y)) \approx (\phi^*(x), y)$ for any $x \in X_n$, $y \in Y_m$, and $\phi : \underline{m} \to \underline{n}$ in Δ and where $X_n \times \hat{\Delta}^n$ is given the product topology and X_n is considered a discrete space.*

Examples/Remarks 5.2.1   (i) Let T be a topological space, $Sing_* T = \{Sing_n T\}$ where $Sing_n T = \{$continuous maps $\hat{\Delta}^n \to T\}$. A map $f : \underline{n} \to \underline{m}$ determines a linear map $\hat{\Delta}^n \to \hat{\Delta}^m$ and hence induces a map $\hat{f} : Sing_m T \to Sing_n T$. So $Sing_* T : \Delta \to $ sets is a simplicial set. Call $Sing_* T$ a Kan complex.

(ii) For any simplicial set X_*, $|X_*|$ is a CW-complex with X_n in one-one correspondence with n-cells in $|X_*|$.

(iii) For any simplicial sets X_*, Y_*, $|X_*| \times |Y_*| \cong |X_* \times Y_*|$ where the product is such that $(X_* \times Y_*)_n = X_n \times Y_n$.

Definition 5.2.4 *Let $\mathcal{A}$ be a small category. The Nerve of $\mathcal{A}$, written $N\mathcal{A}$, is the simplicial set whose n-simplices are diagrams*

$$\mathcal{A}_n = \{A_0 \xrightarrow{f_1} A_1 \longrightarrow \cdots \xrightarrow{f_n} A_n\}$$

where the A_i are $\mathcal{A}$-objects and the f_i are $\mathcal{A}$-morphisms. The classifying space of $\mathcal{A}$ is defined as $|N\mathcal{A}|$ and is denoted by $B\mathcal{A}$.

5.2.1   Properties of $B\mathcal{A}$

(i) $B\mathcal{A}$ is a CW-complex whose n-cells are in one-one correspondence with the diagrams $\mathcal{A}_n$ above. (See 5.2.4(ii)).

(ii) From Examples/Remarks 5.2.1(iii), we have, for small categories $\mathcal{C}$, $\mathcal{D}(I)$ $B(\mathcal{C} \times \mathcal{D}) \approx B\mathcal{C} \times B\mathcal{D}$ where $B\mathcal{C} \times B\mathcal{D}$ is given the compactly generated topology (see [88, 198]). In particular we have the homeomorphism (I) if either $B\mathcal{C}$ or $B\mathcal{D}$ is locally compact (see [198]).

(iii) Let F, G be functors, $\mathcal{C} \to \mathcal{D}$ ($\mathcal{C}$, $\mathcal{D}$ small categories). A natural transformation of functors $\eta : F \to G$ induces a homotopy $B\mathcal{C} \times I \to B\mathcal{D}$ from $B\mathcal{C}$ to $B\mathcal{D}$ (see [165, 198]).

(iv) If $F : \mathcal{C} \to \mathcal{D}$ has a left or right adjoint, then F is a homotopy equivalence (see [165, 198]).

(v) If $\mathcal{C}$ is a category with initial or final object, then $B\mathcal{C}$ is contractible (see [165, 198]).

Examples 5.2.1   (i) A discrete group G can be regarded as a category with one object G whose morphisms can be identified with the elements of G.

The nerve of G, written N_*G, is defined as follows: $-N_n(G) = G^n$, with face maps δ_i given by:

$$\delta_i(g_1, \ldots, g_n) = \begin{cases} (g_2, \ldots, g_n) & i = 0 \\ (g_1, \ldots, g_i g_{i+1}, \ldots, g_n) & 1 \leq i < n - 1 \\ (g_1, \ldots, g_n) & i = n - 1 \end{cases}$$

and degeneracies s_i given by

$$s_i(g_1, g_2, \ldots, g_n) = (g_1, \ldots, g_i, 1, g_{i+1}, \ldots, g_n).$$

The classifying space BG of G is defined as $|N_*(G)|$ and it is a connected CW-complex characterized up to homotopy type by the property that $\pi_1(BG, *) = G$ and $\pi_n(BG, *) = 0$ for all $n > 0$ where $*$ is some base point of BG. Note that BG has a universal covering space usually denoted by EG. (See [135]).

Note that the term classifying space of G comes from the theory of fiber bundles. So, if X is a finite cell complex, the set $[X, BG]$ of homotopy classes of maps $X \to BG$ gives a complete classification of the fiber bundles over X with structure group G.

(ii) Let G be a topological group (possibly discrete) and X a topological G-space. The translation category $\underline{X}$ of X is defined as follows: $-ob(\underline{X}) =$ elements of X; $\text{Hom}_{\underline{X}}(x, x') = \{g \in G | gx = x'\}$. Then the nerve of $\underline{X}$ is the simplicial space equal to $G^n \times X$ in dimension n. $B\underline{X} = |\text{Nerve of}\underline{X}|$ is the Borel space $EG \times_G X$ (see [135]).

(iii) Let $\mathcal{C}$ be a small category, $F : \mathcal{C} \to \underline{Sets}$ a functor, $\mathcal{C}_F$ a category defined as follows

$$ob\mathcal{C}_F = \{(C, x) | C \in ob\mathcal{C}, \ x \in F(C)\}.$$

A morphism from (C, x) to (C', x') is a morphism $f : C \to C'$ such that $f_*(x) = x'$.

The homotopy colimit of F is defined as hocolim $F := B\mathcal{C}_F$. This construction is also called the Bousfield - Kan construction. If the functor F is trivial, we have $B\mathcal{C}_F = B\mathcal{C}$ (see [135]).

$(5.2)^B$ Higher K-theory of exact categories - definitions and examples

In 1.4, we discussed K_0 of exact categories $\mathcal{C}$, providing copious examples. In this section, we define $K_n(\mathcal{C})$ for all $n \geq 0$ with the observation that this definition generalizes to higher dimensions the earlier ones at the zero-dimensional level.

Definition 5.2.5 *Recall from 1.4 that an exact category is a small additive category $\mathcal{C}$ (which is embeddable as a full subcategory of an Abelian category $\mathcal{A}$) together with a family $\mathcal{E}$ of short exact sequences $0 \to C' \xrightarrow{i} C \xrightarrow{j} C'' \to 0 (I)$ such that $\mathcal{E}$ is the class of sequences, (I) is $\mathcal{C}$ that are exact in $\mathcal{A}$, and $\mathcal{C}$ is closed under extensions (i.e., for any exact sequence $0 \to C' \xrightarrow{i} C \xrightarrow{j} C'' \to 0$) in $\mathcal{A}$ with C', C'' in $\mathcal{C}$, we have $C \in \mathcal{C}$.*

In the exact sequence (I) above, we shall refer to i as inflation or admissible monomorphism, j as a deflation or admissible epimorphism; and to the pair (i, j) as a conflation.

Let $\mathcal{C}$ be an exact category. We form a new category $Q\mathcal{C}$ whose objects are the same as objects of $\mathcal{C}$ such that for any two objects $M, P \in ob(Q\mathcal{C})$, a morphism from M to P is an isomorphism class of diagrams $M \xleftarrow{j} N \xrightarrow{i} P$ where i is an admissible monomorphism and j is an admissible epimorphism in $\mathcal{C}$, that is, i and j are part of some exact sequences $0 \to N \xrightarrow{i} P \twoheadrightarrow P' \to 0$ and $0 \to N' \to N \xrightarrow{j} M \to 0$, respectively.

Composition of arrows $M \leftarrow N \rightarrowtail P$ *and* $P \leftarrow R \rightarrowtail T$ *is defined by the following diagram, which yields an arrow*

$$M \leftarrow N \times_P T \rightarrowtail T$$

$$\text{in} \quad \mathcal{QC}$$

$$
\begin{array}{ccc}
M & & \\
\uparrow & & \\
N & \rightarrowtail & P \\
\uparrow & & \uparrow \\
N \times_P T & \rightarrow R & \rightarrowtail T
\end{array}
$$

Definition 5.2.6 *For all* $n \geq 0$, *define*

$$K_n(C) := \pi_{n+1}(B\mathcal{QC}, o),$$

see [165].

Examples 5.2.2 (i) For any ring A with identity, the category $\mathcal{P}(A)$ of finitely generated projective modules over A is exact and we shall write $K_n(A)$ for $K_n(\mathcal{P}(A))$.

Note that for all $n \geq 1$, $K_n(A)$ coincides with the groups $\pi_n(BGL(A)^+)$ defined in 5.1.3.

(ii) Let A be a left Noetherian ring. Then $\mathcal{M}(A)$, the category of finitely generated (left)-A modules, is an exact category, and we denote $K_n(\mathcal{M}(A))$ by $G_n(A)$. The inclusion functor $\mathcal{P}(A) \rightarrow \mathcal{M}(A)$ induces a homomorphism $K_n(A) \rightarrow G_n(A)$.

If A is regular, then $K_n(A) \approx G_n(A)$ (see Remarks and Examples 6.1.1(i)).

(iii) Let X be a scheme (see [79]), $\mathcal{P}(X)$ the category of locally free sheaves of O_X-modules of finite rank (or equivalent category of finite dimensional (algebraic) vector bundles on X.) Then $\mathcal{P}(X)$ is an exact category and we write $K_n(X)$ for $K_n(\mathcal{P}(X))$ (see [164, 165]).

If $X = \text{Spec}(A)$ for some commutative ring A, then we have an equivalence of categories:

$$\mathcal{P}(X) \rightarrow \mathcal{P}(A) : E \rightarrow \Gamma(X, E) = \{A - \text{modules of global sections}\}$$

with inverse equivalence $\mathcal{P}(A) \rightarrow \mathcal{P}(X)$ given by

$$P \rightarrow \widetilde{P} : U \rightarrow O_X(U) \otimes_A P.$$

So,

$$K_n(A) \approx K_n(X).$$

(iv) If X is a Noetherian scheme, then the category $\mathcal{M}(X)$ of coherent sheaves of O_X-modules is exact. We write $G_n(X)$ for $K_n(\mathcal{M}(X))$. If $X = \mathrm{Spec}(A)$, then we have an equivalence of categories $\mathcal{M}(X) \approx \mathcal{M}(A)$ and $G_n(X) \approx G_n(A)$ (see [165]).

(v) Let R be a commutative ring with identity, Λ an R-algebra finitely generated as an R-module, $\mathcal{P}_R(A)$ the category of left Λ-lattices. Then $\mathcal{P}_R(\Lambda)$ is an exact category and we write $G_n(R, \Lambda)$ for $K_n(\mathcal{P}_R(\Lambda))$. If $\Lambda = RG$, G a finite group, write $G_n(R, G)$ for $G_n(R, RG)$. If R is regular, then $G_n(R, \Lambda) \approx G_n(\Lambda)$ (see [106]).

(vi) Let G be a finite group, S a G-set, $\underline{S}$ the translation category of S (or category associated to S) (see 1.1.5). Then, the category $[\underline{S}, \mathcal{C}]$ of functors from $\underline{S}$ to an exact category $\mathcal{C}$ is also an exact category. We denote by $K_n^G(S, \mathcal{C})$ the Abelian group $K_n([\underline{S}, \mathcal{C}])$. As we shall see later, $K_n^G(-, \mathcal{C}) : \underline{GSet} \to \underline{Ab}$ is a 'Mackey' functor, for all $n \geq 0$ (see theorem 10.1.2).

If $S = G/G$ and $\mathcal{C}_G$ denotes the category of representations of G in $\mathcal{C}$, then $[G/G, \mathcal{C}] \approx \mathcal{C}_G$. In particular, $[G/G, \mathcal{P}(R)] \approx \mathcal{P}(R)_G \approx \mathcal{P}_R(RG)$, and so, $K_n^G[G/G, \mathcal{P}(R)] \approx K_n(\mathcal{P}(R)_G) \approx G_n(RG)$ if R is regular. As explained in Example 1.4.1(vi), when $R = \mathbb{C}$, $K_0(\mathcal{P}(\mathbb{C})_G) \approx G_0(\mathbb{C}, G) \approx G_0(\mathbb{C}G) =$ Abelian group of characters $\chi : G \to \mathbb{C}$.

We shall discuss relative generalizations of this in chapter 10.

(vii) Let X be a compact topological space, $F = \mathbb{R}$ or $\mathbb{C}$. Then the category $VB_F(X)$ of vector bundles of X is an exact category and we can write $K_n(VB_F(X))$ as $K_n^F(X)$.

(viii) Let X be an H-space; m, n positive integers; M_m^n an n-dimensional mod$-m$ Moore space, that is, the space obtained form S^{n-1} by attaching an $n-$cell via a map of degree m (see [30, 158]). Write $\pi_n(X, \mathbb{Z}/m)$ for $[M_m^n, X]$, the set of homotopy classes of maps from M_m^n to X. If $X = BQ\mathcal{C}$ where $\mathcal{C}$ is an exact category, write $K_n(\mathcal{C}, \mathbb{Z}/m)$ for $\pi_{n+1}(BQ\mathcal{C}, \mathbb{Z}/m)$, $n \geq 1$, and call this group the mod-m higher K-theory of $\mathcal{C}$. This theory is well-defined for $\mathcal{C} = \mathcal{P}(A)$ where A is any ring with identity and we write $K_n(A, \mathbb{Z}/m)$ for $K_n(\mathcal{P}(A), \mathbb{Z}/m)$. If X is a scheme, write $K_n(X, \mathbb{Z}/m)$ for $K_n(\mathcal{M}(A), \mathbb{Z}/m)$, while for a Noetherian scheme X we shall write $G_n(X, \mathbb{Z}/m)$ for $K_n(\mathcal{M}(X), \mathbb{Z}/m)$. For the applications, it is usual to consider $m = l^s$ where l is a prime and s a positive integer (see chapter 8).

(ix) Let G be a discrete Abelian group, $M^n(G)$ the space with only one non-zero reduced integral cohomology group $\tilde{H}^n(M^n(G))$. Suppose that $\tilde{H}^n(M^n(G)) = G$. If we write $\pi_n(X, G)$ for $[M^n(G), X]$, and we put $G = \mathbb{Z}/m$, we recover (viii) above since $M_m^n = M^n(\mathbb{Z}/m)$. If $G = \mathbb{Z}$, then $M^n(\mathbb{Z}) = S^n$, and so, $\pi_n(X, \mathbb{Z}) = [S^n, X] = \pi_n(X)$.

(x) With notations as in (ix), let $M_{l\infty}^{n+1} = \lim_{\to s} M_{ls}^{n+1}$. For all $n \geq 0$, we shall denote $[M_{l\infty}^{n+1}, BC]$ ($\mathcal{C}$ an exact category) by $K_n^{pr}(\mathcal{C}, \hat{\mathbb{Z}}_l)$ and call this group the profinite (higher) K-theory of $\mathcal{C}$. By way of notation, we shall write $K_n^{pr}(A, \hat{\mathbb{Z}}_l)$ if $\mathcal{P} = \mathcal{M}(A)$, A any ring with identity; $G_n^{pr}(A, \tilde{\mathbb{Z}}_l)$ if $\mathcal{C} = \mathcal{M}(A)$, A any Noetherian ring; $K_n^{pr}(X, \tilde{\mathbb{Z}}_l)$ if $\mathcal{C} = \mathcal{P}(X)$, X any scheme; and $G_n^{pr}(X, \tilde{\mathbb{Z}}_l)$ if $\mathcal{C} = \mathcal{M}(X)$, X a Noetherian scheme. For a comprehensive study of these constructions and applications especially to orders and groupings, see chapter 8.

$(5.2)^C$ K-groups as homotopy groups of spectra

5.2.1 The importance of spectra for this book has to do with the fact that higher K-groups are often expressed as homotopy groups of spectra $\underline{E} = \{E_i\}$ whose spaces $E_i \approx \Omega^k E_{i+k}$ (for k large) are infinite loop spaces. (It is usual to take $i = 0$) and consider E_0 as an infinite loop space.) Also, to each spectrum can be associated a generalized cohomology theory and vice-versa. Hence, Algebraic K-theory can always be endowed with the structure of a generalized cohomology theory. We shall come across these notions copiously later.

Definition 5.2.7 *A spectrum $\underline{E} = \{E_i\}$ for $i \in \mathbb{Z}$ is a sequence of based space E_n and based homeomorphisms $E_i \approx \Omega E_{i+1}(I)$. If we regard $E_i = 0$ for negative i, call $\underline{E}$ a connective spectrum.*

A map $f : \underline{E} = \{E_i\} \to \{F_i\} = \underline{F}$ of spectra is a sequence of based continuous maps strictly compatible with the given homeomorphism (I). The spectra form a category, which we shall denote by $\mathcal{S}$pectra.

From the adjunction isomorphism $[\Sigma X, Y] = [X, \Omega Y]$ for spaces X, Y, we have $\pi_n(\Omega E_i) \cong \pi_{n+1}(E_1)$, and so, we can define the homotopy group of a connective spectrum $\underline{E}$ as $\pi_n(\underline{E}) = \pi_n(E_0) = \pi_{n+1}(E_1) = ... = \pi_{n+i}(E_i) = \cdots = \lim_i \pi_{n+i}(E_i)$.

5.2.2 Each spectrum $\underline{E} = \{E_n\}$ gives rise to an extraordinary cohomology theory E^n in such a way that if X_+ is a space obtained from X by adjoining a base point, $E^n(X) = [X_+, E_n]$ and conversely. This cohomology theory is also denoted by $H^n(X, E)$. One can also associate to $\underline{E}$ a homology theory defined by $E_n(X) = \lim_{k \to +\infty} \pi_{n+k}(E_k \wedge X_+)$, which is also denoted $H_n(X, \underline{E})$.

Let $X_+ = (X, x)$, $Y_+ = (Y, y)$ be two pointed spaces. Recall that the smash product is defined by

$$X \wedge Y = X \times Y / (\{x\} \times Y \cup X \times \{y\}).$$

Now, if $\underline{E}$ is a spectrum and X_+ a pointed space, we define a smash product spectrum $X \wedge \underline{E}$ by $(X \wedge \underline{E})_n = X \wedge E_n$.

Recall also that for a pointed space X, the reduced cone

$$cone(X) := X \times [0,1]/(X \times \{1\}) \cup \{x\} \times [0,1].$$

If A is a subspace of X, then the spectrum $\underline{E}$ defines a homology theory

$$H_n(X, A; E) := \pi_n(X_+ \bigcup_{A^+} cone(A_+) \bigwedge \underline{E}).$$

We shall use these ideas when we discuss equivariant homology theories in Chapter 14.

Examples 5.2.3   (i) *Eilenberg - Maclane Spectrum*

Let $E_s = K(A, s)$ where each $K(A, s)$ is an Eilenberg - Maclane space (where A is an Abelian group and $\pi_n(K(A, s)) = \delta_{ns}(A)$. By adjunction isomorphism, we have $K(A, n) \approx \Omega K(A, n+1)$, and get the Eilenberg - Maclane spectrum whose associated cohomology theory is ordinary cohomology with coefficients in A, otherwise defined by means of singular chain complexes.

(ii) *The suspension spectrum*

Let X be a based space. The nth space of the suspension spectrum $\Sigma^\infty X$ is $\Omega^\infty \Sigma^\infty (\sum^n X)$ and the homotopy groups $\pi_n(\Sigma^\infty X) = \lim_{k \to \infty} \pi_{n+k}(\Sigma^k X)$. When $X = S^0$, we obtain the sphere spectrum $\Sigma^\infty(S^0)$ and $\pi_n(\Sigma^\infty(S^0)) = \lim_{k \to \infty} \pi_{n+k}(S^k)$, called the stable n-stem and denoted by π_n^S.

Note that there is an adjoint pair $(\Sigma^\infty, \Omega^\infty)$ of functors between spaces and spectra and we can write $\Sigma^\infty X = \{X, \Sigma X, \Sigma^2 X, ...\}$. Also, if $\underline{E}$ is an Ω-spectrum, $\Omega^\infty \underline{E}$ is an infinite loop space (indeed, every infinite loop space is the initial space of an Ω-spectrum), and $\pi_n(\underline{E}) = [\Sigma^\infty S^n, \underline{E}] = \pi_n(\Omega^\infty \underline{E})$.

5.2.2 Higher K-groups as homotopy groups of spectra

(i) We now have another way of defining $K_n(\mathcal{C})$ when $\mathcal{C}$ is an exact category. We could also obtain $K_n(\mathcal{C})$ via spectra. For example, we could take the Ω-spectrum (see 1.2) $\underline{BQC} = \{\Omega BQ\mathcal{C}, BQ\mathcal{C}, BQ^2\mathcal{C}, ...\}$ where $Q^i\mathcal{C}$ is the multi-category defined in [225] and $\pi_n(\underline{BQC}) = K_n(\mathcal{C})$.

(ii) Let R be any ring with identity. The suspension ΣR of R is defined as $\Sigma R = \Sigma \mathbb{Z} \otimes_{\mathbb{Z}} R$, where $\Sigma \mathbb{Z} = \mathcal{C}\mathbb{Z}/J\mathbb{Z}$; where $\mathcal{C}\mathbb{Z}$, the cone of $\mathbb{Z}$, is the set of infinite matrices with integral coefficients having only a finite number of nontrivial elements in each row and column; $J\mathbb{Z}$ is the ideal of $\mathcal{C}\mathbb{Z}$ consisting of all entries having only finitely many nontrivial coefficients.

Note that $K_n(R) \cong K_{n+1}(\Sigma R)$ for all $n \geq 0/$.

The K-theory spectrum of R is the Ω-spectrum $\underline{K}_R$ whose n^{th} space is $(\underline{K}_R)_n = K_o(\Sigma^n R) \times BGL(\Sigma^n R)^+$ for all $n \geq 0$ and $K_i(R) \simeq\simeq \pi_i(\underline{K}_R)$ for all $i \geq 0$.

We shall sometimes write $\mathbb{K}(R)$ or $\underline{K}(R)$ for $\underline{K}_R$.

(iii) For any CW-complex X, let $X(n)$ be the nth connected cover of X, that is, the fiber of the nth Postnikov section $X \to X[n]$ of X (see [7]). So, $X(n)$ is n-connected and $\pi_i(X(n)) \simeq \pi_i(X)$ for all $i \geq n$.

Now, for any ring R, the zero-connected K-theory spectrum of R is the Ω-spectrum $\underline{X}_R$ whose n^{th} space is $(\underline{X}_R)_n = BGL(\sum^n R)^+(n)$ for all $n \geq 0$. Then, for all integers $i \geq 1$, $K_i(R) \simeq \pi_i(\underline{X}_R)$.

5.3   Higher K-theory of symmetric monoidal categories – definitions and examples

5.3.1 A symmetric monoidal category is a category $\mathcal{S}$ equipped with a functor $\perp: \mathcal{S} \times \mathcal{S} \to \mathcal{S}$ and a distinguished object 0 such that $\perp$ is coherently associative and commutative in the sense of Maclane (that is, satisfying properties and diagrams in definition 1.3.1). Note that $B\mathcal{S}$ is an H-space (see [62]).

Examples 5.3.1   (i) Let $(Iso\,\mathcal{S})$ denote the subcategory of isomorphisms in $\mathcal{S}$, that is, $ob(Iso\,\mathcal{S}) = ob\mathcal{S}$; morphisms are isomorphisms in $\mathcal{S}$. $\pi_0(Iso\,\mathcal{S})=$ set of isomorphism classes of objects of $\mathcal{S}$. Then $\mathcal{S}^{iso} := \pi_0(Iso\,\mathcal{S})$ is monoid.

$Iso(\mathcal{S})$ is equivalent to the disjoint union $\coprod Aut_\mathcal{S}(\mathcal{S})$, and $B(Iso\,\mathcal{S})$ is homotopy equivalent to $\coprod B(Aut_\mathcal{S}(S))$, $S \in \mathcal{S}^{iso}$.

(ii) If $\mathcal{S} = FSet$ in (1), $Aut_{FSet}(S) \simeq \sum_n$ (symmetric group of degree n). $Iso(FSet)$ is equivalent to the disjoint union $\coprod \Sigma_n$. $B(Iso(FSet))$ is homotopy equivalent to $\coprod B\sum_n$.

(iii) $B(Iso\,\mathcal{P}(R))$ is equivalent to disjoint union $\coprod B\,Aut(P)$ $P \in \mathcal{P}(R)$.

(iv) Let $\mathcal{F}(R) =$ category of finitely generated free R-modules $(Iso\,\mathcal{F}(R)) = \coprod GL_n(R)$, and $B(Iso\,(\underline{F}(R)))$ is equivalent to disjoint union $\coprod BGL_n(R)$. If R satisfies the invariant basis property, then $Iso(\mathcal{F}(R))$ is a full subcategory of $Iso(\mathcal{P}(R))$, and $Iso(\mathcal{F}(R))$ is cofinal in $Iso(\mathcal{P}(R))$.

5.3.2 Suppose that every map in $\mathcal{S}$ is an isomorphism and every translation $S \perp: Aut_\mathcal{S}(T) \to Aut_\mathcal{S}(S \perp T)$ is an injection. We now define a category $\mathcal{S}^{-1}\mathcal{S}$ such that $K(\mathcal{S}) = B(\mathcal{S}^{-1}\mathcal{S})$ is a 'group completion' of $B\mathcal{S}$.

Recall that a group completion of a homotopy commutative and homotopy associative H-space X is an H-space Y together with an H-space map $X \to Y$ such that $\pi_0(Y)$ is the group completion of (that is, the Grothendieck group associated to) the monoid $\pi_0(X)$ (see 1.2.1), and the homology ring $H_*(Y, R)$ is isomorphic to the localization $\pi_0(X)^{-1}H_*(X, R)$ of $H_*(X, R)$.

Definition 5.3.1 *Define $\mathcal{S}^{-1}\mathcal{S}$ as follows:*

$$ob(\mathcal{S}^{-1}\mathcal{S}) = \{(S, T) | S, T \in ob\mathcal{S}\}.$$

$$mor_{\mathcal{S}^{-1}\mathcal{S}}((S_1, T_1), (S_1', T_1^{1'})) = \begin{cases} \text{equivalence classes of composites} \\ (S_1, T_1) \xrightarrow{S_\perp} (S \perp S_1, S \perp T_1) \xrightarrow{(f,g)} ((S_1', T_1')). \end{cases}$$

(i) *The composite*

$$(S_1, T_1) \xrightarrow{S_\perp} (S \perp S_1, S \perp T_1) \xrightarrow{(f,g)} ((S_1', T_1'))$$

is said to be equivalent to

$$(S_1, T_1) \xrightarrow{T_\perp} (T \perp S_1, T \perp T_1) \xrightarrow{(f',g')} ((S_1', T_1'))$$

if there exists is an isomorphism $\alpha : S \cong T$ in $\mathcal{S}$ such that composition with $\alpha \perp S_1$, $\alpha \perp T_1$ send f' and g' to f.

(ii) *Since we have assumed that every translation is an injection in 5.3.2, it means that $\mathcal{S}^{-1}\mathcal{S}$ determines its objects up to unique isomorphism.*

(iii) *$\mathcal{S}^{-1}\mathcal{S}$ is a symmetric monoidal category with $(S, T) \perp (S', T') = (S \perp S', T \perp T')$, and the functor $\mathcal{S} \to \mathcal{S}^{-1}\mathcal{S} : S \to (0, S)$ is monoidal. Hence, $B(\mathcal{S}^{-1}\mathcal{S})$ is an H-space (see [62]).*

(iv) *$B\mathcal{S} \to B(\mathcal{S}^{-1}\mathcal{S})$ is an H-space map and $\pi_0(\mathcal{S}) \to \pi_0(\mathcal{S}^{-1}\mathcal{S})$ is a map of Abelian monoids.*

(v) *$\pi_0(\mathcal{S}^{-1}\mathcal{S})$ is an Abelian group.*

Examples 5.3.2 (i) If $\mathcal{S} = \coprod GL_n(R) = Iso\mathcal{F}(R)$, then $B(\mathcal{S}^{-1}\mathcal{S})$ is a group completion of $B\mathcal{S}$ and $B(\mathcal{S}^{-1}\mathcal{S})$ is homotopy equivalent to $\mathbb{Z} \times BGL(R)^+$. See [62] for a proof. See theorem 5.3.1 for a more general formulation of this.

(ii) For $\mathcal{S} = IsoFSet$, $B\mathcal{S}^{-1}\mathcal{S}$ is homotopy equivalent to $\mathbb{Z} \times B\sum^+$ where Σ is the infinite symmetric group (see [62]).

Definition 5.3.2 *Let $\mathcal{S}$ be a symmetric monoidal category in which every morphism is an isomorphism. Define $K_n^\perp(\mathcal{S}) := \pi_n(B(\mathcal{S}^{-1}\mathcal{S}))$. Note that $K_0^\perp$ as defined above coincides with $K_0^\perp(\mathcal{S})$ as defined in 1.3.1. This is because $K_0^\perp(\mathcal{S}) = \pi_0(B(\mathcal{S}^{-1}\mathcal{S}))$ is the group completion of the Abelian monoid $\pi_0(\mathcal{S}) = \mathcal{S}^{iso}$. For a proof, see [62].*

Remarks 5.3.1 Suppose that S is a symmetric monoidal category, which has a countable sequence of objects $S_1, S_2, ...,$ such that $S_{n+1} = S_n \perp T_n$ for some $T_n \in S$ and satisfying the cofinality condition; that is, for every $S \in S$, there exists an S' and an n such that $S \perp S' \cong S_n$. If this situation arises, then we can form $Aut(S) = co\lim_{n\to\infty} Aut_S(S_n)$.

Theorem 5.3.1 [62] *Suppose that $S = Iso(S)$ is a symmetric monoidal category whose translations are injections, and that the conditions of Remarks 5.3.1 are satisfied so that the group $Aut(S)$ exists. Then the commutator subgroup E of $Aut(S)$ is a perfect normal subgroup; $K_1(S) = Aut(S)/E$, and $B Aut(S)^+$ is the connected component of the identity in the group completion of $B(S^{-1}S)$. Hence, $B(S^{-1}S) \cong K_0(S) \times BAut(S)^+$.*

Example 5.3.1 Let R be a commutative ring with identity. We saw in 1.3.4(viii) that $(S = \mathcal{P}ic(R), \otimes)$ is a symmetric monoidal category. Since $\pi_0(S)$ is a group, S and $S^{-1}S$ are homotopy equivalent (see [62]). Hence, we get $K_0\mathcal{P}ic(R) = \mathcal{P}ic(R)$, $K_1(\mathcal{P}ic(R)) = U(R)$, units of R, and $K_n(\mathcal{P}ic(R)) = 0$ for all $n \geq 2$ (see [240]).

5.4 Higher K-theory of Waldhausen categories – definitions and examples

Definition 5.4.1 *A category with cofibrations is a category C with zero object together with a subcategory $co(C)$ whose morphisms are called cofibrations written $A \rightarrowtail B$ and satisfying axioms*

(C1) Every isomorphism in C is a cofibration.

(C2) If $A \rightarrowtail B$ is a cofibration, and $A \to C$ a C-map, then the pushout $B \cup_A C$ exists in C.

$$A \rightarrowtail B$$
$$\downarrow \quad\quad \downarrow$$
$$C \rightarrowtail B \cup_A C$$

- *Hence, coproducts exist in C, and each cofibration $A \rightarrowtail B$ has a cokernel $C = B/A$.*
- *Call $A \rightarrowtail B \twoheadrightarrow B/A$ a cofibration sequence.*

(C3) For any object A, the unique map $0 \to A$ is a cofibration.

Definition 5.4.2 *A Waldhausen category (or W-category for short) $\mathcal{C}$ is a category with cofibrations together with a subcategory $w(\mathcal{C})$ of weak equivalences (w.e. for short) containing all isomorphisms and satisfying the Gluing axiom for weak equivalences: For any commutative diagram*

$$
\begin{array}{ccc}
C & \leftarrow A \rightarrowtail B \\
\downarrow \sim & \downarrow \sim & \downarrow \sim \\
C' & \leftarrow A' \rightarrowtail B'
\end{array}
$$

in which the vertical maps are weak equivalences and the two right horizontal maps are cofibrations, the induced map $B \cup_A C \to B' \cup_{A'} C'$ is also a weak equivalence.

We shall sometimes denote $\mathcal{C}$ by $(\mathcal{C}, w)$.

Definition 5.4.3 *A Waldhausen subcategory $\mathcal{A}$ of a W-category $\mathcal{C}$ is a subcategory which is also W-category such that*

(a) *The inclusion $\mathcal{A} \subseteq \mathcal{C}$ is an exact functor.*

(b) *The cofibrations in $\mathcal{A}$ are the maps in $\mathcal{A}$ which are cofibrations in $\mathcal{C}$ and whose kernel lies in $\mathcal{A}$.*

(c) *The weak equivalences in $\mathcal{A}$ are the weak equivalences of $\mathcal{C}$ that lie in $\mathcal{A}$.*

Definition 5.4.4 *A W-category $\mathcal{C}$ is said to be saturated if whenever (f, g) are composable maps and fg is a w.e. Then f is a w.e. if and only if g is*

- The cofibrations sequences in a W-category $\mathcal{C}$ form a category $\mathcal{E}$. Note that $ob(\mathcal{E})$ consists of cofibrations sequences $E : A \rightarrowtail B \twoheadrightarrow C$ in $\mathcal{C}$. A morphism $E \to E' : A' \rightarrowtail B' \twoheadrightarrow C'$ in $\mathcal{E}$ is a commutative diagram (I)

$$
\begin{array}{ccc}
A & \rightarrowtail B \twoheadrightarrow C \\
\downarrow & \downarrow & \downarrow \\
A' & \rightarrowtail B' \twoheadrightarrow C'
\end{array}
$$

To make $\mathcal{E}$ a W-category, we define a morphism $E \to E'$ in $\mathcal{E}$ to be a cofibration if $A \to A'$, $C \to C'$, and $A' \cup_A B \to B'$ are cofibrations in $\mathcal{C}$, while $E \to E'$ is a w.e. if its component maps $A \to A'$, $B \to B'$, and $C \to C'$ are w.e. in $\mathcal{C}$.

5.4.1 Extension axiom A W-category $\mathcal{C}$ is said to satisfy extension axiom if for any morphism $f : E \to E'$ as in 5.4.4., maps $A \to A'$, $C \to C'$, being w.e. in $\mathcal{C}$, implies that $B \to B'$ is also a w.e.

Examples 5.4.1 (i) Any exact category $\mathcal{C}$ is a W-category where cofibrations are the admissible monomorphisms and w.e. are isomorphisms.

(ii) If $\mathcal{C}$ is any exact category, then the category $Ch_b(\mathcal{C})$ of bounded chain complexes in $\mathcal{C}$ is a W-category where $w.e.$ are quasi-isomorphisms (that is, isomorphisms on homology), and a chain map $\underline{A.} \to \underline{B.}$ is a cofibration if each $A_i \to B_i$ is a cofibration (admissible monomorphisms) in $\mathcal{C}$.

(iii) Let $\mathcal{C}$ be the category of finite, based CW-complexes. Then $\mathcal{C}$ is a W-category where cofibrations are cellular inclusion and $w.e.$ are homotopy equivalences.

(iv) If $\mathcal{C}$ is a W-category, define $K_0(\mathcal{C})$ as the Abelian group generated by objects of $\mathcal{C}$ with relations

(i) $A \xrightarrow{\sim} B \Rightarrow [A] = [B]$.

(ii) $A \rightarrowtail B \twoheadrightarrow C \Rightarrow [B] = [A] + [C]$.

Note that this definition agrees with the earlier $K_0(\mathcal{C})$ given in 1.4.2 for an exact category.

5.4.2 In order to define the K-theory space $K(\mathcal{C})$ such that $\pi_n(K(\mathcal{C})) = K_n(\mathcal{C})$ for a W-category C, we construct a simplicial W-category $S_*\mathcal{C}$, where $S_n\mathcal{C}$ is the category whose objects $A.$ are sequences of n cofibrations in $\mathcal{C}$, that is,

$$A. : 0 = A_0 \rightarrowtail A_1 \rightarrowtail A_2 \rightarrowtail \dots \rightarrowtail A_n$$

together with a choice of every subquotient $A_{i,j} = A_j/A_i$ in such a way that we have a commutative diagram

$$
\begin{array}{ccccccc}
 & & & & & & A_{n-1,n} \\
 & & & & & & \uparrow \\
 & & & & & & \big\uparrow \\
 & & A_{23} & \rightarrowtail & \dots & \rightarrowtail & A_{2n} \\
 & & \uparrow & & & & \uparrow \\
 & A_{12} \rightarrowtail & A_{13} & \rightarrowtail \dots & \rightarrowtail & A_{1n} \\
 & \uparrow & \uparrow & & & & \uparrow \\
A_1 \rightarrowtail & A_2 \rightarrowtail & A_3 & \rightarrowtail \dots & \rightarrowtail & A_n
\end{array}
$$

By convention, put $A_{jj} = 0$ and $A_{0j} = A_j$.

A morphism $A. \to B.$ is a natural transformation of sequences.

A weak equivalence in $S_n(\mathcal{C})$ is a map $A. \to B.$ such that each $A_i \to B_i$ (and hence each $A_{ij} \to B_{ij}$) is a $w.e.$ in $\mathcal{C}$. A map $A. \to B.$ is a cofibration if for every $0 \le i < j < k \le n$, the map of cofibration sequence is a cofibration in $\mathcal{E}(\mathcal{C})$.

$$
\begin{array}{ccccc}
A_{ij} & \rightarrowtail & A_{ik} & \twoheadrightarrow & A_{jk} \\
\downarrow & & \downarrow & & \downarrow \\
B_{ij} & \rightarrowtail & B_{ik} & \twoheadrightarrow & B_{jk}
\end{array}
$$

For $0 < i \le n$, define exact functors $\delta_i : S_n(\mathcal{C}) \to S_{n+1}(\mathcal{C})$ by omitting A_i from the notations and re-indexing the A_{jk} as needed. Define $\delta_0 : S_n\mathcal{C} \to$

$S_{n+1}(\mathcal{C})$ where δ_0 omits the bottom row. We also define $s_i : S_n(\mathcal{C}) \to S_{n+1}(\mathcal{C})$ by duplicating A_i and re-indexing (see [224]).

We now have a simplicial category $n \to wS_n\mathcal{C}$ with degreewise realization $n \to B(wS_n\mathcal{C})$, and denote the total space by $|wS.\mathcal{C}|$ (see [224]).

Definition 5.4.5 *The K-theory space of a W-category $\mathcal{C}$ is $K(\mathcal{C}) = \Omega|wS.\mathcal{C}|$. For each $n \geq 0$, the K-groups are defined as $K_n(\mathcal{C}) = \pi_n(K\mathcal{C})$.*

5.4.3 By iterating the $S.$ construction, one can show (see [224]) that the sequence

$$\{\Omega|wS.\mathcal{C}|, \Omega|wS.S.\mathcal{C}|, ..., \Omega|wS.^n\mathcal{C}|\}$$

forms a connective spectrum $\mathbb{K}(\mathcal{C})$ called the K-theory spectrum of $\mathcal{C}$. Hence, $K(\mathcal{C})$ is an infinite loop space (see 1.2.2).

Examples 5.4.2   (i) Let $\mathcal{C}$ be an exact category, $Ch_b(\mathcal{C})$ the category of bounded chain complexes over $\mathcal{C}$. It is a theorem of Gillet - Waldhausen that $K(\mathcal{C}) \cong K(Ch_b(\mathcal{C}))$, and so, $K_n(\mathcal{C}) \cong K_n(Ch_b(\mathcal{C}))$ for every $n \geq 0$ (see [216]).

(ii) **Perfect Complexes** Let R be any ring with identity and $\mathcal{M}'(R)$ the exact category of finitely presented R-modules. (Note that $\mathcal{M}'(R) = \mathcal{M}(R)$ if R is Noetherian.) An object $M.$ of $Ch_b(\mathcal{M}'(R))$ is called a perfect complex if $M.$ is quasi-isomorphic to a complex in $Ch_b(\mathcal{P}(\mathcal{R}))$. The perfect complexes form a Waldhausen subcategory $Perf(R)$ of $Ch_b(\mathcal{M}'(R))$. So, we have

$$K(R) \cong K(Ch_b(\mathcal{P}(R)) \cong K(Perf(R)).$$

(iii) **Derived Categories** Let $\mathcal{C}$ be an exact category and $H^b(\mathcal{C})$ the (bounded) homotopy category of $\mathcal{C}$, that is, stable category of $Ch_b(\mathcal{C})$ (see [98]). So, $ob(H^b(\mathcal{C})) = Ch_b(\mathcal{C})$ and morphisms are homotopy classes of bounded complexes. Let $A(\mathcal{C})$ be the full subcategory of $H^b(\mathcal{C})$ consisting of acyclic complexes (see [98]). The derived category of $D^b(\mathcal{C})$ of $\mathcal{E}$ is defined by $D^b(\mathcal{C}) = H^b(\mathcal{C})/A(\mathcal{C})$. A morphism of complexes in $Ch_b(\mathcal{C})$ is called a quasi-isomorphism if its image in $D^b(\mathcal{C})$ is an isomorphism. We could also define unbounded derived category $D(\mathcal{C})$ from unbounded complexes $Ch(\mathcal{C})$. Note that there exists a faithful embedding of $\mathcal{C}$ in an Abelian category $\mathcal{A}$ such that $\mathcal{C} \subset \mathcal{A}$ is closed under extensions and the exact functor $\mathcal{C} \to \mathcal{A}$ reflects the exact sequences. So, a complex in $Ch(\mathcal{C})$ is acyclic if and only if its image in $Ch(\mathcal{A})$ is acyclic. In particular, a morphism in $Ch(\mathcal{C})$ is a quasi-isomorphism if and only if its image in $Ch(\mathcal{A})$ is a quasi-isomorphism. Hence, the derived category $D(\mathcal{C})$ is the category obtained from $Ch(\mathcal{C})$ by formally inverting quasi-isomorphisms.

(iv) **Stable derived categories and Waldhausen categories** Now let
$\mathcal{C} = \mathcal{M}'(R)$. A complex $M.$ in $\mathcal{M}'(R)$ is said to be compact if the
functor $Hom(M., -)$ commutes with arbitrary set-valued coproducts.
Let $\underline{Comp(R)}$ denote the full subcategory of $D(\mathcal{M}'(R))$ consisting of
compact objects. Then we have

$$\underline{Comp(R)} \subset D^b(\mathcal{M}'(R)) \subset D(\mathcal{M}'(R)).$$

Define the stable derived category of bounded complexes $\underline{D}^b(\mathcal{M}'(R))$ as
the quotient category of $D^b(\mathcal{M}'(R))$ with respect to $\underline{Comp(R)}$. A morphism of complexes in $Ch_b(\mathcal{M}'(R))$ is called a stable quasi-isomorphism
if its image in $\underline{D}^b(\mathcal{M}'(R))$ is an isomorphism. The family of stable
quasi-isomorphism in $\mathcal{A} = Ch_b(\mathcal{M}'(R))$ is denoted $\omega\mathcal{A}$.

(v) **Theorem** [56]

(1) $\omega(Ch_b(\mathcal{M}'(R)))$ forms a set of weak equivalences and satisfies the
saturation and extension axioms.

(2) $Ch_b(\mathcal{M}'(R))$, together with the family of stable quasi-isomorphisms,
is a Waldhausen category.

Exercises

5.1 Let X, Y be connected CW-complexes, N, N' perfect normal subgroups
of $\pi_1(X)$, and $\pi_1(Y)$, respectively. Show that

(a) $(X \times Y)^+ \rightarrow X^+ \times Y^+$ is a homotopy equivalence.

(b) If $\widetilde{X}$ is the universal covering space of X, which corresponds to the
subgroup N, show that $\widetilde{X}^+$ is up to homotopy the universal covering
space of X^+.

5.2 Let A be a ring with identity. Show that

(a) $K_3(A) \simeq H_3(St(A))$.

(b) $K_n(R) \simeq \pi_n(BSt(R)^+)$ for all $n \geq 3$.

5.3 Show that the functors $K_n^{k-v} : \mathcal{R}ing \rightarrow \mathbb{Z}\text{-}\mathcal{M}od$ is homotopy invariant,
i.e., for any ring R with identity $K_n^{k-v}(R) \cong K_n^{k-v}(R[t])$ for all $n \geq 1$.

5.4

(a) Let $\mathcal{A}$ be a small category. Show that the classifying space $B\mathcal{A}$ of $\mathcal{A}$ is a CW-complex whose n-cells are in one-one correspondence with the diagrams $\mathcal{A}_n$ in definition 5.2.4.

(b) If $\mathcal{C}$ is a category with initial or final object, show that $B\mathcal{C}$ is contractible.

5.5 Let $\mathcal{S} = \coprod GL_n(R) = Iso\mathcal{F}(R)$ where $\mathcal{F}(R)$ is the category of finitely generated free R-modules (R a ring with identity). Show that $B(\mathcal{S}^{-1}\mathcal{S})$ is a group completion of $B\mathcal{S}$, and $B(\mathcal{S}^{-1}\mathcal{S})$ is homotopy equivalent to $\mathbb{Z} \times BGL(R)^+$.

5.6 Let R be a ring with identity, $\mathcal{M}'(R)$ the exact category of finitely p-resented R-modules, and $Ch_b(\mathcal{M}'(R))$ the category of bounded chain complexes over $\mathcal{C}$. Show that $Ch_b(\mathcal{M}'(R))$ together with the family of stable quasi-isomorphisms in $Ch_b(\mathcal{M}'(R))$ is a Waldhausen category.

Chapter 6

Some fundamental results and exact sequences in higher K-theory

---

6.1 Some fundamental theorems

$(6.1)^A$ Resolution theorem

Let $\mathcal{P} \subset \mathcal{H}$ be full exact subcategories of an Abelian category $\mathcal{A}$, both closed under extensions and inheriting their exact structure from $\mathcal{A}$. Suppose that

(i) Every object M of $\mathcal{H}$ has a finite $\mathcal{P}$-resolution.

(ii) $\mathcal{P}$ is closed under kernels in $\mathcal{H}$, that is, if $L \to M \to N$ is an exact sequence in $\mathcal{H}$ with $M, N \in \mathcal{P}$, then L is also in $\mathcal{P}$. Then $K_n\mathcal{P} \cong K_n\mathcal{H}$ for all $n \geq 0$.

(See [165] for the proof of this result.)

Remarks and Examples 6.1.1   (i) Let R be a regular Noetherian ring. Then by taking $\mathcal{H} = \mathcal{M}(R)$, $\mathcal{P} = \mathcal{P}(R)$ in $(6.1)^A$, we have $K_n(R) \cong G_n(R)$ for all $n \geq 0$.

(ii) Let R be any ring with identity and $\mathcal{H}(R)$ the category of all R-modules having finite homological dimension (that is, having finite resolution by finitely generated projective R-modules), $\mathcal{H}_s(R)$ the subcategory of modules in $\mathcal{H}(R)$ having resolutions of length less than or equal to s. Then by $(6.1)^A$, applied to $\mathcal{P}(R) \subseteq \mathcal{H}_s \subseteq \mathcal{H}(R)$, we have

$$K_n(R) \cong K_n(\mathcal{H}(R)) \cong K_n(\mathcal{H}_s(R))$$

for all $s \geq 1$.

(iii) Let $T = \{T_i\}$ be an exact connected sequence of functors from an exact category $\mathcal{C}$ to an Abelian category (that is, given an exact sequence $0 \to M' \to M \to M'' \to 0$ in $\mathcal{C}$, there exists a long exact sequence

$\cdots \to T_2 M'' \to T_1 M' \to T_1 M'' \to$). Let $\mathcal{P}$ be the full subcategory of T-acyclic objects (that is, objects M such that $T_n(M) = 0$ for all $n \geq 1$) and assume that for each $M \in \mathcal{C}$, there is a map $P \to M$ such that $P \in \mathcal{P}$ and that $T_n M = 0$ for n sufficiently large. Then $K_n \mathcal{P} \cong K_n \mathcal{C}$ for all $n \geq 0$ (see [165]).

(iv) As an example of (iii), let A, B be Noetherian rings, $f : A \to B$ a homomorphism, B a flat A-module. Then we have homomorphism of K-groups. $(B \otimes_A ?)_* : G_n(A) \to G_n(B)$ (since $(B \otimes_A ?)_*$ is exact.) Let B be of finite tor-dimension as a right A-module. Then, by applying (iii) above to $\mathcal{C} = \mathcal{M}(A)$, $T_i(M) = Tor_i^A(B, M)$, and taking $\mathcal{P}$ as the full subcategory of $\mathcal{M}(A)$ consisting of M such that $T_i M = 0$ for $i > 0$, we have $K_n(\mathcal{P}) \simeq G_n(A)$.

(v) Let $\mathcal{C}$ be an exact category and $Nil(\mathcal{C})$ the category, whose objects are pairs (M, ν) whose $M \in \mathcal{C}$ and ν is a nilpotent endomorphism of M. Let $\mathcal{C}_0 \subset \mathcal{C}$ be an exact subcategory of $\mathcal{C}$ such that every object of $\mathcal{C}$ has a finite $\mathcal{C}_0$-resolution. Then every object of $Nil(\mathcal{C})$ has a finite $Nil(\mathcal{C}_0)$ resolution, and so, by $(6.1)^A$,

$$K_n(Nil(\mathcal{C}_0)) \approx K_n(Nil(\mathcal{C})).$$

$(6.1)^B$ Additivity theorem (for exact and Waldhausen categories)

6.1.1 Let $\mathcal{A}, \mathcal{B}$ be exact categories. A sequence of functors $F' \to F \to F''$ from $\mathcal{A}$ to $\mathcal{B}$ is called an exact sequence of exact functors if

$$0 \to F'(A) \to F(A) \to F''(A) \to 0$$

is an exact sequence in $\mathcal{B}$ for every $A \in \mathcal{A}$.

Let $\mathcal{A}, \mathcal{B}$ be Waldhausen categories. If $F'(A) \rightarrowtail F(A) \twoheadrightarrow F''(A)$ is a cofibration sequence in $\mathcal{B}$, and for every cofibration $A \rightarrowtail A'$ in $\mathcal{A}$, $F(A) \cup_{F'(A)} F'(A') \to F(A')$ is a cofibration in $\mathcal{B}$, say that $F' \rightarrowtail F \twoheadrightarrow F''$ a short exact sequence or a cofibration sequence of exact functors.

Additivity theorem 6.1.1 *Let $F' \rightarrowtail F \twoheadrightarrow F''$ be a short exact sequence of exact functors from $\mathcal{A}$ to $\mathcal{B}$ where both $\mathcal{A}$ and $\mathcal{B}$ are either exact categories or Waldhausen categories. Then $F_* \simeq F'_* + F''_* : K_n(\mathcal{A}) \to K_n(\mathcal{B})$.*

Remarks and Examples 6.1.2   (i) It follows from theorem 6.1.1 that if $0 \to F_1 \to F_2 \to \cdots \to F_s \to 0$ is an exact sequence of functors $\mathcal{A} \to \mathcal{B}$, then

$$\sum_{k=0}^{s}(-1)^k F_k = 0 : K_n(\mathcal{A}) \to K_n(\mathcal{B})$$

for all $n \geq 0$ (see [165]).

(ii) Let X be a scheme, $E \in \mathcal{P}(X)$ (see example 5.2.2(iii)). Then we have an exact functor $(E \otimes ?) : \mathcal{P}(X) \to \mathcal{P}(X)$, which induces a homomorphism $K_n(X) \to K_n(X)$. Hence, we obtain the homomorphism

$$K_0(X) \otimes K_n(X) \to K_n(X) : (E) \otimes y \to (E \otimes ?)_* y, \quad y \in K_n(X)$$

making each $K_n(X)$ a $K_0(X)$-module.

(iii) **Flasque categories** An exact (or Waldhausen) category is called Flasque if there is an exact functor $\infty : \mathcal{A} \to \mathcal{A}$ and a natural isomorphism $\infty(A) \cong A \coprod \infty(A)$, that is, $\infty \cong 1 \coprod \infty$ where 1 is the identity functor. By theorem 6.1.1, $\infty_* = 1_* \coprod \infty_*$ and hence the identity map $1_* : K(\mathcal{A}) \to K(\mathcal{A})$ is null homotopic. Hence $K(\mathcal{A})$ is contractible, and so, $\pi_n(K(\mathcal{A})) = K_n(\mathcal{A}) = 0$ for all n.

$(6.1)^C$ Devissage

Devissage theorem 6.1.1 [165] *Let $\mathcal{A}$ be an Abelian category, $\mathcal{B}$ a non-empty full subcategory closed under sub-objects and finite products in $\mathcal{A}$. Suppose that every object M of $\mathcal{A}$ has a finite filtration*

$$0 = M_0 \subset M_1 \subset ... \subset M_n = M$$

such that $M_i/M_{i-1} \in \mathcal{B}$ for each i, then the inclusion $Q\mathcal{B} \to Q\mathcal{A}$ is a homotopy equivalence. Hence, $K_i(\mathcal{B}) \cong K_i(\mathcal{A})$.

Corollary 6.1.1 [165] Let $\underline{a}$ be a nilpotent two-sided ideal of a Noetherian ring R. Then for all $n \geq 0$, $G_n(R/\underline{a}) \cong G_n(R)$.

Examples 6.1.1 (i) Let R be an Artinian ring with maximal ideal $\underline{m}$ such that $\underline{m}^r = 0$ for some r. Let $k = R/\underline{m}$ (for example, $R \equiv \mathbb{Z}/p^r, k \equiv \mathbb{F}_p$). In Devissage theorem 6.1.1, put $\mathcal{B}$ = category of finite-dimensional k-vector spaces and $\mathcal{A} = \mathcal{M}(R)$. Then we have a filtration $0 = \underline{m}^r M \subset \underline{m}^{r-1} M \subset ... \underline{m} M \subset M$ for any $M \in \mathcal{M}(R)$. Hence by Devissage theorem 6.1.1, $G_n(R) \approx K_n(k)$.

(ii) Let X be a Noetherian scheme, $i : Z \subset X$ the inclusion of a closed subscheme. Then $\mathbb{Z}$ is an Abelian subcategory of $\mathcal{M}(X)$ via the direct image $i : \mathcal{M}(Z) \subset \mathcal{M}(X)$. Let $\mathcal{M}_Z(X)$ be the Abelian category of O_X-modules supported on Z, $\underline{a}$ an ideal sheaf in O_X such that $O_X/\underline{a} = O_Z$. Then every $M \in M_Z(X)$ has a finite filtration $M \supset M\underline{a} \supset M\underline{a}^2 ...$ and so, by Devissage, $K_n(\mathcal{M}_Z(X)) \approx K_n(\mathcal{M}(Z)) \approx G_n(Z)$.

6.2 Localization

$(6.2)^A$ Localization sequence plus examples

6.2.1 A full subcategory $\mathcal{B}$ of an Abelian category $\mathcal{A}$ is called a Serre subcategory if whenever

$$0 \to M' \to M \to M'' \to 0$$

is an exact sequence in $\mathcal{A}$; then $M \in \mathcal{B}$ if and only if $M', M'' \in \mathcal{B}$. Given such a $\mathcal{B}$, we can construct a quotient Abelian category $\mathcal{A}/\mathcal{B}$ as follows:

$$ob(\mathcal{A}/\mathcal{B}) = ob(\mathcal{A}).$$

Then, $\mathcal{A}/\mathcal{B}(M, N)$ is defined as follows: If $M' \subseteq M, N' \subseteq N$ are sub-objects such that $M/M' \in ob(\mathcal{B})$, $N' \in ob(\mathcal{B})$, then there exists a natural isomorphism $\mathcal{A}(M, N) \to \mathcal{A}(M', N/N')$.

As M', N' range over such pairs of objects, the group $\mathcal{A}(M', N/N')$ forms a direct system of Abelian groups and we define

$$\mathcal{A}/\mathcal{B}(M, N) = \varinjlim_{(M', N')} \mathcal{A}(M', N/N').$$

Note: Let $T : \mathcal{A} \to \mathcal{A}/\mathcal{B}$ be the quotient functor: $M \to T(M)$. Then

(i) $T : \mathcal{A} \to \mathcal{A}/\mathcal{B}$ is an additive functor.

(ii) If $\mu \in \mathcal{A}(M, N)$, then $T(\mu)$ is null if and only if $Ker(\mu) \in ob(\mathcal{B})$, and $T(\mu)$ is an epimorphism if and only if $Coker(\mu) \in ob(\mathcal{B})$.

(iii) $\mathcal{A}/\mathcal{B}$ is an additive category such that $T : \mathcal{A} \to \mathcal{A}/\mathcal{B}$ is an additive functor

Localization theorem 6.2.1 [165] *If $\mathcal{B}$ is a Serre subcategory of an Abelian category $\mathcal{A}$, then there exists a long exact sequence*

$$\cdots \to K_n(\mathcal{B}) \to K_n(\mathcal{A}) \to K_n(\mathcal{A}/\mathcal{B}) \to K_{n-1}(\mathcal{B}) \to \cdots$$

$$\cdots \to K_0(\mathcal{B}) \to K_0(\mathcal{A}) \to K_0(\mathcal{B}) \to 0. \qquad (I)$$

Examples 6.2.1   (i) Let A be a Noetherian ring, $S \subset A$ a central multiplicative system; $\mathcal{A} = \mathcal{M}(A), \mathcal{B} = \mathcal{M}_S(A)$, the category of finitely generated S-torsion A-modules; $\mathcal{A}/\mathcal{B} \simeq \mathcal{M}(A_S) = $ category of finitely generated A_S-modules.

Let T be the quotient functor $\mathcal{M}(A) \to \mathcal{M}(A)/\mathcal{M}_S(A)$;
$u : \mathcal{M}(A)/\mathcal{M}_S(A) \to \mathcal{M}(A_S)$ is an equivalence of categories such that $u.T \simeq L$, where $L : \mathcal{M}(A) \to \mathcal{M}(A_S)$. We thus have an exact sequence $K_{n+1}(\mathcal{M}(A_S)) \to K_n(\mathcal{M}_S(A) \to K_n(\mathcal{M}(A)) \to K_n(\mathcal{M}(A_S)) \to K_{n-1}(\mathcal{M}_S(A))$, that is,

$$\cdots \to K_n(\mathcal{M}_S(A)) \to G_n(A) \to G_n(A_S) \to K_{n-1}(\mathcal{M}_S(A)) \to \cdots$$

(ii) Let $A = R$ in (i) be a Dedekind domain with quotient field $F, S = R-0$. Then, one can show that

$$\mathcal{M}_S(R) = \bigcup_{\underline{m}} \mathcal{M}(R/\underline{m}^k)$$

as $\underline{m}$ runs through all maximal ideals of R. So,

$$K_n(\mathcal{M}_S(R)) \simeq \oplus_{\underline{m}} \lim_{k \to \infty} G_n(R/\underline{m}^k)$$
$$= \oplus_{\underline{m}} G_n(R/\underline{m}) = \oplus_{\underline{m}} K_n(R/\underline{m}).$$

So, using theorem 6.2.1, we have an exact sequence

$$\to K_{n+1}(F) \to \oplus_{\underline{m}} K_n(R/\underline{m}) \to K_n(R) \to K_n(F)$$
$$\to \oplus_{\underline{m}} K_{n-1}(R/\underline{m}) \cdots$$
$$\to \oplus_{\underline{m}} K_2(R/\underline{m}) \to K_2(R) \to K_2(F)$$
$$\to \oplus_{\underline{m}} K_1(R/\underline{m}) \to K_1(R) \to K_1(F)$$
$$\to \oplus_{\underline{m}} K_0(R/\underline{m}) \to K_0(R) \to K_0(F),$$

that is,

$$\cdots \to \cdots \to \oplus K_2(R/m) \to K_2(R) \to K_2(F) \to \oplus(R/m)^*$$
$$\to R^* \to F^* \to \oplus \mathbb{Z} \to \mathbb{Z} \oplus Cl(R) \to \mathbb{Z} \to 0.$$

(iii) Let R in (i) be a discrete valuation ring (for example, ring of integers in a p-adic ring) with unique maximal ideal $\underline{m} = sR$. Let $F = $ quotient field of R. Then $F = R[\frac{1}{s}]$, residue field $= R/m = k$. Hence, we obtain the following exact sequence

$$\to K_n(k) \to K_n(R) \to K_n(F) \to K_{n-1}(k) \cdots$$

$$\to K_2(k) \to K_2(R) \to K_2(F) \to K_1(k) \cdots \to K_0(F) \to 0. \qquad (II)$$

Gersten's Conjecture says that the above sequence breaks up into split short exact sequences

$$0 \to K_n(R) \xrightarrow{\alpha_n} K_n(F) \xrightarrow{\beta_n} K_{n-1}(k) \to 0.$$

For this to happen, one must have that, for all $n \geq 1$, $K_n(k) \to K_n(R)$ is a zero map, and that there exists a map $K_{n-1}(k) \xrightarrow{\eta_n} K_n(F)$ such that $K_n(F) \simeq K_n(R) \oplus K_{n-1}(k)$, that is, $\beta_n \eta_n = 1_{K_{n-1}(k)}$.

True for $n = 0$:

$$K_0(R) \simeq K_0(F) \simeq \mathbb{Z}.$$

True for $n = 1$:

$$K_1(F) \simeq F^*, \quad K_1(R) = R^*, \quad F^* = R^* \times \{s^n\}.$$

True for $n = 2$:

$$0 \to K_2(R) \xrightarrow{\alpha_2} K_2(F) \xrightarrow{\beta_2} K_1(k) \to 0.$$

Here, β_2 is the tame symbol. If the characteristic of F is equal to the characteristic of k, then Gersten's conjecture is also known to be true. When k is algebraic over F_p, then Gersten's conjecture is also true. It is not known (whether the conjecture is true) in the case when $\mathrm{Char}(F) = 0$ or $\mathrm{Char}(k) = \mathrm{p}$.

(iv) Let R be a Noetherian ring, $S = \{s^n\}$ a central multiplicative system $\mathcal{B} = \mathcal{M}_S(R), \mathcal{A} = \mathcal{M}(R)$.

$$\mathcal{A}/\mathcal{B} = \mathcal{M}(R_S).$$

Then theorem 6.2.1 gives

$$\cdots \to G_{n+1}(R_S) \to K_n(\mathcal{M}_S(R)) \to G_n(R) \to G_n(R_S) \to K_{n-1}(\mathcal{M}_S(R)).$$

Note that

$$K_n(\mathcal{M}_S(R)) = K_n(\bigcup_{n=1}^{\infty} \mathcal{M}(R/s^n R)).$$

Now, by Devissage,

$$G_n(R/s^n R) \simeq G_n(R/sR).$$

Hence,

$$K_n(\bigcup_{n=1}^{\infty} \mathcal{M}(R/s^n R)) = \lim_{n \to \infty} G_n(R/s^n R) = G_n(R/sR).$$

So, we have

$$\cdots G_{n+1}\left(R(\frac{1}{s})\right) \to G_n(R/sR) \to G_n(R) \to G_n\left(R(\frac{1}{s})\right) \to G_{n-1}(R/sR) \to .$$

(v) Let R be the ring of integers in a p-adic field F, Γ a maximal R-order in a semi-simple F-algebra Σ. If $S = R - 0$, then $F = R_S$,

$$\mathcal{B} = \mathcal{M}_S(\Gamma), \quad \mathcal{A} = \mathcal{M}(\Gamma), \quad \mathcal{A}/\mathcal{B} = \mathcal{M}(\Sigma).$$

Then sequence (6.2.1) yields an exact sequence

$$\ldots \to K_n(\Gamma) \to K_n(\Sigma) \to K_{n-1}(\mathcal{M}_S(\Gamma)) \to K_{n-1}(\Gamma) \to K_{n-1}(\Sigma).$$

One can see from (iv) that if $\underline{m} = \pi R$ is the unique maximal ideal of R, then

$$K_n(\mathcal{M}_S(\Gamma)) = \lim_{n\to\infty} G_n(\Gamma/\pi^n\Gamma) = G_n(\Gamma/\pi\Gamma) \simeq K_n(\Gamma/rad\Gamma)$$

(see theorem 7.1.1). Here, $\Sigma = \Gamma_S$ where $S = \{\pi^i\}$. We have also used above the Corollary to Devissage, which says that if $\underline{a}$ is a nilpotent ideal in a Noetherian ring R, then $G_n(R) \simeq G_n(R/\underline{a})$.

(vi) Let R be the ring of integers in an algebraic number field F, Λ any R-order in a semi-simple F-algebra Σ. Let $S = R = 0$. Then we have the following exact sequence

$$\ldots \to K_n(\mathcal{M}_S(\Lambda)) \to G_n(\Lambda) \to G_n(\Sigma) \to K_{n-1}(\mathcal{M}_S(\Lambda)) \to \ldots$$

One can show that $K_n(\mathcal{M}_S(\Lambda)) \simeq \oplus G_n(\Lambda/p\Lambda)$ where p runs through all the prime ideals of R. For further details about how to use this sequence to obtain finite generation of $G_n(\Lambda)$, and the fact that $SG_n(\Lambda)$ is finite (see 7.1.13 on page 137).

(vii) Let X be a Noetherian scheme, U an open subscheme of X, $Z = X - U$, the closed complement of U in X. Put $\mathcal{A} = \mathcal{M}(X) =$ category of coherent O_X-modules, $\mathcal{B}$, the category of coherent O_X-modules whose restriction to U is zero (that is, category of coherent modules supported on Z). $\mathcal{A}/\mathcal{B}$ is the category of coherent O_U-modules. Then we have the following exact sequence

$$\ldots G_n(Z) \to G_n(X) \to G_n(U) \to G_{n-1}(Z) \ldots \to G_0(Z) \to G_0(X) \to G_0(U) \to 0.$$

So far, our localization results have involved mainly the G_n-theory, which translates into K_n-theory when the rings involved are regular. We now obtain localization for the K_n-theory.

Theorem 6.2.1 *Let S be a central multiplicative system for a ring R, $\mathcal{H}_S(R)$ the category of S-torsion finitely generated R-modules of finite projective dimension. If S consists of nonzero divisors, then there exits an exact sequence*

$$\ldots \to K_{n+1}(R_S) \to K_n(\mathcal{H}_S(R)) \xrightarrow{\eta} K_n(R) \xrightarrow{\alpha} K_n(R_S) \to \ldots$$

For the proof, see [60].

Remarks 6.2.1 It is still an open problem to understand $K_n(\mathcal{H}_S(R))$ for various rings R.

If R is regular (for example, $R = \mathbb{Z}$, ring of integers in a number field, Dedekind domains, maximal orders), the $\mathcal{M}_S(R) = \mathcal{H}_S(R)$, and

$$K_n(\mathcal{H}_S(R)) = K_n(\mathcal{M}_S(R)); \quad G_n(R) = K_n(R).$$

So, we recover the G-theory. If R is not regular, then $K_n(\mathcal{H}_S(R))$ is not known in general.

Definition 6.2.1 *Let $\alpha : A \to B$ be a homomorphism of rings A, B. Suppose that s is a central non-zero divisor in B. Call α an analytic isomorphism along s if $A/sA \simeq B/\alpha(s)B$.*

Theorem 6.2.2 *If $\alpha : A \to B$ is an analytic isomorphism along $s \in S = \{s^i\}$ where s is a central non-zero divisor, then $H_S(A) = H_S(B)$.*

$\mathcal{P}F$ follows by comparing localization sequences for $A \to A[\frac{1}{s}]$ and $B \to B[\frac{1}{s}]$ (see [240]).

$(6.2)^B$ Fundamental theorem for higher K-theory

6.2.2 Let $\mathcal{C}$ be an exact category, $\text{Nil}(\mathcal{C})$ the category of nilpotent endomorphism in $\mathcal{C}$, i.e., $\text{Nil}(\mathcal{C}) = \{(M,\nu)|M \in \mathcal{C}, \nu$ being a nilpotent endomorphism of $M\}$. Then we have two functors: $Z : \mathcal{C} \to \text{Nil}(\mathcal{C})Z(M) = (M,0)$ (where '0' = zero endomorphism) and $F : \text{Nil}(\mathcal{C}) \to \mathcal{C} : F(M,\nu) = M$ satisfying $FZ = 1_\mathcal{C}$. Hence we have a split exact sequence $0 \to K_n(\mathcal{C}) \xrightarrow{Z} K_n(\text{Nil}(\mathcal{C})) \to \text{Nil}_n(\mathcal{C}) \to 0$, which defines $\text{Nil}_n(\mathcal{C})$ as the cokernel of Z.

Hence, $K_n(\text{Nil}(\mathcal{C})) \simeq K_n(\mathcal{C}) \oplus \text{Nil}_n(\mathcal{C})$.

6.2.3 Let R be a ring with identity, $\mathcal{H}(R)$ the category of R-modules of finite homological dimension, $\mathcal{H}_S(R)$ the category of S-torsion objects of $\mathcal{H}(R)$. $\mathcal{M}_S(R)$ is the category of finitely generated S-torsion R-modules. One can show (see [88, 165]) that if $S = T_+ = \{t^i\}$, a free Abelian monoid on one generator t, then there exist isomorphisms $\mathcal{M}_{T_+}(R[t]) \simeq \text{Nil}(\mathcal{M}(R)), \mathcal{H}_{T_+}(R[t]) \simeq \text{Nil}(\mathcal{H}(R))$ and $K_n(\mathcal{H}_{T_+}(R[t]) \simeq K_n(R) \oplus \text{Nil}_n(R)$ where we write $\text{Nil}_n(R)$ for $\text{Nil}_n(\mathcal{P}(R))$.

Moreover, the localization sequence (theorem 6.2.1) breaks up into short exact sequences

$$0 \to K_n(R[t]) \to K_n(R[t,t^{-1}]) \xrightarrow{\partial} K_{n-1}(\text{Nil}(R)) \to 0$$

(see [88] for a proof).

Theorem 6.2.3 Fundamental theorem of higher K-theory *Let R be a ring with identity. Define for all*

$n \geq 0, NK_n(R) := Ker(K_n(R[t]) \xrightarrow{\tilde{i}_+} K_0(R))$ *where $\tilde{i}_+$ is induced by the augmentation $t = 1$.*

Then there are canonical decompositions for all $n \geq 0$

(i) $K_n(R[t]) \simeq K_n(R) \oplus NK_n(R)$.

(ii) $K_n(R[t, t^{-1}]) \cong K_n(R) \oplus NK_n(R) \oplus NK_n(R) \oplus K_{n-1}(R)$.

(iii) $K_n(Nil(R)) \cong K_n(R) \oplus NK_{n+1}(R)$.

The above decompositions are compatible with a split exact sequence

$$0 \to K_n(R) \to K_n(R[t]) \oplus K_n(R[t^{-1}]) \to K_n(R[t, t^{-1}]) \to K_{n-1}(R) \to 0.$$

We close this subsection with fundamental theorem for G-theory.

Theorem 6.2.4 *Let R be a Noetherian ring. Then*

(i) $G_n(R[t]) \simeq G_n(R)$.

(ii) $G_n(R[t, t^{-1}]) \simeq G_n(R) \oplus G_{n-1}(R)$.

(See [88, 165] for proof of the above results.)

6.3 Some exact sequences in the K-theory of Waldhausen categories

6.3.1 Cylinder functors

A Waldhausen category has a cylinder functor if there exists a functor $T : Ar\mathcal{A} \to \mathcal{A}$ together with three natural transformations p, j_1, j_2 such that, to each morphism $f : A \to B$, T assigns an object Tf of $\mathcal{A}$, and $j_1 : A \to Tf$, $j_2 : B \to Tf$, $p : Tf \to B$, satisfying certain properties (see [56, 224]).

Cylinder Axiom. For all f, $p : Tf \to B$ is in $w(\mathcal{A})$.

6.3.1 Let $\mathcal{A}$ be a Waldhausen category. Suppose that $\mathcal{A}$ has two classes of weak equivalences $\nu(\mathcal{A}), \omega(\mathcal{A})$ such that $\nu(\mathcal{A}) \subset \omega(\mathcal{A})$. Assume that $\omega(\mathcal{A})$ satisfies the saturation and extension axioms and has a cylinder functor T that satisfies the cylinder axiom. Let $\mathcal{A}^\omega$ be the full subcategory of $\mathcal{A}$ whose objects are those $A \in \mathcal{A}$ such that $0 \to A$ is in $\omega(\mathcal{A})$. Then $\mathcal{A}^\omega$ becomes a Waldhausen category with $co(\mathcal{A}^\omega) = co(\mathcal{A}) \cap \mathcal{A}^\omega$ and $\nu(\mathcal{A}^\omega) = \nu(\mathcal{A}) \cap \mathcal{A}^\omega$.

Theorem 6.3.1 Waldhausen fibration sequence [224]. *With the notations and hypothesis of 6.4.1, suppose that $\mathcal{A}$ has a cylinder functor T that is a cylinder functor for both $\nu(\mathcal{A})$ and $\omega(\mathcal{A})$. Then the exact inclusion functors $(\mathcal{A}^\omega, \nu) \to (\mathcal{A}, \omega)$ induce a homotopy fiber sequence of spectra*

$$K(\mathcal{A}^\omega, \nu) \to K(\mathcal{A}, \nu) \to K(\mathcal{A}, \omega)$$

and hence a long exact sequence

$$K_{n+1}(\mathcal{A}^\omega) \to K_n(\mathcal{A}) \to K_n(\mathcal{A}, \nu) \to K_n(\mathcal{A}, \omega) \to$$

The next result is a long exact sequence realizing the cofiber of the Cartan map as K-theory of a Waldhausen category (see [56]).

Theorem 6.3.2 [56] *Let R be a commutative ring with identity. The natural map $K(\mathcal{P}(R)) \to K(\mathcal{M}'(R))$ induced by $\mathcal{P}(R) \hookrightarrow \mathcal{M}'(R)$ fits into a cofiber sequence of spectra $K(R) \to K(\mathcal{M}'(R)) \to K(\mathcal{A}, \omega)$ where $(\mathcal{A}, \omega)$ is the Waldhausen category of bounded chain complexed over $M'(R)$ with weak equivalences being quasi-isomorphisms. In particular, we have a long exact sequence*

$$\cdots \to K_{n*1}(\mathcal{A}, \omega) \to K_n(R) \to G'_n(R) \to K_{n-1}(\mathcal{A}, \omega) \to \cdots$$

where

$$G'_n(R) = K_n(\mathcal{M}'(R))$$

(see chapter 13 for applications to orders).

We close this subsection with a generalization of the localization sequence (theorem 6.2.1). In theorem 6.4.3 below, the requirement that S contains no zero divisors is removed.

Theorem 6.3.3 [216] *Let S be a central multiplicatively closed subset of a ring R with identity, Perf(R, S) the Waldhausen subcategory of Perf(R) consisting of perfect complexes M such that $S^{-1}M$ is an exact complex. The $K(\text{Perf}(R, S)) \to K(R) \to K(S^{-1}R)$ is a homotopy fibration. Hence there is a long exact sequence*

$$\cdots K_{n+1}(S^{-1}R) \xrightarrow{\delta} K_n(\text{Perf}(R, S)) \to K_n(R) \to K_n(S^{-1}R) \to \cdots$$

6.4   Exact sequence associated to an ideal; excision; and Mayer - Vietoris sequences

6.4.1 Let Λ be a ring with identity, $\underline{a}$ a 2-sided ideal of Λ. Define $F_{\Lambda,\underline{a}}$ as the homotopy fiber of $BGL(\Lambda)^+ \to B\overline{GL}(\Lambda/\underline{a})^+$ where $\overline{GL}(\Lambda/\underline{a} =$

image $(GL(\Lambda) \to GL(\Lambda/\underline{a}))$. Then $F_{\Lambda,\underline{a}}$ depends not only on $\underline{a}$ but also on Λ.

If we denote $\pi_n(F_{\Lambda,\underline{a}})$ by $K_n(\Lambda, \underline{a})$, then we have a long exact sequence

$$\to K_n(\Lambda, \underline{a}) \to K_n(\Lambda) \to K_n(\Lambda/\underline{a}) \to K_{n-1}(\Lambda, \underline{a}) \to \qquad (I)$$

from the fibration $F_{\Lambda,\underline{a}} \to BGL(\Lambda)^+ \to \overline{BGL}(\Lambda/\underline{a})^+$.

Definition 6.4.1 *Let B be any ring without unit and $\widetilde{B}$ the ring with unit obtained by formally adjoining a unit to B, i.e., $\widetilde{B} =$ set of all $(b, s) \in B \times \mathbb{Z}$ with multiplication defined by $(b, s)(b', s') = (bb' + sb' + s'b, ss')$.*

Define $K_n(B)$ as $K_n(\widetilde{B}, B)$. If Λ is an arbitrary ring with identity containing B as a two-sided ideal, then B is said to satisfy excision for K_n if the canonical map $K_n(B) := K_n(\widetilde{B}, B) \to K_n(\Lambda, B)$ is an isomorphism for any ring Λ containing B. Hence, if in 6.5.1 $\underline{a}$ satisfies excision, then we can replace $K_n(\Lambda, \underline{a})$ by $K_n(\underline{a})$ in the long exact sequence (I). We denote $F_{\tilde{\underline{a}},\underline{a}}$ by $F_{\underline{a}}$.

6.4.2 We now present another way to understand $F_{\underline{a}}$ (see [33]). Let $\Gamma_n(\underline{a}) := \text{Ker}(GL_n(\underline{a} \oplus \mathbb{Z}) \to GL_n(\mathbb{Z}))$ and write $\Gamma(\underline{a}) = \lim_{\longrightarrow} \Gamma_n(\underline{a})$. Let Σ_n denote the $n \times n$ permutation matrix. Then Σ_n can be identified with the nth symmetric group. Put $\Sigma = \lim_{\longrightarrow} \Sigma_n$. Then Σ acts on $\Gamma_{\underline{a}}$ by conjugation, and so, we can form $\widetilde{\Gamma}(\underline{a}) = \Gamma(\underline{a}) \rtimes \Sigma$. One could think of $\widetilde{\Gamma}(a)$ as the group of matrices in $GL_n(\underline{a} \oplus \mathbb{Z})$ whose image in $GL_n(\mathbb{Z})$ is a permutation matrix. Consider the fibration $B\Gamma\underline{a} \to B\widetilde{\Gamma}(\underline{a}) \to B(\Sigma)$. Note that $B(\Sigma), B\widetilde{\Gamma}(\underline{a})$ has associated $+-$ construction that are infinite loop spaces. Define $F_{\underline{a}}$ as the homotopy fiber

$$F_{\underline{a}} \to B\widetilde{\Gamma}(\underline{a})^+ \to B\Sigma^+.$$

Then, for any ring Λ (with identity) containing $\underline{a}$ as a two-sided ideal, we have a map of fibrations

$$
\begin{array}{ccc}
F_{\underline{a}} & \xrightarrow{f_{\Lambda\underline{a}}} & F_{\Lambda,\underline{a}} \\
\downarrow & & \downarrow \\
B\widetilde{\Gamma}(\underline{a})^+ & \longrightarrow & BGL(\Lambda)^+ \\
\downarrow & & \downarrow \\
B\Sigma^+ & \longrightarrow & BGL(\Lambda/\underline{a})^+
\end{array}
$$

Definition 6.4.2 *Let $\underline{a}$ be a ring without unit, $S \subseteq \mathbb{Z}$ a multiplicative subset. Say that $\underline{a}$ is an S-excision ideal if for any ring Λ with unit containing $\underline{a}$ as a two-sided ideal, then $f_{\Lambda,\underline{a}}$ induces an isomorphism $\pi_*(F_{\underline{a}}) \otimes S^{-1}\mathbb{Z} \approx \pi_*(F_{\Lambda,\underline{a}}) \otimes S^{-1}\mathbb{Z}$.*

Theorem 6.4.1 [33] *Let $\underline{a}$ be a ring without unit and $S \subseteq \mathbb{Z}$ a multiplicative set such that $\underline{a} \otimes S^{-1}\mathbb{Z} = 0$ or $\underline{a} \otimes S^{-1}\mathbb{Z}$ has a unit. Then $\underline{a}$ is an S-excision ideal and*

$$H_*(F_{\underline{a}}, S^{-1}\mathbb{Z}) \cong H_n(\Gamma(\underline{a}); \quad S^{-1}\mathbb{Z}).$$

Examples and applications 6.4.1   (i) If $\underline{a}$ is a two-sided ideal in a ring Λ with identity such that $\Lambda/\underline{a}$ is annihilated by some $s \in \mathbb{Z}$, then the hypothesis of 4.7.5 is satisfied by $S = \{s^i\}$ and $\underline{a}$ is an S-excision ideal.

(ii) Let R be the ring of integers in a number field F, Λ an R-order in a semi-simple F-algebra, Γ a maximal R-order containing Λ. Then there exists $s \in \mathbb{Z}$, $s > 0$ such that $s\Gamma \subset \Lambda$, and so, $\underline{a} = s\Gamma$ is a 2-sided ideal in both Λ and Γ. Since s annihilates $\Lambda/\underline{a}$ (also $\Gamma/\underline{a}$), $\underline{a}$ is an S-excision ideal, and so, we have a long exact Mayer - Vietoris sequence

$$\to K_{n+1}(\Gamma/\underline{a})\left(\frac{1}{s}\right) \to K_n(\Lambda)\left(\frac{1}{s}\right) \to$$

$$\to K_n(\Lambda/\underline{a})\left(\frac{1}{s}\right) \oplus K_n(\Gamma)\left(\frac{1}{s}\right) \to K_n(\Gamma/\underline{a})\left(\frac{1}{s}\right) \to$$

where we have written $A(\frac{1}{s})$ for $A \otimes \mathbb{Z}(\frac{1}{s})$ for any Abelian group A. Also see [237].

(iii) Let Λ be a ring with unit and $K_n(\Lambda, \mathbb{Z}/r)$ K-theory with mod-r coefficients (see examples 5.2.2(viii)). Let $S = \{s \in \mathbb{Z} | (r, s) = 1\}$. Then multiplication by $s \in S$ is invertible on $K_n(\Lambda, \mathbb{Z}/r)$. Hence, for an S-excision ideal $\underline{a} \subset \Lambda$, $\pi_*(F_{\Lambda,\underline{a}}) \otimes S^{-1}\mathbb{Z} \simeq \pi_*(F_{\underline{a}}) \otimes S^{-1}\mathbb{Z}$ implies that $\pi_*(F_{\Lambda,\underline{a}}; \mathbb{Z}/r) \cong \pi_*(F_{\underline{a}}, \mathbb{Z}/r)$.

If we write $\mathbb{Z}_{(r)}$ for $S^{-1}\mathbb{Z}$ in this situation, we have that $K_n(\Lambda, \mathbb{Z}/r)$ satisfies excision on the class of ideals $\underline{a}$ such that $\underline{a} \otimes \mathbb{Z}_{(r)} = 0$ or $\underline{a} \otimes \mathbb{Z}_{(r)}$ has a unit.

Exercises

6.1 Prove Additivity theorem 6.1.1.

Let $F' \rightarrowtail F \twoheadrightarrow F''$ be a short exact sequence of exact functors from $\mathcal{A}$ to $\mathcal{B}$ where both $\mathcal{A}$ and $\mathcal{B}$ are either exact categories or Waldhausen categories. Show that $F_* \simeq F'_* + F''_* : K_n(A) \to K_n(B)$.

6.2 Let A, B be rings, $S = \{s^i\}$ where s is a central non-zero divisor in B. Let $\alpha : A \to B$ be an analytic isomorphism along s. Show that $H_S(A) = H_S(B)$.

6.3 Let $\mathcal{C}$ be an exact category, $\mathcal{E}(\mathcal{C})$ the category of all sequences $E : A \rightarrowtail B \twoheadrightarrow B/A = C$. Show that the functors $s, t, q : \mathcal{E}(\mathcal{C}) \to \mathcal{C}$ given by $s(E) = A, t(E) = B, q(E) = C$ are exact. Show that the exact functor $(s, q) : \mathcal{E}(\mathcal{C}) \twoheadrightarrow \mathcal{C} \times \mathcal{C}$ induces a homotopy equivalence $K(\mathcal{E}) \to K(\mathcal{C}) \times K(\mathcal{C})$.

Chapter 7

Some results on higher K-theory of orders, grouprings, and modules over 'EI' categories

7.1 Some finiteness results on K_n, G_n, SK_n, SG_n of orders and groupings

Recall that if R is a Dedekind domain with quotient field F, and Λ is any R-order in a semi-simple F-algebra Σ, then $SK_n(\Lambda) := \mathrm{Ker}\,(K_n(\Lambda)) \to K_n(\Sigma)$ and $SG_n(\Lambda) := \mathrm{Ker}\,(G_n(\Lambda)) \to G_n(\Sigma)$. Also, any R-order in a semi-simple F-algebra Σ can be embedded in a maximal R-order Γ, which has well-understood arithmetic properties relative to Σ. More precisely, if $\Sigma = \prod_{i=1}^{r} M_{n_i}(D_i)$, then Γ is Morita equivalent to $\prod_{i=1}^{r} M_{n_i}(\Gamma_i)$ where Γ_i are maximal orders in the division algebra D_i, and so, $K_n(\Gamma) \approx \oplus K_n(\Gamma_i)$ while $K_n(\Sigma) \approx \prod_{i=1}^{r} K_n(D_i)$. So, the study of K-theory of maximal orders in a semi-simple algebras can be reduced to the K-theory of maximal orders in division algebras.

Note also that the study of $SK_n(\Lambda)$ facilitates the understanding of $K_n(\Lambda)$ apart from the various topological applications known for $n = 0, 1, 2$ where $\Lambda = \mathbf{Z}G$ for some groups G that are usually fundamental groups of some spaces (see $(2.3)^C$, $(3.2)^B$). Also $SK_n(\Lambda)$ is involved in the definition of higher class groups, which generalizes to higher K-groups the notion of class groups of orders and groupings (see 7.4).

In this section, we shall prove several finiteness results on higher K-theory of R-orders where R is the ring of integers in a number field or p-adic field. We shall focus first on results on maximal orders in semi-simple algebras in the following subsection $(7.1)^A$.

$(7.1)^A$ Higher K-theory of maximal orders

7.1.1 Let L be a p-adic field, R the integers of L, Γ the maximal order in a semi-simple L-algebra Σ, πR the radical of R, and $S = \{\pi^i\}$ $i \leq 0$; then, $\Sigma = S^{-1}\Gamma$. Let $\mathcal{M}(\Gamma)$ and $\mathcal{M}(\Sigma)$ be the categories of finitely generated Γ-modules and Σ-modules, respectively. Since Γ and Σ are regular, we have $K_i(\Gamma) \simeq K_i(\mathcal{M}(\Gamma))$ and $K_i(\Sigma) \simeq K_i(\mathcal{M}(\Sigma))$, 6.1.2 (i).

Now, suppose $\mathbf{M}_S(\Gamma)$ is the category of finitely generated S-torsion Γ-modules. Then Quillen's location sequence (theorem 6.2.1) yields

$$\cdots \to K_{n+1}(\Sigma) \to K_n(\mathbf{M}_S(\Gamma)) \to K_n(\Gamma) \to K_n(\Sigma) \to K_{n-1}(\mathbf{M}_S(\Gamma)) \to \cdots$$

We now prove the following.

Theorem 7.1.1 *Let R, L, Σ be as in 7.1.1 above and $\mathbf{m}$ the radical of Γ. Then*

(i) $K_n(\mathbf{M}_S(\Gamma)) \simeq K_n(\Gamma/\mathbf{m})$ *for all $n \geq 0$.*

(ii) (a) $K_n(\Gamma) \to K_n(\Sigma)$ *has finite kernel and co-kernel for all $n \geq 1$.*

     (b) $0 \to K_{2n}(\Gamma) \to K_{2n}(\Sigma) \to K_{2n-1}(\Gamma/\mathbf{m}) \to K_{2n-1}(\Gamma) \to K_{2n-1}(\Sigma) \to 0$ *is exact for all $n \geq 2$.*

     (c) $0 \to K_2(\Gamma) \to K_2(\Sigma) \to K_1(\Gamma/\mathbf{m}) \to K_1(\Gamma) \to K_1(\Sigma) \to K_0(\Gamma/\mathbf{m}) \to K_0(\Gamma) \to K_0(\Sigma)$ *is exact.*

PROOF

(i) Note that $\mathbf{M}(\Gamma/\pi\Gamma) \subset \mathbf{M}_S(\Gamma)$, and every object M of $\mathbf{M}_S(\Gamma)$ has a finite filtration $0 = \pi^n M \subset \pi^{n-1}M \subset \cdots \subset \pi^0 M = M$ with quotients in $\mathbf{M}(\Gamma/\pi\Gamma)$. So, by theorem 6.1.1 (Devissage), $K_i(\mathbf{M}_S(\Gamma)) \simeq K_i(\mathbf{M}(\Gamma/\pi\Gamma))$. Now, $\mathrm{rad}(\Gamma/\pi\Gamma)$ is nilpotent in $\Gamma/\pi\Gamma$; so, by corollary 6.1.1, we have

$$K_i(\mathcal{M}(\Gamma/\pi\Gamma)) \simeq K_i(\mathcal{M}((\Gamma/\pi\Gamma)/\mathrm{rad}(\Gamma/\pi\Gamma))).$$

However, $(\Gamma/\pi\Gamma)/\mathrm{rad}(\Gamma/\pi\Gamma) \simeq \Gamma/\mathrm{rad}\Gamma$, and $\Gamma/\mathrm{rad}\Gamma$ is regular. So

$$K_i(\mathcal{M}_S(\Gamma)) \simeq K_i(\Gamma/\mathbf{m}).$$

(ii) By (i), Quillen's localization sequence theorem 6.1.1 becomes

$$\cdots \to K_{n+1}(\Sigma) \to K_n(\Gamma/\mathbf{m}) \to K_n(\Gamma) \to K_n(\Sigma) \to K_{n-1}(\Gamma/\mathbf{m}) \to \cdots$$

$$(7.1)$$

Now, for any ring R, $K_i(M_t(R)) \simeq K_i(R)$, where $M_t(R)$ is the ring of $t \times t$ matrices over R. (This is because the Morita equivalence of $\mathcal{P}(R)$ and $\mathcal{P}(M_t(R))$ yields

$$K_i(\mathcal{P}(R)) \simeq K_t\left(\mathcal{P}(M_t(R))\right).$$

Now, $\Gamma/\mathbf{m}$ is a finite direct product of matrix algebras over finite field-

s. So, by example 5.1.1(v), we have $K_n(\Gamma/\mathbf{m}) = 0$ if n is even ≥ 2, and $K_n(\Gamma/\mathbf{m})$ is finite if n is odd ≥ 1. So (a), (b), (c) follow from sequence 7.1 above.

$\square$

Corollary 7.1.1 Suppose Σ is a direct product of matrix algebras over fields; then

$$0 \to K_{2n}(\Gamma) \to K_{2n}(\Sigma) \to K_{2n-1}(\Gamma/\mathbf{m}) \to 1$$

is exact for all $n \leq 1$.

PROOF Suppose $\Sigma = \Pi M_{n_i}(L_i)$, say; then,

$$\Gamma = \Pi M_{n_t}(R_i) \qquad \text{and} \qquad \Gamma/\mathbf{m} = \Pi M_{n_t}(R_i/\mathbf{m}_i),$$

where R_i is the maximal order in the field L_i, and $\mathbf{m}$ = radical of R_i. So, by Morita duality, the sequence of theorem 7.1.2 ii(b) reduces to the required form.
$\square$

Corollary 7.1.2 Let $R, L, \Gamma, \Sigma, \mathbf{m}$ be as in 7.1.1 above. Then for all $n \geq 1$ the transfer map $K_1(\Gamma/\mathbf{m}) \to K_1(\Gamma)$ is non-zero if Σ is a product of matrix algebras over division rings.

PROOF Suppose $K_i(\Gamma/\mathbf{m} \to K_1(\Gamma)$ is zero; then, from the sequence in theorem 7.1.2(ii)(c), the kernel of $K_1(\Gamma) \to K_1(\Sigma)$ would be zero, contradicting the last statement of theorem 3.2.5(ii).
$\square$

Remark 7.1.1 It follows from the $K_2 - K_1$ localization exact sequence in theorem 7.1.1(c) that a non-commutative analogue of Dennis - Stein $K_2 - K_1$ short exact sequence for a discrete valuation ring does not hold for a maximal order Γ in a semi-simple algebra Σ that is a product of matrix algebras over division rings.

Remarks/notations 7.1.1 Our next aim is to obtain explicit computation of the transfer map $K_n(\Gamma/\mathbf{m}) \to K_n(\Gamma)$ when Σ is a product of matrix algebras over division rings. It suffices to do this for a maximal order Γ in a central division algebra over a p-adic field K. Let $\overline{\Gamma}$ be the residue field of Γ and k the residue field of K.

We shall prove the following result, due to M. Keating (see [99]).

Theorem 7.1.2 *For all $n \geq 1$, there are exact sequences*

(a) $0 \to K_{2n}(\Gamma) \to K_{2n}(D) \to K_{2n-1}(k) \to 0$.

(b) $0 \to K_{2n-1}(\overline{\Gamma})/K_{2n-1}(k) \to K_{2n-1}(\Gamma) \to K_{2n-1}(D) \to 0$.

The proof of theorem 7.1.2 will be in several steps.

Lemma 7.1.1 *Every inner automorphism of a ring R induces the identity on $K_n(R)$ for all $n \geq 1$.*

PROOF   See [99].   ▯

Lemma 7.1.2 *For all $n \geq 1$, the reduction map $K_n(\Gamma) \to K_n(\overline{\Gamma})$ is surjective.*

PROOF   The proof makes use of the fact that Quillen K-theory K_n and Karoubi - Vilamayor K_n^h coincide for the regular rings Γ and $\overline{\Gamma}$ (see theorem 5.1.3(ii)). Hence it suffices to prove the result for K_n^h: Also, for any ring R, $K_n^h(R) \simeq K_1^h(\Omega^{n-1}R)$ where $\Omega^{n-1}R$, the iterated loop ring of R is the polynomial ring $\Omega^{n-1}R = t_1 \ldots t_{n-1}(1 - t_{n-1})R[t_1 \ldots t_n]$. Also, $K_1^n(R) = GL(R)/Uni(R)$ where $Uni(R)$ is the subgroup of $GL(R)$ generated by the unipotent matrices. Moreover, there exists a natural splitting $K_1^h(R+\Omega^{n-1}R) = K_1^h(R) \oplus K_1^h(\Omega^{n-1}R)$.

Now, since $(\overline{\Gamma} + \Omega^{n-1}\overline{\Gamma})$ is a local ring, $GL(\overline{\Gamma} + \Omega^{n-1}\overline{\Gamma})$ is generated by units of $\overline{\Gamma}$ and the elementary matrices $E(\overline{\Gamma} + \Omega^{n-1}\overline{\Gamma})$.

Since Γ is complete in the p-adic topology, units of $\overline{\Gamma}$ can be lifted to Γ. Hence $GL(\Gamma + \Omega^{n-1}\Gamma)$ maps onto $GL(\overline{\Gamma} + \Omega^{n-1}\overline{\Gamma})$; hence the lemma.

Proof of theorem 7.1.2 Let δ_n be the natural map (reduction map) $K_{2n-1}(\Gamma) \to K_{2n-1}(\overline{\Gamma})$ and τ_n the transfer map $K_{2n-1}(\overline{\Gamma}) \to K_{2n-1}(\Gamma)$ to be defined below. Then it suffices to prove that $\mathrm{Im}\tau_n \cong K_{2n-1}(\overline{\Gamma})/K_{2n-1}(k)$.

Let $|k| = q$, $|\overline{\Gamma}| = q^t$, say. There is a primitive $(q^t - 1)$-th root ω of unity in Γ such that $\overline{\Gamma} = k(\overline{\omega})$.

Let R be a maximal order of K and π a uniformizing parameter in R. Then, Γ can be expressed as a twisted group-ring $\Gamma = R[\omega, \hat{\pi}]$, where $\hat{\pi}\omega\hat{\pi}^{-1} = \omega^{qr}$, and $\hat{\pi}^t = \pi(r,t) = 1$ where π is the uniformizing parameter of R.

Now, $K_{2n-1}(\overline{\Gamma})$ is a cyclic group of order $q^{tn} - 1$. Let γ be a generator of $K_{2n-1}(\overline{\Gamma})$. Since by lemma 7.1.2, the natural map $K_{2n-1}(\Gamma) \to K_{2n-1}(\overline{\Gamma})$ is surjective, there exists an element $\hat{\gamma}$ in $K_{2n-1}(\Gamma)$ that maps onto γ. Let α

be an automorphism of $K_{2n-1}(\Gamma)$ induced by conjugation by $\widehat{\pi}$. Define τ_n as a mapping of sets from $K_{2n-1}(\overline{\Gamma})$ to $K_{2n-1}(\Gamma)$ by $\tau_n(\gamma^j) = (\hat{\gamma})^j (\alpha \hat{\gamma})^{-1}$ for $j = 0, 1, \ldots, q^t - 2$ and extend to the whole of $K_{2n-1}(\overline{\Gamma})$ by "linearity".

Let F be the Frobenius automorphism of $\overline{\Gamma}$. Then conjugation by $\widehat{\pi}$ induces F^r on $\overline{\Gamma}$, and so, the composition of τ_n with δ_n is $1 - F_*^n$ where F_* is the map induced on the $K_n(\overline{\Gamma})$ by Frobenius automorphism. Since F^n generates $\mathrm{Gal}(\overline{\Gamma}/k)$, and we can identify $K_{2n-1}(k)$ with the fixed subgroup of $K_{2n-1}(\overline{\Gamma})$, we now have that $\delta_n(\mathrm{Im}\tau_n) \cong K_{2n-1}(\overline{\Gamma})/K_{2n-1}(k)$. Also it follows from lemma 7.1.1 and the definition of τ_n that $\mathrm{Im}\tau_n$ is contained in $SK_{2n-1}(\Gamma)$.

Recall from example 6.2.1(v) and 7.1.1 that we have a localization sequence

$$\ldots K_{2n-1}(D) \to K_n(\mathbf{M}_S(\Gamma)) \to K_n(\Gamma) \to K_n(D) \to K_{n-1}(\mathbf{M}_S(\Gamma)) \to \cdots$$
(II)

and that $K_n(\mathbf{M}_S(\Gamma)) \overset{\alpha}{\cong} K_n(\Gamma/\mathbf{m}) = K_n(\overline{\Gamma})$ where $\underline{\mathbf{m}}$ is the maximal ideal of Γ and the isomorphism α is induced by the functor F from $\mathbf{M}_S(\Gamma)$ to $\mathbf{M}_S(\overline{\Gamma})$, which associates a module M to the direct sum of its composition factors. Note that in II above $S = R - 0$.

Now, let L be a maximum subfield of D, R' the maximal order in L, with residue class field $\overline{R}'$. The inclusion of L in D induces a commu-

tative diagram
$$\begin{array}{ccc} K_{2n}(D) & \longrightarrow & K_{2n-1}(\overline{\Gamma}) \\ \uparrow & & \uparrow \psi \\ K_{2n}(L) & \overset{\beta}{\longrightarrow} & K_{2n-1}(\overline{R}') \end{array}$$
where ψ is induced by the functor

$\Psi(V) = F\left(D \underset{R'}{\otimes} V\right)$, V an R-module, and $\overline{R}' = R'/\mathbf{m}'$ where $\underline{\mathbf{m}}'$ is the maximal ideal in R'.

Suppose that $\underline{\mathbf{m}}'\Gamma = \Gamma\widehat{\pi}^e$. Then $\Psi(\overline{R}') = \Gamma/\Gamma\widehat{\pi} \oplus \cdots \oplus \Gamma\widehat{\pi}^{e-1}/\Gamma\widehat{\pi}^e$, and we have an isomorphism of $\overline{\Gamma}$-modules $\Gamma\widehat{\pi}^i/\Gamma\widehat{\pi}^{i+1} \cong [F^{ri}] cl(\widehat{\pi}^{-i}x\widehat{\pi})$ where $cl(-)$ means residue class, F is the Frobenius automorphism of $\overline{\Gamma}/k$, and $[F^{ri}]$ is the functor $\mathcal{P}(\overline{\Gamma}) \to \mathcal{P}(\overline{\Gamma})$ given by $[F^{ri}] P = P$ as additive group but $r([F^{ri}])p = [F^{ri}] (F^{ri})^{-1} rp$ for $r \in \overline{\Gamma}$, $p \in P$. Since Ψ is additive, we see that $cl(-) = (1 + [F^r] + \cdots + [F^{r(e-1)}]) \left(\overline{\Gamma} \underset{\overline{R}'}{\otimes} -\right)$. Hence $\Psi = 1 + F_*^n + \cdots + F_*^{r(e-1)}$. Since $(r,t) = 1$, $e = (\overline{R}'; k)$ divides $t = (\overline{R}'; k)$, F^r generates $\mathrm{Gal}(\overline{R}'/k)$, and so, $\mathrm{Im}(\Psi) = K_{2n-1}(k)$. Since β is surjective, we see that $K_{2n-1}(k) \subset \mathrm{Im}\left(K_{2n}(D) \to K_{2n-1}(\overline{\Gamma})\right)$, and so, the proof of theorem 7.1.2 is complete. $\qquad \Box$

Remark 7.1.2 In the notation of the proof of theorem 7.1.2, it follows that $|SK_{2n-1}(\Gamma)| = (q^{nt} - 1)/(q^n - 1)$.

Our next result is a K-theoretic characterization of p-adic semi-simple algebras due to A. Kuku (see [107]). In what follows (theorem 7.1.3), L is a p-adic field with ring of integers R.

Theorem 7.1.3 *Let Γ be the maximal R-order in a semi-simple L-algebra Σ. Then the following are equivalent:*

(i) Σ is unramified over its center.

(ii) For all $n \geq 1$, $SK_{2n-1}(\Gamma) = 0$.

(iii) There exists $n \geq 1$ such that $SK_{2n-1}(\Gamma) = 0$.

Remark Since Γ is Morita equivalent to $\prod M_{n_i}(\Gamma_i)$ and $\Sigma = \prod M_{n_i}(D_j)$, say, where Γ_i is a maximal order in some division algebra D_i over L, it suffices to prove the following in order to prove theorem 7.1.3.

Lemma 7.1.3 *Let Γ be the maximal order in a central division algebra D over a p-adic field F.*

   Then the following are equivalent:

(i) $D = F$.

(ii) For all $n \geq 1$, $SK_{2n-1}(\Gamma) = 0$.

(iii) There exists $n \geq 1$ such that $SK_{2n-1}(\Gamma) = 0$.

PROOF   (i)$\Rightarrow$(ii): Suppose that $D = F$, and R is the ring of integers of F, then $\Gamma = R$, and Quillen's localization sequence yields an exact sequence

$$\cdots \rightarrow K_{2n}(R) \rightarrow K_{2n-1}(R/\mathrm{rad}R) \rightarrow SK_{2n-1}(R) \rightarrow 0$$

and so, the result follows from [105] 1.3. (Also see examples 6.2.1(iii).)
(ii)$\Rightarrow$(iii) is trivial. (iii)$\Rightarrow$(i): Suppose $(D : F) = t^2$, say. Let $\overline{\Gamma} = \Gamma/\mathrm{rad}\Gamma$, $\overline{R} = R/\mathrm{rad}R$.
Then $(\overline{\Gamma} : \overline{R}) = t$.   Now, by theorem 7.1.2 $|SK_{2n-1}(\overline{\Gamma})| = |K_{2n-1}(\overline{\Gamma})|/|K_{2n-1}(\overline{R})|$. Also, $K_{2n-1}(\overline{\Gamma})$ has order $p^{lnt} - 1$ if $|\overline{R}| = p^l$ and $K_{2n-1}(\overline{R})$ has order $p^{ln} - 1$. So, $SK_{2n-1}(\Gamma) = 0$ if and only if $(p^{lnt} - 1)/(p^{ln} - 1) = 1$ if and only if $t = 1$ iff $t^2 = 1$ iff $D = F$.   ☐

Remark 7.1.3 If L is a p-adic field with integers R, G a finite group of order prime to p, and LG splits, then RG is a maximal order, and so, $SK_{2n-1}(RG) = 0$ for all $n \geq 1$ by 7.1.3.

7.1.2 We now observe that the global version of theorem 7.1.2 holds under suitable hypothesis (see [99]). So, let K be a global field with integers R, L a finite extension of K. For each finite prime $\underline{q}$ of L, let $l(\underline{q})$ be the residue field at $\underline{q}$. We assume the following hypothesis, which is known to hold for function fields and also for number fields (see [196, 99]) by Soule's work.
   The homomorphisms

$$K_{2n}(L) \rightarrow \coprod_{\underline{q}} K_{2n-1}(l(\underline{q})) \qquad n > 0 \tag{T}$$

are surjective where $\underline{q}$ runs through all finite primes of L.

7.1.3 In the notation of 7.1.2, let D be a central simple K-algebra, Γ a maximal order in D. For each prime p of K, Γ has a unique two-sided maximal ideal $\underline{m}$ above $\underline{p}$. There is a local division algebra $D(\underline{p})$ such that $\widehat{D}_{\underline{p}} = M_s\left(D(\underline{p})\right)$ for some $s = s(\underline{p})$ and $\Gamma/\underline{m} \cong M_s(d(\underline{p}))$ where $d(\underline{p})$ is the residue field of $\widehat{D}(\underline{p})$. Also, the Quillen localization sequence (see 6.2.3) yields

$$\cdots \to K_{n+1} \overset{\eta}{\to} \coprod_{\underline{p}} K_n(d(\underline{p})) \to K_n(\Gamma) \to K_n(D) \to \cdots$$

where $K_n\left(M_S(\Gamma)\right) \cong \coprod K_n(d(\underline{p}))$ and $S = R - 0$.

We now state the global version of theorem 7.1.2, also due to M.E. Keating [99].

Theorem 7.1.4 *Assume hypothesis T. Then there exist exact sequences*

$$0 \to K_{2n}(\Gamma) \to K_{2n}(D) \to \coprod_{\underline{p}} K_{2n-1}(k(\underline{p})) \to 0$$

and

$$0 \to \coprod_{\underline{p}} \left(K_{2n-1}(d(\underline{p}))/K_{2n-1}(k(\underline{p}))\right) \to K_{2n-1}(\Gamma) \to K_{2n-1}(D) \to 0$$

for all $n \geq 1$.

The proof of 7.1.4 uses the following lemma.

Lemma 7.1.4 *Let L be a maximal subfield of D. Then for each prime $\underline{q}$ in L above $\underline{p}$, the natural transformation of localization sequences induces a map*

$$\Phi = \Phi(\underline{q}, \underline{p}) : K_{2n-1}(l(\underline{q})) \to K_{2n-1}(d(\underline{p}))$$

where $l(\underline{q})$ is the residue field of L at $\underline{q}$. Suppose that one of the following two conditions hold:

(i) For each prime $\underline{q}$ of L above $\underline{p}$, $\left(\widehat{L}_{\underline{q}} : \widehat{K}_{\underline{p}}\right)^2 = \left(D(\underline{p}), K_{\underline{p}}\right)$.

(ii) D splits at $\underline{p}$ and L is unramified with $\left(\widehat{L}_{\underline{q}} : \widehat{K}_{\underline{p}}\right)$ constant at the primes $\underline{q}$ above $\underline{p}$. Then $\operatorname{im}\Phi \cong K_{2n-1}\left(k(\underline{p})\right)$.

PROOF   Left as an exercise.   □

Sketch of proof of theorem 7.1.4. Condition (ii) of lemma 7.1.4 is satisfied almost always by any maximal cyclic subfield of D. Now, by [9], we can find such a subfield that satisfies condition (i) at any given finite set of primes of K. Here we can find by hypothesis T a pair of subfields L and L' of D so that $\coprod_{\underline{p}} K_{2n-1}(k(\underline{p})) \subset \eta \left(\mathrm{im} K_{2n}(L) + \mathrm{im} K_{2n}(L') \right)$ where the images are computed in $K_{2n}(D)$.

On the other hand, local consideration show that $SK_{2n-1}(\Gamma)$ maps surjectively onto $\coprod_{\underline{p}} K_{2n-1}\left(d(\underline{p})/K_{2n-1}(k(\underline{p})) \right)$. Hence the result.

Note. In view of Soule's proof that hypothesis T also holds for number fields (see [196]), it follows that T holds for global fields since it was earlier known to hold for function fields.

Next, we record for later use the following results of A.A. Suslin and A.V. Yufryakov (see [204]).

Theorem 7.1.5 [204] *Let F be a local field of char p (i.e., a complete discretely valued field with finite residue field of char p (e.g., a p-adic field). Let R be the ring of integers of F, Γ a maximal order in a central division algebra D of degree d^2 over F. Then for all $n > 0$, $K_n(\Gamma) \otimes \widehat{\mathbb{Z}}_p$, and $K_n(D) \otimes \widehat{\mathbb{Z}}_p$ decompose into a direct sum of a torsion group and a uniquely divisible subgroup.*

Moreover $K_n(\Gamma) \otimes \widehat{\mathbb{Z}}_q \simeq K_n(\overline{\Gamma}) \otimes \widehat{\mathbb{Z}}_q$, $q \neq p$ for all $n \geq 1$ where $\overline{\Gamma}$ is the residue class field of Γ.

PROOF See [204]. ⬜

The next result due to X. Guo and A. Kuku (see [72]) shows that the kernel of the reduction map of a maximal order in a central division algebra over number fields is finite.

Theorem 7.1.6 *Let F be a number field and D a central division algebra of dimension m^2 over F. Let R be the ring of integers of F, and Γ a maximal R-order in D. For any place v of F, let k_v be the residue ring of R at v. Then, the residue ring of Γ_v is a matrix ring over d_v, where d_v is a finite field extension of k_v and the kernel of the reduction map*

$$K_{2n-1}(\Gamma) \overset{(\pi_v)}{\to} \prod_{\text{finite } v} K_{2n-1}(d_v)$$

is finite. Hence the kernel of

$$K_{2n-1}(\Gamma) \overset{\varphi}{\to} \prod_{\text{finite } v} K_{2n-1}(\widehat{\Gamma}_v)$$

is finite.

PROOF   It is well known that the residue ring of $\widehat{\Gamma}_v$ is a matrix ring over d_v, where d_v is a finite field extension of k_v (see [99]).   □

By theorem 7.1.11 $K_{2n-1}(\Gamma)$ is finitely generated. So, it suffices to prove that the kernel of the reduction map is a torsion group in order to show that it is finite.

Let

$$i : K_{2n-1}(R) \to K_{2n-1}(\Gamma)$$

be the homomorphism induced by inclusion, and let

$$tr : K_{2n-1}(\Gamma) \to K_{2n-1}(R)$$

be the transfer homomorphism. Then

$$i \circ tr(x) = x^{m^2}$$

for any $x \in K_{2n-1}(\Gamma)$ by lemma 7.2.2(b).

So, if there is a torsion-free element $x \in \text{Ker}(\pi_v)$, then $tr(x)$ is a torsion-free element in $K_{2n-1}(R)$. Consider the following commutative diagram

$$
\begin{array}{ccc}
K_{2n-1}(R) & \xrightarrow{(\pi_v')} & \prod_{\text{finite } v} K_{2n-1}(k_v) \\
\downarrow{\scriptstyle \iota} & & \downarrow{\scriptstyle (\iota_v)} \\
K_{2n-1}(\Gamma) & \xrightarrow{(\pi_v)} & \prod_{\text{finite } v} K_{2n-1}(d_v)
\end{array}
$$

By Theorem 1 of [8], the kernel of (π_v') is finite. So $(\iota_v) \circ (\pi_v') \circ tr(x)$ is torsion-free. But $x \in \text{Ker}(\pi_v)$, and so, $(\iota_v) \circ (\pi_v') \circ tr(x)$ must be 0 since, from the above diagram,

$$(\iota_v) \circ (\pi_v') \circ tr(x) = (\pi_v)(x^{m^2}) = 0.$$

This is a contradiction. Hence $\text{Ker}(\pi_v)$ is finite.

The last statement follows from the following commutative diagram

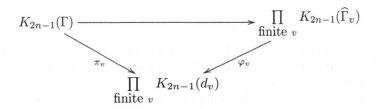

and the fact that $\text{Ker}(\pi_v)$ is finite (as proved above).

We close this subsection with a brief discussion on Wild kernels for higher K-theory of division and semi-simple algebras. First, we define these notions that generalize the notion of Wild kernels for number fields and use 7.1.6 to prove that Wild kernels are finite Abelian groups.

Definition 7.1.1 *Recall that if F is a number field and R the ring of integers of F, then the wild kernel $WK_n(F)$ is defined by*

$$WK_n(F) := Ker\left(K_n(F) \to \prod_{\text{finite } v} K_n\left(\widehat{F_v}\right)\right)$$

where v runs through all the finite places v of F, and $\widehat{F_v}$ is the completion of F at v (see [12]). It is proved in [12] that $WK_n(F)$ is finite, and in [12], Banaszak et al. conjectured that for all number fields F and all $n \geq 0$ we should have $WK_n(F)_l = div(K_n(F))_l$.

Now, suppose D is a central division algebra over F and Γ a maximal R-order in D. Following X. Guo and A. Kuku (see [72]) we define wild kernel

$WK_n(D)$ *of D by $WK_n(D) := Ker\left(K_n(D) \to \prod_{\text{finite } v} K_n\left(\widehat{D_v}\right)\right)$ and show*

that $WK_n(D)$ is finite for all $n \geq 0$ (see theorem 7.1.7 and proposition 7.1.1). We also define pseudo-wild kernel $W'K_n(D)$ of D by

$$W'K_n(D) = Ker\left(K_n(D) \to \prod_{\text{non complex } v} K_n\left(\widehat{D_v}\right)\right)$$

with the observation that $W'K_n(D)$ is a subgroup of $WK_n(D)$. Note that these definitions extend to $WK_n(\Sigma)$, $W'K_n(\Sigma)$ where Σ is a semi-simple F-algebra. These results and some others connected to this topic are due to X. Guo and A. Kuku (see [72]).

Theorem 7.1.7 *Let F be a number field, D a central division algebra over F. Then the wild kernel $WK_{2n-1}(D)$ is finite.*

PROOF By theorem 7.1.4, the following sequence is exact

$$0 \to \oplus_{\text{finite } v} K_{2n-1}(d_v)/K_{2n-1}(k_v) \to K_{2n-1}(\Gamma) \to K_{2n-1}(D) \to 0 \quad \text{(I)}$$

where Γ is a maximal order in D, k_v is the residue ring of the ring of integers R of F at v, and the residue ring of $\widehat{\Gamma}_v$ is a matrix ring over d_v, and d_v is a finite field extension of k_v.

Since $K_{2n-1}(d_v)/K_{2n-1}(k_v)$ is trivial for almost all v, it follows that

$$\oplus_{\text{finite } v} K_{2n-1}(d_v)/K_{2n-1}(k_v)$$

is a finite group. Also, it follows from theorem 7.1.11 that $K_{2n-1}(\Gamma)$ is finitely generated. So, $K_{2n-1}(D)$ is finitely generated, which implies that $WK_{2n-1}(D)$ is finitely generated. So it suffices to prove that $WK_{2n-1}(D)$ is a torsion group.

If $x \in WK_{2n-1}(D) \subset K_{2n-1}(D)$ is torsion free, then from I above, we can find an element $x_1 \in K_{2n-1}(\Gamma)$ such that the image of x_1 under the homomorphism

$$i : K_{2n-1}(\Gamma) \to K_{2n-1}(D)$$

is x, and x_1 is also torsion free. By theorem 7.1.6, the kernel of the composite of the following maps

$$K_{2n-1}(\Gamma) \to \prod_{\text{finite } v} K_{2n-1}(\Gamma_v) \to \prod_{\text{finite } v} K_{2n-1}(d_v)$$

is finite. If x_2 is the image of x_1 in $\prod_{\text{finite } v} K_{2n-1}(\Gamma_v)$, then x_2 is torsion free.
Consider the following commutative diagram II with the maps of elements illustrated in diagram III:

$$
\begin{array}{ccc}
K_{2n-1}(\Gamma) & \longrightarrow & \prod\limits_{\text{finite } v} K_{2n-1}(\Gamma_v) \\
\downarrow & & \downarrow \\
K_{2n-1}(D) & \longrightarrow & \prod\limits_{\text{finite } v} K_{2n-1}(D_v)
\end{array}
\qquad \text{(II)}
$$

$$
\begin{array}{ccc}
x_1 & \longrightarrow & x_2 \\
\downarrow & & \downarrow \\
x & \longrightarrow & x_3 \, ,
\end{array}
\qquad \text{(III)}
$$

where x_3 is the image of x_2 in $\prod_{\text{finite } v} K_{2n-1}(D_v)$. Since D is ramified at finitely many places of F, $k_v = d_v$ for almost all v. So, $K_{2n-1}(\Gamma_v) \simeq K_{2n-1}(D_v)$ for almost all v by theorem 7.1.3. Hence the kernel of the right vertical arrow in diagram II is finite. So, x_3 is torsion free. However $x \in WK_{2n-1}(D)$, and so, $x_3 = 0$. This is a contradiction. Hence $WK_{2n-1}(D)$ is finite. □

Proposition 7.1.1 Let F be a number field and D a central division algebra over F. Then, for all $n \geq 0$, the wild kernel $WK_{2n}(D)$ is contained in the image of $K_{2n}(\Gamma) \to K_{2n}(D)$. In particular, $WK_{2n}(D)$ is finite.

PROOF   Consider the following commutative diagram

$$
\begin{array}{ccccccc}
0 & \longrightarrow & WK_{2n}(D) & \longrightarrow & K_{2n}(D) & \overset{f}{\longrightarrow} & \prod_v K_{2n}(D_v) \\
& & \downarrow & & =\downarrow & & \downarrow{\scriptstyle\tau} \\
0 & \longrightarrow & K_{2n}(\Gamma) & \longrightarrow & K_{2n}(D) & \underset{g}{\longrightarrow} & \prod_v K_{2n-1}(d_v),
\end{array}
$$

where the middle vertical arrow is an identity. By this commutative diagram,

$$g = \tau \circ f$$

which implies $\operatorname{Ker} f \subset \operatorname{Ker} g$. So, $WK_{2n}(D) \subset K_{2n}(\Gamma)$.
Let

$$tr : K_{2n}(\Gamma) \to K_{2n}(R)$$

be the transfer homomorphism, and let

$$i : K_{2n}(R) \to K_{2n}(\Gamma)$$

be the homomorphism induced by the inclusion. Then for any $x \in K_{2n}(\Lambda)$,

$$i \circ tr(x) = x^{m^2},$$

where m^2 is the dimension of D over F. Since $K_{2n}(R)$ is a torsion group, $K_{2n}(\Gamma)$ is also a torsion group. So, it must be finite, which implies $WK_{2n}(D)$ is finite. $\square$

Theorem 7.1.8 *Let Σ be a semi-simple algebra over a number field F. Then the wild kernel $WK_n(\Sigma)$ is contained in the torsion part of the image of the homomorphism*

$$K_n(\Gamma) \to K_n(\Sigma),$$

where Γ is a maximal order of Σ. In particular, $WK_n(\Sigma)$ is finite.

PROOF   Assume $\Sigma = \prod_{i=1}^{k} M_{n_i}(D_i)$, where D_i is a finite dimensional F-division algebra with center E_i. Let Γ be a maximal order of Σ. We know that Γ is Morita equivalent to $\prod_{i=1}^{k} M_{n_i}(\Gamma_i)$, where Γ_i is maximal order of D_i. So $WK_n(\Sigma) = \prod_{i=1}^{k} (WK_n(D_i))$ and $K_n(\Gamma) = \prod_{i=1}^{k} K_n(\Gamma_i)$. This theorem follows from theorems 7.1.7 and 7.1.1. $\square$

Remark 7.1.4   It was also proved in [72] by Guo/Kuku that

(i) $WK_n(\Sigma)/W'K_n(\Sigma)$ is a finite 2-group with 8-rank 0 if $n \equiv 0, 4, 6 \pmod{8}$ (see [72], theorem 3.1).

(ii) $\mathrm{div}(K_2(D)) \subseteq WK_2(D)$, and that when the index of D is square free, then

    (a) $\mathrm{div}K_2(D) \simeq \mathrm{div}K_2(F)$.

    (b) $WK_2(D) \simeq WK_2(F)$ and $|WK_2(D)/\mathrm{div}(K_2(D))| \leq 2$ (see [72], theorem 3.2).

(iii) If $(D : F) = m^2$, then

    (a) $\mathrm{div}(K_n(D))_l = WK_n(D)_l$ for all odd primes l and $n \leq 2$.

    (b) If l does not divide m, then $\mathrm{div}K_3(D)_l = WK_3(D)_l = 0$.

    (c) If $F = Q$ and l does not divide m, then $\mathrm{div}K_n(D)_l \subset WK_n(D)_l$ for all n (see [72], theorem 3.4).

$(7.1)^B$ K_n, G_n, SK_n, SG_n of arbitrary orders

In this subsection, we obtain some finiteness results on K_n, G_n, SK_n, SG_n of arbitrary orders. These results are all due to A.O. Kuku up to theorem 7.1.14 (see [108, 110, 112, 113]). Thereafter, we focus on some vanishing results on $SG_n(\Lambda)$ due to R. Laubenbacher and D. Webb (see [131]).

Let R be the ring of integers in an algebraic number field F, Λ any R-order in a semi-simple F-algebra Σ. Then, the inclusion map $\Lambda \rightarrow \hat{\Lambda}_{\underline{p}}$ induces a group homomorphism $K_n(\Lambda) \xrightarrow{\hat{\varphi}} K_n\left(\hat{\Lambda}_{\underline{p}}\right)$ and hence a map $SK_n\Lambda \xrightarrow{\langle\hat{\varphi}\rangle} \prod SK_n\left(\hat{\Lambda}_{\underline{p}}\right)$. Note that a similar situation holds with SK_n replaced by SG_n.

Theorem 7.1.9 $SK_n\left(\hat{\Lambda}_{\underline{p}}\right) = 0$ *for almost all* $\underline{p}$ *and* $\langle\hat{\varphi}\rangle$ *is surjective.*

PROOF It is well known that $\hat{\Lambda}_{\underline{p}}$ is maximal for almost all $\underline{p}$. It is also well known that Σ has only finitely many non-split completions, i.e., $\hat{\Sigma}_{\underline{p}}$ splits for almost all $\underline{p}$. In any case, $\hat{\Lambda}_{\underline{p}}$ is maximal in a split semi-simple algebra $\hat{\Sigma}_{\underline{p}}$ for almost all $\underline{p}$. So, for almost all $\underline{p}$, $SK_n\left(\hat{\Lambda}_{\underline{p}}\right) = 0$ by theorem 7.1.3 and corollary 7.1.1. So, $\prod SK_n\left(\hat{\Lambda}_{\underline{p}}\right) = \prod_{i=1}^{m} SK_n\left(\hat{\Lambda}_{\underline{p}_i}\right)$ for some finite number m of $\underline{p}_i$'s. $\qquad\square$

7.1.4 We now show that $\langle\hat{\varphi}\rangle$ is surjective. Let $S = R - 0$, $\hat{S}_{\underline{p}} = \hat{R}_{\underline{p}} - 0$. Then Quillen's localization sequence for $K_n(\Lambda)$ and $K_n\left(\hat{\Lambda}_{\underline{p}}\right)$ and the above result yield the following diagram:

$$
\begin{array}{ccccccc}
0 \to & \mathrm{Coker}\big(K_{n+1}(\Lambda) \xrightarrow{\delta_{n+1}} K_{n+1}(\Sigma)\big) \to & K_n\left(\mathcal{H}_S(\Lambda)\right) & \to & SK_n(\Lambda) & \to 0 \\
 & \downarrow {\scriptstyle \rho_n} & \downarrow {\scriptstyle \beta_n} & & \downarrow {\scriptstyle <\hat{\varphi}>} & \\
0 \to & \prod_{\underline{p}} \mathrm{Coker}\big(K_{n+1}(\widehat{\Lambda}_{\underline{p}}) \to K_{n+1}(\widehat{\Sigma}_{\underline{p}})\big) \to & \prod K_n\left(\mathcal{H}_{\hat{S}_{\underline{p}}}\left(\widehat{\Lambda}_{\underline{p}}\right)\right) & \to & \prod_{i=1}^{m} SK_n\left(\widehat{\Lambda}_{\underline{p}_i}\right) & \to 0
\end{array}
$$

We now show that β_n maps $K_n\left(\mathcal{H}_S(\Lambda)\right)$ isomorphically onto $\oplus K_n\left(\mathcal{H}_{\hat{S}_{\underline{p}}}\left(\widehat{\Lambda}_{\underline{p}}\right)\right)$. If $M \in \mathcal{H}_{\hat{S}_{\underline{p}}}\left(\widehat{\Lambda}_{\underline{p}}\right)$, then $M \simeq \bigoplus_{\underline{p}} M_{\underline{p}}$ where $\underline{p}$ runs through the prime ideals of R and $M_{\underline{p}} = 0$ except for a finite number of summands. Moreover, $M_{\underline{p}} \in \mathcal{H}_{\hat{S}_{\underline{p}}}\left(\widehat{\Lambda}_{\underline{p}}\right)$. So, $\mathcal{H}_S \simeq \bigoplus_{\underline{p}} \mathcal{H}_{\hat{S}_{\underline{p}}}\left(\widehat{\Lambda}_{\underline{p}}\right)$. Now,

$$
\bigoplus_{\underline{p}} \mathcal{H}_{S_{\underline{p}}}\left(\widehat{\Lambda}_{\underline{p}}\right) = \lim_{m \to \infty} \bigoplus_{i=1}^{m} \mathcal{H}_{\hat{S}_{\underline{p}_i}}\left(\widehat{\Lambda}_{\underline{p}_i}\right).
$$

So, by Quillen's results (see [165]).

$$
K_n\left(\mathcal{H}_S(\Lambda)\right) \simeq \lim_{m \to \infty} \bigoplus_{i=1}^{m}\left(\mathcal{H}_{\hat{S}_{\underline{p}_i}}\left(\widehat{\Lambda}_{\underline{p}_i}\right)\right) \simeq \oplus_{\underline{p}} K_n\left(\mathcal{H}_{\hat{S}_{\underline{p}}}\left(\widehat{\Lambda}_{\underline{p}}\right)\right).
$$

So, β_n really maps onto $\bigoplus_{\underline{p}} K_n \mathcal{H}_{\hat{S}_{\underline{p}}}\left(\widehat{\Lambda}_{\underline{p}}\right)$, and by diagram chasing, it is clear that ρ_n also maps into $\mathrm{Coker}\big(K_{n+1}(\widehat{\Lambda}_{\underline{p}}) \to K_{n+1}(\widehat{\Sigma}_{\underline{p}})\big)$. The surjectivity of $\langle\hat{\varphi}\rangle$ will follow from the following.

Theorem 7.1.10 *For all $n \geq 1$, there exists an exact sequence*

$$
0 \to Coker\rho_n \to SK_n(\Lambda) \xrightarrow{\langle\hat{\varphi}\rangle} \bigoplus_{i=1}^{m} SK_n\left(\widehat{\Lambda}_{\underline{p}_i}\right) \to 0.
$$

PROOF   The result follows by applying the Snake Lemma to the commutative diagram

$$
\begin{array}{ccccccc}
0 \to & \mathrm{Coker}\big(K_{n+1}(\Lambda) \to K_{n+1}(\Sigma)\big) \to & K_n\left(\mathcal{H}_S(\Lambda)\right) & \to & SK_n(\Lambda) & \to 0 \\
 & \downarrow {\scriptstyle \rho_n} & \downarrow {\scriptstyle \beta_n} & & \downarrow {\scriptstyle \langle\hat{\varphi}\rangle} & \\
0 \to & \bigoplus_{\underline{p}} \mathrm{Coker}\big(K_{n+1}(\widehat{\Lambda}_{\underline{p}}) \to K_{n+1}(\widehat{\Sigma}_{\underline{p}})\big) \to & \bigoplus_{\underline{p}} K_n\left(\mathcal{H}_{\hat{S}_{\underline{p}}}\left(\widehat{\Lambda}_{\underline{p}}\right)\right) & \to & \bigoplus_{i=1}^{m} SK_n\left(\widehat{\Lambda}_{\underline{p}_i}\right) & \to 0
\end{array}
$$

$\square$

Remark 7.1.5 It should be noted that theorem 7.1.10 holds with SK_n replaced by SG_n, where ρ_n is now a map from $\mathrm{Coker}\left(G_{n+1}(\Lambda) \to G_{n+1}(\Sigma)\right)$ to $\oplus_{\underline{p}}\mathrm{Coker}\left(G_{n+1}(\widehat{\Lambda}_{\underline{p}}) \to G_{n+1}(\widehat{\Sigma}_{\underline{p}})\right)$. Note that the proof is similar where β_n

is now isomorphism $K_n \left(\mathcal{M}_S(\Lambda) \right) \simeq \oplus_{\underline{p}} G_n \left(\widehat{\Lambda}_{\underline{p}} / \underline{p} \widehat{\Lambda}_{\underline{p}} \right)$. Details are left to the reader.

Theorem 7.1.11 *Let R be the ring of integers in a number field F, Λ an R-order in a semi-simple F-algebra Σ. Then for all $n \geq 1$*

(i) $K_n(\Lambda)$ is a finitely generated Abelian group.

(ii) $SK_n(\Lambda)$ is a finite group.

(iii) $SK_n(\widehat{\Lambda}_{\underline{p}})$ is finite (or zero) for any prime ideal $\underline{p}$ of R.

PROOF

(i) Note that $K_n(\Lambda) = \lim_{m \to \infty} K_{n,m}(\Lambda)$ where $K_{n,m}(\Lambda) = \pi_n(BGL_m^+(\Lambda)) \simeq \pi_n(BE_m^+(\Lambda))$ since $BE_m^+(\Lambda)$ is the universal covering space of $BGL_m^+(\Lambda)$ (see [25, 220]).

Now, by the stability result of Suslin [202], $K_{n,m}(\Lambda) = K_{n,m+1}(\Lambda)$ if $m \leq \max(2n+1, n+3)$ since Λ satisfies SR_3. Now $E_m(\Lambda)$ is an Arithmetic group since $SL_m(\Lambda)/E_m(\Lambda)$ is finite (see [189]). So, by a result of Borel ([27], theorem 11.4.4), $H_n(E_m(\Lambda))$ is finitely generated. Now $H_n(E_m(\Lambda)) \cong H_n(BE_m(\Lambda)) = H_n(BE_m^+(\Lambda))$, and moreover, $BE_m^+(\Lambda)$ is simply connected H-space for $m \geq 3$. Moreover, by ([197], 9.6.16), $\pi_n(BE_m^+(\Lambda))$ is finitely generated iff $H_n(BE_m^+(\Lambda))$ is finitely generated. So $K_n(\Lambda) = \pi_n(BE_m^+(\Lambda))$ is finitely generated. The proof for $n = 1$ is well known (see [20]).

(ii) It follows from (i) that $SK_n(\Lambda)$ is finitely generated as a subgroup of $K_n(\Lambda)$, and so, we only have to show that $SK_n(\Lambda)$ is torsion.

If Γ is a maximal R-order containing Λ, then, since any R-order is a $\mathbb{Z}$-order, there exists $s \in \mathbb{Z}$ such that $s\Gamma \subset \Lambda$. If we put $\underline{q} = s\Gamma$, then $\underline{q}$ is a two-sided ideal of Λ and Γ such that s annihilates $\Lambda/\underline{q}$ and $\Gamma/\underline{q}$, and so, we obtain a Cartesian square

$$
\begin{array}{ccc}
\Lambda & \longrightarrow & \Gamma \\
\downarrow & & \downarrow \\
\Lambda/\underline{q} & \longrightarrow & \Gamma/\underline{q}
\end{array}
$$

which by [33, 237] leads to an exact Mayer - Vietoris sequence:

$$
\begin{aligned}
\to K_{n+1} \left(\Gamma/\underline{q} \right)_{1/s} \to K_n(\Lambda)_{1/s} \to K_n(\Gamma)_{1/s} \oplus K_n(\Lambda(\underline{q}) \\
\to K_n(\Gamma/\underline{q})_{1/s} \to K_{n-1}(\Lambda)_{1/s} \to \cdots
\end{aligned}
\tag{I}
$$

where, for any Abelian group Λ we write $A_{1/s}$ for $A \otimes \mathbb{Z}[1/s]$.

If we write $A = A_s \oplus A_{s'} \oplus A_f$ where s, s', f denote, respectively, s-torsion, s'-torsion, and free parts of A, we have $A_{1/s} \simeq A_{s'} \oplus (A_f)_s$.

Now, $\Lambda/\underline{q}$, $\Gamma/\underline{q}$ are finite rings, and so, $K_n(\Lambda/\underline{q})_{1/s}$ and $K_n(\Gamma/\underline{q})_{1/s}$ are isomorphic, respectively, to $K_n\left((\Lambda/\underline{q})/\mathrm{rad}(\Lambda/\underline{q})\right)_{1/s}$ and $K_n\left((\Gamma/\underline{q})/\mathrm{rad}(\Gamma/\underline{q})\right)_{1/s}$. Now, since $(\Lambda/\underline{q})/\mathrm{rad}(\Lambda/\underline{q})$ and $(\Gamma/\underline{q})/\mathrm{rad}(\Gamma/\underline{q})$ are finite semi-simple rings, computing their K-groups reduces to computing K-groups of finite fields. Hence $K_{2n}(\Lambda/\underline{q})_{1/s}$ and $K_{2n}(\Gamma/\underline{q})_{1/s} = 0$ for all $n \geq 1$, and the sequence I reduces to

$$\cdots K_{2n+1}(\Gamma/\underline{q})_{1/s} \to K_{2n}(\Lambda)_{1/s} \to K_{2n}(\Gamma)_{1/s} \to 0 \qquad \text{(II)}$$

and

$$\cdots 0 \to K_{2n-1}(\Lambda)_{1/s} \xrightarrow{\delta} K_{2n-1}(\Gamma)_{1/s} \oplus K_{2n-1}(\Lambda/\underline{q})_{1/s} \to K_{2n-1}(\Gamma/\underline{q})_{1/s} \qquad \text{(III)}$$

It follows from the sequence II above that the canonical map $K_{2n}(\Lambda) \xrightarrow{\beta_{2n}} K_{2n}(\Gamma)$ is a monomorphism mod s-torsion. Now, $SK_{2n}(\Gamma) = 0$ since Γ is regular. So, $SK_{2n}(\Lambda) \simeq \mathrm{Ker}\beta_{2n}$ is torsion. But $SK_{2n}(\Lambda)$ is also finitely generated as a subgroup of $K_{2n}(\Lambda)$ by (i) above. Hence $SK_{2n}(\Lambda)$ is finite.

Now, let $\eta : K_{2n-1}(\Gamma)_{1/s} \oplus K_{2n-1}(\Lambda/\underline{q})_{1/s} \to K_{2n-1}(\Gamma)_{1/s}$ be the projection onto the first factor $K_{2n-1}(\Gamma)_{1/s}$. Then, $\mathrm{Ker}(K_{2n-1}(\Lambda)_{1/s} \xrightarrow{\rho} K_{2n-1}(\Gamma)_{1/s})$ is contained in $\mathrm{Ker}\,\eta$ since we have a commutative diagram

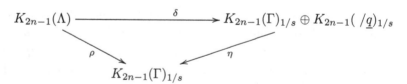

and so, $\rho = \eta\delta$, whence the sequence $0 \to \mathrm{Ker}\delta \to \mathrm{Ker}\rho \to \mathrm{Ker}\eta$ is exact where $\mathrm{Ker}\delta = 0$. Now $\mathrm{Ker}\eta \simeq K_{2n-1}(\Lambda/\underline{q})_{1/s}$ is torsion since $K_{2n-1}(\Lambda) \to K_{2n-1}(\Gamma)$ is a monomorphism mod torsion. But $SK_{2n-1}(\Gamma)$ is finite since $SK_{2n-1}(\Gamma) \simeq SG_{2n-1}(\Lambda)$ and $G_{2n-1}(\Gamma/\underline{q})$ is finite (see theorem 7.1.12(ii)). Hence $SK_{2n-1}(\Lambda)$ is torsion. But $SK_{2n-1}(\Lambda)$ is also finitely generated by (i) above. Hence $SK_{2n-1}(\Lambda)$ is finite.

(iii) It is well known that $\widehat{\Lambda}_{\underline{p}}$ is maximal for almost all $\underline{p}$. It is also well known that Σ has only finitely many non-split extension, i.e., $\widehat{\Sigma}_{\underline{p}}$ splits for almost all p. In other words, $\widehat{\Lambda}_{\underline{p}}$ is maximal in a split semi-simple $\widehat{\Sigma}_{\underline{p}}$ for almost all $\underline{p}$. So, for almost all $\underline{p}$, $SK_{2n-1}\left(\widehat{\Lambda}_{\underline{p}}\right) = 0$ (see theorem

7.1.3) and $SK_{2n}\left(\widehat{\Lambda}_{\underline{p}}\right) = 0$ (see corollary 7.1.1). Also, it was shown in theorem 7.1.13 that there exists a surjection

$$SK_n(\Lambda) \rightarrow \oplus_{i=1}^{m} SK_n\left(\widehat{\Lambda}_{\underline{p}_i}\right).$$

Hence the result.

Remarks 7.1.1 Let R be the ring of integers in a number field F, $\widehat{R}_p$ the completion of R at a prime ideal $\underline{p}$ of R, G a finite group. It follows from theorem 7.1.11 that, for all $n \geq 1$, $K_n(RG)$ is finitely generated; $SK_n(RG)$ and $SK_n(\widehat{R}_pG)$ are finite groups.

To facilitate other computations, we now show that if A is any finite ring, then $K_n(A)$, $G_n(A)$ are finite Abelian groups.

Theorem 7.1.12 *Let A be any finite ring with identity. Then for all $n \geq 1$,*

(i) *$K_n(A)$ is a finite group.*

(ii) *$G_n(A)$ is a finite group, and $G_{2n}(A) = 0$.*

PROOF

(i) First note that for $m \geq 1$, $E_m(R)$ is a finite group, and so, for all $n \geq 1$, $H_n(E_m(R))$ is finite. (Homology of a finite group is finite.) Now, put $K_{n,m}(R) = \pi_n\left(BGL_m^+(R)\right)$. Then, by the stability result of Suslin [202], $K_{n,m}(R) \simeq K_{n,m+1}(R)$ if $m \geq \max(2n+1, s.r.R+1)$, where $s.r.R + 1$ is the stable range of R (see [20]). Also for $n \geq 2$, $\pi_n\left(BGL_m^+(R)\right) \simeq \pi_n\left(BE_m^+(R)\right)$ since $BE_m^+(R)$ is the universal covering space of $BGL_m^+(R)$ (see [220], proof of 4.12, or [136]). Now, $H_n(BE_m(R)) \simeq H_n(BE_m(R)^+)$ by the property of the plus construction since $BE_m(R) \rightarrow BE_m^+(R)$ is acyclic. Moreover, $BE_m^+(R)$ is simply connected. So $K_n(R) = \pi_n\left(BE_m^+(R)\right)$ is finite since, by [197], 9.6.16, $\pi_n\left(BE_m^+(R)\right)$ is finite if and only if $H_n\left(BE_m^+(R)\right)$ is finite.

(ii) Note that $\overline{A} = A/\mathrm{rad}A$ is a semi-simple ring and hence regular. Also $\overline{A}$ is a finite direct product of matrix algebras over finite fields. So, by Devissage, $G_n(A) \simeq G_n(\overline{A}) \simeq K_n(\overline{A})$, which reduces to computing K_n of finite fields. Hence the result.

We next prove some finiteness results for G_n and SG_n of orders (see theorem 7.1.13 below). These results are analogues to theorem 7.1.11 for K_n, SK_n of orders.

Theorem 7.1.13 *Let R be the ring of integers in a number field F, Λ any R-order in a semi-simple F-algebra Σ, Γ a maximal order in Σ containing Λ, $\alpha_r : G_r(\Gamma) \to G_r(\Lambda)$ the map induced by the functor $\mathcal{M}(\Gamma) \to \mathcal{M}(\Lambda)$ given by restriction of scalars. Then, for all $n \geq 1$,*

  (i) *$\alpha_{2n-1} : G_{2n-1}(\Gamma) \to G_{2n-1}(\Lambda)$ has finite kernel and cokernel.*

  (ii) *$\alpha_{2n} : G_{2n}(\Gamma) \to G_{2n}(\Lambda)$ is injective with finite cokernel.*

  (iii) *$G_n(\Lambda)$ is finitely generated.*

  (iv) *$SG_{2n-1}(\Lambda)$ is finite and $SG_{2n}(\Lambda) = 0$.*

  (v) *$SG_{2n-1}(\Lambda_{\underline{p}})$, $SG_{2n-1}(\widehat{\Lambda}_{\underline{p}})$ are finite of order relatively prime to the rational prime p lying below $\underline{p}$ and $SG_{2n}(\Lambda_{\underline{p}}) = SG_{2n}(\widehat{\Lambda}_{\underline{p}}) = 0$.*

PROOF

(i) Since Λ is also a $\mathbb{Z}$-order, there exists a non-zero integer s such that $\Lambda \subset \Gamma \subset \frac{1}{s}\Lambda$ (see [171]). If $S = \{1, s, s^2, \ldots\}$, we have $\Lambda_S = R\left[\frac{1}{s}\right] \otimes \Lambda = R\left[\frac{1}{s}\right] \otimes \Gamma = \Gamma_S$.

Now, since $G_{2n}(\Gamma/s\Gamma) = G_{2n}(\Lambda/s\Lambda) = 0$ (see theorem 7.1.12), we have the following commutative diagram where the top and bottom sequences are exact and ρ is an isomorphism:

$$\cdots \longrightarrow G_{2n-1}(\Gamma/s\Gamma) \xrightarrow{\gamma_{2n-1}} G_{2n-1}(\Gamma) \xrightarrow{\delta_{2n-1}} G_{2n-1}(\Gamma_S) \longrightarrow 0$$
$$\downarrow \qquad\qquad \downarrow{\alpha_{2n-1}} \qquad\qquad \downarrow{\rho}$$
$$\cdots \longrightarrow G_{2n-1}(\Lambda/s\Lambda) \xrightarrow{\gamma'_{2n-1}} G_{2n-1}(\Lambda) \xrightarrow{\delta'_{2n-1}} G_{2n-1}(\Lambda_S) \longrightarrow 0$$

From the right-hand commutative square, we have an exact sequence

$$0 \to \ker\alpha_{2n-1} \to \ker\delta_{2n-1} \to \ker\delta'_{2n-1} \to \text{Coker}\alpha_{2n-1} \to 0.$$

Now, $\text{Ker}\delta_{2n-1} = \text{Im}\gamma_{2n-1}$ is finite since $G_{2n-1}(\Gamma/s\Gamma)$ is a finite group by theorem 7.1.12. Hence $\text{Ker}\alpha_{2n-1}$ is finite as a subgroup of $\text{Ker}\delta_{2n-1}$. Also $\text{Ker}\delta'_{2n-1} = \text{Im}\gamma'_{2n-1}$ is finite since $G_{2n-1}(\Gamma/s\Gamma)$ is finite. So, $\text{Coker}\alpha_{2n-1}$ is finite.

(ii) We also have the following commutative diagram where the rows are exact:

$$0 \longrightarrow G_{2n}(\Gamma) \xrightarrow{\delta_{2n}} G_{2n}(\Gamma_S) \xrightarrow{\nu_{2n}} G_{2n-1}(\Gamma/s\Gamma) \longrightarrow \cdots$$
$$\downarrow{\alpha_{2n}} \qquad\qquad \downarrow{\rho} \qquad\qquad \downarrow$$
$$0 \longrightarrow G_{2n}(\Lambda) \xrightarrow{\delta'_{2n}} G_{2n}(\Lambda_S) \xrightarrow{\nu'_{2n}} G_{2n-1}(\Lambda/s\Lambda) \longrightarrow \cdots$$

The result follows from the following exact sequence associated with the left-hand commutative square:

$$0 \to \mathrm{Ker}\alpha_{2n} \to \mathrm{Ker}\delta_{2n} \to \mathrm{Ker}\delta'_{2n} \to \mathrm{Coker}\alpha_{2n} \to \mathrm{Coker}\delta_{2n} \to \cdots .$$

where $\mathrm{Ker}\alpha_{2n} = 0$ since $\mathrm{Ker}\delta_{2n} = 0$ and $\mathrm{Coker}\delta_{2n} \approx \mathrm{Im}\nu_{2n}$ is finite as a subgroup of the finite group $G_{2n-1}(\Gamma/s\Gamma)$.

(iii) Since $\forall n \geq 1$, $K_n(\Gamma)$ is finitely generated; then, $\mathrm{Im}\alpha_n$ has a finite index in $G_n(\Lambda)$, and so, $G_n(\Lambda)$ is finitely generated.

(iv),(v) Let X be the set of prime ideals of R. Then $\forall n \geq 1$ and we have the following exact sequences (see example 6.2.1(v)):

$$\cdots \to \oplus_{\underline{p} \in X} G_r(\Lambda/\underline{p}\Lambda) \to SG_r(\Lambda) \to 0. \qquad (i)$$

$$\cdots \to G_r(\Lambda_{\underline{p}}/\underline{p}\Lambda_{\underline{p}}) \to SG_r(\Lambda_{\underline{p}}) \to 0. \qquad (ii)$$

$$\cdots \to G_r(\widehat{\Lambda}_{\underline{p}}/\underline{p}\widehat{\Lambda}_{\underline{p}}) \to SG_r(\widehat{\Lambda}_{\underline{p}}) \to 0. \qquad (iii)$$

Note that $G_r(\Lambda/\underline{p}\Lambda) \simeq G_r(\Lambda_{\underline{p}}/\underline{p}\Lambda_{\underline{p}}) \simeq K_r(\Lambda_{\underline{p}}/\mathrm{rad}\Lambda_{\underline{p}})$ is a finite group since $\Lambda_{\underline{p}}/\underline{p}\Lambda_{\underline{p}} \simeq \Lambda/\underline{p}\Lambda$ is a finite ring and $(\Lambda_{\underline{p}}/\mathrm{rad}\Lambda_{\underline{p}}) \simeq (\Lambda_{\underline{p}}/\underline{p}\Lambda_{\underline{p}})/\mathrm{rad}(\Lambda_{\underline{p}}/\underline{p}\Lambda_{\underline{p}})$. Furthermore, if $r = 2n$, $G_{2n}(\Lambda/\underline{p}\Lambda_{\underline{p}}) \simeq G_{2n}(\widehat{\Lambda}/\underline{p}\widehat{\Lambda}_{\underline{p}}) = 0$ by theorem 7.1.12(ii). If $r = 2n - 1$, then $\underset{\underline{p} \in X}{\oplus} G_r(\Lambda/\underline{p}\Lambda)$ is a torsion group, and so, from (i), $SG_{2n-1}(\Lambda)$ is torsion. But $SG_{2n-1}(\Lambda)$ is finitely generated by (iii). Hence $SG_{2n-1}(\Lambda)$ is finite.

Now, $\Lambda_{\underline{p}}/\mathrm{rad}\Lambda_{\underline{p}} \simeq \widehat{\Lambda}_{\underline{p}_{l_i}}/\mathrm{rad}\widehat{\Lambda}_{\underline{p}} \simeq \prod_{i=1}^{k} M_{n_i}(F_{q_i})$ where F_{q_i} is a finite field of order $q_i = p^{l_i}$, say for the rational prime p lying below $\underline{p}$. So, $K_{2n-1}(\Lambda_{\underline{p}}/\mathrm{rad}\Lambda_{\underline{p}}) \simeq \prod K_{2n-1}(F_{q_i}) \simeq \prod$ (cyclic groups of order $(q_i^n - 1)$) (see example 5.1.1(v)). So, $|K_{2n-1}(\Lambda_{\underline{p}}/\mathrm{rad}\Lambda_{\underline{p}})| \equiv -1 \bmod p$, and so, it follows from (ii) and (iii) above that $SG_{2n-1}(\Lambda_{\underline{p}})$ and $SG_{2n-1}(\widehat{\Lambda}_{\underline{p}})$ are finite groups of order relatively prime to p.

$\square$

Remarks 7.1.2 (i) Note that theorem 7.1.13 holds for $\Lambda = RG$, $\Lambda_{\underline{p}} = R_{\underline{p}}G$, $\widehat{\Lambda}_{\underline{p}} = \widehat{R}_{\underline{p}}G$ where G is a finite group, $R_{\underline{p}}$ is localization of R at $\underline{p}$, and $\widehat{R}_{\underline{p}} = $ completion of R at $\underline{p}$.

(ii) One can also prove easily that for all $n \geq 1$, $G_n(\Lambda_{\underline{p}})$ is a finitely generated Abelian group. Hence $G_n(R_{\underline{p}}G)$ is finitely generated.

(iii) One important consequence of theorem 7.1.13(iii) is the following result, which says that G_n of arbitrary finite algebras (i.e., R-algebra finitely generated as R-modules) are finitely generated (see theorem 7.1.14 below).

(iv) We shall also prove in 7.5 that $G_n(RV)$ is finitely generated if V is a virtually infinite cyclic group.

Theorem 7.1.14 *Let R be the ring of integers in a number field, A any finite algebra (i.e., A an R-algebra finitely generated as an R-module). Then $G_n(A)$ is finitely generated.*

PROOF It is well known (see [20] III 8.10) that there exists a nilpotent ideal N in A such that $A/N = T \times \Lambda$ where T is a semi-simple ring and Λ is an R-order in a semi-simple F-algebra. Hence $G_n(A) \simeq G_n(A/N) \simeq G_n(T) \times G_n(\Lambda)$. Now, $G_n(\Lambda)$ is finitely generated by theorem 7.1.13(iii). Note that T is a finite ring since T is torsion and finitely generated. Moreover, $G_n(T)$ is finite by theorem 7.1.12(ii). Hence the result. ⬚

7.1.5 Our next aim is to prove the following result on the vanishing of $SG_n(\Lambda)$ under suitable hypothesis on Λ. This result due to R. Lauberbacher and D. Webb (see [131]) has the interesting consequence that if R is the ring of integers in a number field F, and G a finite group, the $SG_n(RG) = 0$ for all $n \geq 1$. We also have some other consequences due to A. Kuku, i.e., $SG_n(\widehat{\Lambda}_{\underline{p}}) = 0$ and $SG_n(\widehat{R}_{\underline{p}}G) = 0$, for any prime ideal $\underline{p}$ of R (see [117] or chapter 8).

Theorem 7.1.15 [131] *Let R be a Dedekind domain with quotient field F, Λ any R-order in a semi-simple F-algebra. Assume that*

(i) $SG_1(\Lambda) = 0$.

(ii) $G_n(\Lambda)$ is finitely generated for all $n \geq 1$.

(iii) $R/\underline{p}$ is finite for all primes $\underline{p}$ of R.

(iv) If ζ is an ℓ^s-th root of unity for any rational prime ℓ and positive integer s, $\tilde{R}$ the integral closure of R in $F(\zeta)$, then $SG_1(\tilde{R} \otimes_R \Lambda) = 0$.

Then $SG_n(\Lambda) = 0$ for all $n \geq 1$.

The proof of theorem 7.1.15 will depend on the next two results – lemma 7.1.5 and theorem 7.1.16 below.

7.1.6 In the notation of theorem 7.1.15, let $S = R - 0$, $\mathcal{M}_S(\Lambda)$ the category of finitely generated S-torsion Λ-modules, $\underline{m}$ a maximal two-sided ideal of Λ, $k(\underline{m})$ a finite field extension of $R/R \cap \underline{m}$ such that $\Lambda/\underline{m}$ is a full matrix ring over $k(\underline{m})$. Recall that for any exact category $\mathcal{C}$, $K_n(\mathcal{C}, \mathbb{Z}/s)$ is the mod-s K-theory of $\mathcal{C}$ as defined in example 5.2.2(viii). These groups are also discussed further in chapter 8.

We now prove the following.

Lemma 7.1.5 *In the notation of 7.1.6, there is an isomorphism*

$$\prod_{\underline{m} \subset \Lambda max\ ideal} K_n(k(\underline{m}), \mathbb{Z}/\ell^\nu) \overset{\cong}{\to} K_n(\mathcal{M}_S(\Lambda); \mathbb{Z}/\ell^\nu).$$

PROOF Using Devissage argument, one has that $K_n(\mathcal{M}_S(\Lambda)) \simeq \oplus K_n$ (End(T)) where T runs through all isomorphism classes of simple Λ-modules. Now, let T be a simple Λ-module. Then, $ann_\Lambda(T)$ is a two-sided ideal of Λ, which contains a product $\underline{m}_1 \cdot \underline{m}_2 \cdots \underline{m}_r = \underline{m}_1 \cap \underline{m}_2 \cap \cdots \cap \underline{m}_r$ of maximal two-sided ideals $\underline{m}_1, \underline{m}_2, \cdots, \underline{m}_r$ of Λ. Hence T is a simple module over $\Lambda/\underline{m}_1 \cdot \underline{m}_2 \cdots \underline{m}_r \cong \Lambda/\underline{m}_1 \times \Lambda/\underline{m}_2 \times \cdots \Lambda/\underline{m}_r$. Hence T is a simple module over $\Lambda/\underline{m}_i$ for some i. If $\underline{m}$ is a maximal two-sided ideal of Λ, then $\Lambda/\underline{m}$ is a simple finite-dimensional algebra over $R/(R \cap \underline{m})$.

Since the residue fields of R are finite, it follows that $\Lambda/\underline{m}$ is a full matrix ring over a finite field extension $k(\underline{m})$ of $R/R \cap \underline{m}$, and the result follows from the fact that K-theory is Morita-invariant. □

Theorem 7.1.16 *In the notation of theorem 7.1.1, we have that for all odd $n \geq 1$ and rational primes ℓ*

$$SG_n(\Lambda, \mathbb{Z}/\ell^\nu) = Ker(G_n(\Lambda, \mathbb{Z}/\ell^\nu) \to G_n(\Sigma, \mathbb{Z}/\ell^\nu)) = 0$$

where $\nu \geq 2$ if $\ell = 2$.

PROOF Recall that Quillen's localization sequence

$$\cdots G_{n+1}(\Sigma) \to G_n(\mathcal{M}_S(\Lambda)) \to G_n(\Lambda) \to G_n(\Sigma) \to \cdots$$

is induced by the sequence of exact categories

$$\mathcal{M}_S(\Lambda) \to \mathcal{M}(\Lambda) \to \mathcal{M}(\Sigma). \tag{I}$$

Since R is contained in the centre of Λ, we obtain a commutative diagram of exact functors and exact categories

$$
\begin{array}{ccccc}
\mathcal{M}_S(\Lambda) \times \mathcal{P}(R) & \longrightarrow & \mathcal{M}(\Lambda) \times \mathcal{P}(R) & \longrightarrow & \mathcal{M}(\Lambda) \times \mathcal{P}(R) \\
\downarrow & & \downarrow & & \downarrow \\
\mathcal{M}_S(\Lambda) & \longrightarrow & \mathcal{M}(\Lambda) & \longrightarrow & \mathcal{M}(\Sigma)
\end{array} \tag{II}
$$

where the vertical functors are given by $- \otimes_R -$ and are biexact. Since $\ell^\nu \neq 2$, the vertical functors induce pairing on K-groups with or without coefficients.

From [131] corollary 7.14, we now have a commutative diagram

$$
\begin{array}{ccc}
G_2(\Sigma, \mathbb{Z}/\ell^\nu) \times K_{n-1}(R, \mathbb{Z}/\ell^\nu) & \longrightarrow & G_{n+1}(\Sigma, \mathbb{Z}/\ell^\nu) \\
\downarrow{\scriptstyle \partial_2 \times id} & & \downarrow{\scriptstyle \partial_{n+1}} \\
\prod_{\underline{m} \subset \Lambda} K_1\left(k(\underline{m}); \mathbb{Z}(\ell^\nu)\right) \times K_{n-1}(R, \mathbb{Z}/\ell^\nu) & \longrightarrow & \prod_{\underline{m} \subset \Lambda} K_n\left(k(\underline{m}); \mathbb{Z}/\ell^\nu\right)
\end{array} \qquad \text{(III)}
$$

So, in order to prove theorem 7.1.16, it suffices to show that ∂_{n+1} is surjective.

Now, consider the following commutative diagram of exact functors

$$
\begin{array}{ccc}
\prod_{\underline{m}} \mathcal{M}(\Lambda/\underline{m}) \times \mathcal{P}(R) & \longrightarrow & \prod_{\underline{m}} \mathcal{M}(\Lambda/\underline{m}) \\
\downarrow & & \downarrow \\
\mathcal{M}_S(\Lambda) \times \mathcal{P}(R) & \longrightarrow & \mathcal{M}_S(\Lambda)
\end{array} \qquad \text{(IV)}
$$

where the top arrow is given by $(\oplus_i \mathcal{M}_i, P) \to \oplus_i M \otimes_R P)$ and the vertical functors induce isomorphisms in K-theory (by Devissage); then the bottom horizontal map in III is induced by the top horizontal map in IV.

If $\mathrm{char}(F) = \ell$, then $\mathrm{char}(k(\underline{m})) = \ell$ for all $\underline{m} \subset \Lambda$. Since $k(\underline{m})$ is finite, then $K_n(k(\underline{m}))$ contains no ℓ-torsion for n odd, and so, it follows from the Bockstein sequence (see chapter 8)

$$
K_n(k(\underline{m}) \xrightarrow{\ell^\nu} K_n(k(\underline{m}) \to K_n(k(\underline{m}), \mathbb{Z}/\ell^\nu) \to K_{n-1}(k(\underline{m}) \to \cdots
$$

that $K_n(k(\underline{m}), \mathbb{Z}/\ell^\nu) = 0$ for all odd n. Hence, in this case, ∂_{r+1} is trivially surjective.

Now, suppose $\mathrm{char}(F) \neq \ell$. We shall prove the theorem under the assumption that R contains a primitive ℓ^ν-th root of unity ζ. Let $\underline{p} \subset R$ be a prime ideal such that $\mathrm{char}(R/\underline{p}) \neq \ell$. It was shown in [30] theorem 2.6 that $\underset{n \geq 0}{\oplus}; K_n(R/\underline{p}; \mathbb{Z}/\ell^\nu)$ is a graded $\mathbb{Z}/\ell^\nu$-algebra isomorphic to $\Lambda(\alpha_{\underline{p}}) \otimes_{\mathbb{Z}/\ell^\nu} P(\beta_{\underline{p}})$ where $\Lambda(\alpha_p)$ is the exterior algebra generated over $\mathbb{Z}/\ell^\nu$ by a generator $\alpha_{\underline{p}}$ of the cyclic group $K_1(R/\underline{p}; \mathbb{Z}/\ell^\nu) \cong (R/\underline{p})^*/\ell^\nu(R/\underline{p})^*$ and $P(\beta_{\underline{p}})$ is the polynomial algebra generated over $\mathbb{Z}/\ell^\nu$ by an element $\beta_{\underline{p}} \in K_2(R/\underline{p}; \mathbb{Z}/\ell^\nu)$ mapped by the Bockstein to the image of $\zeta \in R^* \cong K_1(R)$ in $K_1(R/\underline{p})$ (see [30]). In the case where $\mathrm{char}(R/\underline{p}) = \ell$, we have that $K_n(R/\underline{p}; \mathbb{Z}/\ell^\nu) = 0$ for all odd n (see [30]).

If k is a finite field extension of R/p, then, if $\mathrm{char}(R/p) \neq \ell$,

$$
\oplus_{n \geq 0} K_n(k; \mathbb{Z}/\ell^\nu) \cong \bigwedge(\alpha') \otimes_{\mathbb{Z}/\ell^\nu} P(\beta_p)
$$

where α' is a generator of $K_1(k; \mathbb{Z}/\ell^\nu) \cong k^*/\ell^\nu k^*$ [131]. Squares in an exterior algebra are zero, and so, for odd n, $K_n(R/p; \mathbb{Z}/\ell^\nu)$, resp. $K_n(k; \mathbb{Z}/\ell^\nu)$, is a cyclic $\mathbb{Z}/\ell^\nu$-module, generated by $\alpha_p \cdot \beta_p^{(n-1)/2}$, resp. $\alpha' \cdot \beta_p^{(n-1)/2}$. As shown above, $\oplus_{n \leq 0} K_n(\underline{m}); \mathbb{Z}/\ell^\nu) \cong \bigwedge(\alpha_{\underline{m}}) \otimes_{\mathbb{Z}/\ell^\nu} P(\beta)$, where $\alpha_{\underline{m}}$ is a generator of $K_1(k(\underline{m}); \mathbb{Z}/l^\nu)$, and $\beta_{\underline{m}}$ is the image of the Bott element $\beta \in K_2(R; \mathbb{Z}/l^\nu)$ (see [131]) under the map

$$K_2(R; \mathbb{Z}/l^\nu) \longrightarrow K_2(R/p; \mathbb{Z}/l^\nu) \stackrel{\cong}{\longrightarrow} K_2(k(\underline{m}); \mathbb{Z}/l^\nu).$$

Our aim is to show that ∂_{n+1} in square III is onto. So, let $\Sigma_{\underline{m}} y_{\underline{m}}(\beta)_{\underline{m}}^{(n-1)/2}$ be a typical element of $\coprod K_n(k(\underline{m}); \mathbb{Z}/l^\nu)$. From diagram (IV) we see that this element is the image of $(\sigma_{\underline{m}} y_{\underline{m}})(\beta)^{(n-1)/2}$ in $\coprod_{\underline{m}} K_1(k(\underline{m}); \mathbb{Z}/\ell^\nu) \times K_{n-1}(R; \mathbb{Z}/\ell^\nu)$ under the bottom horizontal map in square III, which, therefore, is onto. By assumption, ∂_2 is onto, hence so is $\partial_2 \times id$. It follows that ∂_{n+1} is onto. This proves the theorem under the assumption that R contains a primitive ℓ^νth root of unity.

In the general case, let ζ be a primitive ℓ^νth root of unity in an algebraic closure of R, and let $\tilde{F} = F(\zeta)$ and $\tilde{R}$ be the integral closure of R in $\tilde{\Lambda} = \tilde{R} \otimes_R \Lambda$, and $\tilde{\Sigma} = \tilde{F} \otimes_F \Sigma$. Restriction of scalars induces a map of localization sequences

$$
\begin{array}{ccc}
G_{n+1}(\tilde{\Sigma}; \mathbb{Z}/\ell^\nu) & \xrightarrow{\tilde{\partial}_{n+1}} & \coprod_{\tilde{p} \subset \tilde{M}} K_n(k(\tilde{\underline{m}}); \mathbb{Z}/\ell^\nu) \\
\downarrow & & \downarrow \\
G_{n+1}(\Sigma; \mathbb{Z}/\ell^\nu) & \longrightarrow & \coprod_{p \subset M} K_n(k(\underline{m}); \mathbb{Z}/\ell^\nu)
\end{array}
$$

It follows from the ideal theory of orders (see [38]) that for given $\tilde{\underline{m}}$ and $\underline{m}$, the corresponding component of f is zero unless $\tilde{\underline{m}}$ lies over $\underline{m}$.
In this case, it is the transfer map $K_n(k(\tilde{\underline{m}}); \mathbb{Z}/\ell^\nu) \to K_n(k(\underline{m}); \mathbb{Z}/\ell^\nu)$ induced by the field extension $k(\tilde{\underline{m}}/k$.

Since n is odd, the map $K_n(k(\tilde{m}); \mathbb{Z}/\ell^\nu) \to K_n(k(\underline{m}); \mathbb{Z}/\ell^\nu)$ is onto ([196], Lemma 9). Therefore, f is onto. Since ∂_{n+1} is also onto, this completes the proof. $\Box$

Proof of theorem 7.1.15. If n is even, then it follows from lemma that $K_n(\mathcal{M}_S(\Lambda)) = 0$. Therefore $SG_n(\Lambda) = 0$ also.

So, assume that n is odd. By hypothesis, $G_n(\Lambda)$ is finitely generated, and so, $SG_n(\Lambda)$ is finite by hypothesis (iii). Let $x \in SG_n(\Lambda)$. We shall show that the ℓ-primary component x_i of x is zero for all primes ℓ. Let ν be an integer large enough so that ℓ^ν annihilates the ℓ-primary component of $G_n(\Lambda)$. (If $\ell = 2$, also choose $\nu \geq 2$.) Since the Bockstein sequence for K-theory with

$\mathbb{Z}/\ell^\nu$-coefficients is functorial, we obtain a commutative diagram

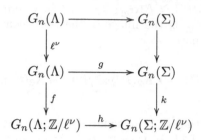

Write $G_n(\Lambda) = B \oplus C \oplus D$, where B is the torsion-free part of $G_n(\Lambda)$, C is the prime-to-ℓ torsion, and D is the ℓ-primary part. Then the kernel of f equals $(\ell^\nu \cdot B) \oplus C$.

Write $x = y + x_i$ with $x_i \in D$, and $y \in C$. Then, since $g(x) = 0$, we have

$$0 = hf(x) = hf(x_i + y) = hf(x_i) + hf(y) = hf(x_i).$$

But h is one-to-one by theorem 7.1.16, and hence $f(x_i) = 0$. Since $x_i \in D$, this implies that $x_i = 0$.

This completes the proof of Theorem 7.1.15.

Corollary 7.1.3 Let R be the ring of integers in a number field F, G a finite group. Then $SG_n(RG) = 0$ for all $n > 0$.

PROOF   It was proved in [100] that $SG_1(RG) = 0$. Also, by theorem 7.1.13, $G_n(RG)$ is finitely generated for all $n \geq 1$. So, the hypothesis of 7.1.15 are satisfied for $\Lambda = RG$ and the result follows.   □

Remarks 7.1.3   (i) In [117] A.O   Kuku showed that in the notation of 7.1.1, $SG_n(\widehat{\Lambda_p}) = 0$ for any prime ideal p of R if Λ satisfies the hypothesis of 7.1.15. Hence $SG_n(\widehat{R_p}G) = 0$ (see chapter 8).

(ii) Recall that 7.1.16 was proved for odd n. In [117] A. Kuku showed that for all even n, $SG_n(\Lambda, \mathbb{Z}/\ell^\nu) = 0$ in the notation of 7.1.16 if Λ satisfies the hypothesis of 7.1.15 (see chapter 8). Hence $SG_n(RG, \mathbb{Z}/\ell^\nu) = 0$ for all even n.

(iii) In [117] A. Kuku showed also that if Λ satisfies the hypothesis of 7.1.15, then for all $n \geq 1$ $SK_n(\Lambda) \cong \mathrm{Ker}(K_n(\Lambda) \to G_n(\Lambda))$, i.e., $SK_n(\Lambda)$ is isomorphic to the kernel of the Cartan map (see chapter 8).

Before we close the section, we present the following computation (theorem 7.1.17) due to A. Kuku that, if G is a finite p-group, then $SK_{2n-1}(\mathbb{Z}G)$, $SK_{2n-1}(\widehat{\mathbb{Z}}_\ell G)$ are finite p-groups for all $n \geq 1$ (see [121]). Recall that we showed already in 7.1.8 (ii) that $SK_n(\mathbb{Z}G)$ is finite for all $n \geq 1$, and for any finite group G.

Theorem 7.1.17 *Let G be a finite p-group. Then*

(a) $SK_{2n-1}(\mathbb{Z}G)$ *is a finite p-group.*

(b) *For any rational prime ℓ, $SK_{2n-1}(\widehat{\mathbb{Z}}_\ell G)$ is a finite p-group (or zero) for all $n \geq 1$.*

PROOF

(a) Let Γ be a maximal order containing $\mathbb{Z}G$.
If $|G| = p^s$ and we write $\mathbf{b} = p^s\Gamma$, then the Cartesian square

$$
\begin{array}{ccc}
\mathbb{Z}G & \longrightarrow & \Gamma \\
\downarrow & & \downarrow \\
\mathbb{Z}G/\mathbf{b} & \longrightarrow & \Gamma/\mathbf{b}
\end{array}
$$

yields a long Mayer - Vietoris sequence

$$\cdots \to K_{n+1}(\Gamma/\mathbf{b})\left(\tfrac{1}{p}\right) \to K_n(\mathbb{Z}G)\left(\tfrac{1}{p}\right) \to K_n(\Gamma)\left(\tfrac{1}{p}\right) \oplus K_n(\mathbb{Z}G/\mathbf{b})\left(\tfrac{1}{p}\right)$$
$$\to K_n(\Gamma/b)\left(\tfrac{1}{p}\right) \to K_{n-1}(\mathbb{Z}G)\left(\tfrac{1}{p}\right) \to \cdots$$

$$(I)$$

(see [33, 237]).

Since p^s annihilates $(\mathbb{Z}G)/\mathbf{b}$ and $\Gamma/\mathbf{b}$ then $(\mathbb{Z}G)/\mathbf{b}$ and $\Gamma/\mathbf{b}$ are $\mathbb{Z}/p^s$-algebras. Moreover, $I = rad((\mathbb{Z}G)/\mathbf{b})$, $J = rad(\Lambda/\mathbf{b})$ are nilpotent in the finite rings $(\mathbb{Z}G)/\mathbf{b}$ and $\Gamma/\mathbf{b}$, respectively, and so, by [236] 5.4, we have for all $n \geq 1$ that $K_n(\mathbb{Z}G)/\mathbf{b}, I)$ and $K_n(\Gamma/\mathbf{b}, J)$ are p-groups. By tensoring the exact sequences II and III below by $\mathbb{Z}\left(\tfrac{1}{p}\right)$

$$\cdots \to K_{n+1}((\mathbb{Z}G)/\mathbf{b})/I) \to K_n((\mathbb{Z}G)/\mathbf{b}, I) \to K_n((\mathbb{Z}G)/\mathbf{b}) \to \quad (II)$$

$$\cdots \to K_{n+1}((\Gamma/\mathbf{b})/J) \to K_n((\Gamma)/\mathbf{b}, J) \to K_n((\Gamma)/\mathbf{b}) \to \cdots \quad (III)$$

we have $K_n(\Gamma/\mathbf{b})\left(\tfrac{1}{p}\right) \cong K_n((\Gamma/\mathbf{b})/J)\left(\tfrac{1}{p}\right)$ and $K_n((\mathbb{Z}G/\mathbf{b})\left(\tfrac{1}{p}\right) \cong K_n((\mathbb{Z}G/\mathbf{b})/I)\left(\tfrac{1}{p}\right)$.

Now, $(\mathbb{Z}G/\mathbf{b})/I$ and $(\Gamma/\mathbf{b})/J$ are finite semi-simple rings and hence direct products of matrix algebras over finite fields. So, by Quillen's results, $K_{2r}(\mathbb{Z}G/\mathbf{b})\left(\tfrac{1}{p}\right) \simeq K_{2r}((\mathbb{Z}G/\mathbf{b})/I)\left(\tfrac{1}{p}\right) = 0$ and $K_{2r}((\Gamma/\mathbf{b}))\left(\tfrac{1}{p}\right) \cong K_{2r}((\Gamma/\mathbf{b})/J)\left(\tfrac{1}{p}\right) = 0$. So, the exact sequence I above becomes

$$0 \to K_{2r-1}(\mathbb{Z}G)\left(\frac{1}{p}\right) \to K_{2r-1}(\Gamma)\left(\frac{1}{p}\right) \oplus K_{2r-1}(\mathbb{Z}G/\mathbf{b})\left(\frac{1}{p}\right) \to \cdots$$

which shows that $K_{2r-1}(\mathbb{Z}G) \to K_{2r-1}(\Gamma)$ is a monomorphism *mod p*-torsion: i.e., $Ker(K_{2r-1}(\mathbb{Z}G) \to K_{2r-1}(\Gamma))$ is a p-torsion group. It is also finite since it is finitely generated as a subgroup of $K_{2r-1}(\mathbb{Z}G)$, which is finitely generated (see theorem 7.1.13). Hence $\mathrm{Ker}(K_{2r-1}(\mathbb{Z}G) \xrightarrow{\beta} K_{2r-1}(\Gamma))$ is a finite p-group.

Now, the exact sequence associated to composite $\alpha = \gamma\beta$ in the commutative diagram

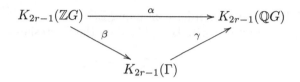

is

$$0 \to Ker\beta \to SK_{2r-1}(\mathbb{Z}G) \to SK_{2r-1}(\Gamma) \to \cdots$$

where $Ker\beta$ is a finite p-group.

Now, $\Gamma = \mathbb{Z} \oplus (\oplus_{i=1}^t M_{n_i}(\mathbb{Z}[\omega_i]))$. But it is a result of Soule [196] that if F is a number field and O_F the ring of integers of F, then $SK_n(O_F) = 0 \ \forall n \geq 1$. Hence $SK_n(\Gamma) = 0 \ \forall n \geq 1$. So, $SK_{2r-1}(\mathbb{Z}G) \simeq Ker\beta$ is a finite p-group.

(b) Let $S = \mathbb{Z} - 0$, $\widehat{S}_\ell = \widehat{\mathbb{Z}_\ell} - 0$. Then by applying the Snake lemma to the following commutative diagram

$$
\begin{array}{ccccccccc}
0 \to & \dfrac{K_{2n}(\mathbb{Q}G)}{Im(K_{2n}(\mathbb{Z}G))} & \longrightarrow & K_{2n-1}(\mathcal{H}_S(\mathbb{Z}G) & \longrightarrow & SK_{2n-1}(\mathbb{Z}G) \to 0 \\
& \downarrow & & \downarrow\scriptstyle{\wr\wr} & & \downarrow \\
0 \to & \underset{\ell}{\oplus}\dfrac{K_{2n}(\widehat{Q}_\ell G)}{Im(K_{2n}(\widehat{\mathbb{Z}}_\ell G))} & \longrightarrow & \underset{\ell}{\oplus} K_{2n-1}(H_{\widehat{S}_i}(\widehat{\mathbb{Z}}_\ell G) & \longrightarrow & \underset{\ell}{\oplus} SK_{2n-1}(\widehat{\mathbb{Z}}_\ell G) \to 0
\end{array}
$$

where $K_{2n-1}(\mathcal{H}_S(\mathbb{Z}G)) \xrightarrow{\sim} \underset{\ell}{\oplus} K_{2n-1}(H_{\widehat{S}_i}(\widehat{\mathbb{Z}}_\ell G))$ is an isomorphism, we obtain a surjective map

$$SK_{2n-1}(\mathbb{Z}G) \twoheadrightarrow \underset{\ell}{\oplus}(\widehat{S}K_{2n-1}(\widehat{\mathbb{Z}}_\ell G). \tag{I}$$

Now, it is well known that for $\ell \neq p$, $\widehat{\mathbb{Z}}_\ell G$ is a maximal order in a split semi-simple algebra $\widehat{Q}_\ell G$. Now, when $\widehat{\mathbb{Z}}_\ell G$ is a maximal order in $\widehat{Q}_\ell G$, we have from theorem 7.1.3 that $SK_{2n-1}(\widehat{\mathbb{Z}}_\ell G) = 0$ iff $\widehat{Q}_\ell G$ splits. Hence $\bigoplus_\ell SK_{2n-1}(\widehat{\mathbb{Z}}_\ell G)$ in I reduces to only $SK_{2n-1}(\widehat{\mathbb{Z}}_p G)$.

But by (a), $SK_{2n-1}(\mathbb{Z}G)$ is a finite p-group. Hence so is $SK_{2n-1}(\widehat{\mathbb{Z}}_p G)$.

$\square$

Remarks 7.1.4 Even though we do not have finite generation results for $K_n(\Lambda)$, $G_n(\Lambda)$, of p-adic orders Λ, the following results, theorem 7.1.18 concerning finite kernel and cokernel, are quite interesting.

Theorem 7.1.18 *Let R be the ring of integers in a p-adic field F, Λ any R-order in a semi-simple F-algebra Σ, Γ a maximal R-order containing Λ. Then, for all $n \geq 2$,*

    (i) The canonical map $K_n(\Gamma) \to K_n(\Sigma)$ has finite kernel and cokernel.

    (ii) The canonical map $G_n(\Lambda) \to G_n(\Sigma)$ has finite kernel and cokernel.

    (iii) $\alpha_n : G_n(\Gamma) \to G_n(\Lambda)$ has finite kernel and cokernel where α_n is the map induced by the functor $\underline{M}(\Gamma) \to \underline{M}(\Lambda)$ given by restriction of scalars.

PROOF   Left as an exercise.   ⧠

7.2   Ranks of $K_n(\Lambda)$, $G_n(\Lambda)$ of orders and grouprings plus some consequences

---

$(7.2)^A$ Ranks of K_n and G_n of orders Λ

7.2.1 Let R be the ring of integers in a number field F, Λ any R-order in a semi-simple F-algebra Σ. We proved in theorem 7.1.11(i) that for all $n \geq 1$, $K_n(\Lambda)$ is a finitely generated Abelian group, and in 7.1.13(iii) that $G_n(\Lambda)$ is also finitely generated.

The aim of this section is to obtain information about the ranks of these groups. More precisely, we show in theorem 7.2.1 that if Γ is a maximal R-order containing Λ, then for all $n \geq 2$, $\mathrm{rank}K_n(\Lambda) = \mathrm{rank}G_n(\Lambda) = \mathrm{rank}(K_n(\Gamma)) = \mathrm{rank}(K_n(\Sigma))$. It follows that if $\Lambda = RG$ (G a finite group), then $\mathrm{rank}K_n(RG) = \mathrm{rank}G_n(RG) = \mathrm{rank}K_n(\Gamma) = \mathrm{rank}K_n(FG)$. These results are due to A.O. Kuku (see [115]).

We also prove an important consequence of the results above, namely that for all $n \geq 1$, $K_{2n}(\Lambda)$, $G_{2n}(\Lambda)$ are finite groups. This is also due to A.O. Kuku (see [121]).

Theorem 7.2.1 *Let R be the ring of integers in a number field F, Λ any R-order in a semi-simple F-algebra Σ, Γ a maximal R-order containing Λ. Then for all $n \geq 2$ $rankK_n(\Lambda) = rankK_n(\Gamma) = rankG_n(\Lambda) = rankK_n(\Sigma)$.*

The proof of theorem 7.2.1 will be in several steps.

Theorem 7.2.2 *Let R be the ring of integers in a number field F, and Γ a maximal R-order in a semi-simple F-algebra Σ. Then the canonical map $K_n(\Gamma) \to K_n(\Sigma)$ has finite kernel and torsion cokernel for all $n \geq 2$. Hence, $rankK_n(\Gamma) = rankK_n(\Sigma)$.*

PROOF   Since Γ, Σ are regular, we have $K_n(\Gamma) \simeq G_n(\Gamma)$ and $K_n(\Sigma) \simeq G_n(\Sigma)$. So, we show that $G_n(\Gamma) \to G_n(\Sigma)$ has finite kernel and torsion cokernel. Now, $SG_n(\Gamma) = SK_n(\Gamma)$ is finite for all $n \geq 1$ (being finitely generated and torsion, see theorem 7.1.11). Also the localization sequence of Quillen yields

$$\cdots \to G_{n+1}(\Gamma) \to G_{n+1}(\Sigma) \to \bigoplus_{\underline{p}} G_n(\Gamma/\underline{p}\Gamma) \cdots ,$$

where $\underline{p}$ runs through the prime ideals of R. Now, for all $n \geq 1$, each $G_n(\Gamma/\underline{p}\Gamma)$ is finite since $\overline{\Gamma}/\underline{p}\Gamma$ is finite (see theorem 7.1.12(ii)). So, $\oplus_{\underline{p}}G_n(\Gamma/\underline{p}\Gamma)$ is torsion. Hence, $G_{n+1}(\Sigma)/\mathrm{Im}(G_{n+1}(\Gamma))$ is torsion, as required.   ⬚

Lemma 7.2.1 (Serre) *Let $A \to B \oplus K \to C \oplus L \to D$ be an exact sequence of Abelian groups. If A, B, C and D are finite (resp. torsion), then the kernel and cokernel of $K \to L$ are both finite (resp. torsion).*

Theorem 7.2.3 *Let R be the ring of integers in a number field F, Λ any R-order in a semi-simple F-algebra, Γ a maximal order containing Λ. Then, for all $n \geq 1$, the map $G_n(\Gamma) \to G_n(\Lambda)$ induced by the functor $\mathcal{M}(\Gamma) \to \mathcal{M}(\Lambda)$ given by restriction of scalars has finite kernel and cokernel. Hence, $rankG_n(\Gamma) = rankK_n(\Gamma) = rankG_n(\Lambda)$.*

PROOF   There exists a non-zero element $s \in R$ such that $\Lambda \subset \Gamma \subset \Lambda(1/s)$. Let $S = \{s^i\}$, $i \geq 0$. Then $\Lambda_S = \Lambda \otimes_R R_S \simeq \Gamma \otimes_R R_S = \Gamma_S$. We show that for all $n \geq 1$, $\alpha_n : G_n(\Gamma) \to G_n(\Lambda)$ has finite kernel and cokernel.

Consider the following commutative diagram of exact sequences:

$$\cdots \to G_n(\Gamma/s\Gamma) \xrightarrow{\beta_n} G_n(\Gamma) \longrightarrow G_n(\Gamma_S) \longrightarrow G_{n-1}(\Gamma/s\Gamma) \to \cdots \quad \text{(I)}$$

$$\cdots \to G_n(\Lambda/s\Lambda) \xrightarrow{\beta'_n} G_n(\Lambda) \xrightarrow{\rho'_n} G_n(\Lambda_S) \longrightarrow G_{n-1}(\Lambda/s\Lambda)$$

with vertical maps σ_n and δ, where δ is an isomorphism.

From I we extract the Mayer - Vietoris sequence

$$G_n(\Gamma/s\Gamma) \to G_n(\Lambda/s\Lambda) \oplus G_n(\Gamma) \to G_n(\Lambda) \to G_{n-1}(\Gamma/s\Gamma). \quad \text{(II)}$$

Now, since $\Gamma/s\Gamma$ and $\Lambda/s\Lambda$ are finite and $n \geq 1$, all the groups in II above are finite (see theorem 7.1.12(ii)) except $G_n(\Gamma)$ and $G_n(\Lambda)$. The result now follows from Lemma 7.2.1.   ⬚

Theorem 7.2.4 *Let R be the ring of integers in a number field F, Λ any R-order in a semi-simple F-algebra, Γ a maximal order containing Λ. Then, for all $n \geq 1$ the map $K_n(\Lambda) \to K_n(\Gamma)$ has finite kernel and cokernel. Hence $\mathrm{rank}K_n(\Lambda) = \mathrm{rank}K_n(\Gamma)$.*

To be able to prove theorem 7.2.4, we first prove the following.

Theorem 7.2.5 *Let R be the ring of integers in a number field F, Λ any R-order in a semi-simple F-algebra, Γ a maximal order containing Λ. Then, for all $n \geq 1$, the map $K_n(\Lambda) \to K_n(\Gamma)$ (induced by the inclusion map $\Lambda \to \Gamma$) is an isomorphism mod torsion.*

PROOF   First note that since every R-order is a $\mathbb{Z}$-order, there exists a non-zero integer s such that $\Lambda \subset \Gamma \subset \Lambda(1/s)$. Put $\underline{q} = s\Gamma$. Then we have a Cartesian square

$$
\begin{array}{ccc}
\Lambda & \longrightarrow & \Gamma \\
\downarrow & & \downarrow \\
\Lambda/\underline{q} & \longrightarrow & \Gamma/\underline{q}
\end{array}
\qquad\qquad (\mathrm{I})
$$

Now, by tensoring I with $\mathbb{Z}(1/s)$, if we write $A(1/s)$ for $A \otimes \mathbb{Z}(1/s)$ for any Abelian group A, we have long exact Mayer - Vietoris sequence (see [33, 237])

$$
\cdots K_{n+1}(\Gamma/\underline{q})\left(\frac{1}{s}\right) \xrightarrow{\rho} K_n(\Lambda)\left(\frac{1}{s}\right) \xrightarrow{\eta} K_n(\Gamma)\left(\frac{1}{s}\right) \oplus K_n(\Lambda/\underline{q})\left(\frac{1}{s}\right)
$$

$$
\xrightarrow{\alpha} K_n(\Gamma/\underline{q})\left(\frac{1}{s}\right) \to K_{n-1}(\Lambda)\left(\frac{1}{s}\right) \to \cdots \qquad (\mathrm{II})
$$

Now, $\Lambda/\underline{q}$ and $\Gamma/\underline{q}$ are finite rings, and so, $K_n(\Lambda/\underline{q})$ and $K_n(\Gamma/\underline{q})$ are finite groups (see theorem 7.1.12). The result is now immediate from Lemma 7.2.1. □

Proof of theorem 7.2.4.   Let $\alpha_n : K_n(\Lambda) \to K_n(\Gamma)$ denote the map. By Theorem 7.2.5, the kernel and cokernel of α_n are torsion. Also, for all $n \geq 1$, $K_n(\Lambda)$ are finitely generated (see theorem 7.1.11). Hence, the kernel and cokernel of α_n are finitely generated and hence finite. So, $\mathrm{rank}K_n(\Gamma) = \mathrm{rank}K_n(\Lambda)$.

As a fallout from the above, we now have the following result, which also proves that $SK_n(\Lambda)$ is finite for any R-order Λ.

Theorem 7.2.6 *Let R be the ring of integers in a number field F, Λ any R-order in a semi-simple F-algebra Σ. Then the canonical map $K_n(\Lambda) \to K_n(\Sigma)$ has finite kernel and torsion cokernel.*

PROOF   From the commutative diagram

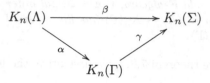

we have an exact sequence

$$0 \to \ker \alpha \to SK_n(\Lambda) \to SK_n(\Gamma) \to \frac{K_n(\Gamma}{\mathrm{Im}K_n(\Lambda)}$$

$$\to \frac{K_n(\Sigma}{\mathrm{Im}K_n(\Lambda)} \to \frac{K_n(\Sigma}{\mathrm{Im}(K_n(\Gamma))} \to 0.$$

Now, by theorem 7.2.4, $\ker\alpha$ is finite, and by theorem 7.1.13, $SK_n(\Gamma) = SG_n(\Gamma)$ is finite for all $n \geq 1$. Hence, from the exact sequence I above, $SK_n(\Lambda)$ is finite.

Also, by theorem 7.2.4, $K_n(\Gamma)/\mathrm{Im}K_n(\Lambda)$ is finite, and by Theorem 7.2.2, $K_n(\Sigma)/\mathrm{Im}(K_n(\Gamma))$ is torsion. Hence the result.   ▯

Remarks 7.2.1   (i) The above results hold for $\Lambda = RG$ where G is any finite group.

(ii) The ranks of $K_n(R)$ and $K_n(F)$ are well known and are due to Borel (see [26, 27]). More precisely, let r_1 be the number of embeddings of F in $\mathbb{R}$ and r_2 the number of distinct conjugate pairs of emebeddings of F in $\mathbb{C}$ with image not contained in $\mathbb{R}$. Then

$$\mathrm{rank}K_n(F) = \begin{cases} 1 & \text{if } n = 0, \\ \infty & \text{if } n = 1, \\ 0 & \text{if } n = 2k \ k > 0, \\ r_1 + r_2 & \text{if } n = 4k + 1, \\ r_2 & \text{if } n = 4k + 3, \end{cases}$$

$$\mathrm{rank}K_n(R) = \begin{cases} 1 & \text{if } n = 0, \\ r_1 + r_2 - 1 & \text{if } n = 1, \\ r_1 + r_2 & \text{if } n = 4k + 1, \\ r_2 & \text{if } n = 4k + 3, \\ 0 & \text{if } n = 2k \ k > 0, \end{cases}$$

It means that if Σ is a direct product of matrix algebras over fields and Γ is a maximal order in Σ, then $\mathrm{rank}K_n(\Gamma) = \mathrm{rank}K_n(\Sigma)$ is completely determined since $\Sigma = \prod M_{n_i}(F_i)$ and $\Gamma = \prod M_{n_i}(R_i)$ where R_i is the ring of integers in F_i. Also, by theorem 7.2.1, this is equal to rank $G_n(\Lambda)$ as well as rank $K_n(\Lambda)$ if Λ is any R-order contained in Γ.

However, if Σ does not split, there exists a Galois extension E of F that splits Σ, in which case we can reduce the problem to that of computation of ranks of K_n of fields.

$(7.2)^B$ $K_{2n}(\Lambda)$, $G_{2n}(\Lambda)$ are finite for all $n \geq 1$ and for all R-orders Λ

Theorem 7.2.7 *Let R be the ring of integers in a number field F, Λ any R-order in a semi-simple F-algebra Σ. Then $K_{2n}(\Lambda)$, $G_{2n}(\Lambda)$ are finite groups for all $n \geq 1$. Hence, $K_{2n}(RG)$, $G_{2n}(RG)$ are finite groups for all $n \geq 1$.*

To prove 7.2.7 we shall first prove the following.

Theorem 7.2.8 *Let R be the ring of integers in a number field F, Γ a maximal R-order in a semi-simple F-algebra Σ. Then $K_{2n}(\Gamma)$ is a finite group for all $n \geq 1$.*

Remarks 7.2.2 Since Γ is Morita equivalent to $\prod_{i=1}^{r} M_{n_i}(\Gamma_i)$, and $\Sigma = \prod_{i=1}^{r} M_{n_i}(D_i)$, say, where Γ_i is a maximal R-order in a division algebra D_i, it suffices to prove that $K_{2n}(\Gamma)$ is finite if Γ is a maximal order in a central division algebra D over a number field F. To accomplish this, we first prove the following result 7.2.2.

Lemma 7.2.2   *(a) Let D be a division algebra of dimension m^2 over its center F. For $n \geq 0$, let*

$$i_n : K_n(F) \to K_n(D)$$

be the homomorphism induced by the inclusion map $i : F \hookrightarrow D$; and

$$tr_n : K_n(D) \to K_n(F)$$

the transfer map. Then for all $n \geq 0$, each of $i_n \circ tr_n$ and $tr_n \circ i_n$ is multiplication by m^2.

  (b) Let R be the ring of integers in a number field F and D a central division algebra over F of dimension m^2, Γ a maximal R-order in D, $i_n^1 : K_n(R) \to K_n(\Gamma)$ the homomorphism induced by the inclusion $i^1 : R \hookrightarrow \Gamma$, and $tr_n^1 : K_n(\Gamma) \to K_n(R)$ the transfer map. Then, for all $n \geq 0$, $i_n^1 \circ tr_n^1$ and $tr_n^1 \circ i_n^1$ are multiplication by m^2.

PROOF

  (a) Every element d of D acts on the vector space D of dimension m^2 over F via left multiplication, i.e., there is a natural inclusion

$$t : D \to M_{m^2}(F).$$

This inclusion induces the transfer homomorphism of K-groups

$$t_n : K_n(D) \to K_n(M_{m^2}(F)) \simeq K_n(F).$$

The composition of t with $i : F \hookrightarrow D$, namely,

$$F \xrightarrow{i} D \xrightarrow{t} M_{m^2}(F)$$

is diagonal, i.e.,

$$t \circ i(x) = diag(x, x, \ldots, x).$$

So, by Lemma 1 of [65], $t_n \circ i_n$ is multiplication by m^2.

The composition

$$D \xrightarrow{t} M_{m^2}(F) \xrightarrow{i} M_{m^2}(D)$$

is not diagonal. But we will prove that it is equivalent to the diagonal map. By the Noether - Skolem Theorem, there is an inner automorphism φ such that the following diagram commutes.

$$
\begin{array}{ccc}
D & \xrightarrow{t} & M_{m^2}(F) \\
{\scriptstyle diag}\downarrow & & \downarrow{\scriptstyle i} \\
M_{m^2}(D) & \xrightarrow[\varphi]{} & M_{m^2}(D),
\end{array}
$$

where $diag$ is the diagonal map. By the Lemma 2 of [65], the induced homomorphism $K_n(\varphi)$ is an identity. So $i_n \circ tr_n$ is multiplication by m^2, also by Lemma 1 of [65].

(b) Proof of (b) is similar to that of (a) above with appropriate modification, which involves the use of a Noether - Skolem theorem for maximal orders, which holds since any R-automorphism of a maximal order can be extended to an F-automorphism of D.

$\Box$

Proof of 7.2.8. As observed in 7.2.2, it suffices to prove the result for Γ, a maximal order in a central division F-algebra D of dimension m^2, say. Now, we know that for all $n \geq 1$, $K_{2n}(\Gamma)$ is finitely generated. So, it suffices to show that $K_{2n}(\Gamma)$ is torsion. So, let $tr : K_{2n}(\Gamma) \to K_{2n}(R)$ be the transfer map and $i : K_{2n}(R) \to K_{2n}(\Gamma)$ the map induced by the inclusion $R \hookrightarrow \Gamma$. Let $x \in K_{2n}(\Gamma)$. Then $i \circ tr(x) = x^{m^2}$ (I). Since $K_{2n}(R)$ is torsion (see [196]), $tr(x)$ is torsion, and so, from (I), x is torsion. Hence $K_{2n}(\Gamma)$ is torsion. But $K_{2n}(\Gamma)$ is finitely generated. Hence $K_{2n}(\Gamma)$ is finite.

Proof of theorem 7.2.7. Let Γ be a maximal order in Σ containing Λ. By theorem 7.1.12, $K_{2n}(\Lambda)$ is finite, and so, $\mathrm{rank}K_{2n}(\Gamma) = 0$. Also, by 7.2.1, $\mathrm{rank}K_{2n}(\Lambda) = \mathrm{rank}G_{2n}(\Lambda) = \mathrm{rank}K_{2n}(\Gamma) = 0$. Hence $K_{2n}(\Lambda)$, $G_{2n}(\Lambda)$ are finite groups for all $n \geq 1$.

7.3   Decomposition of $G_n(RG)$ $n \geq 0$, G finite Abelian group; Extensions to some non-Abelian groups, e.g., quaternion and dihedral groups

7.3.1 The aim of this section is, first, to obtain decompositions for $G_n(RG)$ $n \geq 0$, R a left Noetherian ring with identity, G a finite Abelian group as a generalization of Lenstra's decomposition theorem 2.4.2, and then to extend the decomposition to some non-Abelian groups. Note that the results in $(7.3)^A$, $(7.3)^B$, and $(7.3)^C$ below are all due to D. Webb (see [231]).

Recall from 2.4 that if C is a finite cyclic group $< t >$, say, $\mathbb{Z}(C) :=$ $\mathbb{Z}C/(\Phi_{|C|}(t)) \cong \mathbb{Z}[\zeta_{|C|}]$ and $\mathbb{Z} < C >= \mathbb{Z}(C) \left(\frac{1}{|C|}\right)$ – all in the notation of 2.4. If R is an arbitrary ring, $R(C) = R \otimes_{\mathbb{Z}} \mathbb{Z}(C), R < C >= R \otimes \mathbb{Z} < C >$.

If G is a finite Abelian group, $QG \simeq \prod_{C \in X(G)} Q(C), \Gamma = \prod_{C \in X(G)} \mathbb{Z}(C)$ is a maximal $\mathbb{Z}$-order in QG containing $\mathbb{Z}G$. (Here $X(G)$ is the set of cyclic quotients of G.) If we write $A = \prod_{C \in X(G)} \mathbb{Z} < C >$, and R is an arbitrary ring,

$$R \otimes \Gamma = \prod_{C \in X(G)} R(C), R \otimes A = \prod_{C \in X(G)} R < C > .$$

Note that $M \in \mathcal{M}(\mathbb{Z}G)$ implies that $R \otimes M \in \mathcal{M}(RG)$ and that we have a functor $(I).\mathcal{M}(R \otimes \Gamma) \xrightarrow{res} \mathcal{M}(RG)$ induced by restriction of s-calars. Also flatness of A over Γ yields a functor (extension of scalars) $\mathcal{M}(R \otimes \Gamma) \xrightarrow{ext} \mathcal{M}(R \otimes A)(II)$.

In this section we shall prove among other results the following.

Theorem 7.3.1 [231] *If G is a finite Abelian group and R a Noetherian ring, then for all $n \geq 0$, $G_n(RG) \simeq \bigoplus_{C \in X(G)} G_n(R < C >)$, i.e., $G_n(RG) \simeq$ $G_n(R \otimes A)$.*

$(7.3)^A$ Lenstra functor and the decomposition

7.3.2 Let $S = \mathbb{Z} - 0$. The proof of theorem 7.3.1 above involves the definition of Lenstra functor $L : \mathcal{M}_S(\Gamma) \to \mathcal{M}_S(\Gamma)$, which is a homotopy e-quivalence of classifying spaces. The functor $res : \mathcal{M}(\Gamma) \to \mathcal{M}(\mathbb{Z}G)$ induces a functor $res : \mathcal{M}_S(\Gamma) \to \mathcal{M}_S(\mathbb{Z}G)$, and $ext : \mathcal{M}_S(\Gamma) \to \mathcal{M}(A)$ induces $ext : \mathcal{M}_S(\Gamma) \to \mathcal{M}_S(A)$, and we shall see that L carries the homotopy fiber of res into the homotopy fiber of ext (a topological analogue of Lenstra's observation that L carries the relations $\mathcal{R}$, in the Heller - Reiner presentation $G_0(\mathbb{Z}G) \simeq G_0(\Gamma)/\mathcal{R}_1$ (see [82]), into the relation $\mathcal{R}_2$ in the presentation

$G_0(A) \cong G_0(\Gamma)/\mathcal{R}_2$ arising in the localization sequence $\Gamma \to A$. This enables one to map the homotopy fiber sequence

$$\Omega BQ\mathcal{M}(\mathbb{Z}G) \longrightarrow \Omega BQ\mathcal{M}(QG) \longrightarrow BQ\mathcal{M}_S(\mathbb{Z}G)$$

to the sequence

$$\Omega BQ\mathcal{M}(A) \longrightarrow \Omega BQ\mathcal{M}(QG) \longrightarrow BQ\mathcal{M}_S(A)$$

in such away that $\Omega BQ\mathcal{M}(\mathbb{Z}G) \to \Omega BQ\mathcal{M}(A)$ is a weak equivalence and thus results in the isomorphism $G_n(\mathbb{Z}G) \simeq G_n(A)$. We now go into more details.

7.3.3 Let G be a finite group, $G(p)$ its Sylow-p-subgroup. If G is also A-belian, let $G(p') = \bigoplus_{\substack{q \text{ prime} \\ q \neq p}} G(q)$ so that we have the primary decomposition $G \simeq G(p) \times G(p')$. For any set $\mathcal{P}$ of primes, let $G(\mathcal{P}) := \prod_{p \in \mathcal{P}} G(p)$, the $\mathcal{P}$-torsion part of G.

If C is a cyclic quotient of G, write $\mathcal{P}(C)$ for the set of all rational primes dividing $|C|$. If $\mathcal{P}' \subset \mathcal{P}(C)$, the inclusion $C(\mathcal{P}') \overset{i_{\mathcal{P}'}}{\hookrightarrow} C$ induces a map of grouprings $i_{\mathcal{P}'} : BC(\mathcal{P}') \to BC$ where B is any ring with identity.

Similarly, the projection $C \overset{r_{\mathcal{P}'}}{\twoheadrightarrow} C(\mathcal{P}')$ induces $\hat{r}_{\mathcal{P}'} : BC \twoheadrightarrow BC(\mathcal{P}')$. Note that $\hat{r}_{\mathcal{P}'}, \hat{i}_{\mathcal{P}'} = $ identity on $BC(\mathcal{P}')$. The map $\hat{i}_{\mathcal{P}'} : BC(\mathcal{P}') \twoheadrightarrow B(C)$ induces a map $i_S : BC(\mathcal{P}') \to B(C)$ such that the diagram

$$\begin{array}{ccc} BC(\mathcal{P}') & \overset{\hat{i}_{\mathcal{P}'}}{\to} & BC \\ \downarrow & & \downarrow \\ B(C(\mathcal{P}')) & \overset{i_{\mathcal{P}'}}{\to} & B(C). \end{array}$$

commutes.

7.3.4 Note that in 7.3.3, the map $\hat{r}_{\mathcal{P}'}$ does not descend like $\hat{i}_{\mathcal{P}'}$. However, if we put $\mathcal{P}' = \mathcal{P}(C) - \{p\}$, we see that $\hat{r}_{\mathcal{P}'}$ descends modulo p. Indeed let $i = i_{\mathcal{P}(C)-\{p\}} : \mathbb{F}_p(C(p')) \to \mathbb{F}_p(C)$. Then there exists a left inverse $r : \mathbb{F}_p(C) \twoheadrightarrow \mathbb{F}_p(C(\mathcal{P}'))$ of i such that $\text{Ker}(r) = $ radical $J(\mathbb{F}_p(C))$ – a nilpotent ideal of $\mathbb{F}_p((C))$.

Now, put $R_p = R/pR$. Then we have induced maps $i : R_p(C(p')) \to R_p(C)$ and $r : R_p(C) \to R_p(C(p'))$ such that $ri = id$ and $\text{Ker}(r)$ is nilpotent.

Let $\hat{i} = \hat{i}_{\mathcal{P}(C)-\{p\}} : R_p C(p') \to R_p C$, $\hat{r} = \hat{r}_{\mathcal{P}(C)-\{p\}} : R_p C \to R_p C(p')$, $\nu_C : R_p G \twoheadrightarrow R_p C \twoheadrightarrow R_p(C)$; $\nu_C^* : \mathcal{M}(R_p(C)) \to \mathcal{M}(R_p G)$. We now have the following lemma.

Lemma 7.3.1 *In the diagrams below, $\hat{i}^*$ and $\hat{r}^*$ are homotopy inverse. So also are i^* and r^*. Moreover, the diagrams*

$$\begin{array}{ccc} BQ\mathcal{M}(R_p(C)) & \overset{\nu_C^*}{\to} & BQ\mathcal{M}(R_p G) \\ \downarrow{i^*} & & \downarrow{\hat{i}^*} \\ BQ\mathcal{M}(R_p(C)) & \overset{\nu_{C(p')}^*}{\to} & BQ\mathcal{M}(R_p G(p')) \end{array} \qquad (I)$$

$$BQ\mathcal{M}(R_p(C(p'))) \xrightarrow{\nu_C^*} BQ\mathcal{M}(R_pG)$$

$$\uparrow r^* \qquad\qquad\qquad \uparrow \hat{r}^* \qquad\qquad (II)$$

$$BQ\mathcal{M}(R_p(C(p'))) \xrightarrow{\nu_{C(p')}^*} BQ\mathcal{M}(R_pG(p'))$$

commute, the first strictly and the second up to homotopy.

PROOF $\hat{r}$ is left inverse to $\hat{i}$, i.e., $\hat{r}\hat{i} = id$. To show that $\hat{i}^*$ and $\hat{r}^*$ are homotopy inverses, it suffices to show that either is a homotopy equivalence. But $\hat{r}$ has a nilpotent kernel, and so, by Devissage, r^* is a homotopy equivalence. Similarly, i^* and r^* are homotopy inverses since r has a nilpotent kernel. $\qquad\qquad\qquad\qquad\qquad\qquad\qquad\qquad\qquad\qquad\qquad\Box$

Since the diagram

$$R_pG \twoheadrightarrow R_pC \twoheadrightarrow R_p(C)$$

$$\uparrow i \qquad\quad \uparrow i \qquad\quad \uparrow i$$

$$R_pG(p') \twoheadrightarrow R_pC(p') \twoheadrightarrow R_p(C(p'))$$

commutes, then homotopy-commutativity of the diagram (II) follows from strict commutativity of (I).

7.3.5 Note that $R_p \otimes \Gamma \cong \prod\limits_{C \in X(G)} R_p(C)$, and so, we have an identification

$$BQ\mathcal{M}(R_p \otimes \Gamma) \xrightarrow{\sim} \prod\limits_{C \in X(G)} BQ\mathcal{M}(R_p(C)).$$

Now define a functor

$$E : \prod\limits_{C \in X(\pi)} \mathcal{M}(R_p(C)) \to \prod\limits_{C' \in X(G(p'))} \mathcal{M}(R_p(C'))$$

by

$$(M, (C)) \to (i_{P(0)-p}^* M, (C(p')));$$

here $(M, (C))$ denotes the vector

$$(0, \ldots, 0, M, 0 \ldots, 0) \in \prod\limits_{C \in X(G)} \mathcal{M}(R_p(C))$$

with M in the Cth component. The same notation is used for elements of $\prod\limits_{C' \in X(G(p'))} \mathcal{M}(R_p(C'))$. Note that addition is by direct sum componentwise.

Lemma 7.3.2 *There is a homotopy equivalence*

$$\alpha : BQ\mathcal{M}(R_pG) \longrightarrow \prod\limits_{C' \in X(G(p'))} BQ\mathcal{M}(R_p(C'))$$

such that the diagram

$$BQ\mathcal{M}(R_p \otimes \Gamma) \xrightarrow{\quad res_p \quad} BQ\mathcal{M}(R_pG)$$

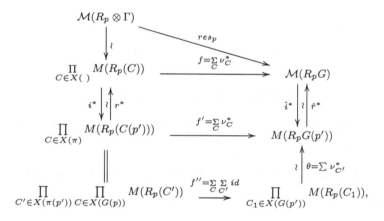

$$\prod_{C \in X(C)} BQ\mathcal{M}(R_p(C)) \xrightarrow{\quad E \quad} \prod_{C \in X(C)} BQ\mathcal{M}(R_p(C'))$$

commutes. Hence, res_p is given up to a canonical equivalence by $(M, (C)) \mapsto (i^*_{\mathcal{P}(C)-p}M, (C(p')))$.

PROOF   Consider the diagram

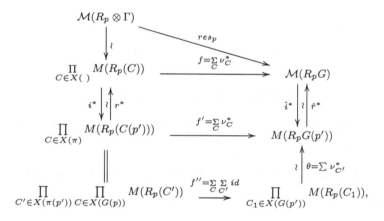

where i^* denotes the product of maps $i^*_{\mathcal{P}(C)-\{p\}}$ and r^* the product of the maps $r^*_{\mathcal{P}(C)-\{p\}}$, θ is induced by the isomorphism $\mathbb{F}_pG(p') \cong \prod_{C' \in X(G(p'))} \mathbb{F}_p(C')$ of [231], (1.9), the equality in the left column is obtained by writing $C' = C(p'), C_1 = C(p)$, and the map f'' is the sum (over C') of the maps $\prod_{C \in X(G(p))} M(R_p(C')) \to M(R_p(C'))$ given by $(M_{C_1}) \to \sum_{C_1} M_{C_1}$. By lemma 7.3.1, the upper square commutes, while the lower by definition. Moreover, E is just $f'' \circ i^*$. Define α as the right-hand column: $\alpha = \theta^{-1} \circ \hat{i}^*$. Then, $\alpha \circ res_p = f'' \circ i^* = E$, as desired.   ▯

Next, define $F: \prod_{C \in X(\pi)} \mathcal{M}(R_p(C)) \to \prod_{C' \in X(G(p'))} \mathcal{M}(R_p(C'))$ by

$$(M, (C)) \longmapsto \begin{cases} (M, (C)) & \text{if} \quad p \nmid |C| \\ 0 & \text{if} \quad p \mid |C| \end{cases}$$

Lemma 7.3.3 *There is a homotopy equivalence $\beta : BQ\mathcal{M}(R_p \otimes A) \to$*

$\prod_{C' \in X(G(p'))} BQ\mathcal{M}(R_p(C'))$ *such that the diagram*

$$
\begin{array}{ccc}
BQ\mathcal{M}(R_p \otimes \Gamma) & \xrightarrow{\;ext_p\;} & BQ\mathcal{M}(R_p \otimes A) \\
\Big\downarrow{\wr} & & {\wr}\Big\downarrow{\beta} \\
\prod_{C \in X(\pi)} BQ\mathcal{M}(R_p(C)) & \xrightarrow{\;\varepsilon\;} & \prod_{C \in X(G(p'))} BQ\mathcal{M}(R_p(C'))
\end{array}
$$

commutes. Thus, ext_p may be replaced, up to a canonical homotopy equivalence, by F.

PROOF In the diagram

$$
\begin{array}{ccc}
\mathcal{M}(R_p \otimes F) & \xrightarrow{\;ext_p\;} & \mathcal{M}(R_p \otimes A) \\
\Big\downarrow{\wr} & & {\wr}\Big\downarrow \\
\prod_{C \in X(\pi)} \mathcal{M}(R_p(C)) & \xrightarrow{\;\prod_p ext_p\;} & \prod_{C \in X(\pi)} \mathcal{M}(R_p\langle C\rangle) \\
& \searrow{F} & \Big\| \\
& & \prod_{C' \in X(\pi(p'))} \mathcal{M}(R_p(\;)),
\end{array}
$$

the equality in the right-hand column is justified by the observation that $\mathbb{F}_p\langle C\rangle = \mathbb{F}_p(C)$ if $p \nmid |C|$, i.e., if $C \in X(G(p'))$ (since $|C|$ is already a unit in $\mathbb{F}_p$), while $\mathbb{F}_p < C >= 0$ if $p \mid |C|$ (since then p is both zero and a unit). Define β as the right-hand column; then we have the required result. $\square$

7.3.6 Identifying $\mathcal{M}(R \otimes \Gamma)$ with $\prod_{C \in X(\pi)} \mathcal{M}(R(C))$ as above, define the Lenstra functor $L : \mathcal{M}(R \otimes \Gamma) \to \mathcal{M}(R \otimes \Gamma)$ by $L(M,(C)) = \sum_{\mathcal{P}' \subseteq \mathcal{P}(C)} (i_{\mathcal{P}'}^* M, (C(\mathcal{P}')))$; as above, $(M,(C))$ denotes the vector $(0,\ldots,0,M, 0,\ldots,0)$ with M in the Cth component, and the sum is meaningful since one can add in each factor $\mathcal{M}(R(C))$ by direct sum. The same formula defines Lenstra functors

$$L : \mathcal{M}_S(R \otimes \Gamma) \to \mathcal{M}_S(R \otimes \Gamma) \quad \text{and} \quad L : \mathcal{M}(R \otimes QG) \to \mathcal{M}(R \otimes QG)$$

Lemma 7.3.4 *The Lenstra functors are homotopy equivalences.*

PROOF We show this for $\mathcal{M}(R \otimes \Gamma)$, the proofs in the other cases being identical. Since $BQ\mathcal{M}(R \otimes \Gamma)$ is a CW-complex, it suffices to show that L

is a weak equivalence, i.e., that it induces an automorphism of $G_n(R \otimes \Gamma) = \pi_{n+1}(BQ\mathcal{M}(R \otimes \Gamma))$ for all $n \geq 0$. On $G_n(R \otimes \Gamma) \cong \oplus G_n(R(C))$, L_* is given by $L_*[x,(C)] = \sum_{\mathcal{P}' \subseteq \mathcal{P}(C)} [i_{\mathcal{P}}^*(x), (C(\mathcal{P}'))]$, where $[x,(C)]$ denotes the vector $(0, \ldots, 0, x, 0, \ldots, 0)$ with x in the Cth component. Define an endomorphism $\tilde{L}$ of $\bigoplus_{p \in X(C)} G_n(R(C))$ by

$$\tilde{L}[x,(C)] = \sum_{\mathcal{P}' \subseteq \mathcal{P}(C)} (-1)^{|\mathcal{P}(C) - \mathcal{P}'|}[i_{\mathcal{P}'}^*(x), (C(\mathcal{P}'))].$$

A purely formal Möbius inversion argument (see [92] or [241] for details) establishes that $\tilde{L}$ and L_* are inverse isomorphisms. So, L is a homotopy equivalence.   ▯

Lemma 7.3.5 *There is a homotopy equivalence* $\lambda : \mathcal{M}_S(RG) \to \mathcal{M}_S(R \otimes A)$ *such that*

$$\begin{array}{ccc} \mathcal{M}_S(R \otimes \Gamma) & \xrightarrow{\ L\ } & \mathcal{M}_S(R \otimes \Gamma) \\ \Big\downarrow{res^{tor}} & & \Big\downarrow{ext^{tor}} \\ \mathcal{M}_S(RG) & \xrightarrow{\ \lambda\ } & \mathcal{M}_S(R \otimes A) \end{array}$$

commutes.

PROOF   It suffices, by Devissage, to define equivalences $\lambda_p : \mathcal{M}(R_pG) \longrightarrow \mathcal{M}(R_p \otimes A)$ such that

$$\begin{array}{ccc} \mathcal{M}(R_p \otimes \Gamma) & \xrightarrow{\ L\ } & \mathcal{M}(R_p \otimes \Gamma) \\ \Big\downarrow{res_p} & & \Big\downarrow{ext_p} \\ \mathcal{M}(R_pG) & \xrightarrow{\ \lambda_p\ } & \mathcal{M}(R_p \otimes A) \end{array}$$

commutes, for each prime p. Consider the canonical homotopy equivalences

$$\alpha : \mathcal{M}(R_pG) \xrightarrow{\ \simeq\ } \prod_{C' \in X(G(p'))} \mathcal{M}(R_p(C'))$$

and $$\beta : \mathcal{M}(R_p \otimes A) \longrightarrow \prod_{C \in X(G(p'))} \mathcal{M}(R_p(C'))$$

of lemma 7.3.2 and lemma 7.3.3, relative to which res_p and ext_p can be described by $res_p(M,(C)) = (i_{\mathcal{P}(C)-\{p\}}^* M, (C(p'))$,

$$ext_p(M,(C)) = \begin{cases} (M,(C)) & \text{if} \ \ p \nmid |C| \\ 0 & \text{if} \ \ p \mid |C| \end{cases}$$

Define a section s of res_p by $s(M, (C')) = (M, (C'))$ and $\lambda_p = \mathrm{ext}_p \circ L \circ s$. We easily check that the diagram commutes, viz.,

$$
\begin{aligned}
\lambda_p \circ \mathrm{res}_p(M, (C)) &= \mathrm{ext}_p \circ L \circ s \circ \mathrm{res}_p(M, (C)) \\
&= \mathrm{ext}_p \circ L(i^*_{PC-\{p\}}(M, (C))) \\
&= \mathrm{ext}_p \left(\sum_{\mathcal{P}' \subseteq \mathcal{P}(C) - \{p\}} (i^*_{\mathcal{P}'} M, (C(\mathcal{P}'))) \right) \\
&= \sum_{\mathcal{P}' \subseteq \mathcal{P}(C) - \{p\}} (i^*_{\mathcal{P}'} M, (C(\mathcal{P}')))
\end{aligned}
$$

By the above description of ext_p, $p \nmid C(\mathcal{P}') \mid$ when $\mathcal{P}' \subseteq \mathcal{P}(C) - \{p\}$. On the other hand,

$$
\begin{aligned}
\mathrm{ext}_p \circ L(M, (C)) &= \mathrm{ext}_p \left(\sum_{\mathcal{P}' \subseteq \mathcal{P}(C)} (i^*_{\mathcal{P}'} M, (C(\mathcal{P}'))) \right) \\
&= \mathrm{ext}_p \left(\sum_{\mathcal{P}' \subseteq \mathcal{P}(C) p \in \mathcal{P}'} (i^*_{\mathcal{P}'} M, (C(\mathcal{P}'))) \right. \\
&\qquad \left. + \sum_{\mathcal{P}' \subseteq \mathcal{P}(C) - \{p\}} (i^*_{\mathcal{P}'} M, (C(\mathcal{P}'))) \right) \\
&= \sum_{\mathcal{P}' \subseteq \mathcal{P}(C) - \{p\}} (i^*_{\mathcal{P}'} M, (C(\mathcal{P}')))
\end{aligned}
$$

since all terms in the first summation vanish under ext_p, as $p \in \mathcal{P}' \Rightarrow p \| C(\mathcal{P}') \|$. Thus, $\lambda_p \circ \mathrm{res}_p = \mathrm{ext}_p \circ L$, as desired. The same Möbius inversion argument as in the proof of lemma 7.3.4 shows that λ_p induces an isomorphism on homotopy and hence is a homotopy equivalence. $\qquad \Box$

Theorem 7.3.2 *There is a weak equivalence*
$\Omega B\mathcal{Q}\mathcal{M}(RG) \to \Omega B\mathcal{Q}\mathcal{M}(R \otimes \Gamma)$.

PROOF   In the diagram

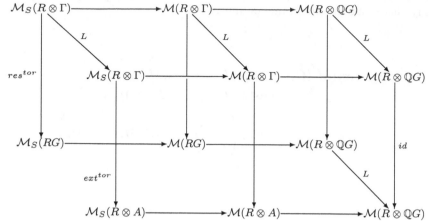

the rows are homotopy fiber sequences by Quillen's localization theorem (see 6.2.1 or [165]). So, one obtains a diagram

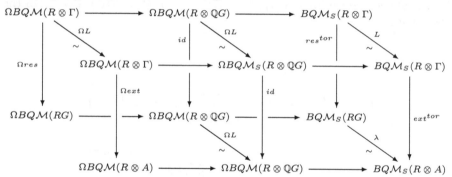

in which the rows are homotopy fiber sequences, all squares not involving λ obviously commute, and the maps L, ΩL, λ are homotopy equivalences. But the right face commutes, by lemma 7.3.5; so the bottom face also commutes. Thus there is an induced map $\rho : \Omega BQM(RG) \to \Omega BQM(R \otimes A)$ on the fiber, which is a weak equivalence, by the five lemma.   ▯

Theorem 7.3.1 now follows since ρ induces an isomorphism on homotopy groups $G_*(RG) \xrightarrow{\sim} G_*(R \otimes A) \cong \prod_{\rho \in C(\pi)} G_*(R < C >)$.

$(7.3)^B$ $G_n(RH)$, H dihedral group or non-Abelian group of order pq

Recall that we defined in 2.4 G-rings, twisted-grouprings, and crossed-product rings, and we discussed some of their properties. We shall copiously use some of these properties for the rest of this section.

We now prove the following result, which is a higher-dimensional analogue of theorem 2.4.6. Recall that a subgroup G^1 of G is cocyclic if G/G^1 is cyclic.

Theorem 7.3.3 *Let $H = G \rtimes G_1$ be the semi-direct product of G and G_1, where G is a finite Abelian group and G_1 any finite group such that the action of G_1 on G stabilizes every cocyclic subgroup of G, so that G acts on each cyclic quotient C of G. Let R be a Noetherian ring.*
Then for all $n \geq 0$, $G_n(RH) \simeq \bigoplus\limits_{C \in X(G)} G_n(R < C > \#G_1)$.

PROOF   By 2.4.7(v), $QH \cong QG\#G_1$, which by 2.4.2 and 2.4.7(iii) is $\prod\limits_{C \in X(G)} \mathbb{Q}(C)\#G$. Let $\Gamma = \prod\limits_{C \in X(G)} \mathbb{Z}(C)\#G_1$, a $\mathbb{Z}$-order containing $\mathbb{Z}H$, and let $A = \prod\limits_{\rho \in X(\pi)} \mathbb{Z} < C > \#G_1$. The maps $i_\mathcal{P}, r_\mathcal{P}, i_\mathcal{P}^*$, and $r_\mathcal{P}^*$ defined in 7.3^A are all G_1-equivariant, and so, one has induced maps

$$\hat{i}_\mathcal{P} : RG(\mathcal{P})\#G_1 \to RG\#G_1 = RH, \quad \hat{r}_\mathcal{P} : RG\#G_1 \twoheadrightarrow RG(\mathcal{P})\#G_1,$$

etc., having the same properties as in lemma 7.3.1. For example, given a cyclic quotient $C \in X(G)$ and prime p,

$$r_{\mathcal{P}(C)-\{p\}} : R_p(C)\#G_1 \to R_p(C(p'))\#G_1$$

has a nilpotent kernel, by lemma 2.4.7(ii). Hence, one can define the Lenstra functor as before, and the proof from this stage is identical to that of theorem 7.3.1.   ☐

Proposition 7.3.1 *Let H be a non-Abelian group of order pq, $p \mid q - 1$. Let G_1 denote the unique subgroup of order p of $Gal(\mathbb{Q}(\zeta_q)/\mathbb{Q})$. Then*

$$G_*(\mathbb{Z}H) \simeq G_*(\mathbb{Z}) \oplus G_* \left(\mathbb{Z} \left[\zeta_p, \frac{1}{p} \right] \right) \oplus G_* \left(\mathbb{Z} \left[\zeta_q, \frac{1}{q} \right]^{G_1} \right).$$

PROOF H is a semi-direct product $G \rtimes G_1$ where G is cyclic of order q and G_1 is cyclic of order p. By theorem 7.3.3, $G_*(\mathbb{Z}H) \simeq G_*(\mathbb{Z}\#G_1) \oplus G_* \left(\mathbb{Z} \left[\zeta_q, \frac{1}{q} \right] \#G_1 \right)$. In the first summand, G_1 acts trivially, and so, this is an ordinary groupring, and $G_*(\mathbb{Z}\#G_1) = G_*(\mathbb{Z}) \oplus G_* \left(\left[\zeta_p, \frac{1}{p} \right] \right)$ by theorem 7.3.1. In $\mathbb{Z} \left[\zeta_q, \frac{1}{q} \right] \#G_1$, G_1 acts faithfully, hence, as the unique p-element subgroup of the Galois group. But $\mathbb{Z} \left[\zeta_q, \frac{1}{q} \right]$ is unramified over $\mathbb{Z} \left[\frac{1}{q} \right]$, and hence over its invariant subring $\mathbb{Z}\left[\zeta_q, \frac{1}{q}\right]^{G_1}$. Now, by 2.4.6(vi)

$$\mathbb{Z}\left[\zeta_q, \frac{1}{q}\right] \# G_1 \simeq M_p\left(\mathbb{Z}\left[\zeta_q, \frac{1}{q}\right]^{G_1}\right),$$

which is Morita equivalent to $\mathbb{Z}\left[\zeta_q, \frac{1}{q}\right]^{G_1}$. Hence

$$G_*\left(\mathbb{Z}\left[\zeta_q, \frac{1}{q}\right] \# G_1\right) \simeq G_*\left(\mathbb{Z}\left[\zeta_q, \frac{1}{q}\right]^{G_1}\right),$$

□

Proposition 7.3.2 For the dihedral group D_{2n} of order $2n$,

$$G_*(\mathbb{Z}D_{2n}) \simeq \bigoplus_{\substack{d|2 \\ d>2}} G_*\left(\mathbb{Z}\left[\zeta_d, \frac{1}{d}\right]_+\right) \oplus G_*\left(\mathbb{Z}\left[\frac{1}{2}\right]\right)^\varepsilon \oplus G_*(\mathbb{Z}),$$

where

$$\varepsilon = \begin{cases} 1 & \text{if } n \text{ is odd,} \\ 2 & \text{if } n \text{ is even,} \end{cases}$$

and $\mathbb{Z}\left[\zeta_d, \frac{1}{d}\right]$ is the complex conjugation-invariant subring of $\mathbb{Z}\left[\zeta_d, \frac{1}{d}\right]$.

PROOF $D_{2n} \simeq G \rtimes G_1$ where G is cyclic of order n and G_1 is of order 2. So, theorem 7.3.3 yields

$$G_*(\mathbb{Z}D_{2n}) = \bigoplus_{C \in X(G)} G_*(\mathbb{Z}<C> \# G_1).$$

Now, let C be a cyclic quotient of order $d > 2$. Then $\mathbb{Z}<C> \# G_1 \simeq \mathbb{Z}\left[\zeta_d, \frac{1}{d}\right] \# G_1$ where G_1 acts by complex conjugation. Again, $\mathbb{Z}\left[\zeta_d, \frac{1}{d}\right]$ is unramified over its invariant subring $\mathbb{Z}\left[\zeta_d, \frac{1}{d}\right]_+$. Hence

$$\mathbb{Z}\left[\zeta_d, \frac{1}{d}\right] \# G_1 \cong M_2\left(\mathbb{Z}\left[\zeta_d, \frac{1}{d}\right]_+\right)$$

by 2.4.6(v) yielding $G_*(\mathbb{Z}<C> \# G_1) \simeq G_*\left(\mathbb{Z}\left[\zeta_d, \frac{1}{d}\right]_+\right)$, as above. For n even, there is a cyclic quotient C of order 2 on which G_1 acts trivially. So, $\mathbb{Z}<C> \# G_1 = \mathbb{Z}\left[\frac{1}{2}\right]G_1 \simeq \mathbb{Z}\left[\frac{1}{2}\right] \times \mathbb{Z}\left[\frac{1}{2}\right]$, yielding a contribution of $G_*\left(\mathbb{Z}\left[\frac{1}{2}\right]\right)^2$. Finally, for the trivial cyclic quotient C, $\mathbb{Z}<C> \# G_1 = \mathbb{Z}G_1$, so that $G_*(\mathbb{Z}<C> \# G_1) \simeq G_*(\mathbb{Z}) \oplus G_*\left(\mathbb{Z}\left[\frac{1}{2}\right]\right)$, by theorem 7.3.1. This completes the proof of proposition 7.3.2.   □

Example 7.3.1 For the symmetric group S_3, $G_3(\mathbb{Z}S_3) \cong \mathbb{Z}/48\mathbb{Z} \oplus \mathbb{Z}/48\mathbb{Z}$.

PROOF By proposition 7.3.1, $G_3(\mathbb{Z}S_3) \cong G_3(\mathbb{Z}) \oplus G_3\left(\mathbb{Z}\left[\zeta_3, \frac{1}{3}\right]_+\right) = G_3(\mathbb{Z}) \oplus G_3\left(\mathbb{Z}\left[\frac{1}{3}\right]\right)$. By the calculation of Lee and Szczarba [132], $G_3(\mathbb{Z}) \cong \mathbb{Z}/48\mathbb{Z}$. By Soulé's theorem [196], $SG_3(\mathbb{Z}) = SG_2(\mathbb{Z}) = 0$, and it follows easily that the localization sequence for $\mathbb{Z} \to \mathbb{Z}\left[\frac{1}{3}\right]$ breaks up, yielding $0 \to G_3(\mathbb{Z}) \to G_3\left(\mathbb{Z}\left[\frac{1}{3}\right]\right) \to G_2(\mathbb{F}_3) \to 0$. But $G_2(\mathbb{F}_3) = 0$, and so, $G_3\left(\mathbb{Z}\left[\frac{1}{3}\right]\right) \cong \mathbb{Z}/48\mathbb{Z}$. $\quad\square$

$(7.3)^C$ $G_n(RH)$, H the generalized quaternion group of order 4.2^s

7.3.7 Let H be the generalized quaternion of order 4.2^s, i.e., the subgroup of the units $\mathbb{H}^\times$ of the Hamilton quaternion algebra generated by $x = e^{\pi i/2^s}$, $y = j$. Equivalently, G has a presentation: $< x, y \mid x^{2^5} = y^2, y^4 = 1, yxy^{-1} = x^{-1} >$. For $n \geq 0$, let $G_1 = \{1, \gamma\}$ be a two-element group acting on $\mathbb{Q}[\zeta_{2^n}]$ by complex conjugation, with fixed field $\mathbb{Q}[\zeta_{2^n}]_+$, the maximal real subfield. Let $c : G_1 \times G_1 \to \mathbb{Q}[\zeta_{2^{s+1}}]^\times$ be the normalized 2-cocycle given by $c(\gamma, \gamma) = -1$, and let $\Sigma = \mathbb{Q}[\zeta_{2^{s+1}}] \#_c G_1$ the crossed-product algebra usually denoted by

$$(\mathbb{Q}[\zeta_{2^{s+1}}]/\mathbb{Q}[\zeta_{2^{s+1}}]_+, c).$$

Let Γ be a maximal $\mathbb{Z}$-order in Σ.

Proposition 7.3.3

$$G_*(\mathbb{Z}H) \simeq \bigoplus_{j=0}^{s} G_*\left(\mathbb{Z}\left[\zeta_{2^j}, \frac{1}{2^j}\right]_+\right) \oplus G_*\left(\Gamma\left[\frac{1}{2^{s+1}}\right]\right) \oplus G_*\left(\mathbb{Z}\left[\frac{1}{2}\right]\right)^2$$

PROOF Let $G = < x >$, a cyclic subgroup of index 2, and let $G_1 = H/G = \{1, \gamma\}$, where γ is the image of y. Using the normalized transversal $G_1 \to H$ given by lifting $\gamma \in G_1$ to $y \in H$, one can see that the extension $1 \to G \to H \to G_1 \to 1$ is determined by the normalized 2-cocycle $z : G_1 \times G_1 \to G$ defined by $z(\gamma, \gamma) = \nu$, where $\nu = x^{2^s} = y^2$ is the element of G of order 2.

By 2.4.7(v) and 7.3.1, $\mathbb{Q}H \simeq \mathbb{Q}G \#_z G_1 \simeq \left(\prod_{j=0}^{s+1} \mathbb{Q}[\zeta_{2^j}]\right) \#_z G_1$. Since the identification $\mathbb{Q}G \xrightarrow{\sim} \prod_{j=0}^{s+1} \mathbb{Q}[\zeta_{2^j}]$ is given by $x \mapsto (\zeta_{2^j})_j$, the image of $z(\gamma, \gamma) = \nu$ is $(1, \ldots, 1, -1)$, so by 2.4.7(iii), $\mathbb{Q}H \xrightarrow{\sim} \prod_{j=0}^{s+1} \mathbb{Q}[\zeta_{2^j}] \#_c G_1$, where all cocycles but the last are trivial. For $j = 0$ or 1, $\mathbb{Q}[\zeta_{2^j}] = Q$, and G_1 is acting trivially.

So, $\mathbb{Q}[\zeta_{2^j}]\#G_1 = \mathbb{Q}G_1 \simeq \mathbb{Q} \times \mathbb{Q}$. Thus $\mathbb{Q}H \xrightarrow{\sim} \mathbb{Q} \times \mathbb{Q}^3 \times \prod_{j=0}^{s} \mathbb{Q}[\zeta_{2^j}]\#G_1 \times \Sigma$ is the Wedderburn decomposition of $\mathbb{Q}H$, where the first factor corresponds to the trivial representation.

Let $B = \mathbb{Z} \times \mathbb{Z}^3 \times \prod_{j=2}^{s} \mathbb{Z}[\zeta_{2^j}]\#G_1, \times \mathbb{Z}[\zeta_{2^{s+1}}]\#_c G_1$, a $\mathbb{Z}$-order in $\mathbb{Q}H$ containing $\mathbb{Z}H$, and let

$$A = \mathbb{Z} \times \mathbb{Z}\left[\frac{1}{2}\right]^3 \times \prod_{j=2}^{s} \mathbb{Z}\left[\zeta_{2^j}, \frac{1}{2^j}\right]\#G_1 \times \mathbb{Z}\left[\zeta_{2^{s+1}}, \frac{1}{2^{s+1}}\right]\#_c G_1.$$

Then we have restriction-of-scalars and extension-of-scalars maps res : $\mathcal{M}(B) \to \mathcal{M}(\mathbb{Z}H)$, ext : $\mathcal{M}(B) \to \mathcal{M}(A)$. Let $B = B_1 \times \cdots \times B_{s+4}$, $A = A_1 \times \cdots \times A_{s+4}$. For each factor B_j of B, there is a unique ring map $\mathbb{Z} = B_0 \xrightarrow{i_j} B_j$. Let i_j denote also the mod-2 reduction $\mathbb{F}_2 = B_0/2B_0 \to B_j/2B_j$.

Next, note that each $B_j/2B_j$ admits a map $r_j : B_j/2B_j \to \mathbb{F}_2$, which is left-inverse to i_j and has a nilpotent kernel. For the $\mathbb{Z}$-factors $B_1, \ldots, B_4$, this is obvious. For the factors $\mathbb{Z}[\zeta_{2^j}]\#G_1$ with $2 \leq j \leq s$, this follows from the fact that 2 is totally ramified in $\mathbb{Z}[\zeta_{2^j}]$ ([241], proposition 7.4.1), so that $2\mathbb{Z}[\zeta_{2^j}] = \mathfrak{p}^{2^{j-1}}$, where $\mathfrak{p}$ is the unique prime over 2, and the residue extension $\mathbb{F}_2 \to \frac{\mathbb{Z}[\zeta_{2^j}]}{\mathfrak{p}}$ is trivial. Then there is map $\frac{\mathbb{Z}[\zeta_{2^j}]}{2\mathbb{Z}[\zeta_{2^j}]}\#G_1 \twoheadrightarrow \frac{\mathbb{Z}[\zeta_{2^j}]}{\mathfrak{p}}\#G_1 \simeq \mathbb{F}_2 G_1$ with nilpotent kernel. By decomposing this map with the augmentation $\mathbb{F}_2 G_1 \twoheadrightarrow \mathbb{F}_2$, we have the desired map $\frac{\mathbb{Z}[\zeta_{2^j}]}{2\mathbb{Z}[\zeta_{2^j}]}\#\Gamma \xrightarrow{r_j} \mathbb{F}_2$. It is clear that r_j is left-inverse to i_j since i_j is just the structure map of $\mathbb{F}_2$-algebra $\frac{\mathbb{Z}[\zeta_{2^j}]}{2\,\mathbb{Z}[\zeta_{2^j}]}\#G_1$. Note that the same argument works for the last factor $B_{s+4} = \mathbb{Z}[\zeta_{2^{s+1}}]\#_c G_1$ since the cocycle c becomes trivial modulo 2. $\quad\square$

Let $\hat{i} : \mathbb{F}_2 \to \mathbb{F}_2 H$ denote the structure map, $\hat{r} : \mathbb{F}_2 H \twoheadrightarrow \mathbb{F}_2$ the augmentation, so that $\hat{r}\hat{i} = id$; $\hat{r}$ has nilpotent kernel by [230] 1.6.
The following lemma is analogous to lemma 7.3.1.

Lemma 7.3.6 *Fix a factor B_j of B; then i_j^* and r_j^* are homotopy inverses, as are $\hat{i}^*$ and $\hat{r}^*$. Moreover, the diagrams*

$$\begin{array}{ccc}
BQ\mathcal{M}(B_j/2B_j) & \longrightarrow & BQ\mathcal{M}(\mathbb{F}_2 H) \\
i_j^* \downarrow & & \hat{i}^* \downarrow \\
BQ\mathcal{M}(\mathbb{F}_2) & \xrightarrow{id} & BQ\mathcal{M}(\mathbb{F}_2)
\end{array}
\qquad
\begin{array}{ccc}
BQ\mathcal{M}(B_j/2B_j) & \longrightarrow & BQ\mathcal{M}(\mathbb{F}_2 H) \\
r_j^* \uparrow & & \hat{r}^* \uparrow \\
BQ\mathcal{M}(\mathbb{F}_2) & \xrightarrow{id} & BQ\mathcal{M}(\mathbb{F}_2)
\end{array}$$

commute, the first strictly, the second up to homotopy.

PROOF   Identical to that of lemma 7.3.1; the left diagram is obviously strictly commutative, since both vertical maps are the $\mathbb{F}_2$-algebra structure maps. $\quad\square$

Now, identify $\mathcal{M}(B/2B)$ with $\prod\limits_{j=1}^{s+4} \mathcal{M}(B_j/2B_j)$ in the obvious way. Define a

functor $E : \prod\limits_{j=1}^{s+4} \mathcal{M}(B_j/2B_j) \to \mathcal{M}(\mathbb{F}_2)$ by $(M,j) \to (i_j^* M, 1)$, where, as usual,

(M,j) denotes the vector $(0, \ldots, 0, M, 0, \ldots, 0)$ with M the jth entry.

Lemma 7.3.7 *The diagram*

$$
\begin{array}{ccc}
BQ\mathcal{M}(B/2B) & \xrightarrow{\ res_2\ } & BQ\mathcal{M}(\mathbb{F}_2 H) \\
\Big\downarrow{\wr} & & \Big\downarrow{\hat{\imath}^*} \\
\prod\limits_{j=1}^{s+4} BQ\mathcal{M}(B_j/2B_j) & \xrightarrow{\ E\ } & BQ\mathcal{M}(\mathbb{F}_2)
\end{array}
$$

commutes; thus res_2 is given up to canonical homotopy equivalences by
$(M,j) \mapsto (i_j^* M, 1)$.

PROOF   Consider the diagram

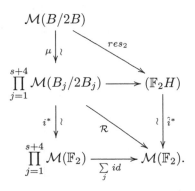

where i^* is the product of the maps $i_j^* : M(B_j/2B_j) \to M(\mathbb{F}_2)$. The square commutes, by (4.3), and the lower triangle commutes by definition of E. Thus $\hat{\imath}^* \circ res_2 = E \circ \mu$, as claimed. $\Box$

7.3.8 Define $F : \prod\limits_{j=1}^{s+1} BQ\mathcal{M}(B_j/2B_j) \to \mathcal{M}(\mathbb{F}_2, j)$ by

$$
(M,j) \longrightarrow \begin{cases} (M,j) & \text{if } j = 1 \\ 0 & \text{if } j > 1. \end{cases}
$$

Lemma 7.3.8 *The diagram*

$$
\begin{array}{ccc}
BQ\mathcal{M}(B/2B) & \xrightarrow{\;ext_2\;} & BQ\mathcal{M}(B/2B) \\
\;\;\downarrow{\scriptstyle\iota} & & \;\;\Vert{\scriptstyle\beta} \\
\overset{s+4}{\underset{j=1}{\Pi}}\, BQ\mathcal{M}(B_j/2B_j) & \xrightarrow{\;F\;} & BQ\mathcal{M}(\mathbb{F}_2)
\end{array}
$$

commutes.

PROOF   The identification β comes from the fact that for $j > 1$, $A_j/2A_j = 0$, since 2 is a unit; the assertion is then obvious. $\square$

7.3.9 Define the Lenstra functor $L : \mathcal{M}(B) \to \mathcal{M}(B)$ (via the identification $\mathcal{M}(B) \simeq \overset{s+4}{\underset{j=1}{\Pi}} \mathcal{M}(B_j)$), by

$$
L(M,j) = \begin{cases} (M,j) + (i_j^* M, 1) & \text{for } j > 1 \\ (M,j) & \text{for } j = 1. \end{cases}
$$

The same formula defines functors $L : \mathcal{M}_S(B) \to \mathcal{M}_S(B)$, $L : \mathcal{M}(\mathbb{Q}H) \to \mathcal{M}(\mathbb{Q}H)$. These functors are homotopy equivalences by the Whitehead theorem since the induced maps on homotopy groups are isomorphisms; for example, the induced map $\pi_*(BQ\mathcal{M}(B)) = \overset{s+4}{\underset{j=1}{\oplus}} G_{*-1}(B_j)$ has matrix

$$
\begin{bmatrix}
1 & i_2^* & i_3^* & \cdots & i_{s+4}^* \\
  & 1 & 0 & \cdots & 0 \\
  & & 1 & & \vdots \\
  & & & \ddots & \vdots \\
0 & & & & \ddots \\
  & & & & 1
\end{bmatrix}
$$

and hence is an isomorphism.

Lemma 7.3.9 *There is a homotopy equivalence $\lambda : M_S(\mathbb{Z}H) \to M_S(\Lambda)$ such that*

$$
\begin{array}{ccc}
\mathcal{M}_S(B) & \xrightarrow{\;L\;} & \mathcal{M}_S(B) \\
{\scriptstyle res^{tor}}\downarrow & & \downarrow{\scriptstyle ext^{tor}} \\
\mathcal{M}_S(\mathbb{Z}H) & \xrightarrow{\;\lambda\;} & \mathcal{M}_S(\Lambda)
\end{array}
$$

commutes.

PROOF   By Devissage, it suffices to define equivalences $\lambda_p : \mathcal{M}(\mathbb{F}_p H) \to \mathcal{M}(A/pA)$ such that the diagram

$$
\begin{array}{ccc}
\mathcal{M}(B/pB) & \xrightarrow{\ L\ } & \mathcal{M}(B/pB) \\
{\scriptstyle res_p}\downarrow & & \downarrow{\scriptstyle ext_p} \\
\mathcal{M}(\mathbb{F}_p H) & \xrightarrow{\ \lambda_p\ } & \mathcal{M}(A/pA)
\end{array}
$$

commutes, for each prime p. For odd p, this is trivial, since both res_p and ext_p are then isomorphisms. So, it suffices to treat the case $p = 2$. By 7.3.7 and 7.3.8, the above diagram can be replaced up to canonical homotopy equivalences by

$$
\begin{array}{ccc}
\displaystyle\prod_{j=1}^{s+4} \mathcal{M}(B_j/2B_j) & \xrightarrow{\ L\ } & \displaystyle\prod_{j=1}^{s+4} \mathcal{M}(B_j/2B_j) \\
{\scriptstyle E}\downarrow & & \downarrow{\scriptstyle F} \\
\mathcal{M}(\mathbb{F}_2) & \xrightarrow{\ \lambda_2\ } & \mathcal{M}(\mathbb{F}_2)
\end{array}
$$

For $j > 0$, $F \circ L(M, j) = F(M, j) + F(i_j^* M, 1) = i_j^* M = E(M, j)$, while for $j = 0$, manifestly $F \circ L(M, 0) = (M, 0) = E(M, 0)$. Thus one can simply take $\lambda_2 = id$. $\square$

We conclude the proof of theorem 7.3.3 with the same formal argument as that of theorem 7.3.2. One obtains the diagram

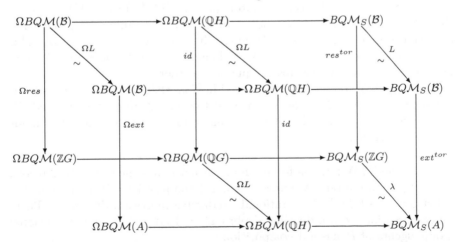

in which the rows are homotopy fiber sequences and the right face commutes; hence the bottom face commutes, inducing a weak equivalence :

$\Omega BQ\mathcal{M}(\mathbb{Z}H) \to \Omega BQ\mathcal{M}(A)$ on the fiber. It follows that

$$G_* (\mathbb{Z}H) \simeq G_*(A) \simeq G_*(\mathbb{Z}) \oplus G_* \left(\mathbb{Z}\left[\frac{1}{2}\right]\right)^3$$

$$\oplus \left(\overset{s}{\underset{j=2}{\oplus}} \ G_* \left(\mathbb{Z}\left[\zeta_{2^j}, \frac{1}{2}\right] \# G_1 \right) \right) \oplus G_* \left(\mathbb{Z}\left[\zeta_{2^{s+1}}, \frac{1}{2^{s+1}}\right] \# G_1 \right).$$

But by remarks 2.4.6(vi), $\mathbb{Z}\left[\zeta_{2^j}, \frac{1}{2^j}\right] \# G_1 \simeq M_2 \left(\mathbb{Z}\left[\zeta_{2^j}, \frac{1}{2^j}\right]_+\right)$, which is Morita equivalent to $\mathbb{Z}\left[\zeta_{2^j}, \frac{1}{2^j}\right]_+$. Similarly, since $\mathbb{Z}[\zeta_{2^{s+1}}] \#_c G_1 \subseteq \Gamma$, 2.4.7(iv) shows that

$$\mathbb{Z}\left[\zeta_{2^{s+1}}, \frac{1}{2^{s+1}}\right] \#_c G_1 = \Gamma\left[\frac{1}{2^{s+1}}\right].$$

Hence,

$$G_*(\mathbb{Z}) \simeq \overset{s}{\underset{j=0}{\oplus}} G_* \left(\mathbb{Z}\left[\zeta_{2^j}, \frac{1}{2^j}\right]_+ \right) \oplus G_* \left(\Gamma\left[\frac{1}{2^{s+1}}\right] \right) \oplus G_* \left(\mathbb{Z}\left[\frac{1}{2}\right] \right)^2$$

and the proof of theorem 7.3.3 is complete.

$(7.3)^D$ $G_n(RH)$, (H a nilpotent group) plus a conjecture of Hambleton, Taylor, and Williams

7.3.10 The aim of this subsection is to discuss a formula, due to Hambleton, Taylor, and Williams, henceforth abbreviated as HTW. In [76], HTW proved that this formula for $G_n(RH)$ holds for H any finite nilpotent group, and then conjectured that the formula should hold for any finite group H. This conjecture is seen from [231, 233, 234] to hold for dihedral groups, non-Abelian groups of order pq, and groups of square-free order.

However, D. Webb and D. Yao proved in [235] that this conjecture fails for the symmetric group S_5. This means that the conjecture must be revised. D. Webb and D. Yao think that it is reasonable to conjecture that 7.3.12 holds for solvable groups.

7.3.11 The HTW formula for any finite group H has the form 7.3.12 below. Let H be a finite group. We first obtain a decomposition of $\mathbb{Q}H$ a follows:

Let $\rho : H \to GL(V_\rho)$ be a rational irreducible representation of H. Then, we can associate to ρ a division algebra $D_\rho = \mathrm{End}_{\mathbb{Q}H}(V_\rho)$, and the rational group algebra of H has a decomposition

$$\mathbb{Q}H \cong \prod_\rho \mathrm{End}_{D_\rho}(V_\rho) \cong \prod_\rho M_{n_\rho}(D_\rho) \tag{I}$$

where ρ ranges over the set $X(H)$ of irreducible rational representations of H.

For such a ρ, let r be the order of the kernel of ρ and s the degree of any of the irreducible complex constituents of $\mathbb{C} \otimes_Q V_\rho$. Put $\omega_\rho = |H|/rs$. Let Γ_ρ be a maximal $\mathbb{Z}(1/\omega_\rho)$-order in D_ρ (or equivalently in $M_{n_\rho}(D_\rho)$ since both are Morita equivalent).

Jacobinski theorem [171] 41.3 provides some useful information about ω_ρ. Firstly, both $\mathbb{Z}[1/\omega_\rho]H$ and $\mathbb{Z}[1/\omega_\rho] \otimes_\mathbb{Z} \Gamma$ (where Γ is a maximal $\mathbb{Z}$-order in $\mathbb{Q}H$ containing $\mathbb{Z}H$) are subrings of $\mathbb{Q}H$, and their projections into the factor $\mathrm{End}_{D_\rho}(V_\rho)$ of $\mathbb{Q}H$ are equal. More generally, suppose that for $\rho_1, \ldots, \rho_t$, each ω_{ρ_i} divides r (where r is a fixed integer). Then the projections of $\mathbb{Z}\left[\frac{1}{r}\right]H$ and $\mathbb{Z}[1/r] \otimes_\mathbb{Z} \Gamma$ into $\prod_{i=1}^{\tau} \mathrm{End}_{D_{\rho_i}}(V_{\rho_i})$ are equal.

We now state the HTW conjecture.

7.3.12 HTW Conjecture. In the notation of 7.3.11, suppose that R is a Noetherian ring. Then for all $n \geq 0$, $G_n(RH) \cong \bigoplus_{\rho \in X(H)} G_n(R \otimes_\mathbb{Z} \Gamma_\rho)$ for any finite group H.

The following theorem 7.3.4, due to HTW, shows that conjecture 7.3.12 is true for finite p-groups.

Theorem 7.3.4 [76] *Let R be a Noetherian ring and H a finite p-group, (p any prime). Then, in the notation of 7.3.11, we have for all $n \geq 0$ that $G_n(RH) \cong \bigoplus_{\rho \in X(H)} G_n(R \otimes_\mathbb{Z} \Gamma_\rho)$.*

Sketch of Proof. The proof uses the result that if H is a finite p-group and R is a Noetherian ring, then we have a split exact sequence

$$0 \to G_n(RH) \to G_n\left(R\left(\frac{1}{p}\right)\right)H) \oplus G_n(R) \to G_n\left(R\left(\frac{1}{p}\right)\right) \to 0 \quad \text{(I)}$$

(see [76] for the proof of (I)).

Now for any p-group H, $\mathbb{Z}(\frac{1}{p})H$ is a maximal $\mathbb{Z}(\frac{1}{p})$-order in $\mathbb{Q}H$ (see [171] 41.1) and hence is a product of maximal $\mathbb{Z}(\frac{1}{p})$-orders in the factors in the decomposition of $\mathbb{Q}H$. But each of this is Morita equivalent to a maximal $\mathbb{Z}(\frac{1}{p})$-order in D_ρ (see [171] 21.7). If ρ is non-trivial, let Γ_ρ denote a maximal $\mathbb{Z}(\frac{1}{p})$-order in D_ρ. It is Morita equivalent to a maximal order in $M_{n_\rho}(D_\rho)$. If ρ is trivial put $\Gamma_\rho = \mathbb{Z}$.

Since G_n preserves products and Morita equivalences yields isomorphisms of G_n-groups, we have the required result by applying (I).

Remarks 7.3.1   (i) HTW observe in [76] that the Lenstra - Webb theorem for $G_n(RH)$, H Abelian, can be derived from theorem 7.3.5 as follows: Let $H = H' \oplus H(p)$ where $H(p)$ is the Sylow-p-subgroup of H. Then

$G_n(RH) = \underset{\rho}{\oplus} \, G_n(RH') \otimes_{\mathbb{Z}} \Gamma_\rho$ where the sum runs over the irreducible rational representations of $H(p)$. Moreover $G_n((RH) \otimes_{\mathbb{Z}} \Gamma_\rho) \equiv G_n((R \otimes_{\mathbb{Z}} \Gamma_\rho))H$, and so, we can proceed by induction on the order of H.

(ii) The following generalization of the above discussion is also due to HTW. So, let P be a p-group that is normal in H. Then, one can obtain a Mayer - Vietoris sequence

$$\ldots G_n(R[H/P]) \to G_n(R(1/p)[H/P] \oplus G_n(RH) \to G_n(R(1/p)H) \to \ldots$$
$$(II)$$

One could write $\mathbb{Z}(1/p)G = \mathbb{Z}(1/p)[H/P] \to \mathbb{Z}[1/p]G$. This map sends $g \in H/P$ to $1/\,|\,P\,|\, \underset{h \in \pi^{-1}(g)}{\sum} h$ where $\pi : H \to H/P$ is the projection.

From this splitting and the $M - V$ sequence (II), we have

$$G_n(RH) = G_n(R[H/P] \oplus G_n(R \otimes_{\mathbb{Z}} A).$$

Now, the quotient H/P acts on the irreducible complex representation of P. Suppose that the isotropy group of each non-trivial irreducible representation is a p-group. Then Clifford's theorem shows that ω_ρ is a power of p for each irreducible rational representation that does not factor through H/P. We can then identify A above with a piece of the maximal $\mathbb{Z}(1/p)$-order of $\mathbb{Q}H$. Hence the conjecture holds for $\mathbb{Q}H$ if it holds for $\mathbb{Q}[H/P]$. HTW claim that this situation holds for the alternating and symmetric group on four letters and certain metacyclic groups H (e.g., H having a cyclic normal subgroup P of order p^r, say, such that the composite $H/P \to \mathrm{Aut}(P) \to \mathrm{Aut}(\mathbb{Z}/p\mathbb{Z})$ has p-torsion kernel.

(iii) HTW also observes that conjecture 7.3.12 holds for all finite nilpotent groups. This is because if the conjecture holds for two groups H, H' whose orders are relatively prime, then the conjecture holds for $H \oplus H'$ (see [76]).

(iv) In [234], D. Webb proves that conjecture 7.3.12 holds for finite groups of square-free order by methods already discussed in $(7.3)^A$. For details, see [234].

(v) However, D. Webb and D. Yao proves that the conjecture fails for the symmetric group S_5. We close this section with a brief discussion of the counterexample.

7.3.13 First observe that if $R = \mathbb{Z}$, conjecture 7.3.12 becomes

$$G_n(\mathbb{Z}H) \cong \oplus G_n(\Gamma_\rho).$$

Also note that $\omega_\rho = |\, im\rho \,|/s$ (in the notation of 7.3.11).

The group S_5 provides a counterexample because the rank of $G_n(\mathbb{Z}S_5)$ predicted by the HTW conjecture is different from the rank of $G_1(\mathbb{Z}S_5)$ (based on the the formula of rank $G_1(\mathbb{Z}G)$ for any finite group G by M.E. Keating, see [100]).

7.3.14 Counterexample to HBT conjecture Now consider the case of $G = S_n$. It is well known ([90], theorem (4.12)) that the irreducible rational representations of S_n are in bijective correspondence with the partitions of n. As λ ranges over the partitions of n, the Specht modules S^λ furnish a complete set of irreducible rational representations. Moreover, the Specht modules are absolutely irreducible. The dimensions of the Specht modules are given by the *Hook length formula* ([90] theorem (20.1)):

$$\dim(S^\lambda) = n!/\Pi(\text{hook lengths}).$$

The meaning of this is the following: given a partition λ of n, consider the associated Young diagram. The *hook length* of an entry in the diagram is the number of entries in the *hook* consisting of all entries below and to the right of the given entry (in the same row or column as the given entry) in the diagram. For example, in the Young diagram

$$O \; X \; *$$
$$O \; *$$

associated to the partition (3.2) of 5, the hook length of the entry labeled X is 3, its hook consisting of the X and the asterisks. The Specht modules $S^{(3,2)}$ has dimension $5!/3 \cdot 4 \cdot 2 = 5$. From the hook length formula, it is easy to determine the degrees of the irreducible rational representations of S_5; they are tabulated below:

Partition λ : (5) (4,1) (3,2) (3,1^2) (2^2,1) (2,1^3) (1^5)

$\dim(S^\lambda)$:   1   4   5   6   5   4   1

Thus, the rational group algebra decomposes as

$$QS_5 = Q \times M_4(Q) \times M_5(Q) \times M_6(Q) \times M_5(Q) \times M_4(Q) \times Q;$$

the first factor (associated to the partition (5)) corresponds to the one-dimensional trivial representation, and the last factor (associated to (1^5)) corresponds to one-dimensional parity representation. Since each representation except the trivial representation $S^{(5)}$ and the parity representation $S(1^5)$ is faithful, each of the integers $| im\rho_\lambda |$, except for these two, is just 5!, while $| im\rho_5 | = 1$ and $| im\rho_{(1^5)} | = 2$. Since each irreducible representation is absolutely irreducible, each integer g_λ defined above coincides with $\dim(S^\lambda)$. From this, one easily tabulates the integers w_λ appearing in the conjecture; for example, $w_{(4,1)} = 5!/\dim(S^{(4,1)}) = 5!/4 = 2 \cdot 3 \cdot 5$. The result is:

λ :   (5) (4,1) (3,2) (3,1^2) (2^2,1) (2,1^3) (1^5).

w_λ :   1   2.3.5 2^3.3 2^2.5 2^3.3   2.3.5   2.

Since $G_1(\mathbb{Z}[1/w_{\rho_\lambda}])$ is the group of units of $\mathbb{Z}[1/w_{\rho_\lambda}]$, its rank is the number of prime divisors of w_{ρ_λ}. Thus, the conjecture 7.3.12 predicts that the rank of $G_1(\mathbb{Z}S_5)$ should be 13.

7.3.15 Keating's computation In [100], Keating computated $G_1(\mathbb{Z}G)$ for any finite group G. His formula for the rank of $G_1(\mathbb{Z}G)$ is the following. If the decomposition of the rational group algebra is $\mathbb{Q}G \cong \Sigma_1 \times \cdots \times \Sigma_k$, let $\wp_i$ be the maximal $\mathbb{Z}$-order in the center of Σ_i. Let r_i be the rank of the group of units of $\wp_i$, and let v_i be the number of primes of $\wp_i$ that divide $|G|$. Finally, let ε denote the number of isomorphism classes of simple $\mathbb{Z}G$ modules annihilated by $|G|$. Then Keating's rank formula is:

$$\text{rank}(G_1(\mathbb{Z}G)) = r_1 + \cdots + r_k + v_1 + \cdots + v_k - \varepsilon. \qquad (1.2)$$

In the case $G = S_5$, each $\wp_i$ is just $\mathbb{Z}$, so each r_i is 0 and each $v_i = 3$. To determine ε, it suffices to determine the number of simple FS_5-modules, where F is the finite field of order 2, 3, or 5. For this, recall that for any field F, there is a natural bilinear form on the permutation module M^λ over F arising from the Young subgroup S_λ of S_5; the Specht module S^λ is a subspace of M^λ, and one defines $D^\lambda = S_\lambda/(S^\lambda \cap S^{\lambda\perp})$. For a prime p, a partition $\lambda = (\lambda_1, \lambda_2, \lambda_3, \dots)$ of n is p-singular if there is some i for which $0 \neq \lambda_{i+1} = \lambda_{i+2} = \cdots = \lambda_{i+p}$; λ is p-regular otherwise. By [90] theorem (11.5), the simple FS_5-modules are precisely the spaces D^λ as λ ranges over all p-regular partitions of 5, where p is the characteristic of F. Since the 2-regular partitions of 5 are (5), (4,1), and (3,2), there are precisely three simple F_2S_5-modules. Similarly, there are five simple F_3S_5-modules and six simple F_5S_5-modules. Thus $\varepsilon = 14$.

Substituting into (2.1), it follows that the rank of $G_1(\mathbb{Z}G_5)$ is 7. Since this disagrees with the prediction of 7.3.12, the conjecture cannot hold in general.

It may be reasonable to conjecture that 7.3.12 holds for solvable groups.

7.4 Higher dimensional class groups of orders and grouprings

$(7.4)^A$ Generalities on higher class groups

7.4.1 In 2.3, we introduced class groups of Dedekind domains, orders, and grouprings and reviewed some of their properties.

Now, if R is a Dedekind domain with quotient field F and Λ any R-order in a semi-simple F-algebra Σ, the higher class group $C\ell_n(\Lambda)$ $n \geq 0$ are defined as

$$Cl_n(\Lambda) := Ker(SK_n(\Lambda) \to \bigoplus_p SK_n(\hat{\Lambda}_p)) \qquad (I)$$

where p runs through all the prime ideals of R and $Cl_n(\Lambda)$ coincides with the usual class group $Cl(\Lambda)$ at zero-dimensional level. Our attention in this section is focused on $Cl_n(\Lambda)$ for R-orders Λ when R is the ring of integers in a number field, and we assume in the ensuing discussion that our R-orders are of this form.

The groups $Cl_1(\Lambda)$, $Cl_1(RG)$, which are intimately connected with White-head groups and Whitehead torsion, have been extensively studied by R. Oliv-er (see [159]). It is classical that $Cl_0(\Lambda)$, $Cl_1(\Lambda)$ are finite groups. However, it follows from some results of this author that $Cl_n(\Lambda)$ is finite for all $n \geq 1$ (see theorem 7.1.11(ii)). If Γ is maximal R-order, it follows that $Cl_n(\Gamma) = 0$ for all $n \geq 1$. This result is due to M.E. Keating (see [99]).

7.4.2 A lot of work has been done, notably by R. Oliver, on $Cl_1(\Lambda)$ and $Cl_1(\mathbb{Z}G)$ where G is a finite group, in connection with his intensive study of $SK_1(\Lambda)$ and $SK_1(\mathbb{Z}G)$, $SK_1(\hat{\mathbb{Z}}_pG)$, etc. (see [159]). We note in particular the following properties of $Cl_1(\Lambda)$, where R is the ring of integers in a number field and Λ any R-order in a semi-simple F algebra.

(i) $Cl_1(\Lambda)$ is finite.

(ii) If G is any Abelian group, $Cl_1(RG) = SK_1(RG)$.

(iii) $Cl_1(\mathbb{Z}G) \neq 0$ if G is a non-Abelian p-group.

(iv) $Cl_1(\mathbb{Z}G) = 0$ if G is Dihedral or quaternion 2-group.

For further information on computations of $Cl_1(\mathbb{Z}G)$, see [159].

We now endeavor to obtain information on $Cl_n(\Lambda)$ for all $n \geq 1$.

We first see that for all $n \geq 1$, $Cl_1(\Lambda)$ is a finite group. This follows from some earlier results of the author. We state this result formally.

Theorem 7.4.1 *Let R be the ring of integers in a number field F, Λ any R-order in a semi-simple F-algebra Σ. Then $Cl_n(\Lambda)$ is a finite group for all $n \geq 1$.*

PROOF   It suffices to show that $SK_n(\Lambda)$ is finite (see theorem 7.1.11(ii)).   □

We next present a fundamental sequence involving $Cl_n(\Lambda)$ in the following.

Theorem 7.4.2 *Let R be the ring of integers in a number field F, Λ any R-order in a semi-simple F-algebra Σ. If $\underline{p}$ is any prime = maximal ideal of*

R, write $\hat{\Lambda}_{\underline{p}} = \hat{R}_{\underline{p}} \otimes \Lambda, \hat{\Sigma}_{\underline{p}} = \hat{F}_{\underline{p}} \otimes \Sigma$ where $\hat{R}_{\underline{p}}, \hat{F}_{\underline{p}}$ are completions of R and F, respectively, at $\underline{p}$. Then we have the following exact sequence:

$$0 \to K_{n+1}(\Sigma)/Im(K_{n+1}(\Lambda)) \to$$

$$\bigoplus_{\underline{p} \in max(R)} (K_{n+1}(\hat{\Sigma}_{\underline{p}})/Im(K_{n+1}(\hat{\Lambda}_{\underline{p}})) \to C\ell_n(\Lambda) \to 0 \qquad (I)$$

PROOF   See theorem 7.1.10.   □

Lemma 7.4.1 *In the exact sequence*

$$0 \to C\ell_n(\Lambda) \to SK_n(\Lambda) \to \bigoplus_{\underline{p}} SK_n(\hat{\Lambda}_p) \to 0,$$

$SK_n(\hat{\Lambda}_{\underline{p}}) = 0$ *for almost all* p, *i.e.,* $\bigoplus_{p} SK_n(\hat{\Lambda}_p)$ *is a finite direct sum.*

PROOF   See theorem 7.1.9.   □

Remarks 7.4.1   (i) In view of theorem 7.1.10, there exists a finite set $\mathcal{P}(\Lambda)$ of prime ideals of R such that, for $\underline{p} \in \mathcal{P}(\Lambda)$, $\hat{\Lambda}_{\underline{p}}$ is maximal and $\hat{\Sigma}_{\underline{p}}$ splits, in which case $SK_n(\hat{\Lambda}_{\underline{p}}) = 0$ for all $n \geq 1$. We shall often write $\mathcal{P}$ for $\mathcal{P}(\Lambda)$ when the context is clear, as well as $\check{\mathcal{P}} = \check{\mathcal{P}}(\Lambda)$ for the set of rational primes lying below the prime ideals in $\mathcal{P} = \mathcal{P}(\Lambda)$.

(ii) If $\Lambda = RG$ where G is the finite group, then the prime ideals $\underline{p} \in \mathcal{P}$ lies above the prime divisors of $|G|$. In particular, if $R = \mathbb{Z}$, then $\check{\mathcal{P}}$ consists of the prime divisors of $|G|$.

(iii) If Γ is a maximal order containing Λ such that p does not divide $[\Gamma : \Lambda] :=$ the index of Λ in Γ, then $p \in \check{\mathcal{P}}$ (see [159]).

Theorem 7.4.3 *Let R be the ring of integers in a number field F, Λ any R-order in a semi-simple F-algebra, Γ a maximal R-order containing Λ. Then for all $n \geq 1$, there exists an exact sequence*

$$0 \longrightarrow \frac{K_{n+1}(\Gamma)}{Im K_{n+1}(\Lambda)} \longrightarrow \bigoplus_{\underline{p} \in \mathcal{P}} \frac{K_{n+1}(\hat{\Gamma}_{\underline{p}})}{Im K_{n+1}(\hat{\Lambda}_{\underline{p}})} \longrightarrow C\ell_n(\Lambda) \longrightarrow 0.$$

Moreover, $\frac{K_{n+1}(\Gamma)}{Im(K_{n+1}\Lambda)}$ *and* $\bigoplus_{\underline{p} \in \mathcal{P}} \frac{K_{n+1}(\hat{\Gamma}_{\underline{p}})}{Im K_{n+1}(\hat{\Lambda}_{\underline{p}})}$ *are finite groups.*

PROOF   From theorem 7.4.2 we have a commutative diagram of exact sequences.

$$
\begin{array}{ccccccccc}
0 & \longrightarrow & \dfrac{K_{n+1}(\Sigma)}{Im K_{n+1}(\Lambda)} & \longrightarrow & \underset{p}{\oplus}\, \dfrac{K_{n+1}(\hat{\Sigma}_p)}{Im K_{n+1}(\hat{\Lambda}_p)} & \longrightarrow & C\ell_n(\Lambda) & \longrightarrow & 0 \\
 & & \downarrow & & \downarrow & & \downarrow & & \\
0 & \longrightarrow & \dfrac{K_{n+1}(\Sigma)}{Im(K_{n+1}(\Gamma))} & \longrightarrow & \underset{p}{\oplus}\, \dfrac{K_{n+1}(\hat{\Sigma}_p)}{Im K_{n+1}(\hat{\Gamma}_p)} & \longrightarrow & 0 & &
\end{array}
$$

where $\underline{p}$ ranges over all ideals of R. Taking kernels of vertical arrows, we have the required sequence.

$$
0 \longrightarrow \frac{K_{n+1}(\Gamma)}{Im(K_{n+1}(\Lambda))} \longrightarrow \underset{\underline{p}}{\oplus}\, \frac{K_{n+1}(\hat{\Gamma}_{\underline{p}})}{Im K_{n+1}(\hat{\Lambda}_{\underline{p}})} \longrightarrow C\ell_n(\Lambda) \longrightarrow 0 \qquad (I)
$$

To be able to replace the middle term of (I) by $\underset{\underline{p}\in\mathcal{P}}{\oplus}\, \frac{K_{n+1}(\Gamma_{\underline{p}})}{Im K_{n+1}(\Lambda_{\underline{p}})}$, it suffices to show that $\underset{\underline{p}\notin\mathcal{P}}{\oplus}\, \frac{K_{n+1}(\hat{\Gamma}_{\underline{p}})}{Im K_{n+1}(\hat{\Lambda}_{\underline{p}})} = 0$ for all $n \geq 1$.

Recall from remarks 7.4.1 that for $\underline{p}\in\mathcal{P}$, $\hat{\Lambda}_{\underline{p}}$ is maximal, and $\hat{\Sigma}_{\underline{p}}$ splits. Now, suppose that $n = 2r$(even). Then by theorem 7.1.1, $SK_{2n}(\hat{\Lambda}_{\underline{p}}) = 0$, and so,

$$
\frac{K_{2r+1}(\hat{\Sigma}_{\underline{p}})}{Im K_{2r+1}(\hat{\Gamma}_{\underline{p}})} = 0 \qquad (II)
$$

from the localization sequence

$$
\cdots \longrightarrow K_{2r+1}(\hat{\Gamma}_{\underline{p}}) \longrightarrow K_{2r+1}(\hat{\Sigma}_{\underline{p}}) \longrightarrow SK_{2r}(\hat{\Lambda}_{\underline{p}}) = 0.
$$

Now suppose that $n = 2r - 1$(odd); then by theorem 7.1.3 we also have $SK_{2r-1}(\hat{\Gamma}_{\underline{p}}) = 0$ since $\hat{\Sigma}_{\underline{p}}$ splits. So,

$$
\frac{K_{2r}(\hat{\Sigma}_{\underline{p}})}{Im K_{2r}(\hat{\Lambda}_{\underline{p}})} = 0 \qquad (III)
$$

from the localization sequence

$$
\longrightarrow K_{2r}(\hat{\Lambda}_{\underline{p}}) \longrightarrow K_{2r}(\hat{\Sigma}_{\underline{p}}) \longrightarrow SK_{2r-1}(\hat{\Lambda}_{\underline{p}}) = 0.
$$

Hence, from (II) and (III), we have that for all $n \geq 1$

$$
\underset{\underline{p}\notin\mathcal{P}}{\oplus}\, \frac{K_{n+1}(\hat{\Sigma}_{\underline{p}})}{Im K_{n+1}(\hat{\Lambda}_{\underline{p}}))} = 0. \qquad (IV)
$$

Now, from the commutative diagram

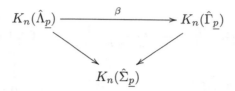

we have an exact sequence

$$\cdots \to SK_n(\Gamma_{\underline{p}}) \to \frac{K_n(\hat{\Gamma}_{\underline{p}})}{ImK_n(\hat{\Lambda}_{\underline{p}})} \to \frac{K_n(\hat{\Sigma}_{\underline{p}})}{ImK_n(\hat{\Lambda}_{\underline{p}})} \to \frac{K_n(\hat{\Sigma}_{\underline{p}})}{ImK_n(\hat{\Gamma}_{\underline{p}})} \to 0$$

Now, for all $n \geq 1$, $SK_n(\hat{\Gamma}_{\underline{p}}) = 0$ (see theorems 7.1.1 and 7.1.3). Also from (IV) above, $\frac{K_{n+1}(\hat{\Sigma}_{\underline{p}})}{Im(K_{n+1})(\hat{\Lambda}_{\underline{p}})} = 0$ for all $n \geq 1$.

Hence, from (V), $\frac{K_{n+1}(\hat{\Gamma}_{\underline{p}})}{Im(K_{n+1}(\hat{\Lambda}_{\underline{p}}))} = 0$ for all $n \geq 1$ and $\underline{p} \in \mathcal{P}$.

That $\underset{\underline{p} \in \mathcal{P}}{\oplus} \frac{K_{n+1}(\hat{\Gamma}_{\underline{p}})}{ImK_{n+1}(\hat{\Lambda}_{\underline{p}})}$ and $\frac{K_{n+1}(\Gamma)}{ImK_{n+1}(\Lambda)}$ are finite groups follows from theorem 7.2.4 and the exact sequence in the statement of theorem 7.4.3. ⟦

$(7.4)^B$ Torsion in odd dimensional higher class groups

7.4.3 There have been considerable research efforts in recent years to understand torsion in higher class groups of orders. Theorem 7.4.11, due to Kolster and Laubenbacher [102] provides information on torsion in odd-dimensional class groups $C\ell_{2n-1}(\Lambda)(n \geq 1)$ of arbitrary orders Λ in semi-simple algebras over number fields. One important consequence of this result is that the only p-torsion possible in $C\ell_{2n-1}(RG)$ (G-finite group) are for those primes p dividing the order of G. These considerations lead to computations of $C\ell_{2n-1}(\mathbb{Z}S_r)$ when S_r is the symmetric group of degree r and $C\ell_{2n-1}(\mathbb{Z}D_{2r})$, where D_{2r} is the dihedral group of order $2r$. We also express $C\ell_{2n-1}(\Lambda)$ as a homomorphic image of $\prod_{i=1}^{r} H^0(E_i, \hat{\mathbb{Q}}_p/\hat{\mathbb{Z}}_p(n))$ under the assumption that local Quillen - Lichtenbaum conjecture holds. Finally, we indicate some connections of higher class groups to homogeneous functions.

The results in this subsection are all due to M. Kolster and R. Laubenbacher (see [102]).

Definition 7.4.1 *For any ring A, we shall write $K_n^c(A)$ for the quotient of $K_n(A)$ modulo its maximal divisible subgroups.*

Now, let F be a number field with ring of integers R, D a division algebra over F, and Γ a maximal R-order in D. If $\mathcal{P}$ is a set of prime ideals in R, we define $C\ell_n(\Gamma, \mathcal{P})$ by

$$C\ell_n(\Gamma, \mathcal{P}) = Coker\left(K_{n+1}^c(D) \to \bigoplus_{\underline{p} \in \mathcal{P}} K_{n+1}^c(D_{\underline{p}}) \oplus \bigoplus_{\underline{p} \notin \mathcal{P}} K_{n+1}^c(D_{\underline{p}})/im(K_{n+1}^c(\Gamma_{\underline{p}}))\right).$$

We shall write $\check{\mathcal{P}}$ for the set of rational primes lying below the prime ideals in $\mathcal{P}$.

If Λ is an R-order in a semi-simple F-algebra Σ, we shall write $\mathcal{P}(\Lambda)$ for the set of prime ideals $\underline{p}$ of R such that $\hat{\Lambda}_{\underline{p}}$ is not a maximal order in $\hat{\Sigma}_{\underline{p}}$. Note that $\mathcal{P}(\Lambda)$ is a finite set, with r elements, say. We shall sometimes write $\mathcal{P}$ for $\mathcal{P}(\Lambda)$ and $\check{\mathcal{P}}$ for $\check{\mathcal{P}}(\Lambda)$ when the context is clear.

Lemma 7.4.2 *For all $n \geq 1$ there is an isomorphism*

$$C\ell_n(\Lambda) \cong Coker\left(\bigoplus_{\underline{p} \in \mathcal{P}(\Lambda)} K_{n+1}^c(\hat{\Lambda}_{\underline{p}}) \longrightarrow \bigoplus_{i=1}^{r} C\ell_n(\Gamma_i, \mathcal{P}(\Lambda))\right).$$

PROOF See [102]. The proof makes use of the fact that $C\ell_n(\Lambda) \simeq Coker\left(K_{n+1}(\hat{\Lambda}) \oplus K_{n+1}(\Sigma) \to K_{n+1}(\hat{\Sigma})\right)$ where the term $K_{n+1}(\hat{\Lambda}) \oplus K_{n+1}(\Sigma) \to K_{n+1}(\hat{\Sigma})$ is part of the long exact Mayer - Vietoris sequence associated with the arithmetic sequence

$$\begin{array}{ccc} \Lambda & \to & \Sigma \\ \downarrow & & \downarrow \\ \hat{\Lambda} & \to & \hat{\Sigma} \end{array}$$

where $\hat{\Lambda} = \Pi\hat{\Lambda}_{\underline{p}}$ and $\hat{\Sigma} = \prod_{\underline{p}} (\hat{\Sigma}_{\underline{p}}, \hat{\Lambda}_{\underline{p}})$ are the adele rings of Λ and Σ, respectively. Details are left to the reader (see [102]). $\quad\square$

Theorem 7.4.4 *Let F be a number field with ring of integers R, Λ any R-order in a semi-simple F-algebra Σ. Then, for all $n \geq 1$, p-torsion in $C\ell_{2n-1}(\Lambda)$ can only occur for primes p in $\check{\mathcal{P}}(\Lambda)$, i.e., $C\ell_{2n-1}(\Lambda)(q) = 0$ for $q \notin \check{\mathcal{P}}(\Lambda)$.*

PROOF In view of lemma 7.4.2, it suffices to show that in the situation of general $\mathcal{P}$ in the definition of $C\ell_n(\Gamma, \mathcal{P})$ in 7.4.1, if $q \notin \check{\mathcal{P}}$, then $C\ell_{2n-1}(\Gamma, \mathcal{P})(q) = 0$. If $\underline{p}$ is a prime ideal of R, let $\overline{\Gamma}_{\underline{p}}$ denote the residue class field of $\hat{\Gamma}_{\underline{p}}$. For any prime $q \neq char(\overline{\Gamma}_{\underline{p}})$, Suslin's rigidity theorem yields an isomorphism $K_{2n}^c(\hat{\Gamma}_{\underline{p}}) \otimes \hat{\mathbb{Z}}_q \cong K_{2n}(\overline{\Gamma}_{\underline{p}}) \otimes \hat{\mathbb{Z}}_q$ (see [204], lemma 2). This is true in particular if $\underline{p} \in \mathcal{P}(\Lambda)$ and $q \notin \check{\mathcal{P}}(\Lambda)$. Hence the q torsion in $C\ell_{2n-1}(\Gamma, \mathcal{P})$ coincides with the q-torsion in the cokernel of the map $K_{2n}^c(D) \to \bigoplus_{\underline{p}} K_{2n}^c(\hat{D}_{\underline{p}})/K_{2n}^c(\hat{\Gamma}_{\underline{p}})$. It is therefore sufficient to show that this cokernel is zero.

Consider the following commutative diagram of localization sequences:

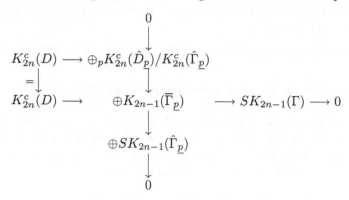

It follows from a result of Keating (see 7.1.16) that $SK_{2n-1}(\Gamma)$ $\cong \oplus\limits_{p} SK_{2n-1}(\Gamma_{\underline{p}})$, under the hypothesis that certain transfer maps in the localization sequence of a number ring are zero. This fact was proved later by Soulé [196] (see [102] for the correction of an error in [196]). The result now follows from the Snake lemma.   ⟋

Corollary 7.4.1 Let G be any finite group. Then for all $n \geq$, the only possible p-torsion in $C\ell_{2n-1}(O_F G)$ is for those p dividing the order of G.

7.4.4 Local Quillen - Lichtenbaum (Q - L) conjecture Let R be the ring of integers in a number field E. This conjecture states that for $p = char(k_{\mathbf{q}})$ where $k_{\mathbf{q}} = R/\mathbf{q}$, there is an isomorphism $K_{2n}(\hat{R}_{\mathbf{q}})(p) \cong H^0(\hat{E}_{\mathbf{q}}, \hat{Q}_p/\hat{\mathbb{Z}}_p(n))^*$. Here, for any Abelian group A, A^* denotes the dual $Hom(A, Q/\mathbb{Z})$. Also, $\mathbf{q}$ is non-zero.

Note. Local (Q - L) conjecture has been proved for p odd if $E_{\mathbf{q}}/\hat{Q}_p$ is unramified.

Theorem 7.4.5 [102] *Let R be the ring of integers in a number field F, Λ any R-order in a semi-simple F-algebra $\Sigma \cong \prod\limits_{i=1}^{r} M_{n_i}(D_i)$, D_i skew fields with center E_i. Then, for each odd prime $p \in \check{P}$, there is a surjection*

$$\prod_{i=1}^{r} H^0(E_i, \hat{Q}_p/\hat{\mathbb{Z}}_p(n)) \longrightarrow C\ell_{2n-1}(\Lambda)(p)$$

in the following case where $| \mathcal{P} | = r$.

   (i) *All the skewfields $D_i = E_i$ are commutative, and the local Q - L conjecture holds for all local fields $\hat{E}_{i\mathbf{q}}$ where $\mathbf{q}$ lies above p.*

   (ii) *For all $\mathbf{q} \in \mathcal{P}$ and all i, the local degree $deg(\hat{D}_{i_{\mathbf{q}}})$ is not divisible by the residue characteristic, and the local Q - L conjecture holds for all local fields $\hat{E}_{i\mathbf{q}}$, $\mathbf{q}$ above p.*

We now discuss some applications of the above results on odd-dimensional class groups to some grouprings.

Remarks 7.4.2 Let G be a finite group, R the ring of integers in a number field F. Let C_F be a set of irreducible F-characters. Under the assumptions of theorem 7.4.5, we obtain a surjection

$$\prod_{\chi \in C_F} H^0(F\chi, Q_p/\mathbb{Z}_p(n)) \longrightarrow Cl_{2n-1}(RG)(p)$$

for each odd prime p dividing G.

Now, observe that for a skew field D occurring in the decomposition of FG, the local degrees are prime to the residue characteristic except in the case of dyadic prime where the local degree may be 2 (see [159]). For example, if the Sylow-2-subgroups of G is elementary Abelian, then the latter case does not occur.

Theorem 7.4.6 [102] *Let S_r be the symmetric group on r letters, and let $n \geq 0$. Then $Cl_{4n+1}(\mathbb{Z}S_r)$ is a finite 2-torsion group, and the only possible odd torsion in $Cl_{4n-1}(\mathbb{Z}S_r)$ that can occur are for odd primes p such that $\frac{p-1}{2}$ divides n.*

PROOF Follows from theorem 7.4.5 (2) since the simple factors of QS_r are matrix rings over Q (see [39] theorem 75.19). ▯

Theorem 7.4.7 [102] *Let D_{2r} be the Dihedral group of order $2r$. If the local Q - L conjecture is true, then $Cl_{4n+1}(\mathbb{Z}D_{2r})$ is a finite 2-torsion group.*

PROOF Follows from theorem 7.4.5 (1) since all skew fields appearing as simple components of $\mathbb{Z}D_{2r}$ are commutative (see [38] example 7.3.9) ▯

Remarks 7.4.3 In theorems 7.4.6 and 7.4.7 we can replace $\mathbb{Z}$ by the ring of integers in a number field F that is unramified at all primes dividing $| G |$.

Remarks 7.4.4 We now provide an application of odd-dimensional higher class groups to homogeneous functions. We saw earlier in theorem 7.4.5 that there exists an epimorphism $\prod_{\chi \in C_F} H^0(F(\chi), \hat{Q}_p/\mathbb{Z}_p(n)) \to Cl_{2n-1}(RG)(p)$. We provide an alternative description of $\prod_{\chi \in C_F} H^0(F(\chi), Q_p/\mathbb{Z}_p(d))$ where p divides $| G |$ and $d \geq 1$. First we have the following.

Theorem 7.4.8 *Let G be a finite group and $d \geq 1$, F a number field and $\Omega_F = Gal(\overline{F}/F)$ the absolute Galois group of F. Then there is an isomorphism*

$$Hom_{\Omega_F}(RG, \hat{Q}_p/\hat{Z}_p(d)) \cong \prod_{\chi \in C_F} H^0(F(\chi), \hat{Q}_p/\hat{Z}_p(d))$$

PROOF See [102].   ▯

Definition 7.4.2 *Let G be a finite Abelian group, d a non-negative integer. A function $f : G \to \mathbb{Q}/\mathbb{Z}$ is homogeneous of degree d if $f(nx) = n^d f(x)$ for all $x \in G$ and all $n \in \mathbb{N}$ such that $(n, o(x)) = 1$. Denote by $hmg^d(G)$ the (finite) Abelian group (under pointwise addition) of all homogeneous functions of degree d on G.*

A subgroup H of G is called cocyclic if the quotient G/H is cyclic. Let $\phi : H \to \mathbb{Q}/\mathbb{Z}$ be a character. Then the induced character

$$\phi_H : G \longrightarrow \mathbb{Q}/\mathbb{Z}$$

defined by

$$\phi_H(x) = \begin{cases} \phi(x) & if \quad x \in H \\ 0 & otherwise, \end{cases}$$

is homogeneous of degree 1, and is called a cocyclic function. Let $Coc(G)$ denote the subgroup of $Hmg(G) = Hmg^1(G)$ generated by all cocyclic functions.

Theorem 7.4.9 *Let G be an Abelian group of odd order. Then there exists a surjection from $Hmg(\hat{G})$ onto $C\ell_1(\mathbb{Z}G)$ with kernel $Coc(\hat{G})$, i.e., we have an exact sequence*

$$0 \longrightarrow Coc(\hat{G}) \longrightarrow Hmg(\hat{G}) \longrightarrow C\ell_1(\mathbb{Z}G) \longrightarrow 0.$$

PROOF See [42].   ▯

For further information, see exercise 7.13 and [102].

$(7.4)^C$ Torsion in even-dimensional higher class groups $C\ell_{2r}(\Lambda)$ of Orders

7.4.5 In this subsection we now turn our attention to even-dimensional higher class groups $C\ell_{2r}(\Lambda)$ with the aim of obtaining information on p-torsion. Even though an analogue of theorem 7.4.4 for general R-orders is yet to be proved (if it is true), we do have a result due to Guo and Kuku that provides an analogue of theorem 7.4.4 for Eichler orders in quaternion algebras and hereditary orders in semi-simple F-algebras where F is a number field. Guo and Kuku proved the result simultaneously for "generalized Eichler orders" that combine the properties of Eichler orders and hereditary orders (see [74]).

Definition 7.4.3 *Let R be a ring. For each ideal I of R, let $(I)^{m \times n}$ denote the set of all $m \times n$ matrices with entries in I. If $\{I_{ij} : 1 \leq i, j \leq r\}$ is a set of ideals in R, we write*

$$\Lambda = \begin{pmatrix} (I_{11}) & (I_{12}) & \ldots & (I_{1r}) \\ (I_{21}) & (I_{22}) & \ldots & (I_{2r}) \\ \ldots & \ldots & \ldots & \ldots \\ (I_{r1}) & (I_{r2}) & \ldots & (I_{rr}) \end{pmatrix}^{(n_1, \ldots, n_r)}$$

to indicate that Λ is the set of all matrices $(T_{ij})_{1 \leq i, j \leq r}$, where for each pair (i, j) the matrix T_{ij} ranges over all elements of $(I_{ij})^{n_i \times n_j}$.

Definition 7.4.4 *Let $\Sigma \simeq M_n(D)$, where D is a finite-dimensional division algebra over a number field F with integers R. We call an order Λ in Σ generalized Eichler order if each $\hat{\Lambda}_{\mathbf{p}}$ has the form*

$$\begin{pmatrix} (\Delta) & (\mathbf{p}^{k_\mathcal{P}}) & (\mathbf{p}^{k_\mathcal{P}}) & \ldots & (\mathbf{p}^{k_\mathcal{P}}) \\ (\Delta) & (\Delta) & (\mathbf{p}^{k_\mathcal{P}}) & \ldots & (\mathbf{p}^{k_\mathcal{P}}) \\ (\Delta) & (\Delta) & (\Delta) & \ldots & (\mathbf{p}^{k_\mathcal{P}}) \\ \ldots & \ldots & \ldots & \ldots & \ldots \\ (\Delta) & (\Delta) & (\Delta) & \ldots & (\Delta) \end{pmatrix}^{n_1, \ldots, n_r}$$

where $k_{\mathbf{p}} \geq 1$ and Δ is the unique maximal order in $D_{(\mathbf{p})}$ satisfying $D \otimes_F \hat{F}_{\mathbf{p}} \simeq M_m(D_{(\mathbf{p})})$. If $\Sigma \simeq \underset{i}{\oplus} M_{n_i}(D_i)$ is a semi-simple algebra, then an order Λ in Σ is called a generalized Eichler order if $\Lambda \simeq \underset{i}{\oplus} \Lambda_i$, where Λ_i is a generalized Eichler order in $M_{n_i}(D_i)$. Note that if all $k_{\mathbf{p}} = 1$, then Λ is hereditary, and if Σ is a quaternion algebra, then Λ is an Eichler order as defined in [222].

Definition 7.4.5 *Let R be a Dedekind domain with quotient field F. An R-order Λ is said to be hereditary if every left Λ-lattice is projective. Note that a maximal order is hereditary but not conversely.*

7.4.6 Let F be a number field with ring of integers R, D a division algebra over F, Γ a maximal R-order in D. We shall adopt the notations in $(7.4)^B$. So, for a set $\mathcal{P}$ of prime ideals of R, let $C\ell_n(\Gamma, \mathcal{P})$ be as defined in 7.4.1. We have the following lemma.

Lemma 7.4.3 *Let R be the ring of integers in a number field F, Λ any R-order in a semi-simple F-algebra Σ. Then the even-dimensional higher class group $C\ell_{2n}(\Lambda)$ is a homomorphic image of*

$$Coker \left(\bigoplus_{\mathbf{p} \in \mathcal{P}(\Lambda)} (K_{2n+1}^c(\hat{\Lambda}_{\mathbf{p}}) \longrightarrow K_{2n+1}^c(\hat{\Sigma}_{\mathbf{p}})) \right)$$

PROOF   Let Γ be a maximal R-order containing Λ. Then, by lemma 7.4.2,

$$Cl_{2n}(\Lambda) \simeq Coker\left(\bigoplus_{\mathbf{p}\in\mathcal{P}(\Lambda)} K^c_{2n+1}(\hat{\Lambda}_{\mathbf{p}}) \longrightarrow Cl_{2n}(\Gamma,\mathcal{P})\right)$$

By theorem 7.1.1 (ii) (b),

$$\bigoplus_{\underline{\mathbf{p}}\in\mathcal{P}(\Lambda)} K^c_{2n+1}(\hat{\Sigma}_{\mathbf{p}})/im(K^c_{2n+1}(\hat{\Gamma}_{\mathbf{p}})) = 0.$$

So,

$$Cl_{2n}(\Gamma,\mathcal{P}) = Coker\left(K^c_{2n+1}(\Sigma) \longrightarrow \bigoplus_{\underline{\mathbf{p}}\in\mathcal{P}(\Lambda)} K^c_{2n+1}(\hat{\Sigma}_{\mathbf{p}})\right).$$

Hence $Cl_{2n}(\Lambda)$ is a homomorphic image of

$$Coker\left(\bigoplus_{\mathbf{p}\in\mathcal{P}(\Lambda)} K^c_{2n+1}(\hat{\Lambda}_{\mathbf{p}}) \longrightarrow \bigoplus_{\mathbf{p}\in\mathcal{P}(\Lambda)} K^c_{2n+1}(\hat{\Sigma}_{\mathbf{p}})\right).$$

□

Proposition 7.4.1 ([101] Theorem A and 2.2) Let R and S be rings, and U an $R - S$-bimodule. Then the natural homomorphisms

$$K_n\left(\begin{pmatrix} R & 0 \\ 0 & S \end{pmatrix}\right) \longrightarrow K_n\left(\begin{pmatrix} R & U \\ 0 & S \end{pmatrix}\right)$$

are isomorphisms. Let $U^* = Hom(U,S)$, where U is considered as a right S-module. If R is the endomorphism ring $End(U)$ of the right S-module U, then the natural homomorphisms

$$K_n\left(\begin{pmatrix} R & U \\ 0 & S \end{pmatrix}\right) \longrightarrow K_n\left(\begin{pmatrix} R & U \\ U^* & S \end{pmatrix}\right)$$

are surjective.

Lemma 7.4.4 *Let R be a ring*

$$R_1 = \begin{pmatrix} (R) & (0) & (0) & \cdots & (0) \\ (R) & (R) & (0) & \cdots & (0) \\ (R) & (R) & (R) & \cdots & (0) \\ \cdots & \cdots & \cdots & \cdots & \cdots \\ (R) & (R) & (R) & \cdots & (R) \end{pmatrix}^{n_1,\ldots,n_r}$$

and

$$R_2 = \begin{pmatrix} (R) \ (R) \ (R) \ \ldots \ (R) \\ (R) \ (R) \ (R) \ \ldots \ (R) \\ (R) \ (R) \ (R) \ \ldots \ (R) \\ \ldots \ \ldots \ \ldots \ \ldots \ \ldots \\ (R) \ (R) \ (R) \ \ldots \ (R) \end{pmatrix}^{n_1,\ldots,n_r}.$$

Then the natural homomorphisms

$$K_n(R_1) \longrightarrow K_n(R_2)$$

are surjective.

PROOF   We will prove this lemma by induction. If $r = 1$, then $R_1 = R_2$. This is the trivial case.

Now suppose that the lemma holds for $r - 1$. Let

$$R_3' = \begin{pmatrix} (R) \ (0) \ \ldots \ (0) \\ (R) \ (R) \ \ldots \ (0) \\ \ldots \ \ldots \ \ldots \ \ldots \\ (R) \ (R) \ \ldots \ (R) \end{pmatrix}^{n_1,\ldots,n_{r-1}}$$

and

$$R_4' = \begin{pmatrix} (R) \ (R) \ \ldots \ (R) \\ (R) \ (R) \ \ldots \ (R) \\ \ldots \ \ldots \ \ldots \ \ldots \\ (R) \ (R) \ \ldots \ (R) \end{pmatrix}^{n_1,\ldots,n_{r-1}}.$$

By induction hypothesis, the homomorphisms

$$K_n(R_3') \longrightarrow K_n(R_4')$$

are surjective. Let

$$R_3 = \begin{pmatrix} R_3' & 0 \\ 0 & (R)^{(n_r)} \end{pmatrix} \simeq R_3' \times (R)^{(n_r)}$$

and

$$R_4 = \begin{pmatrix} R_4' & 0 \\ 0 & (R)^{(n_r)} \end{pmatrix} \simeq R_4' \times (R)^{(n_r)}.$$

Since $K_n(R_3) \simeq K_n(R_3') \oplus K_n((R)^{(n_r)})$ and $K_n(R_4) \simeq K_n(R_4') \oplus K_n((R)^{(n_r)})$, the homomorphisms

$$K_n(R_3) \longrightarrow K_n(R_4)$$

are surjective. By proposition 7.4.1, the homomorphisms

$$K_n(R_3) \longrightarrow K_n(R_1)$$

are surjective. Let

$$
R_5 = \begin{pmatrix}
(R) & (R) & \cdots & (R) & (0) \\
(R) & (R) & \cdots & (R) & (0) \\
\cdots & \cdots & \cdots & \cdots & \cdots \\
(R) & (R) & \cdots & (R) & (0) \\
(R) & (R) & \cdots & (R) & (R)
\end{pmatrix}^{n_1,\ldots,n_r}.
$$

By proposition 7.4.1, the homomorphisms

$$
K_n(R_4) \longrightarrow K_n(R_5)
$$

and

$$
K_n(R_5) \longrightarrow K_n(R_2)
$$

are surjective. Hence the compositions

$$
f_{n*} : K_n(R_3) \longrightarrow K_n(R_4) \longrightarrow K_n(R_5) \longrightarrow K_n(R_2)
$$

are surjective. Let g_{n*} be the composition

$$
K_n(R_3) \longrightarrow K_n(R_1) \longrightarrow K_n(R_2).
$$

Although f_{n*} and g_{n*} are obtained in different ways, they are both induced by the same natural ring inclusion

$$
R_3 \longrightarrow R_2.
$$

By the functoriality of K-theory, $f_{n*} = g_{n*}$. Hence the maps g_{n*} are surjective, which implies that the maps

$$
K_n(R_1) \longrightarrow K_n(R_2)
$$

are surjective. $\Box$

7.4.7 Let $A \simeq M_n(D)$ be a simple algebra and Λ a generalized Eichler order in A. The local order $\Lambda_{\underline{p}}$ is either maximal or isomorphic to some

$$
\begin{pmatrix}
(\Delta) & (\underline{p}^{k_{\mathcal{P}}}) & (\underline{p}^{k_{\mathcal{P}}}) & \cdots & (\underline{p}^{k_{\mathcal{P}}}) \\
(\Delta) & (\Delta) & (\underline{p}^{k_{\mathcal{P}}}) & \cdots & (\underline{p}^{k_{\mathcal{P}}}) \\
(\Delta) & (\Delta) & (\Delta) & \cdots & (\underline{p}^{k_{\mathcal{P}}}) \\
\cdots & \cdots & \cdots & \cdots & \cdots \\
(\Delta) & (\Delta) & (\Delta) & \cdots & (\Delta)
\end{pmatrix}^{n_1,\ldots,n_r}
$$

where $k_p \geq 1$ and Δ is the unique maximal order in D_p. By the Skolem - Noether theorem this isomorphism is given by an inner automorphism. Hence there is an element $a_{\underline{p}} \in A_{\underline{p}}$ such that

$$
\hat{\Lambda}_{\underline{p}} = a_{\underline{p}} \begin{pmatrix}
(\Delta) & (\underline{p}^{k_{\mathcal{P}}}) & (\underline{p}^{k_{\mathcal{P}}}) & \cdots & (\underline{p}^{k_{\mathcal{P}}}) \\
(\Delta) & (\Delta) & (\underline{p}^{k_{\mathcal{P}}}) & \cdots & (\underline{p}^{k_{\mathcal{P}}}) \\
(\Delta) & (\Delta) & (\Delta) & \cdots & (\underline{p}^{k_p}) \\
\cdots & \cdots & \cdots & \cdots & \cdots \\
(\Delta) & (\Delta) & (\Delta) & \cdots & (\Delta)
\end{pmatrix}^{n_1,\ldots,n_r} a_{\underline{p}}^{-1}.
$$

We now define $\hat{\Gamma}'_{\underline{p}} = \hat{\Lambda}_{\underline{p}}$, if $\Lambda_{\underline{p}}$ is maximal, and

$$\hat{\Gamma}'_{\underline{p}} = a_{\underline{p}} \begin{pmatrix} (\Delta) \, (\Delta) \, (\Delta) \, \cdots \, (\Delta) \\ (\Delta) \, (\Delta) \, (\Delta) \, \cdots \, (\Delta) \\ (\Delta) \, (\Delta) \, (\Delta) \, \cdots \, (\Delta) \\ \cdots \quad \cdots \quad \cdots \quad \cdots \quad \cdots \\ (\Delta) \, (\Delta) \, (\Delta) \, \cdots \, (\Delta) \end{pmatrix}^{n_1,\dots,n_r} a_{\underline{p}}^{-1}$$

otherwise. By Theorem 5.3 in [171] there is a global maximal Γ, so that $\hat{\Gamma}_{\underline{p}} = \hat{\Gamma}'_{\underline{p}}$ for all $\underline{p}$.

Let

$$I = \Pi p^{k_{\mathcal{P}}}$$

throughout this section, where $\underline{p}$ runs through all $\underline{p}$ at which $\hat{\Lambda}_{\underline{p}}$ is not maximal, i.e., through $\underline{p} \in \mathcal{P}(\Lambda)$.

Lemma 7.4.5 *For all $n \geq 1$, the natural homomorphisms*

$$K_n(\hat{\Lambda}_{\underline{p}}/I\hat{\Gamma}_{\underline{p}}) \longrightarrow K_n(\hat{\Gamma}_{\underline{p}}/I\hat{\Gamma}_{\underline{p}})$$

are surjective.

PROOF If $\Lambda_{\underline{p}}$ is maximal, then the lemma obviously holds. So, we suppose that $\Lambda_{\underline{p}}$ is not maximal. Without loss of the generality, we assume

$$\hat{\Lambda}_{\underline{p}} = \begin{pmatrix} (\Delta) \, (\underline{p}^{k_{\mathcal{P}}}) \, (\underline{p}^{k_{\mathcal{P}}}) \, \cdots \, (\underline{p}^{k_{\mathcal{P}}}) \\ (\Delta) \, \overline{(\Delta)} \, (\overline{\underline{p}}^{k_{\mathcal{P}}}) \, \cdots \, (\overline{\underline{p}}^{k_{\mathcal{P}}}) \\ (\Delta) \, (\Delta) \, \overline{(\Delta)} \, \cdots \, (\underline{p}^{k_{\mathcal{P}}}) \\ \cdots \quad \cdots \quad \cdots \quad \cdots \quad \cdots \\ (\Delta) \, (\Delta) \, (\Delta) \, \cdots \, (\Delta) \end{pmatrix}^{n_1,\dots,n_r}$$

and

$$\hat{\Gamma}_{\underline{p}} = \begin{pmatrix} (\Delta) \, (\Delta) \, (\Delta) \, \cdots \, (\Delta) \\ (\Delta) \, (\Delta) \, (\Delta) \, \cdots \, (\Delta) \\ (\Delta) \, (\Delta) \, (\Delta) \, \cdots \, (\Delta) \\ \cdots \quad \cdots \quad \cdots \quad \cdots \quad \cdots \\ (\Delta) \, (\Delta) \, (\Delta) \, \cdots \, (\Delta) \end{pmatrix}^{n_1,\dots,n_r}$$

where Δ is the unique maximal order in $\hat{D}_{p)}$. Then

$$\hat{\Lambda}_{\underline{p}}/I\hat{\Gamma}_{\underline{p}} = \begin{pmatrix} (R) \, (0) \, (0) \, \cdots \, (0) \\ (R) \, (R) \, (0) \, \cdots \, (0) \\ (R) \, (R) \, (R) \, \cdots \, (0) \\ \cdots \quad \cdots \quad \cdots \quad \cdots \quad \cdots \\ (R) \, (R) \, (R) \, \cdots \, (R) \end{pmatrix}^{n_1,\dots,n_r}$$

and

$$\hat{\Gamma}_{\underline{p}}/I\hat{\Gamma}_{\underline{p}} = \begin{pmatrix} (R) \ (R) \ (R) \ \dots \ (R) \\ (R) \ (R) \ (R) \ \dots \ (R) \\ (R) \ (R) \ (R) \ \dots \ (R) \\ \dots \ \dots \ \dots \ \dots \ \dots \\ (R) \ (R) \ (R) \ \dots \ (R) \end{pmatrix}^{n_1, \dots, n_r}$$

where $R = \Delta/\underline{p}^{k_\mathcal{P}}$.   ☐

The lemma now follows from lemma 7.4.4.

7.4.8 For any Abelian group G, let $G(\frac{1}{s})$ be the group $G \otimes \mathbb{Z}[\frac{1}{s}]$. For any ring homomorphism

$$f : A \longrightarrow B,$$

we shall write f_* for the induced homomorphism

$$K_n(A)\left(\frac{1}{s}\right) \longrightarrow K_n(B)\left(\frac{1}{s}\right).$$

Lemma 7.4.6 *For all $n \geq 1$, the natural homomorphism*

$$f_{1_*} : K_n(\hat{\Lambda}_{\underline{p}})\left(\frac{1}{s}\right) \longrightarrow K_n(\hat{\Gamma}_{\underline{p}})\left(\frac{1}{s}\right)$$

is surjective, where s is the generator of $I \cap \mathbb{Z}$.

PROOF   The square

$$\begin{array}{ccc} \hat{\Lambda}_{\underline{p}} & \xrightarrow{\ f_1\ } & \hat{\Gamma}_{\underline{p}} \\ {\scriptstyle f_2}\downarrow & & \downarrow{\scriptstyle g_1} \\ \hat{\Lambda}_{\underline{p}}/I\hat{\Gamma}_{\underline{p}} & \xrightarrow[\ g_2\ ]{} & \hat{\Gamma}_{\underline{p}}/I\hat{\Gamma}_{\underline{p}} \end{array}$$

  (I)

has an associated $K_*(\frac{1}{s})$ Mayer - Vietoris sequence

$$\cdots \longrightarrow K_n(\hat{\Lambda}_{\underline{p}})\left(\frac{1}{s}\right) \xrightarrow{(f_{1_*}, f_{2_*})} K_n(\hat{\Gamma}_{\underline{p}})\left(\frac{1}{s}\right) \oplus K_n(\hat{\Lambda}_{\underline{p}}/I\hat{\Gamma}_{\underline{p}})\left(\frac{1}{s}\right)$$

$$\xrightarrow{(g_{1_*}, g_{2_*})} K_n(\hat{\Gamma}_{\underline{p}})/I\hat{\Gamma}_{\underline{p}})\left(\frac{1}{s}\right) \longrightarrow \cdots$$

by [33, 237], where

$$(f_{1_*}, f_{2_*})(x) = (f_{1_*}(x), f_{2_*}(x))$$

for $x \in K_n(\Lambda_{\underline{p}})(\frac{1}{s})$ and

$$(g_{1_*}, g_{2_*})(a, b) = g_{1_*}(a) - g_{2_*}(b)$$

for $a \in K_n(\Gamma_{\underline{p}})(\frac{1}{s})$ and $b \in K_n(\hat{\Gamma}_{\underline{p}}/I\hat{\Gamma}_{\underline{p}})(\frac{1}{s})$.

For any element $x \in K_n(\hat{\Gamma}_{\underline{p}})(\frac{1}{s})$, we can find $y \in K_n(\hat{\Lambda}_{\underline{p}}/I\hat{\Gamma}_{\underline{p}})(\frac{1}{s})$ such that

$$(g_{1_*}, g_{2_*})(x, y) = g_{1_*}(x) - g_{2_*}(y) = 0$$

by lemma 7.4.5. So,

$$(x, y) \in ker(g_{1_*}, g_{2_*}) = im(f_{1_*}, f_{2_*}).$$

Hence $x \in im(f_{1_*})$, which implies that f_{1_*} is surjective. ☐

Corollary 7.4.2 For all $n \geq 1$, the cokernel of

$$K_n(\hat{\Lambda}_{\underline{p}}) \longrightarrow K_n(\hat{\Gamma}_{\underline{p}})$$

has no non-trivial p-torsion, where p is an arbitrary rational prime that does not divide s.

PROOF By lemma 7.4.6, the cokernel of

$$K_n(\hat{\Lambda}_{\underline{p}}) \longrightarrow K_n(\hat{\Gamma}_{\underline{p}})$$

is s-torsion. Hence the result follows. ☐

Corollary 7.4.3 For all $n \geq 0$, the map

$$f_* : K_{2n+1}(\hat{\Lambda}_{\underline{p}}) \left(\frac{1}{s} \right) \longrightarrow K_{2n+1}(\hat{\Sigma}_{\underline{p}}) \left(\frac{1}{s} \right)$$

is surjective, where f_* is induced by the inclusion map $f : \hat{\Lambda}_{\underline{p}} \to \hat{\Sigma}_{\underline{p}}$.

PROOF The map

$$f : K_{2n+1}(\hat{\Lambda}_{\underline{p}}) \left(\frac{1}{s} \right) \longrightarrow K_{2n+1}(\hat{\Sigma}_{\underline{p}}) \left(\frac{1}{s} \right)$$

is the composition

$$K_{2n+1}(\hat{\Lambda}_{\underline{p}}) \left(\frac{1}{s} \right) \xrightarrow{f_{1_*}} K_{2n+1}(\hat{\Gamma}_{\underline{p}}) \left(\frac{1}{s} \right) \xrightarrow{h_*} K_{2n+1}(\hat{\Sigma}_{\underline{p}}) \left(\frac{1}{s} \right)$$

where h_* is induced by the inclusion $h : \hat{\Gamma}_{\underline{p}} \to \hat{\Sigma}_{\underline{p}}$. By theorem 7.1.2(b), h_* is surjective. Since f_{1_*} and h_* are both surjective, f is also surjective. ☐

Theorem 7.4.10 *Let Λ be a generalized Eichler order in a semi-simple algebra A over F. For all $n \geq 1$, the q-primary part of $C\ell_{2n}(\Lambda)$ is trivial for $q \notin \check{\mathcal{P}}(\Lambda)$.*

PROOF   Since Λ can be expressed as the direct sum of generalized Eichler order in the semi-simple components of A, we may assume A is simple. Corollary 7.4.5 implies in this case that

$$Coker(K^c_{2n+1}(\hat{\Lambda}_{\underline{p}}) \longrightarrow K^c_{2n+1}(\hat{A}_{\underline{p}}))_q = 0$$

for $q \notin \check{\mathcal{P}}(\Lambda)$. So the q-primary part of $C\ell_{2n}(\Lambda)$ is trivial for $q \notin \check{\mathcal{P}}(\Lambda)$ by lemma 7.4.3.   ☐

7.5   Higher K-theory of grouprings of virtually infinite cyclic groups

7.5.1 F.T. Farrell and L.E. Jones conjectured in [54] that algebraic K-theory of virtually cyclic subgroups V should constitute "building blocks" for the algebraic K-theory on an arbitrary group G. In [55] they obtained some results on lower K-theory of V. In this section we obtain results on higher K-theory of virtually infinite cyclic groups V in the two cases: (i) when V admits an epimorphism (with finite kernel) to a finite group (see [183]), and (ii) when V admits an epimorphism (with finite kernel) to the finite dihedral group (see [183]). The results in this section are due to A. Kuku and G. Tang (see [123]).

We shall discuss the precise form of this conjecture in chapter 14, which will be devoted to equivariant homology theories. We recall that we briefly discussed its formulation in 4.5.4.

Definition 7.5.1 *A group is called virtually cyclic if it is either finite or virtually infinite cyclic, i.e., contains a finite index subgroup that is infinite cyclic. More precisely, virtually infinite cyclic groups V are of two types, namely,*

   i) The group V that admits an epimorphism (with finite kernel G) to the infinite cyclic group $T =< t >$, i.e., V is the semi-direct product $G \rtimes_\alpha T$, where $\alpha : G \to G$ is an automorphism and the action of T is given by $tgt^{-1} = \alpha(g)$ for all $g \in G$.

   ii) The group V that admits an epimorphism (with finite kernel) to the infinite dihedral group D_∞, i.e., $V = G_0 *_H G_1$ where the groups G_i, $i = 0, 1$, and H are finite and $[G_i : H] = 2$.

$(7.5)^A$ Some preliminary results

In this subsection, we set the stage by proving theorems 7.5.1 and 7.5.2, which constitute generalizations of theorems 1.2 and 1.5 of [55]. Here, we prove the results for an arbitrary R-order Λ in a semi-simple F-algebra Σ (where R is the ring of integers in a number field F) rather than for the special case $\Lambda = \mathbb{Z}G$ (G finite group) treated in [55].

7.5.2 Let R the ring of integers in a number field F, Λ an R-order in a semi-simple F-algebra Σ, and $\alpha : \Lambda \to \Lambda$ is an R automorphism. The α extends to F-automorphism on Σ. Suppose that Γ is a maximal element in the set of all α-invariant R-orders in Σ containing Λ. Let $\max(\Gamma)$ denote the set of all two-sided maximal ideals in Γ and $\max_\alpha(\Gamma)$ the set of all two-sided maximal α-invariant ideals in Γ. Recall that a Λ-lattice in Σ is a Λ-Λ submodule of Σ, which generates Σ as F-vector space. The aim of this section is to prove theorems 7.5.1 and 7.5.2 below.

Theorem 7.5.1 *The set of all two-sided, α-invariant Γ-lattices in Σ is a free Abelian group under multiplication and has $\max_\alpha(\Gamma)$ as a basis.*

PROOF Let $\mathfrak{a}$ be a two-sided, α-invariant Γ-lattice in Σ. Then $\{x \in \Sigma \mid x\mathfrak{a} \subseteq \mathfrak{a}\}$ is an α-invariant R-order containing Γ. Hence, it must be equal to Γ by the maximality of Γ. Similarly $\{x \in \Sigma \mid \mathfrak{a}x \subseteq \mathfrak{a}\} = \Gamma$.

Now, let $\mathfrak{a} \subseteq \Gamma$ be a two-sided, α-invariant Γ-lattice in Σ. Then, $B = \Gamma/\mathfrak{a}$ is a finite ring, and hence, Artinian. So, $\mathrm{rad}B$ is a nilpotent, α-invariant, two-sided ideal in B, and $B/\mathrm{rad}B$ is semi-simple ring. Hence $B/\mathrm{rad}B$ decomposes as a direct sum of simple rings B_i, i.e.,

$$B/\mathrm{rad}B = B_1 \oplus B_2 \otimes \cdots \oplus B_n \tag{I}$$

and $\alpha : B/\mathrm{rad}B \to B/\mathrm{rad}B$ induces a permutation of the factors B_i, i.e.,

$$\alpha(B_i) = B_{\hat{\alpha}(i)} \tag{II}$$

where $\hat{\alpha}$ is a permutation of $\{1, 2, \ldots, n\}$. So, $\mathfrak{a} \in \max_\alpha(\Gamma)$ if and only if both $\mathrm{rad}(\Gamma/\mathfrak{a}) = 0$ and $\hat{\alpha}$ is a cyclic permutation. Hence, $\mathfrak{a} \notin \max_\alpha(\Gamma)$ if and only if there exist a pair of two-sided, α-invariant Γ-lattices $\mathfrak{b}$ and $\mathfrak{c}$ satisfying three properties:

(i) Both $\mathfrak{b}$ and $\mathfrak{c}$ properly contain $\mathfrak{a}$.

(ii) $\mathfrak{b}$ and $\mathfrak{c}$ are both contained in Γ. $\tag{III}$

(iii) $\mathfrak{b}\mathfrak{c} \subseteq \mathfrak{a}$.

Hence, just as in [55], we deduce the following fact:

If $\mathfrak{a}$ is a two-sided, α-invariant Γ-lattice, then a contains a (finite) product of elements from $\max_\alpha(\Gamma)$.   (IV)

If $\mathfrak{a}$ is a two-sided, α-invariant Γ-lattice, then we write

$$\bar{\mathfrak{a}} = \{x \in \Sigma \mid x\mathfrak{a} \subseteq \Gamma\}, \tag{V}$$

which is also a two-sided, α-invariant Γ-lattice.   ▯

We now prove the following.

Lemma 7.5.1 *If* $\mathfrak{p} \in \max_\alpha(\Gamma)$, *then*

$$\bar{\mathfrak{p}} \neq \Gamma. \tag{VI}$$

PROOF   Choose a positive integer $s \in \mathbb{Z}$ such that $s\Gamma \subset \mathfrak{p}$. Applying (i) - (iii) with $\mathfrak{a} = s\Gamma$, we can define elements $\mathfrak{p}_i \in \max_\alpha(\Gamma)$ such that

$$\mathfrak{p}_1 \mathfrak{p}_2 \ldots \mathfrak{p}_n \subset \Gamma.$$

Let us assume that n is the smallest possible integer with this property. Using the characterization of $\max_\alpha(\Gamma)$ given above (III and IV), we see that some $\mathfrak{p}_i$ must be contained in $\mathfrak{p}$ since $\mathfrak{p} \in \max_\alpha(\Gamma)$. And since $\mathfrak{p}_i \in \max_\alpha(\Gamma)$, $\mathfrak{p} = \mathfrak{p}_i$. We can therefore write $\mathfrak{a}\mathfrak{p}\mathfrak{b} \subset s\Gamma$ where $\mathfrak{a} = \mathfrak{p}_1 \ldots \mathfrak{p}_{i-1}$ and $\mathfrak{b} = \mathfrak{p}_{i+1} \ldots \mathfrak{p}_n$. Thus, $\frac{1}{s}\mathfrak{a}\mathfrak{p}\mathfrak{b} \subset \Gamma$, and further, $(\frac{1}{s}\mathfrak{b}\mathfrak{a}\mathfrak{p})\mathfrak{b} \subset \mathfrak{b}$. We have $(\frac{1}{s}\mathfrak{b}\mathfrak{a}\mathfrak{p}) \subset \Gamma$. Hence $\frac{1}{s}\mathfrak{b}\mathfrak{a} \subset \bar{\mathfrak{p}}$, by the definition of $\bar{\mathfrak{p}}$. Since $\mathfrak{b}\mathfrak{a}$ is a product of $n-1$ elements in $\max_\alpha(\Gamma)$, the minimality of n implies that $\mathfrak{b}\mathfrak{a} \subset s\Gamma$, so $\frac{1}{s}\mathfrak{b}\mathfrak{a} \subset \Gamma$. Thus $\bar{\mathfrak{p}} \neq \Gamma$.   ▯

Lemma 7.5.2 *If* $\mathfrak{p} \in \max_\alpha(\Gamma)$, *then* $\bar{\mathfrak{p}}\mathfrak{p} = \Gamma = \mathfrak{p}\bar{\mathfrak{p}}$.

PROOF   Similar to that in step 4, page 21 of [55].   ▯

Lemma 7.5.3 *If* $\mathfrak{p}_1$, $\mathfrak{p}_2 \in \max_\alpha(\Gamma)$, *then* $\mathfrak{p}_1\mathfrak{p}_2 = \mathfrak{p}_2\mathfrak{p}_1$.

PROOF   Similar to that of step 5, in [55].

Note that the proof in ([20] p.158) is easily adapted to yield the following conclusion:

A two-sided, α-invariant Γ-lattice $\mathfrak{a} \subseteq \Gamma$ is uniquely, up to order, a product of elements of $\max_\alpha(\Gamma)$, and we can finish the proof as in [20], p.158.   ▯

Corollary 7.5.1 *If every element of* $\max_\alpha(\Gamma)$ *is a projective right* Γ-*module, then every element in* $\max(\Gamma)$ *is also a projective right* Γ-*module, and consequently,* Γ *is a hereditary ring.*

PROOF   It is the same as the proof of Corollary 1.6 in [55].   ⬚

Theorem 7.5.2 *Let R be the ring of integers in a number field F, Λ any R-order in a semi-simple F-algebra Σ. If $\alpha : \Lambda \to \Lambda$ is an R automorphism, then there exists an R-order $\Gamma \subset \Sigma$ such that*

1) $\Lambda \subset \Gamma$.

2) Γ is α-invariant.

3) Γ is a (right) regular ring. In fact, Γ is a (right) hereditary ring.

PROOF   Let $\mathcal{S}$ be the set consisting of all α-invariant R-orders M of Σ, which contains Λ. Then $\mathcal{S}$ is not empty since $\Lambda \in \mathcal{S}$. Choose Γ to be any maximal member of $\mathcal{S}$. Such a member exists by Zorn's Lemma. Note that this R-order Γ satisfies properties (1) and (2) by definition and is clearly a right Noetherian ring. Hence, it suffices to show that Γ is a right hereditary ring; i.e., that every right Γ-module is either projective or has a length 2 resolution by projective right Γ-modules. To do this, it suffices to show that every maximal two-sided ideal in Γ is a projective right Γ-module. Let $\max(\Gamma)$ denote the set of all two-sided maximal ideals in Γ, and $\max_\alpha(\Gamma)$ the set of all members among the two-sided α-invariant proper ideals in Γ. Note that if $\mathfrak{a} \in \max_\alpha(\Gamma)$, then $\Gamma/\mathfrak{a}$ is a finite ring. To see this, first observe that $\Gamma/\mathfrak{a}$ is finitely generated as an Abelian group under addition. If it were not finite, then there would exist a prime $p \in \mathbb{Z}$ such that the multiples of p in $\Gamma/\mathfrak{a}$ would form a proper two-sided α-invariant proper ideals in $\Gamma/\mathfrak{a}$. But this would contradict the maximality of $\mathfrak{a}$. Also $\Gamma/\mathfrak{a}$ is a (right) Artinian ring since it is a finite ring. But $\mathrm{rad}(\Gamma/\mathfrak{a})$ is an α-invariant two-sided ideal in $\Gamma/\mathfrak{a}$. So the maximality of $\mathfrak{a}$ again shows that $\mathrm{rad}(\Gamma/\mathfrak{a}) = 0$. Hence, $\Gamma/\mathfrak{a}$ is semi-simple ring. We finally remark that $\mathfrak{a}$ is a two-sided Γ-lattice in Σ since $\mathfrak{a}$ has finite index in the lattice Γ.

By corollary 7.5.1, it suffices to show that every element $\mathfrak{p} \in \max_\alpha(\Gamma)$ is a projective right Γ-module. Let $\mathfrak{q}$ be the inverse of $\mathfrak{p}$ given by theorem 7.5.1; i.e., $\mathfrak{q}$ is a two-sided Γ-lattice in Σ, which is α-invariant and satisfies the equations $\mathfrak{pq} = \mathfrak{qp} = \Gamma$. Consequently, there exist elements $a_1, a_2, \ldots, a_n \in \mathfrak{p}$ and $b_1, b_2, \ldots, b_n \in \mathfrak{q}$ such that

$$a_1 b_1 + a_2 b_2 + \cdots + a_n b_n = 1.$$

Now define (right) Γ-module homomorphisms $f : \mathfrak{p} \to \Gamma^n$ and $g : \Gamma^n \to \mathfrak{p}$ by

$$f(x) = (b_1 x, b_2 x, \ldots, b_n x)$$
$$g(y_1, y_2, \ldots, y_n) = a_1 y_1 + a_2 y_2 + \cdots + a_n y_n$$

where $x \in \mathfrak{p}$ and $(y_1, y_2, \ldots, y_n) \in \Gamma^n$. Note that the composite $g \circ f = id_\mathfrak{p}$. Consequently, $\mathfrak{p}$ is a direct summand of Γ^n, which shows that $\mathfrak{p}$ is a right-projective Γ-module.   ⬚

$(7.5)^B$ K-theory for the first type of virtually infinite cyclic groups

In this subsection, we prove that if R is the ring of integers in a number field F, Λ an R-order in a semi-simple F-algebra Σ, and α an automorphism of Λ, then for all $n \geq 0$, $NK_n(\Lambda, \alpha)$ is s-torsion for some positive integer s and that the torsion-free rank of $K_n(\Lambda_\alpha[t])$ is equal to the torsion-free rank of $K_n(\Lambda)$, which is finite by theorem 7.1.11. When $V = G \rtimes_\alpha T$ is a virtually infinite cyclic group of the first type, we show that for all $n \geq 0, G_n(RV)$ is a finitely generated Abelian group and that for all $n < -1$, $K_n(RV) = 0$. We also show that for all $n \geq 0$, $NK_n(RV)$ is $|G|$-torsion.

First, we prove the following result for later use

Theorem 7.5.3 *Let A be a Noetherian ring and α an automorphism of A, and $A_\alpha[t]$ the twisted polynomial ring. Then,*

(i) $G_n(A_\alpha[t]) \cong G_n(A)$ for all $n \geq 0$

(ii) There exists a long exact sequence

$$\ldots \longrightarrow G_n(A) \overset{1-\alpha_*}{\longrightarrow} G_n(A) \longrightarrow G_n(A_\alpha[t, t^{-1}]) \longrightarrow G_{n-1}(A) \longrightarrow \ldots$$

PROOF   (i) follows directly from Theorem 2.18 of (cf. [88], p.194). To prove the long exact sequence in (ii), we denote by $\mathcal{A} = \mathcal{M}(A_\alpha[t])$ the category consisting of finitely generated $A_\alpha[t]$-modules. Consider the Serre subcategory $\mathcal{B}$ of $\mathcal{A} = \mathcal{M}(A_\alpha[t])$, which consists of modules $M \in obj\mathcal{A}$ on which t is nilpotent, i.e.,

$$obj\mathcal{B} = \{M \in obj\mathcal{A} \mid \quad \text{there exists an} \quad m \geq 0 \quad \text{such that} \quad Mt^m = 0\}.$$

Applying the localization theorem to the pair $(\mathcal{A}, \mathcal{B})$, we obtain a long exact sequence

$$\ldots \longrightarrow K_{n+1}(\mathcal{A}/\mathcal{B}) \overset{1-\alpha_*}{\longrightarrow} K_n(\mathcal{B}) \longrightarrow K_n(\mathcal{A}) \longrightarrow K_n(\mathcal{A}/\mathcal{B}) \longrightarrow \ldots$$

By definition and (i) $K_n(\mathcal{A}) = G_n(A_\alpha[t]) \cong G_n(A)$. We will prove that

$$K_n(\mathcal{B}) \cong G_n(A)$$

and

$$K_n(\mathcal{A}/\mathcal{B}) \cong G_n(A_\alpha[T]) := G_n(A_\alpha[t, t^{-1}])$$

for all $n \geq 0$. At first $\mathcal{M}(A) \subseteq \mathcal{M}(A_\alpha[t]/tA_\alpha[t]) \subseteq \mathcal{B}$ (Note that although $t \notin center(A_\alpha[t])$, we have $tA_\alpha[t] = A_\alpha[t]t \lhd A_\alpha[t]$). Using the Devissage theorem one gets

$$K_n(\mathcal{B}) \cong K_n(\mathcal{M}(A_\alpha[t]/tA_\alpha[t]))$$

But

$$0 \longrightarrow tA_\alpha[t] \longrightarrow A_\alpha[t] \xrightarrow{t=0} A \longrightarrow 0$$

is an exact sequence of homomorphisms of rings. So we have

$$A_\alpha[t]/tA_\alpha[t] \cong A$$

as rings, and so, $K_n(\mathcal{M}(A_\alpha[t]/tA_\alpha[t])) \cong G_n(A)$. Thus

$$K_n(\mathcal{B}) \cong K_n(\mathcal{M}(A_\alpha[t]/tA_\alpha[t])) \cong G_n(A).$$

Next we prove that

$$K_n(\mathcal{A}/\mathcal{B}) \cong G_n(A_\alpha[T]).$$

Since $A_\alpha[T]$ is a direct limit of free $A_\alpha[t]$-modules $A_\alpha[t]t^{-n}$, it is a flat $A_\alpha[t]$-module, and this implies that $- \otimes_{A_\alpha} A_\alpha[T]$ is an exact functor from $\mathcal{A}$ to $\mathcal{M}(A_\alpha[T])$ and further induces an exact functor

$$F : \mathcal{A}/\mathcal{B} \longrightarrow \mathcal{M}(A_\alpha[T]).$$

We now prove that F is an equivalence. For any $M \in obj\mathcal{M}(A_\alpha[T])$, pick a generating set $\{x_1, x_2, \ldots, x_l\}$ of the finitely generated $A_\alpha[T]$-module M. Let

$$M_1 = \sum_{i=1}^{l} x_i A_\alpha[t].$$

Then $M_1 \in \mathcal{A}$ and $M_1 \otimes_{A_\alpha[t]} A_\alpha[T] \cong M$. The exact sequence of $A_\alpha[t]$-modules

$$0 \longrightarrow M_1 \longrightarrow M \longrightarrow M/M_1 \longrightarrow 0$$

induces an exact sequence

$$0 \to M_1 \otimes_{A_\alpha[t]} A_\alpha[T] \to M \otimes_{A_\alpha[t]} A_\alpha[T] \to M/M_1 \otimes_{A_\alpha[t]} A_\alpha[T] \to 0.$$

Since $\{x_1, x_2, \ldots, x_l\}$ is a generating set for the $A_\alpha[T]$-module M, then for any $x \in M$, there exist $f_i \in A_\alpha[T]$ $(i = 1, \ldots, l)$ such that $x = \sum_{i=1}^{l} x_i f_i$. Thus, there exists $n \geq 0$ such that $xt^n = \sum_{i=1}^{l} x_i(f_i t^n) \in M_1$, i.e., M/M_1 is t-torsion. Thus $M/M_1 \otimes_{A_\alpha[t]} A_\alpha[T] = 0$. It follows that

$$0 \longrightarrow M_1 \otimes_{A_\alpha[t]} A_\alpha[T] \longrightarrow M \otimes_{A_\alpha[t]} A_\alpha[T] \longrightarrow 0$$

is exact. That is

$$M_1 \otimes_{A_\alpha[t]} A_\alpha[T] \cong M \otimes_{A_\alpha[t]} A_\alpha[T].$$

For any $M, N \in \mathcal{A}$, we have, by definition

$$\mathrm{Hom}_{\mathcal{A}/\mathcal{B}}(M, N) = \varinjlim \mathrm{Hom}_{A_\alpha[t]}(M', N/N')$$

where M/M' and N' are t-torsion. One gets easily:

$$\mathrm{Hom}_{\mathcal{A}/\mathcal{B}}(M, N) = \varinjlim \mathrm{Hom}_{A_\alpha[t]}(M', N/N') = \varinjlim \mathrm{Hom}_{A_\alpha[t]}(M', N/N_t)$$

where $N_t = \{x \in N \mid$ and there exists an $m \geq 0$ such that $xt^m = 0\}$. Define a map

$$\phi : \mathrm{Hom}_{\mathcal{A}/\mathcal{B}}(M, N) \longrightarrow \mathrm{Hom}_{\mathcal{B}}(M \otimes_{A_\alpha[t]} A_\alpha[T], N \otimes_{A_\alpha[t]} A_\alpha[T]).$$

For any $f \in \mathrm{Hom}_{\mathcal{B}}(M \otimes_{A_\alpha[t]} A_\alpha[T], N \otimes_{A_\alpha[t]} A_\alpha[T])$, since M is a finitely generated $A_\alpha[T]$-module, and there exists an $m \geq 0$ such that

$$f(Mt^m \otimes 1) \subseteq N \otimes 1.$$

We can define $Mt^m \overset{\sigma}{\to} N/N_t$, $xt^m \mapsto n$ if $f(xt^m \otimes 1) = n \otimes 1$. This is well defined and σ maps to f under ϕ.

If $\sigma \in \mathrm{Hom}_{A_\alpha[t]}(M', N/N_t)$ is such that its image in $\mathrm{Hom}_{\mathcal{B}}(M \otimes_{A_\alpha[t]} A_\alpha[T])$, $N \otimes_{A_\alpha[t]} A_\alpha[T]$ is zero, then $\sigma \otimes 1 = 0$ implies that $\sigma m \otimes 1 = (\sigma \otimes 1)(m \otimes 1) = 0$ in $N/N_t \otimes_{A_\alpha[t]} A_\alpha[T]$. Hence $\sigma(m) = 0$, and so,

$$\mathrm{Hom}_{\mathcal{A}/\mathcal{B}}(M, N) \cong \mathrm{Hom}_{\mathcal{B}}(M \otimes_{A_\alpha[t]} A_\alpha[T], N \otimes_{A_\alpha[t]} A_\alpha[T]),$$

that is,

$$\mathcal{A}/\mathcal{B} \cong \mathcal{M}(A_\alpha[T]).$$

Hence $K_n(\mathcal{A}/\mathcal{B}) \cong K_n(\mathcal{M}(A_\alpha[T]))$, completing the proof of 2). $\Box$

Theorem 7.5.4 *Let R be the ring of integers in a number field F, Λ any R-order in semi-simple F-algebra Σ, α an automorphism of Λ. Then*

For all $n \geq 0$.

(i) $NK_n(\Lambda, \alpha)$ is s-torsion for some positive integer s. Hence the torsion-free rank of $K_n(\Lambda_\alpha[t])$ is the torsion-free rank of $K_n(\Lambda)$ and is finite. If $n \geq 2$, then the torsion-free rank of $K_n(\Lambda_\alpha[t])$ is equal to the torsion free rank of $K_n(\Sigma)$.

(ii) If G is a finite group of order r, then $NK_n(RG, \alpha)$ is r-torsion, where α is the automorphism of RG induced by that of G.

PROOF   (i) By theorem 7.5.2, we can choose an α-invariant R-order Γ in Σ, which contains Λ and is regular. First note that since every R-order is a $\mathbb{Z}$-order, there is a non-zero integer s such that $\Lambda \subseteq \Gamma \subseteq \Lambda(1/s)$, where $A(1/s)$ denote $A \otimes \mathbb{Z}(1/s)$ for an Abelian group A. Put $\underline{q} = s\Gamma$, and then we have a Cartesian square

$$\begin{array}{ccc} \Lambda & \longrightarrow & \Gamma \\ \downarrow & & \downarrow \\ \Lambda/\underline{q} & \longrightarrow & \Gamma/\underline{q} \end{array} \qquad\qquad (\mathrm{I})$$

Since α induces automorphisms of all the four rings in the square (I) (cf. [55]), we have another Cartesian square

$$
\begin{array}{ccc}
\Lambda_\alpha[t] & \longrightarrow & \Gamma_\alpha[t] \\
\downarrow & & \downarrow \\
(\Lambda/\underline{q})_\alpha[t] & \longrightarrow & (\Gamma/\underline{q})_\alpha[t]
\end{array}
\tag{II}
$$

Note that both $(\Lambda/\underline{q})_\alpha[t]$ and $(\Gamma/\underline{q})_\alpha[t]$ are $\mathbb{Z}/s\mathbb{Z}$-algebra, and so, it follows from (II) that we have a long exact Mayer - Vietoris sequence (cf. [33] or [237])

$$
\cdots \longrightarrow K_{n+1}(\Gamma/\underline{q})_\alpha[t](1/s) \longrightarrow K_n(\Lambda_\alpha[t])(1/s)
$$
$$
\longrightarrow K_n(\Lambda/\underline{q})_\alpha[t](1/s) \oplus K_n(\Gamma_\alpha[t])(1/s)
$$
$$
\longrightarrow K_n(\Gamma/\underline{q})_\alpha[t](1/s) \longrightarrow K_{n-1}(\Lambda_\alpha[t])(1/s) \longrightarrow \cdots
\tag{III}
$$

Since we also have a long exact Mayer - Vietoris sequence

$$
\cdots \longrightarrow K_{n+1}(\Gamma/\underline{q})_\alpha[t](1/s) \longrightarrow K_n(\Lambda_\alpha[t])(1/s)
$$
$$
\longrightarrow K_n(\Lambda/\underline{q})_\alpha[t](1/s) \oplus K_n(\Gamma_\alpha[t])(1/s)
$$
$$
\longrightarrow K_n(\Gamma/\underline{q})_\alpha[t](1/s) \longrightarrow K_{n-1}(\Lambda_\alpha[t])(1/s) \longrightarrow \cdots
\tag{IV}
$$

then, by mapping sequence (III) to sequence (IV) and taking kernels, we obtain another long exact Mayer - Vietoris

$$
\cdots \longrightarrow NK_{n+1}(\Gamma/\underline{q})_\alpha[t](1/s) \longrightarrow NK_n(\Lambda_\alpha[t])(1/s)
$$
$$
\longrightarrow NK_n(\Lambda/\underline{q})_\alpha[t](1/s) \oplus NK_n(\Gamma_\alpha[t])(1/s)
$$
$$
\longrightarrow NK_n(\Gamma/\underline{q})_\alpha[t](1/s) \longrightarrow NK_{n-1}(\Lambda_\alpha[t])(1/s) \longrightarrow \cdots
$$

However, by [55], $\Gamma_\alpha[t]$ is regular since Γ is. So, $NK_n(\Gamma, \alpha) = 0$ by theorem 7.5.3 (i) since $K_n(\Gamma_\alpha[t]) = G_n(\Gamma_\alpha[t]) = G_n(\Gamma) = K_n(\Gamma)$. Both $\Lambda/\underline{q}$ and $\Gamma/\underline{q}$ are finite, and hence quasi-regular. They are also $\mathbb{Z}/s\mathbb{Z}$-algebras, and so, it follows that $(\Lambda/\underline{q})_\alpha[t]$ and $(\Gamma/\underline{q})_\alpha[t]$ are also quasi-regular and $\mathbb{Z}/s\mathbb{Z}$-algebra. We now prove that for a finite $\mathbb{Z}/s\mathbb{Z}$-algebra A, $NK_n(A, \alpha)$ is a s-torsion. Since A is finite, its Jacobson radical $J(A)$ is nilpotent, and by corollary 5.4 of [236], the relative K-groups $K_n(A, J(A))$ are s-torsion for any $n \geq 0$. This implies that

$$
K_n(A)(1/s) \cong K_n(A/J(A))(1/s)
$$

from the relative K-theory long exact sequence tensored with $\mathbb{Z}(\frac{1}{s})$. Similarly, one gets

$$
K_n(A_\alpha[x])(1/s) \cong K_n((A/J(A))_\alpha[x])(1/s).
$$

However, $A/J(A)$ is regular, and so,

$$
K_n(A/J(A))_\alpha[x])(1/s) \cong K_n((A/J(A))(1/s).
$$

by theorem 7.5.3(i).

Hence, we have

$$K_n(A_\alpha[x])(1/s) \cong K_n(A)(1/s).$$

From the finiteness of A one gets that $K_n(A)$ is finite (see theorem 7.1.12). Hence both $K_n(A_\alpha[x])(1/s)$ and $K_n(A)(1/s)$ have the same cardinality. From the exact sequence

$$0 \longrightarrow NK_n(A,\alpha) \longrightarrow K_n(A_\alpha[x]) \to K_n(A) \longrightarrow 0$$

tensored with $\mathbb{Z}(\frac{1}{s})$, we obtain the exact sequence

$$0 \longrightarrow NK_n(A,\alpha)(1/s) \longrightarrow K_n(A_\alpha[x])(1/s) \longrightarrow K_n(A)(1/s) \longrightarrow 0.$$

Hence $NK_n(A,\alpha)(1/s)$ is zero since $K_n(A_\alpha[x])(1/s)$ and $K_n(A)(1/s)$ are isomorphic. Hence, both $NK_{n+1}(\Gamma/\underline{q},\alpha)(1/s)$ and $NK_n(\Lambda/\underline{q},\alpha)(1/s)$ are zero, and so, $NK_n(\Lambda/\underline{q},\alpha)$ is s-torsion.

Since $K_n(\Lambda_\alpha[\bar{t}]) = K_n(\Lambda) \oplus NK_n(\Lambda,\alpha)$ and $NK_(\Lambda,\alpha)$ is torsion, the torsion-free rank of $K_n(\Lambda_\alpha[t])$ is the torsion-free rank of $K_n(\Lambda)$. By theorem 7.1.11 the torsion-free rank of $K_n(\Lambda)$ is finite, and if $n \geq 2$, the torsion-free rank of $K_n(\Sigma)$ is the torsion-free rank of $K_n(\Lambda)$ (see [115]).

(ii) is a direct consequence of (i) since if $\mid G \mid = r$, and we take $\Lambda = RG$, then $r\Gamma \subseteq \Lambda$.   ⬚

Theorem 7.5.5 *Let $V = G \rtimes T$ be the semi-direct product of a finite group G of order r with an infinite cyclic group $T = < t >$ with respect to the automorphism $\alpha : G \to G : g \mapsto tgt^{-1}$. Then,*

(i) $K_n(RV) = 0$ for all $n < -1$.

(ii) The inclusion $RG \to RV$ induces an epimorphism $K_{-1}(RG) \to K_{-1}(RV)$. Hence $K_{-1}(RV)$ is a finitely generated Abelian group.

(iii) For all $n \geq 0$, $G_n(RV)$ is a finitely generated Abelian group.

(iv) $NK_n(RV)$ is r-torsion for all $n \geq 0$.

PROOF

(i) By theorem 7.5.2 there exists an α-invariant regular ring Γ in FG that contains RG. Then, for the integers $s = \mid G \mid$, $RG \subseteq \Gamma \subseteq RG(1/s)$. Put $\underline{q} = s\Gamma$. Then we have a Cartesian square

$$
\begin{array}{ccc}
RG & \longrightarrow & \Gamma \\
\downarrow & & \downarrow \\
RG/\underline{q} & \longrightarrow & \Gamma/\underline{q}
\end{array}
\qquad (VI)
$$

Since α induces automorphisms of all the four rings in the square (VI), we have another Cartesian square

$$
\begin{array}{ccc}
RV & \longrightarrow & \Gamma_\alpha[T] \\
\downarrow & & \downarrow \\
(RG/\underline{q})_\alpha[T] & \longrightarrow & (\Gamma/\underline{q})_\alpha[T]
\end{array}
\qquad \text{(VII)}
$$

(see [55]).

Since lower K-theory has excision property, it follows from [20] that we have lower K-theory exact sequence

$$K_0(RV) \to K_0((RG/\underline{q})_\alpha[T]) \oplus K_0(\Gamma_\alpha[T]) \to K_0((\Gamma/\underline{q})_{\alpha[T]}) \to$$

$$\text{(VIII)}$$

$$\to K_{-1}(RV) \to K_{-1}((RG/\underline{q})_\alpha[T]) \oplus K_{-1}(\Gamma_\alpha[T]) \to \cdots$$

However, $(\Gamma/\underline{q})_\alpha[T]$ and $(RG/\underline{q})_\alpha[T])$ are quasi-regular since $\Gamma/\underline{q}$ and $RG/\underline{q}$ are quasi-regular (see [55]). Also, $\Gamma_\alpha[T]$ is regular since Γ is regular (see [55]).

Hence

$$K_i(\Gamma_\alpha[T]) = K_i((RG/\underline{q})_\alpha[T]) = K_i((\Gamma/\underline{q})_\alpha[T]) = 0$$

for all $i \leq -1$. Thus $K_i(RV) = 0$ for all $i < -1$, from the exact sequence (VIII).

(ii) The proof of (ii) is similar to the proof of a similar statement for $\mathbb{Z}V$ in [55], corollary 1.3, and is omitted.

(iii) Is a direct consequence of theorem 7.5.3 (2) since $G_n(RG)$ is finitely generated for all $n \geq 1$ (see theorem 7.1.13).

(iv) By [55] 1.3.2, we have a Cartesian square

$$
\begin{array}{ccc}
RV & \longrightarrow & \Gamma_\alpha[T] \\
\downarrow & & \downarrow \\
(RG/\underline{q})_\alpha[T] & \longrightarrow & (\Gamma/\underline{q})_\alpha[T]
\end{array}
$$

where $\Gamma_\alpha[T]$ is regular and $(RG/\underline{q})_\alpha[T]$ are quasi-regular (see [55] 1.1 and 1.41).

Moreover, since r annihilates $RG/\underline{q}$ and $\Gamma/\underline{q}$ it also annihilates $(RG/\underline{q})_\alpha[T]$, $(\Gamma/\underline{q})_\alpha[T]$ since for $A = RG/q$, or Γ/q, $A_\alpha[\overline{T}]$ is a direct limit of free $A_\alpha[t]$-module $A_\alpha[t]t^{-n}$. Hence by [236], corollary 3.3(d), $NK_n(A_\alpha[T])$ is r-torsion. (Note that [236], corollary 3.3(d), is valid when p is any integer r.)

We also have by [33, 237] a long exact Mayer-Vietoris sequence

$$\cdots \to NK_{n+1}(\Gamma/\underline{q})_\alpha[T]\left(\frac{1}{r}\right) \to NK_n(RV)\left(\frac{1}{r}\right) \to NK_n(\Gamma/\underline{q})\alpha[T]\left(\frac{1}{r}\right)$$

$$\oplus NK_n(\Gamma)_\alpha[T]\left(\frac{1}{r}\right) \to NK_n(\Gamma/\underline{q})\alpha[T]\left(\frac{1}{r}\right) \to \cdots \qquad (IX)$$

But $NK_n(\Gamma_\alpha[T]) = 0$ since $\Gamma_\alpha[T]$ is regular. Hence, we have $NK_n(RV)(\frac{1}{r}) = 0$ from (IX), and also, $NK_n(RV)$ is r-torsion. ▯

$(7.5)^C$ Nil-groups for the second type of virtually infinite cyclic groups

7.5.3 The algebraic structure of the groups in the second class is more complicated. We recall that a group V in the second class has the form $V = G_0 *_H G_1$ where the groups G_i, $i = 0, 1$, and H are finite, and $[G_i : H] = 2$. We will show that the nil-groups in this case are torsion, too. At first we recall the definition of nil-groups in this case.

Let $\mathcal{T}$ be the category of triples $\mathbf{R} = (R; B, C)$, where B and C are R-bimodules. A morphism in $\mathcal{T}$ is a triple

$$(\phi; f, g) : (R; B, C) \longrightarrow (S; D, E)$$

where $\phi : R \longrightarrow S$ is a ring homomorphism and both $f : B \longrightarrow D$ and $g : C \longrightarrow E$ are $R - S$-bimodule homomorphisms. There is a functor ρ from the category $\mathcal{T}$ to $Rings$ defined by

$$\rho(\mathbf{R}) = R_\rho = \begin{pmatrix} T_R(C \otimes_R B) & C \otimes_R T_R(B \otimes_R C) \\ B \otimes_R T_R(C \otimes_R B) & T_R(B \otimes_R C) \end{pmatrix}$$

where $T_R(B \otimes_R C)(\text{resp.} T_R(C \otimes B))$ is the tensor algebra of $B \otimes_R C$ (resp. $C \otimes_R B$) and $\rho(\mathbf{R})$ is the ring with multiplication given as matrix multiplication and each entry by concatenation. There is a natural augmentation map (cf. [37])

$$\epsilon : R_\rho \longrightarrow \begin{pmatrix} R & 0 \\ 0 & R \end{pmatrix}$$

The nil-group $NK_n(\mathbf{R})$ is defined to be the kernel of the map induced by ϵ on K_n-groups.

We now formulate the nil-groups of interest. Let V be a group in the second class of the form $V = G_0 *_H G_1$ where the groups G_i, $i = 0, 1$ and H are finite, and $[G_i : H] = 2$. Considering $G_i - H$ as the right coset of H in G_i, which is different from H, the free $\mathbb{Z}$-module $\mathbb{Z}[G_i - H]$ with basis $G_i - H$ is a $\mathbb{Z}H$-bimodule that is isomorphic to $\mathbb{Z}H$ as a left $\mathbb{Z}H$-module, but the right action is twisted by an automorphism of $\mathbb{Z}H$ induced by an automorphism of H. Then the nil-groups are defined to be $NK_n(\mathbb{Z}H; \mathbb{Z}[G_0 - H], \mathbb{Z}[G_1 - H])$ using the triple $(\mathbb{Z}G; \mathbb{Z}[G_0 - H], \mathbb{Z}[G_1 - H])$. This inspires us to consider the following general case. Let R, a ring with identity $\alpha : R \to R$, be a ring automorphism. We denote by R^α the $R - R$-bimodule, which is R as a left R-module but with right multiplication given by $a \cdot r = a\alpha(r)$. For any automorphism α and β of R, we consider the triple $\mathbf{R} = (R; R^\alpha, R^\beta)$. We will prove that $\rho(\mathbf{R})$ is in fact a twisted polynomial ring, and this is important for later use.

Theorem 7.5.6 *Suppose that α and β are automorphisms of R. For the triple $\mathbf{R} = (R; R^\alpha, R^\beta)$, let R_ρ be the ring $\rho(\mathbf{R})$, and let γ be a ring automorphism of $\begin{pmatrix} R & 0 \\ 0 & R \end{pmatrix}$ defined by*

$$\gamma : \begin{pmatrix} a & 0 \\ 0 & b \end{pmatrix} \mapsto \begin{pmatrix} \beta(b) & 0 \\ 0 & \alpha(a) \end{pmatrix}.$$

Denote by 1_α (resp. 1_β) the generator of R^α (resp. R^β) corresponding to 1. Then, there is a ring isomorphism

$$\mu : R_\rho \longrightarrow \begin{pmatrix} R & 0 \\ 0 & R \end{pmatrix}_\gamma [x],$$

defined by mapping an element

$$\sum_{i \geq 0} \begin{pmatrix} a_i & 0 \\ 0 & a_i \end{pmatrix} \begin{pmatrix} (1_\beta \otimes 1_\alpha)^i & 0 \\ 0 & (1_\alpha \otimes 1_\beta)^i \end{pmatrix}$$

$$+ \sum_{i \geq 0} \begin{pmatrix} a'_i & 0 \\ 0 & a'_i \end{pmatrix} \begin{pmatrix} 0 & 1_\beta \otimes (1_\alpha \otimes 1_\beta)^i \\ 1_\alpha \otimes (1_\beta \otimes 1_\alpha)^i & 0 \end{pmatrix}$$

to an element

$$\sum_{i \geq 0} \begin{pmatrix} a_i & 0 \\ 0 & b_i \end{pmatrix} x^{2i} + \sum_{i \geq 0} \begin{pmatrix} a'_i & 0 \\ 0 & b'_i \end{pmatrix} x^{2i+1}.$$

PROOF By definition, each element of R_ρ can be written uniquely as

$$\sum_{i \geq 0} \begin{pmatrix} a_i & 0 \\ 0 & a_i \end{pmatrix} \begin{pmatrix} (1_\beta \otimes 1_\alpha)^i & 0 \\ 0 & (1_\alpha \otimes 1_\beta)^i \end{pmatrix}$$

$$+ \sum_{i \geq 0} \begin{pmatrix} a'_i & 0 \\ 0 & a'_i \end{pmatrix} \begin{pmatrix} 0 & 1_\beta \otimes (1_\alpha \otimes 1_\beta)^i \\ 1_\beta \otimes (1_\beta \otimes 1_\alpha)^i & 0 \end{pmatrix}.$$

It is easy to see that μ is an isomorphism from the additive group of R_ρ to $\begin{pmatrix} R & 0 \\ 0 & R \end{pmatrix}[x]$. To complete the proof, we only need to check that μ preserves any product of two elements such as

$$u_i = \begin{pmatrix} a_i & 0 \\ 0 & b_i \end{pmatrix} \begin{pmatrix} (1_\beta \otimes 1_\alpha)^i & 0 \\ 0 & 1_\alpha \otimes 1_\beta)^i \end{pmatrix} \quad i \geq 0$$

and

$$v_i = \begin{pmatrix} a'_j & 0 \\ 0 & b'_j \end{pmatrix} \begin{pmatrix} 0 & (1_\beta \otimes (1_\alpha \otimes 1_\beta)^j \\ 1_\alpha \otimes (1_\beta \otimes 1_\alpha)^j & 0 \end{pmatrix} \quad j \geq 0.$$

We check these case by case. Note that for any a, b $\in$ R, $1_\alpha \cdot a = \alpha(a)1_\alpha$, and $1_\beta \cdot b = \beta(b) \cdot 1_\beta$. Thus,

$$\begin{pmatrix} (1_\beta \otimes 1_\alpha)^i & 0 \\ 0 & 1_\alpha \otimes 1_\beta)^i \end{pmatrix} \begin{pmatrix} a_j & 0 \\ 0 & a_j \end{pmatrix}$$

$$= \begin{pmatrix} (\beta\alpha)^i(a_j) & 0 \\ 0 & (\alpha\beta)^i b_j \end{pmatrix} \begin{pmatrix} (1_\beta \otimes 1_\alpha)^i & 0 \\ 0 & 1_\alpha \otimes 1_\beta)^i \end{pmatrix}.$$

Note that the following equations hold in R_ρ:

$$(1_\beta \otimes 1_\alpha)^i (1_\beta \otimes 1_\alpha)^j = (1_\beta \otimes 1_\alpha)^{i+j}.$$
$$(1_\alpha \otimes 1_\beta)^i (1_\alpha \otimes 1_\beta)^j = (1_\alpha \otimes 1_\beta)^{i+j}.$$
$$(1_\beta \otimes 1_\alpha)^i (1_\beta \otimes (1_\alpha \otimes 1_\beta)^j) = 1_\beta \otimes (1_\alpha \otimes 1_\beta)^{i+j}.$$
$$(1_\alpha \otimes 1_\beta)^i (1_\alpha \otimes (1_\beta \otimes 1_\alpha)^j) = 1_\alpha \otimes (1_\beta \otimes 1_\alpha)^{i+j}.$$

This implies that $\mu(u_i u_j) = \mu(u_i)\mu(u_j)$, and $\mu(u_i v_j) = \mu(u_i)\mu(v_j)$. Similarly, we have

$$\begin{pmatrix} 0 & 1_\beta \otimes (1_\alpha \otimes 1_\beta)^i \\ 1_\alpha \otimes (1_\beta \otimes 1_\alpha)^i & 0 \end{pmatrix} \begin{pmatrix} a'_j & 0 \\ 0 & b'_j \end{pmatrix}$$

$$= \begin{pmatrix} (\beta\alpha)^i(b'_j) & 0 \\ 0 & (\alpha\beta)^i\alpha(a'_j) \end{pmatrix} \begin{pmatrix} 0 & 1_\beta \otimes (1_\alpha \otimes 1_\beta)^i \\ 1_\alpha \otimes (1_\beta \otimes 1_\alpha)^i & 0 \end{pmatrix}.$$

We have also equations in R_ρ:

$$(1_\beta \otimes (1_\alpha \otimes 1_\beta)^i)(1_\alpha \otimes 1_\beta)^j = 1_\beta \otimes (1_\alpha \otimes 1_\beta)^{i+j}.$$
$$(1_\alpha \otimes (1_\beta \otimes 1_\alpha)^i)(1_\beta \otimes 1_\alpha)^j = 1_\alpha \otimes (1_\beta \otimes 1_\alpha)^{i+j}.$$
$$(1_\beta \otimes (1_\alpha \otimes 1_\beta)^i)(1_\alpha \otimes (1_\beta \otimes 1_\alpha)^j) = 1_\beta \otimes (1_\alpha)^{i+j+1}.$$
$$(1_\alpha \otimes (1_\beta \otimes 1_\alpha)^i)(1_\beta \otimes (1_\alpha \otimes 1_\beta)^j) = 1_\alpha \otimes (1_\beta)^{i+j+1}.$$

It follows that $\mu(v_j u_i) = \mu(v_j)\mu(u_i)$, and $\mu(v_j v_i) = \mu(v_j)\mu(v_i)$. Hence μ is an isomorphism. $\qquad\square$

From Theorem 3.1 above, we obtain the following important result.

Theorem 7.5.7 *If R is regular, then $NK_n(R; R^\alpha, R^\beta) = 0$ for all $n \in \mathbb{Z}$. If R is quasi-regular, then $NK_n(R; R^\alpha, R^\beta) = 0$ for all $n \leq 0$.*

PROOF Since R is regular, then $\begin{pmatrix} R & 0 \\ 0 & R \end{pmatrix}$ is regular too. By theorem 7.5.6, R_ρ is a twisted polynomial ring over $\begin{pmatrix} R & 0 \\ 0 & R \end{pmatrix}$, and so, it is regular [57]. By the fundamental theorem of algebraic K-theory, it follows that $NK_n(R; R^\alpha, R^\beta) = 0$ for all $n \in \mathbb{Z}$ (see 6.3). If R is quasi-regular, then R_ρ is quasi-regular also [55]. By the fundamental theorem of algebraic K-theory for lower K-theory, $NK_n(R; R^\alpha, R^\beta) = 0$ for all $n \leq 0$. □

Remark 7.5.1 When $n \leq 1$, the result above is proved in [36] using isomorphism between $NK_n(R; R^\alpha, R^\beta)$ and Waldhausen's groups $\widetilde{\mathrm{Nil}}^W_{n-1}(R; R^\alpha, R^\beta)$ and the fact that $\widetilde{\mathrm{Nil}}^W_{n-1}(R; R^\alpha, R^\beta)$ vanishes for regular rings R [224]. We have given here another proof of the vanishing of lower Waldhausen's groups $\widetilde{\mathrm{Nil}}^W_{n-1}(R; R^\alpha, R^\beta)$ based on the isomorphism of $NK_n(R; R^\alpha, R^\beta)$ and $\widetilde{\mathrm{Nil}}^W_{n-1}(R; R^\alpha, R^\beta)$ for $n \leq 1$.

7.5.4 Now, we specialize to the case that $R = \mathbb{Z}H$, the groupring of a finite group H of order h. Let α and β be automorphisms of R induced by automorphisms of H. Choose a hereditary order Γ as in Theorem 1.6. Then we can define triples in $\mathcal{T}$.

$$\mathbf{R} = (R; R^\alpha, R^\beta).$$
$$\mathbf{\Gamma} = (\Gamma; \Gamma^\alpha, \Gamma^\beta).$$
$$\mathbf{R}/h\Gamma = (R/h\Gamma; (R/h\Gamma)^\alpha (R/h\Gamma)^\beta).$$
$$\mathbf{\Gamma}/h\Gamma = (\Gamma/h\Gamma; (\Gamma/h\Gamma)^\alpha (\Gamma/h\Gamma)^\beta).$$

The triples determine twisted polynomials rings

$$
\begin{array}{lcl}
R_\rho & \text{corresponding to} & \mathbf{R}, \\
\Gamma_\rho & \text{corresponding to} & \mathbf{\Gamma}, \\
(R/h\Gamma)_\rho & \text{corresponding to} & \mathbf{R}/h\Gamma, \\
(\Gamma/h\Gamma)_\rho & \text{corresponding to} & \mathbf{\Gamma}/h\Gamma.
\end{array}
\tag{1}
$$

Hence there is a Cartesian square

$$
\begin{array}{ccc}
RG & \longrightarrow & \Gamma \\
\downarrow & & \downarrow \\
RG/\underline{q} & \longrightarrow & \Gamma/\underline{q}
\end{array}
\tag{2}
$$

which implies that the square

$$\begin{pmatrix} R & 0 \\ 0 & R \end{pmatrix} \longrightarrow \begin{pmatrix} \Gamma & 0 \\ 0 & \Gamma \end{pmatrix} \tag{3}$$

$$\begin{pmatrix} R/h\Gamma & 0 \\ 0 & R/h\Gamma \end{pmatrix} \longrightarrow \begin{pmatrix} \Gamma/h\Gamma & 0 \\ 0 & \Gamma/h\Gamma \end{pmatrix}$$

is a Cartesian square. By theorem 7.5.6, we have the following Cartesian square

$$\begin{array}{ccc} R_\rho & \longrightarrow & \Gamma_\rho \\ \downarrow & & \downarrow \\ (R/h\Gamma)_\rho & \longrightarrow & (\Gamma/h\Gamma)_\rho \end{array} \tag{4}$$

Theorem 7.5.8 *Let V be a virtually infinite cyclic group in the second class having the form $V = G_0 *_H G_1$ where the group G_i, $i = 0, 1$, and H are finite, and $[G_i : H] = 2$. Then the nil-groups*

$$NK_n(\mathbb{Z}H; \mathbb{Z}[G_0 - H], \mathbb{Z}[G_1 - H])$$

defined by the triple $\mathbb{Z}H; \mathbb{Z}[G_0 - H], Z[G_1 - H]$ are $|H|$-torsion when $n \geq 0$ and 0 when $n \leq -1$.

PROOF   The proof is similar to that of theorem 7.5.4(i) using the Cartesian squares (2) and (3) above instead of (I) and (II) used in proof of theorem 7.5.4(i). Details are left to the reader.   ⬚

7.6   Higher K-theory of modules over 'EI' categories

7.6.1 In this section, we study higher K- and G-theory of modules over 'EI' categories. Modules over 'EI' categories constitute natural generalizations for the notion of modules over grouprings, and K-theory of such categories are known to house topological and geometric invariants and are replete with applications in the theory of transformation groups (see [137]). For example, if G is a finite group, and $or(G)$ the orbit category of G (an 'EI' category; see 7.6.2 below), X a G-CW-complex with round structure (see [137]), then the equivariant Riedemester torsion takes values in $Wh(Qor(G))$ where $(Wh(Qor(G))$ is the quotient of $K_1(Qor(G))$ by subgroups of "trivial units" (see [137]). We shall obtain several finiteness results, which are extensions of results earlier obtained for K-theory of grouprings of finite groups. For example, if $\mathcal{C}$ is a finite 'EI' category, and R the ring of integers in a number field F, we show that $K_n(R\mathcal{C}), G_n(R\mathcal{C})$ are finitely generated Abelian groups, $SK_n(R\mathcal{C})$ are finite groups for all $n \geq 0$, and $SG_n(R\mathcal{C}) = 0$ for all $n \geq 1$.

$(7.6)^A$ Generalities on modules over 'EI' categories $\mathcal{C}$

Definition 7.6.1 *An 'EI' category $\mathcal{C}$ is a small category in which every endomorphism is an isomorphism. $\mathcal{C}$ is said to be finite if the set $Is(\mathcal{C})$ of isomorphism classes of $\mathcal{C}$-objects is finite and for any two $\mathcal{C}$-objects X, Y, the set $\mathcal{C}(X, Y)$ of $\mathcal{C}$-morphism from X to Y is finite.*

Examples 7.6.1   (i) Let G be a finite group. Let $ob\mathcal{C} = \{G/H | H \leq G\}$ and morphisms be G-maps. Then $\mathcal{C}$ is a finite EI category called the orbit category of G and denoted $or(G)$. Here $\mathcal{C}(G/H, G/H) \simeq Aut(G/H) \approx N_G(H)/H$ where $N_G(H)$ is the normalizer of H in G (see [137]). We shall denote the group $N_G(H)/H$ by $\overline{N_G(H)}$.

(ii) Suppose that G is a Lie group, $ob\mathcal{C} = \{G/H | H$ compact subgroup of $G\}$ is also called the orbit category of G and denoted $orb(G)$. Here, morphisms are also G-maps.

(iii) Let G be a Lie group. Let $ob\mathcal{C} = \{G/H | H$ compact subgroup of $G\}$, and for $G/H, G/H' \in ob\mathcal{C}$, let $\mathcal{C}(G/H, G/H')$ be the set of homotopy classes of G-maps. Then, $\mathcal{C}$ is an EI category called the discrete orbit category of G and denoted by $or/(G)$.

(iv) Let G be a discrete group, $or_{\mathcal{F}in}(G)$ a category whose objects consist of all G/H, where H runs through finite subgroups of G. Morphisms are G-maps. The $or_{\mathcal{F}in}(G)$ is an EI category.

Note. For further examples of EI categories, see [137].

Definition 7.6.2 *Let R be a commutative ring with identity, $\mathcal{C}$ an EI category. An $R\mathcal{C}$-module is a contravariant functor $\mathcal{C} \to R\text{-}\mathcal{M}od$.*

Note that for any EI category $\mathcal{C}$, the $R\mathcal{C}$-modules form an Abelian category $R\mathcal{C}\text{-}\mathcal{M}od$, i.e., (i) $R\mathcal{C}\text{-}\mathcal{M}od$ has a zero object, and has finite products and coproducts. (ii) Each morphism has a kernel and a cokernel, and (iii) each monomorphism is a kernel and each epimorphism is a cokernel.

Examples 7.6.2   (i) Let G be a discrete group, and $\check{G}$ the groupoid with one object, (i.e., G) and left translation $\ell_g : G \to G : h \to gh$ as morphisms. A left RG-module M is an R-module M together with a group homomorphism $\alpha : G \to Aut_R(M) : g \to \ell_g : M \to M$ where

$\ell_g(m) = gm$. Hence, a left RG-module M uniquely determines an $R\breve{G}$-module and vice-versa, i.e., RG-$\mathcal{M}od = R\breve{G}$-$\mathcal{M}od$.

(ii) If S is a set and $R(S)$ the free R-module generated by S, we have an RC-module $RC(?, x) : C \to R$-$\mathcal{M}od$, $y \to RHom(y, x)$ for any $x \in obC$.

(iii) If G is a finite group and S a G-set, we have an $ROr(G)$-module $orG \to R$-$\mathcal{M}od$: $G/H \to R(S^H) = Rmap(G/H, S)^G$.

Definition 7.6.3 *Let C be an EI category. If we think of obC as an index set, then an obC-set A is a collection $\{A_c | c \in C\}$ of sets indexed by C. We could also think of obC as a category having obC as a set of objects and only the identity as morphisms, in which case we can interpret an obC-set as a functor $obC \to$ sets and a map between obC-sets as a natural transformation. Alternatively an obC-set could be visualized as a pair (L, β) – L a set and $\beta : L \to obC$ a set map. Then $L = \{\beta^{-1}(c) | c \in obC\}$.*

Note that any RC-module M has an underlying obC-set also denoted by M.

Definition 7.6.4 *An RC-module M is free with obC-set $B \in M$ as a base if for any RC-module N, and any map $f : B \to N$ of obC-sets, there is exactly one RC-homomorphism $F : M \to N$ extending f.*

An obC-set (N, β) is said to be finite if N is a finite set. If S is an (N, β)-subset of an RC-module M, define $spanS$ as the smallest RC-submodule of M containing S. Say that M is finitely generated if $M = spanS$ for some finite obC-subset S of M.

If R is a Noetherian ring and C a finite EI category, let $\mathcal{M}(RC)$ be the category of finitely generated RC-modules. Then $\mathcal{M}(RC)$ is an exact category in the sense of Quillen.

An RC-module P is said to be projective if any exact sequence of RC-modules $0 \to M' \to M \to P \to 0$ splits, or equivalently if $Hom_{RC}(P, -)$ is exact, or P is a direct summand of a free RC-module.

Let $\mathcal{P}(RC)$ be the category of finitely generated projective RC-modules. Then $\mathcal{P}(RC)$ is also exact. We write $K_n(RC)$ for $K_n(\mathcal{P}(RC))$.

Finally, let R be a commutative ring with identity, C an EI category, $\mathcal{P}_R(RC)$ the category of finitely generated RC-modules such that for each $X \in obC$, $M(X)$ is projective as R-module. Then $\mathcal{P}_R(RC)$ is an exact category and we write $G_n(R, C)$ for $K_n(\mathcal{P}_R(RC))$. Note that if R is regular, then $G_n(R, C) \cong G_n(RC)$.

7.6.2 Let C_1, C_2 be EI categories, R a commutative ring with identity, B a functor $C_1 \to C_2$. Let $RC_2(??, B(?))$ be the $RC_1 - RC_2$-bimodule: $C_1 \times C_2 \to R$-$\mathcal{M}od$ given by $(X_1, X_2) \to RC_2(X_2, BX_1)$. Define $RC_2 - RC_1$ bimodule analogously. Now define induction functor $ind_B : RC_1$-$\mathcal{M}od \to RC_2$-$\mathcal{M}od$ given by $M \to M \underset{RC_1}{\otimes} RC_2(??, B(?))$.

Also define a restriction functor $res_B : R\mathcal{C}_2\text{-}\mathcal{M}od \to R\mathcal{C}_1\text{-}\mathcal{M}od$ by $N \to N \circ B$.

7.6.3 A homomorphism $R_1 \to R_2$ of commutative rings with identity induces a functor $B : R_1\text{-}\mathcal{M}od \to R_2\text{-}\mathcal{M}od : N \to R_2 \underset{R_1}{\otimes} N$. So, if $\mathcal{C}$ is an EI-category, we have an induced functor $R_1\mathcal{C}\text{-}\mathcal{M}od \to R_2\mathcal{C}\text{-}\mathcal{M}od : M \to B \circ M$ where $(B \cdot M)(X) = B(M(X))$.

We also have an induced exact functor $\mathcal{P}(R_1\mathcal{C}) \to \mathcal{P}(R_2\mathcal{C}) : M \to B \circ M$ and hence a homomorphism $K_n(R_1\mathcal{C}) \to K_n(R_2\mathcal{C})$.

Now, suppose that R is a Dedekind domain with quotient fields F, and $R \hookrightarrow F$ the inclusion map; it follows from above that we have group homomorphisms $K_n(R\mathcal{C}) \to K_n(F\mathcal{C})$ and $G_n(R\mathcal{C}) \to G_n(F\mathcal{C})$.

Now, define $SK_n(R\mathcal{C}) :=$ Kernel of $K_n(R\mathcal{C}) \to K_n(F\mathcal{C})$ and $SG_n(R\mathcal{C}) :=$ Kernel of $G_n(R\mathcal{C}) \to G_n(F\mathcal{C})$.

$(7.6)^B$ $K_n(R\mathcal{C}), SK_n(R\mathcal{C})$

The following splitting theorem for $K_n(R\mathcal{C})$, where R is a commutative ring with identity and $\mathcal{C}$ any EI category, is due to W. Lück (see [137]).

Theorem 7.6.1 *Let R be a commutative ring with identity, $\mathcal{C}$ any EI category. Then*

$$K_n(R\mathcal{C}) \cong \bigoplus_{X \in Is(\mathcal{C})} K_n(R(Aut(X))).$$

PROOF We give a sketch of the proof of this theorem. Details can be found in [137].

Step I: For $X \in ob\mathcal{C}$, define the "splitting functor" $S_X : R\mathcal{C}\text{-}\mathcal{M}od \to R(Aut(X))\text{-}\mathcal{M}od$ by $S_X(M) = M(X)/M'(X)$ where $M'(X)$ is the R-submodule of $M(X)$ generated by the images of the R-homomorphisms $M(f) : M(Y) \to M(X)$ induced by all non-isomorphisms $f : X \to Y$.

Step II: Define the 'extension functor' $E_X : R(Aut(X))\text{-}\mathcal{M}od \to R\mathcal{C}\text{-}\mathcal{M}od$ by

$$(E_X(M) = M \underset{RAut(X)}{\bigotimes} R\mathcal{C}(?, X).$$

Step III: For $\overline{U} \in Is(\mathcal{C})$, the objects $X \in \overline{U}$ constitute a full subcategory of $\mathcal{C}$, which we denote by $\mathcal{C}(U)$. Now define

$$\mathrm{split}K_n(R\mathcal{C}) := \bigoplus_{\overline{U} \in Is(\mathcal{C})} K_n(R\mathcal{C}(U)).$$

Step IV: For each $U \in ob\mathcal{C}$, define the functor $\hat{E}_U : R\mathcal{C}(U)\text{-}\mathcal{M}od \to R\mathcal{C}\text{-}\mathcal{M}od$ by

$$\hat{E}_U(M) = M \bigotimes_{R\mathcal{C}(U)} R\mathcal{C}(?,??)$$

This induces a functor $\mathcal{P}(R\mathcal{C}(U)) \to \mathcal{P}(R\mathcal{C})$ and a homomorphism $K_n(\hat{E}_U) : K_n(R\mathcal{C}(U)) \to K_n(R\mathcal{C})$ and hence a homomorphism

$$E_n(R\mathcal{C}) = \oplus K_n(\hat{E}_U) : \oplus_{\overline{U} \in Is\mathcal{C}} K_n (R\mathcal{C}(U)) \to K_n(R\mathcal{C}),$$

that is, a homomorphism

$$E_n(R\mathcal{C}) : \mathrm{split}K_n(R\mathcal{C}) \to K_n(R\mathcal{C}).$$

Step V: For any $U \in ob(\mathcal{C})$, define a functor $\hat{S}_U : R\mathcal{C}\text{-}\mathcal{M}od \to R\mathcal{C}(U)\text{-}\mathcal{M}od$ by $\hat{S}_U(M) = M \underset{R\mathcal{C}}{\otimes} B$ for the $R\mathcal{C} - R\mathcal{C}(U)$ bimodule B given by $B(X,Y) = R\mathcal{C}(X,Y)$ if $Y \in \overline{U}$ and $B(X,Y) = \{0\}$ if $Y \notin \overline{U}$, where X runs through $ob\mathcal{C}(U)$, and $Y \in ob(\mathcal{C})$. Then each $\hat{S}_U$ induces a homomorphism $K_n(\hat{S}_U) : K_n(R\mathcal{C}) \to K_n(R\mathcal{C}(U))$ and hence a homomorphism

$$S_n(R\mathcal{C}) : K_n(R\mathcal{C}) \to \bigoplus_{\overline{U} \in Is(\mathcal{C})} K_n(R\mathcal{C}(U))$$

i.e.,

$$S_n(R\mathcal{C}) : K_n(R\mathcal{C}) \to \mathrm{split}K_n(R\mathcal{C}).$$

Step VI:

$$E_n(R\mathcal{C}) : K_n(R\mathcal{C}) \to \bigoplus_{\overline{U} \in Is(\mathcal{C})} K_n R\mathcal{C}(U)$$

and

$$S_n(R\mathcal{C}) : \bigoplus_{\overline{U} \in Is(\mathcal{C})} K_n(R\mathcal{C}(U)) \to K_n(R\mathcal{C})$$

are isomorphisms, one the inverse of the other.

Step VII:

$$K_n(R\mathcal{C}(U)) \simeq K_n(R(Aut(X)))$$

(for any $X \in \overline{U}$, via the equivalence of categories $Aut(X)' \to \mathcal{C}(U)$ where for any group G, $\check{G}$ is the groupoid with one object G, and morphisms left translations $1 : G \hookrightarrow G : h \hookrightarrow gh$.

The next result follows from theorem 7.1.11.

Theorem 7.6.2 *Let C be a finite EI category, R the ring of integers in a number field F. Then for any C-object $X, K_n(R(AutX))$ is a finitely generated Abelian group for all $n \geq 1$, and $SK_n(R(Aut(X))$ is a finite group.*

The next result follows from theorem 7.6.1 and theorem 7.6.2.

Corollary 7.6.1 Let R be the ring of integers in a number field F, C any finite EI category. Then for all $n \geq 1$,

(i) $K_n(RC)$ is a finitely generated Abelian group.

(ii) $SK_n(RC)$ is a finite group.

Remarks 7.6.1 Let G be a finite group and $C = or(G)$ the orbit category of G.

It is well known that there is one-one correspondence between $Is(C)$ and the conjugacy classes $con(G)$ of G, i.e., $G/H \simeq G/H'$, if H is conjugate to H'. It is also well known that $C(G/H, G/H) = Aut(G/H) = N_G(H)/H := \overline{N_G H}$ where $N_G H$ is the normalizer of H in G.
So, for any commutative ring R with identity,
$$K_n(Ror(G)) = \bigoplus_{H \in con(G)} K_n R\overline{N_G(H)}.$$

$(7.6)^C$ $G_n(RC), SG_n(RC)$

The following splitting result for $G_n(RC)$ is due to W. Lück see [137].

Theorem 7.6.3 *Let R be a commutative Noetherian ring with identity, C any finite EI category. Then for all $n \geq 1$,*
$$G_n(RC) \cong \bigoplus_{\overline{X} \in Is(C)} G_n(R(Aut(X))).$$

PROOF   We sketch the proof of 7.6.3 and refer the reader to [137] for missing details.

Step I: For each $X \in obC$, define $Res_X : RC\text{-}\mathcal{M}od \to (RAut(X))\text{-}\mathcal{M}od$
$Res_X(M) = M(X)$. Then an RC-module M is finitely generated iff $Res_X(M)$ is finitely generated for all X in obC (see [137]).

Moreover, Res_X induces an exact functor $\mathcal{M}(R\mathcal{C}) \to \mathcal{M}(R(Aut(X)))$, which also induces for all $n \geq 0$ homomorphisms $G_n(R\mathcal{C}) \to G_n(R(Aut(X)))$ and hence homomorphism $Res : G_n(R\mathcal{C}) \to \underset{\overline{X} \in Is(\mathcal{C})}{\oplus} G_n(R(Aut(X)))$. We write $\mathrm{split}G_n(R\mathcal{C})$ for $\underset{\overline{X} \in Is\mathcal{C}}{\oplus} G_n(R(Aut(X)))$.

Step II: For $X \in ob\mathcal{C}$, define a functor $I_X : R(Aut(X)) \to R\mathcal{C}$ by

$$I_X(M) = \begin{cases} M \underset{RAut(X)}{\otimes} R\mathcal{C}(Y,X) \text{ if } \overline{Y} = \overline{X} \\ 0 \qquad\qquad\qquad\qquad \text{ if } \overline{Y} \neq \overline{X} \end{cases}$$

Then we have an induced homomorphism

$$I : \mathrm{split}G_n(R\mathcal{C}) = \bigoplus_{\overline{X} \in Is\mathcal{C}} G_n(R(Aut(X))) \to G_n(R\mathcal{C}).$$

Step III: Res and I are isomorphisms inverse to each other.

$\square$

The following result follows from theorem 7.1.13 and theorem 7.1.15.

Theorem 7.6.4 *Let R be the ring of integers in a number field F,$\mathcal{C}$ a finite EI category. Then, for any $X \in ob\mathcal{C}$, and all $n \geq 1$ $G_n(R(Aut(X)))$ is a finitely generated Abelian group, and $SG_n(R(Aut(X))) = 0$.*

The following result is a consequence of theorem 7.6.3 and theorem 7.6.4.

Corollary 7.6.2 Let R be the ring of integers in a number field F,$\mathcal{C}$ a finite EI-category. Then for all $n \geq 1$,

(i) $G_n(R\mathcal{C})$ is a finitely generated Abelian group.

(ii) $SG_n(R\mathcal{C}) = 0$.

$(7.6)^D$ Cartan map $K_n(R\mathcal{C}) \to G_n(R\mathcal{C})$

7.6.4 In this subsection we briefly discuss Cartan maps and some consequences. Recall that if R is a commutative ring with identity and $\mathcal{C}$ an EI category, then for all $n \geq 0$, the inclusion functor $\mathcal{P}(R\mathcal{C}) \to \mathcal{M}(R\mathcal{C})$ induces a homomorphism called the Cartan map.

First, we observe that if R is regular and $\mathcal{C}$ a finite EI category, then $K_n(R\mathcal{C}) \simeq G_n(R\mathcal{C})$.

We record here the following result whose proof depends on ideas from equivariant K-theory (see theorem 10.4.1).

Theorem 7.6.5 *Let k be a field of characteristic $p, \mathcal{C}$ a finite EI category. Then for all $n \geq 0$, the Cartan homomorphism $K_n(k\mathcal{C}) \to G_n(k\mathcal{C})$ induce isomorphism*

$$\mathbb{Z}\left(\frac{1}{p}\right) \otimes K_n(k\mathcal{C}) \cong \mathbb{Z}\left(\frac{1}{p}\right) \otimes G_n(k\mathcal{C}).$$

The next result is a consequence of theorem 10.4.2. (Also see corollary 10.4.2.)

Corollary 7.6.3 Let R be the ring of integers in a number field F, $\mathbf{m}$ a prime ideal of R lying over a rational prime p. Then for all $n \geq 1$,

(a) the Cartan map $K_n((R/\mathbf{m})\mathcal{C}) \to G_n((R/\mathbf{m})\mathcal{C})$ is surjective.

(b) $K_{2n}(R/\mathbf{m})\mathcal{C})$ is a finite p-group.

$(7.6)^E$ Pairings and module structures

7.6.5 Let $\mathcal{E}, \mathcal{E}_1, \mathcal{E}_2$ be three exact categories, and $\mathcal{E}_1 \times \mathcal{E}_2$ the product category. An exact pairing $\mathcal{E}_1 \times \mathcal{E}_2 \to \mathcal{E} : (M_1, M_2) \to M_1 \circ M_2$ is a covariant functor from $\mathcal{E}_1 \times \mathcal{E}_2$ to $\mathcal{E}$ such that $\mathcal{E}_1 \times \mathcal{E}_2((M_1, M_2), (M_1', M_2')) = \mathcal{E}_1(M_1, M_1') \times \mathcal{E}_2(M_2, M_2') \to \mathcal{E}(M_1 \circ M_2, M_1' \circ M_2')$ is a bi-additive and bi-exact, that is, for a fixed M_2, the functor $\mathcal{E}_1 \to \mathcal{E}$ given by $M_1 \mapsto M_1 \circ M_2$ is additive and exact, and for fixed M_1, the functor $\mathcal{E}_2 \to \mathcal{E} : M_2 \to M_1 \circ M_2$ is additive and exact. It follows from [224] that such a pairing gives rise to a K-theoretic cup product $K_i(\mathcal{E}_j) \times K_j(\mathcal{E}_2) \to K_{i+j}(\mathcal{E})$, and in particular to natural pairing $K_0(\mathcal{E}_1) \circ K_n(\mathcal{E}_2) \to K_n(\mathcal{E})$, which could be defined as follows:

Any object $M_1 \in \mathcal{E}$ induced an exact functor $M_1 : \mathcal{E}_2 \to \mathcal{E} : M_2 \to M_1 \circ M_2$ and hence a map $K_n(M_1) : K_n(\mathcal{E}_2) \to K_n(\mathcal{E})$. If $M_1' \to M_1 \to M_1''$ is an exact sequence in $\mathcal{E}_1$, then we have an exact sequence of exact functors $M_1'^* \to M_1^* \to M_1''^*$ from $\mathcal{E}_2$ to $\mathcal{E}$ such that, for each object $M_2 \in \mathcal{E}_2$, the sequence $M_1'(M_2) \to M_1^*(M_2) \to M_1''^*(M_2)$ is exact in $\mathcal{E}$, and hence, by a result of Quillen (see 6.1.1), induces the relation $K_n(M_1'^*) + K_n(M_1''^*) = K_n(M_1^*)$. So, the map $M_1 \to K_n(M_1) \in Hom(K_n(\mathcal{E}_2), K_n(\mathcal{E}))$ induces a homomorphism $K_0(\mathcal{E}_1) \to Hom(K_n(\mathcal{E}), K_n(\mathcal{E}))$ and hence a pairing $K_0(\mathcal{E}_1) \times K_n(\mathcal{E}) \to K_n(\mathcal{E})$. We could obtain a similar pairing $K_n(\mathcal{E}_1) \times K_0(\mathcal{E}_2) \to K_n(\mathcal{E})$.

If $\mathcal{E}_1 = \mathcal{E}_2 = \mathcal{E}$ and the pairing $\mathcal{E} \times \mathcal{E}$ is naturally associative (and commutative), then the associated pairing $K_0(\mathcal{E}) \times K_0(\mathcal{E}) \to K_0(\mathcal{E})$ turns $K_0(\mathcal{E})$ into an associative (and commutative ring, which may not contain the identity). Suppose that there is a pairing $\mathcal{E} \circ \mathcal{E}_1 \to \mathcal{E}_1$ that is naturally associative with respect to the pairing $\mathcal{E} \circ \mathcal{E} \to \mathcal{E}$, then the pairing

$K_0(\mathcal{E}) \times K_n(\mathcal{E}_1) \to K_n(\mathcal{E}_1)$ turns $K_n(\mathcal{E}_1)$ into a $K_0(\mathcal{E})$-module that may or may not be unitary. However, if $\mathcal{E}$ contains a natural unit, i.e., an object E such that $E \circ M = M \circ E$ are naturally isomorphic to M for each $\mathcal{E}$-object M, then the pairing $K_0(\mathcal{E}) \times K_n(\mathcal{C}_1) \to K_n(\mathcal{E}_1)$ turns $K_n(\mathcal{E}_1)$ into a unitary $K_0(\mathcal{E})$-module.

7.6.6 We now apply the above to the following situation. Let R be a commutative ring with identity, $\mathcal{C}$ a finite EI category.

Let $\mathcal{E} = \mathcal{P}_R(R\mathcal{C})$ be the category of finitely generated $R\mathcal{C}$-modules such that for all, $X \in ob(\mathcal{C})$. $M(X)$ is projective as an R-module. So, $\mathcal{P}_R(R\mathcal{C})$ is an exact category on which we have a pairing

$$\otimes : \mathcal{P}_R(R\mathcal{C}) \times \mathcal{P}_R(R\mathcal{C}) \to \mathcal{P}_R(R\mathcal{C}). \tag{I}$$

If we take $\mathcal{E}_1 = \mathcal{P}(R\mathcal{C})$, then the pairing

$$\otimes : \mathcal{P}_R(R\mathcal{C}) \times \mathcal{P}_R(R\mathcal{C}) \to \mathcal{P}(R\mathcal{C}) \tag{II}$$

is naturally associative with respect to the pairing (I), and so, $K_n(R\mathcal{C})$ is a unitary $(K_0(\mathcal{P}_R(R\mathcal{C})) = G_0(R,\mathcal{C})$-module. Also, $G_n(R,\mathcal{C})$ is a $G_0(R,\mathcal{C})$-module.

7.6.7 Let $\mathcal{C}$ be a finite EI category and $\mathbb{Z}(Is(\mathcal{C}))$ the free Abelian group on $Is(\mathcal{C})$. Note that $\mathbb{Z}(Is(\mathcal{C})) = \underset{Is\mathcal{C}}{\oplus} \mathbb{Z}$. If $\mathbb{Z}^{(Is(\mathcal{C}))}$ is the ring of $\mathbb{Z}$-valued functions on $Is\mathcal{C}$, we can identify each element of $\mathbb{Z}(Is(\mathcal{C}))$ as a function $Is(\mathcal{C}) \to \mathbb{Z}$ via an injective map $\beta : \mathbb{Z}(Is(\mathcal{C})) \to \mathbb{Z}^{Is(\mathcal{C})}$ given by $\beta(X)(Y) = |\mathcal{C}(Y,X)|$ for $X,Y \in ob\mathcal{C}$. Moreover, β identifies $\mathbb{Z}(Is(\mathcal{C}))$ as a subring of $\mathbb{Z}^{(Is(\mathcal{C}))}$. Call $\mathbb{Z}(Is(\mathcal{C}))$ the Burnside ring of $\mathcal{C}$ and denote this ring by $\Omega(\mathcal{C})$. Note that if $\mathcal{C} = orb(G)$, G a finite group, then $\mathbb{Z}(Is(\mathcal{C}))$ is the well-known Burnside ring of π, which is denoted by $\Omega(G)$.

7.6.8 If R is a commutative ring with identity and $\mathcal{C}$ a finite EI category, let $\mathcal{F}(R\mathcal{C})$ be the category of finitely generated free $R\mathcal{C}$-modules. Then, for all $n \geq 1$, the inclusion functor $\mathcal{F}(R\mathcal{C}) \to \mathcal{P}(R\mathcal{C})$ induces an isomorphism $K_n(\mathcal{F}(R\mathcal{C})) \simeq K_n(R\mathcal{C})$ and $K_0(\mathcal{F}(R\mathcal{C})) \simeq \mathbb{Z}(Is\mathcal{C})$ (see [137] 10.42). Now, by the discussion in 4.1, the pairing $K_0(\mathcal{F}(R\mathcal{C})) \times K_n(\mathcal{P}(R\mathcal{C})) \to K_n(\mathcal{P}(R\mathcal{C}))$ makes $K_n(R\mathcal{C})$ a unitary module over the Burnside ring $\mathbb{Z}(is(\mathcal{C})) \simeq K_0(\mathcal{F}(R\mathcal{C}))$.

7.7　Higher K-theory of $\mathcal{P}(A)_G$; A maximal orders in division algebras; G finite group

The initial motivation for the work reported in this section was the author's desire to obtain a non-commutative analogue of a fundamental result of R.G.

Swan in the sense we now describe.

If B is a (not necessarily commutative) regular ring, G a finite group, let $K_0(G, \mathcal{P}(B))$ be the Grothendieck group of the category $[G, \mathcal{P}(B)]$ of G-representations in the category $\mathcal{P}(B)$ of finitely generated projective B-modules (see 1.1). In [209], Swan proved that if R is a semi-local Dedekind domain with quotient field F, then the canonical map $K_0(G, \mathcal{P}(R)) \overset{\delta}{\to} K_0(G, \mathcal{P}(F))$ is an isomorphism. The question arises whether this theorem holds if R is non-commutative. Since the map is always surjective, whether R is commutative or not, the question reduces to asking whether δ is injective if R is a non-commutative semi-local Dedekind domain. We show in theorem 7.7.6 that δ is not always injective via a counterexample of the canonical map $K_0(G, \mathcal{P}(A)) \overset{\delta}{\to} K_0(G, \mathcal{P}(D))$ where A is a maximal order in a central division algebra D over a p-adic field F, and in (2.7) prove that if G is a finite p-group, then the kernel of δ in this example is a finite cyclic p-group.

Now, since $[G, \mathcal{P}(B)]$ is an exact category, and $K_n(G, \mathcal{P}(B)) \cong G_n(BG)$ (see lemma 7.7.4), the K-theory of the category $[G, \mathcal{P}(B)]$ reduces to the K-theory of the groupring BG. First we obtain as much result as we can in the local situation AG where A is a maximal order in a central p-adic division algebra D and then in the global situation of A being a maximal order in a central division algebra over an algebraic number field.

The results of this section are due to A. Kuku (see [110]).

$(7.7)^A$ A transfer map in higher K-theory and non-commutative analogue of a result of R.G. Swan

7.7.1 Let C be a category, G any group (not necessarily finite). A G-object in C, or equivalently, a G-representation in C is a pair (X, α), where X is a C-object and $\alpha : G \to \operatorname{Aut}_C(X)$ is a group homomorphism from G to the group $\operatorname{Aut}_C(X)$ of C-automorphisms of X. The G-objects form a category that can be identified with the category $[G, C]$ of functors from G (considered as a category with one object, with morphism elements of G to C). Note that $[G, C]$ was denoted by C_G or $[G/G, C]$ in 1.1.

Now, let C be an exact category, then $[G, C]$ or equivalently C_G or $[G/G, C]$ is also an exact category (see 10.1.1). If X, Y are pointed spaces, let $\overline{[X, Y]}$ be the set of free homotopy classes of free maps from X to Y. In [191], C. Sherman constructs, for all $n \geq 0$, homomorphisms

$$K_n(G, C) \to [BG, \Omega^{n+1} BQC] \qquad (I)$$

where for any category D, BD is the classifying space of D (see 5.2), and $\Omega^{n+1} BQC$ is the $(n + 1)$th loop space of BQC. Now, put $G = \pi_1(X)$ the fundamental group of X, where X is a connected pointed CW-complex.

Then there exists a canonical pointed homotopy class of maps $\phi : X \to B\pi_1 X$ corresponding to the identity map of $\pi_1 X$ under the isomorphism $[X, B\pi_1 X] \xrightarrow{\sim} \mathrm{Hom}(\pi_1 X, \pi_1 X)$. Composition with ϕ defines a homomorphism

$$[B\pi_1 X, \Omega^{n+1} BQC] \to [X, \Omega^{n+1} BQC]. \tag{II}$$

By composing (II) with (I), we obtain a homomorphism

$$K_n(\pi_1(-), C) \to [-, \Omega^{n+1} BQC], \tag{III}$$

which represents a natural transformation of bifunctors $K_n(\pi_1(\ ), C) \to [-, \Omega^{n+1} BQC]$ from the pointed homotopy category of connected CW-complexes and the category of small categories to the category of groups. For $n = 0$, the map (III) is universal for all natural transformations $K_0(\pi_1(-), C) \to [-, H]$ where H is a connected H-space (see [191]) and agrees with Quillen's construction in [59], section 1, where $C = \mathcal{P}(A)$ is the category of finitely generated projective modules over the ring A. Now, if C, C' are two exact categories and $F : C \to C'$ an exact functor, it follows from the universal properties of the construction that we have a commutative diagram

$$\begin{array}{ccc} K_0(\pi_1(X), C) & \longrightarrow & K_0(\pi_1(X), C') \\ \downarrow & & \downarrow \\ [X, BQC] & \longrightarrow & [X, BQC'], \end{array} \tag{IV}$$

and if we put $X = S^n$ (the nth sphere), then the bottom row of (IV) yields a map $K_n(C) \to K_n(C')$.

7.7.2 We now apply 7.7.1 to the following setup. Let F be a p-adic field (i.e., F is any finite extension of $\hat{Q}_p$, the completion of the field Q of rational numbers at a rational prime p), R the ring of integers of F (i.e., R is the integral closure of $\hat{Z}_p$ in F), A a maximal R-order in a central division F-algebra D, and **m** the unique maximal ideal of A (such that $F \otimes \mathbf{m} = D$). Observe that the restriction of scalars defines an exact functor $\mathcal{P}(A/\mathbf{m}) \to \mathcal{M}(A)$, which induces homomorphisms $K_n(A/\mathbf{m})] \xrightarrow{\beta_n} G_n(A)$. Now, the inclusion functor $\mathcal{P}(A) \to \mathcal{M}(A)$ induces a homomorphism $K_n(A) \xrightarrow{\sigma_n} G_n(A)$ (called the Cartan homomorphism), which is an isomorphism since A is regular. Composing β_n with σ_n^{-1}, we obtain the transfer map $K_n(A/\mathbf{m}) \to K_n(A)$.

Also the exact functor $\mathcal{P}(A/\mathbf{m}) \to \mathcal{M}(A)$ described above induces an exact functor $[G, \mathcal{P}(A/\mathbf{m})] \to [G, \mathcal{M}(A)]$ for any group G, and hence a homomorphism $K_n(G, \mathcal{P}(A/\mathbf{m})) \xrightarrow{\beta_n} K_n(G, \mathcal{M}(A))$ for all $n \geq 0$. Moreover, the inclusion functor $\mathcal{P}(A) \to \mathcal{M}(A)$ induces an exact functor $[G, \mathcal{P}(A)] \to [G, \mathcal{M}(A)]$ and hence a homomorphism $K_n(G, \mathcal{P}(A) \xrightarrow{\hat{\sigma}_n} K_n(G, \mathcal{M}(A))$, which can be shown to be an isomorphism when $n = 0$, by slightly modifying Gersten's proof of a similar result for A, a Dedekind ring R (see [59], theorem 3.1(c)). (The same proof works by putting $A/\mathbf{m}$ for $R/\mathbf{p}$ where $\mathbf{m}$ is the unique prime = maximal ideal of A lying above $\mathbf{p}$, and in this case, only one prime is involved, so it is easier.) Composing $\hat{\sigma}_n^{-1}$ with β_0 yields the map $K_0(G, \mathcal{P}(A/m)) \to K_0(G, \mathcal{P}(A))$. If we now put $G = \pi_1(X)$, where X is a finite connected CW-complex, it follows from 7.7.1 (III) and (IV) that we have a commutative diagram

$$
\begin{array}{ccccc}
K_0(\pi_1(X), \mathcal{P}(A/m)) & \xrightarrow{\hat{\beta}_0} & K_0(\pi_1(X), \mathcal{M}(A)) & \xrightarrow{\hat{\sigma}_0^{-1}} & K_0(\pi_1(X), \mathcal{P}(A)) \\
\downarrow & & \downarrow & & \downarrow \\
[X, BQ\mathcal{P}(A/\mathbf{m})] & \xrightarrow{\delta_n} & [X, BQ\mathcal{M}(A)] & \longrightarrow & [X, BQ\mathcal{P}(A)]
\end{array}
$$

where the vertical arrows on the left and right are universal maps of Quillen (see [191]), and the bottom row yields the transfer map described above when $X = S^n$. We shall need the above in 7.7.4.

7.7.3 Now, let R be a Dedekind domain with quotient field F, A a maximal R-order in a central division algebra over F, and G a finite group. Then the groupring AG is an R-order in the semi-simple F-algebra DG. Note that D is a separable F-algebra, and since FG is also a separable F-algebra if and only if the characteristic of F does not divide the order $|G|$ of G, $DG = D \otimes_F FG$ would be separable if Char $F \nmid |G|$. In particular, DG is separable if F is a p-adic field or an algebraic number field (i.e., any finite extension of the field Q of rational numbers). If $B = R$ or A, Λ a B-algebra finitely generated and projective as a B-module, we shall write $P_B(\Lambda)$ for the category of finitely generated Λ-modules that are B-projective and write $P_B(G)$ for $P_B(BG)$. Note that $[G, \mathcal{P}(B)] \cong \mathcal{P}_B(G)$.

Lemma 7.7.1 *In the notation of 7.7.3,*

(i) $\mathcal{P}_A(G) = \mathcal{P}_R(AG)$.

(ii) *The inclusion $\mathcal{P}_A(G) \to \mathcal{M}(AG)$ induces an isomorphism $K_n(\mathcal{P}_A(G)) \cong G_n(AG)$ for all $n \geq 0$.*

(iii) *If $R/\mathbf{p}$ is finite for every prime ideal $\mathbf{p}$ of R, and $m_\mathbf{p}$ is the unique maximal ideal of A such that $m_\mathbf{p} \supseteq \mathbf{p}A$, then there exist isomorphisms $G_n(AG/\mathbf{p}AG) \cong G_n(AG/m_\mathbf{p}AG) \cong G_n(AG/rad(AG))$ for all $n \geq 0$.*

PROOF

(i) If $M \in P_A(G)$, then $M \in \mathcal{M}(AG)$ and $M \in \mathcal{P}(A)$. But A is an R-order in a separable F-algebra D, and so, A is an R-lattice, i.e., A is R-projective. So, $M \in \mathcal{P}(R)$.

Conversely, suppose that $M \in \mathcal{P}_R(AG)$. Then, $M \in \mathcal{M}(AG)$ and $M \in \mathcal{P}(R)$. But $M \in \mathcal{M}(AG)$ implies that $M \in \mathcal{M}(A)$ and $AG \in \mathcal{M}(A)$. So, $M \in \mathcal{M}(A)$ and $M \in \mathcal{P}(R)$ imply that $M \in \mathcal{P}(A)$ (see [213], p.93, theorem 5.12).

(ii) Let $M \in \mathcal{M}(AG)$. Then, since AG is Noetherian, there exists a resolution $0 \to D_n \to P_{n-1} \to \cdots \to P_0 \to M \to 0$ where each $P_i \in \mathcal{P}(AG)$, and $D_n \in \mathcal{M}(AG)$. Now, $M \in \mathcal{M}(AG)$ implies that $M \in \mathcal{M}(A)$, and since A is regular, each $P_i \in \mathcal{P}(A)$ and $D_n \in \mathcal{P}(A)$ also. So, each $P_i \in \mathcal{P}_A(G)$, and so is D_n. Hence, by the resolution theorem of Quillen (see $(6.1)^A$), $K_n(\mathcal{P}_A(G)) \cong G_n(AG)$. That $K_n(G, \mathcal{P}(A)) \cong G_n(AG)$ follows from the fact that $[G, \mathcal{P}(A)] \cong \mathcal{P}_A(G)$.

(iii) Note that $\mathbf{p}AG \subseteq \mathbf{m_p}AG \subset rad(AG)$, and so, $AG/\mathbf{p}AG$, $AG/\mathbf{m_p}AG$ are finite Artinian rings since they are both finite-dimensional algebras over the finite field $R/\mathbf{p}$. Hence $G_n(AG/\mathbf{p}AG) \cong G_n(((AG)/\mathbf{p}AG)/rad(AG/\mathbf{p}AG)) \cong G_n(AG/rad(AG))$. Similarly, $G_n(AG/\mathbf{m_p}AG) \cong (AG/radAG)$.

$\square$

Remark 7.7.1 (i) Let R be a Dedekind ring with quotient field F such that $R/\mathbf{p}$ is finite for all prime ideals $\mathbf{p}$ of R. Suppose that A is a maximal R-order in a central division algebra D over F, then Quillen's localization sequence connecting $G_n(AG)$ and $G_n(DG)$, $K_n(G, \mathcal{P}(A))$ and $K_n(G, \mathcal{P}(D))$, plus the usual Devissage argument, yields the following commutative diagram where the vertical arrows are isomorphisms:

$$\cdots \to \bigoplus_p K_n(G, \mathcal{P}(A/\mathbf{m_p})) \to K_n(G, \mathcal{P}(A)) \to K_n(G, \mathcal{P}(D)) \to \cdots$$

$$\theta \downarrow \wr \qquad\qquad\qquad \downarrow \wr \qquad\qquad\qquad \downarrow \wr$$

$$\cdots \to \bigoplus_{\mathbf{p}} G_n(AG/\mathbf{p}AG) \to G_n(AG) \to G_n(DG) \to \cdots$$

Note that the isomorphism θ follows from lemma 7.7.1(iii) since $K_n(G, \mathcal{P}(A/\mathbf{m})) \cong G_n(AG/\mathbf{m_p}AG) \cong G_n(AG/\mathbf{p}AG)$. Note also that the canonical map $G_0(AG) \to G_0(DG)$ is always surjective (see [174, 213]).

(ii) Our next aim is to show that a non-commutative analogue of Swan's theorem is not true in general for grouprings whose coefficient rings are non-commutative local Dedekind rings, via the counterexample of AG, where G is a finite group and A a maximal order in a central division algebra over a p-adic field F. Since by 7.7.3 and lemma 7.7.1 $K_n(G, \mathcal{P}(A)) \cong G_n(AG)$, we then try to determine as much of the K-theory of AG as possible.

Theorem 7.7.1 *Let R be the ring of integers in a p-adic field F, A a maximal R-order in a central division algebra D over F, and G any finite group. Then the canonical map $K_0(G, \mathcal{P}(D)) \xrightarrow{\eta_0} K_0(G, \mathcal{P}(D))$ is not always injective.*

PROOF   Suppose that η_0 was injective for all finite groups G, then the transfer map $K_0(G, \mathcal{P}(A/\mathbf{m})) \overset{\delta}{\to} K_0(G, \mathcal{P}(A))$ would be zero where $\mathbf{m} =$ rad A and $A/\mathbf{m}$ is a finite field, and by an argument due to Gersten (see [59], p.224 or p.242, or [191]) one can show that $K_0(G, P(A/\mathbf{m})) \to K_0(G, \mathcal{P}(A))$ is zero for all groups G. This is because there exists a commutative diagram

$$
\begin{array}{ccc}
K_0(G', \mathcal{P}(A/\mathbf{m})) & \longrightarrow & K_0(G', \mathcal{P}(A)) \\
\downarrow & & \downarrow \\
K_0(G, \mathcal{P}(A/\mathbf{m})) & \longrightarrow & K_0(G, \mathcal{P}(A))
\end{array}
\qquad (I)
$$

where G' is a finite group since any representation $\phi : G \to \mathrm{Aut}_{A/\mathbf{m}}(V)$ in $\mathcal{P}_{A/\mathbf{m}}(G)$ factors through a finite quotient G' of G, and so, from the commutative diagram (I), the transfer map $K_0(G, P(A/\mathbf{m})) \to K_0(G, P(A))$ would be zero for all groups G. Now, by the universal property of Quillen K-theory discussed in 7.7.1 and 7.7.2, we have, for any finite connected CW-complex X, a commutative diagram

$$
\begin{array}{ccc}
K_0(\pi_1(X), \mathcal{P}(A/m)) & \longrightarrow & K_0(\pi_1(X), \mathcal{P}(A)) \\
\downarrow & & \downarrow \\
[X, BQ\mathcal{P}(A/m)] & \longrightarrow & [X, BQ\mathcal{P}(A)]
\end{array}
\qquad (II)
$$

and hence a map $K_n(A/m) \to K_n(A)$, which would be zero for all $n \geq 1$ by the universality of the construction, i.e., $K_n(A) \to K_n(D)$ is injective for all $n \geq 1$, contradicting lemma 7.1.3 which says that for all $n \geq 1$, $K_{2n-1}(A) \to K_{2n-1}(D)$ is injective if and only if $D = F$. Hence, the canonical map $K_0(G, P(A)) \to K_0(G, P(D))$ is not always injective.   □

$(7.7)^B$ Higher K-theory of $\mathcal{P}(A)_G$, A a maximal order in a p-adic division algebra

Remark 7.7.2 Theorem 7.7.1 says that there are finite groups G for which η_0 is not injective. We shall exhibit one such class of groups in theorem 7.7.4 below. Meanwhile, we try to obtain more information on the K-theory of AG in this local situation.

7.7.4 Since AG is an R-order in the p-adic semi-simple algebra DG, it follows from theorem 7.1.13(iv) that for all $n \geq 1$, $SG_{2n}(AG) = 0$, and $SG_{2n-1}(AG)$ is a finite group of order relatively prime to p. So, we regard $SG_n(A\pi)$ as known for $n \geq 1$. Before considering the case $n = 0$, we first exploit the above information to obtain information on $SK_n(AG)$. Note that we have the following commutative diagram

$$\cdots \rightarrow K_{n+1}(AG) \xrightarrow{\delta_{n+1}} K_{n+1}(DG) \xrightarrow{\gamma_{n+1}} K_n(\mathcal{H}_S(AG)) \xrightarrow{\rho_n} SK_n(AG) \rightarrow 0$$

$$\downarrow \qquad\qquad \downarrow \wr \qquad\qquad \downarrow \beta_n \qquad\qquad \downarrow \alpha_n$$

$$\cdots \rightarrow G_{n+1}(AG) \xrightarrow{\bar{\delta}_{n+1}} G_{n+1}(DG) \xrightarrow{\bar{\gamma}_{n+1}} G_n((A/\mathbf{m})G) \xrightarrow{\bar{\rho}_n} SG_n(AG) \rightarrow 0$$

$$\text{(III)}$$

from which we obtain the following commutative diagram where the rows are short exact sequences:

$$0 \rightarrow \operatorname{Coker}(\delta_{n+1}) \xrightarrow{\gamma_{n+1}} K_n(\mathcal{H}_S(AG)) \xrightarrow{\rho_n} SK_n(AG) \rightarrow 0$$

$$\downarrow \qquad\qquad\qquad \downarrow \beta_r \qquad\qquad\qquad \downarrow \alpha_r \qquad\qquad \text{(IV)}$$

$$0 \rightarrow \operatorname{Coker}(\bar{\delta}_{n+1}) \xrightarrow{\bar{\gamma}_{n+1}} G_n((A/\mathbf{m})G) \xrightarrow{\bar{\rho}_n} SG_n(AG) \rightarrow 0.$$

Theorem 7.7.2 *In the notation of 7.7.4 we have:*

(i) (a) *The Cartan homomorphisms* $K_n((A/\mathbf{m})G) \xrightarrow{\mu_n} G_n((A/\mathbf{m})G)$ *are surjective for all* $n \geq 1$. *Coker* μ_n *maps onto Coker* β_n *for all* $n \geq 0$. *Hence Coker* $\beta_n = 0$ *for all* $n \geq 1$, *and Coker* β_0 *is a finite p-group.*

(b) *For all* $n \geq 1$, *the Cartan homomorphism* $K_n(AG) \rightarrow G_n(AG)$ *induces a surjection* $SK_n(AG) \rightarrow SG_n(AG)$. *For* $n = 0$, *the Cartan homomorphism* $K_0(AG) \xrightarrow{\chi_0} G_0(AG)$ *induces a homomorphism* $\alpha_0 : SK_0(AG) \rightarrow SG_0(AG)$ *whose cokernel is a finite p-group.*

(ii) $SK_{2n}(AG) = Ker(K_{2n}(AG) \xrightarrow{\chi_{2n}} G_{2n}(AG))$.

PROOF

(i) First note that $A/\mathbf{m}$ is a finite field of characteristic p since A is a local ring, and $A/\mathbf{m}$ is a finite-dimensional skewfield over the finite field $\mathbb{Z}/p$ of order p (Wedderburn theorem!). So, by theorem 10.4.1, for each $n \geq 0$ the Cartan homomorphism $K_n((A/\mathbf{m})G) \xrightarrow{\mu_n} G_n((A/\mathbf{m})G)$ induces an isomorphism $\mathbb{Z}(1/p) \otimes K_n((A/\mathbf{m})G) \cong \mathbb{Z}(1/p) \otimes G_n((A/\mathbf{m})G)$, i.e., the μ_n is an isomorphism mod p-torsion $\forall\, n \geq 0$. Hence the cokernel of μ_n are finite p-groups, for all $n \geq 1$, since $G_n((A/\mathbf{m})G)$ is finite for all $n \geq 1$ (see theorem 7.1.12), $(A/\mathbf{m})G$ being a finite ring. However, by theorem 7.1.12, $G_{2n}((A/\mathbf{m})G) = G_{2n}(AG/rad\ AG) = 0$, and $G_{2n-1}((A/\mathbf{m})G)$ is finite of order relatively prime to p. So, Coker $\mu_{2n} = 0$, and Coker μ_{2n-1} is also zero since $|\text{Coker } \mu_{2n-1}|$ is a power of p and divides $|G_{2n-1}((A/m)G)|$, which is $\equiv 1 \bmod p$, and this is possible if and only if Coker $\mu_{2n-1} = 0$.

That Coker μ_n maps onto Coker β_n, $n \geq 0$ would follow once we show that $\mathcal{P}((A/\mathbf{m})G) \subseteq \mathcal{H}_S(AG)$ (see [20], p.533] for a similar argument for $n = 0$). This we now set out to do. Let $P \in \mathcal{P}(AG/\mathbf{m}AG)$. Then $P \in \mathcal{M}_S(AG)$. Now, for any $P \in \mathcal{P}(AG/\mathbf{m}AG)$, there exists Q such that $P \oplus Q \cong (AG/\mathbf{m}AG)^n$ for some n. We only have to show that $hd_{AG}(AG/\mathbf{m}AG) < \infty$. Note that $\mathbf{m}$ is A-projective since A is hereditary, $\mathbf{m}AG = \mathbf{m} \otimes_A AG \in \mathcal{P}(A)$, and so, is AG-projective. Hence $0 \to \mathbf{m}AG \to AG \to AG/\mathbf{m}AG \to 0$ is a finite $\mathcal{P}(AG)$-resolution of $AG/(\mathbf{m}AG)$. Hence $hd_{AG}(AG/\mathbf{m}AG)$ is finite.

That Coker $\beta_n = 0 \ \forall \ n \geq 1$ follows immediately from above. The statement that Coker β_0 is a finite p-group follows from the well-known result that Coker μ_0 is a finite p-group (see [187], p.132, corollary 1 to theorem 2.5]). That $SK_n(AG) \xrightarrow{\alpha_n} SG_n(AG)$ is surjective $\forall \ n \geq 1$ follows by applying the Snake lemma to diagram (IV) and using the fact that Coker $\beta_n = 0 \ \forall \ n \geq 1$. The last statement follows by applying the Snake lemma to diagram (IV) for $n = 0$ and using the fact that Coker β_0 is a finite p-group.

(ii) Consider the commutative diagram

$$
\begin{array}{ccccccc}
0 & \longrightarrow & SK_r(AG) & \xrightarrow{\nu_r} & K_r(AG) & \xrightarrow{\delta_r} & K_r(DG) & \longrightarrow \cdots \\
& & \downarrow{\alpha_r} & & \downarrow{\chi_r} & & \downarrow{\iota\theta} & \\
0 & \longrightarrow & SG_r(AG) & \xrightarrow{\bar{\nu}_r} & G_r(AG) & \xrightarrow{\bar{\delta}_r} & G_r(DG) & \longrightarrow \cdots
\end{array}
\qquad (V)
$$

where the rows are exact and θ is an isomorphism. If r is even ≥ 2, i.e., $r = 2n, n \geq 1$, then $SG_{2n}(AG) = 0$ for all $n \geq 1$, and so, $\bar{\nu}_r \alpha_r$ is a zero map. By considering the exact sequence associated with the composite $\delta_r = \bar{\delta}_r \chi_r$, we have $SK_{2n}(AG) \cong \text{Ker } \chi_{2n}$, as required.

$\square$

Remarks 7.7.1 Our next aim is to obtain explicit results on $SK_0(G, \mathcal{P}(A)) \cong SG_0(AG)$. Note that we have the following commutative diagrams (VI) and (VII):

$$
\begin{array}{ccccccc}
\longrightarrow K_1(DG) & \xrightarrow{\gamma_1} & K_0(\mathcal{H}_S(AG)) & \xrightarrow{\rho_0} & K_0(AG) & \xrightarrow{\delta_0} & K_0(DG) \\
\| & & \downarrow{\beta_0} & & \downarrow & & \| \\
\longrightarrow G_1(DG) & \xrightarrow{\bar{\gamma}_1} & G_0((A/\mathbf{m})G) & \xrightarrow{\bar{\rho}_0} & G_0(AG) & \xrightarrow{\bar{\delta}_0} & G_0(DG)
\end{array}
\qquad (VI)
$$

and

$$
\begin{array}{ccccccc}
\longrightarrow G_1(DG) & \longrightarrow & G_0(\Gamma/rad\ \Gamma) & \longrightarrow & G_0(\Gamma) & \longrightarrow & G_0(DG) \longrightarrow 0 \\
\| & & \downarrow & & \downarrow & & \| \\
\longrightarrow G_1(DG) & \longrightarrow & G_0((A/\mathbf{m})G) & \longrightarrow & G_0(AG) & \longrightarrow & G_0(DG) \longrightarrow 0
\end{array}
\qquad (VII)
$$

where in (VII), Γ is a maximal R-order containing AG, and in both cases the rows are exact. We now prove the following.

Theorem 7.7.3 *Let R be the ring of integers in a p-adic field F, G a finite group, and A a maximal R-order in a central division algebra D over F. Then*

(i) *The canonical map $K_0(AG) \rightarrow K_0(DG)$ is injective.   Hence, $SK_0(G, \mathcal{P}(A)) \cong Coker(K_0(\mathcal{H}_S(AG))) \rightarrow G_0(AG/(\mathbf{m}AG))$.*

(ii) *If Γ is a maximal R-order containing AG, then $SK_0(\Gamma) = 0$. Let $G_0(\Gamma) \rightarrow G_0(AG)$ be the map induced by restriction of scalars; then $SK_0(G, \mathcal{P}(A)) \cong Coker(K_0(\Gamma)) \rightarrow G_0(AG))$.*

PROOF   Note that AG is an R-order in DG, and so, the generators of $K_0(AG)$ have the form $x = [P] - [Q]$, where P, Q are in $\mathcal{P}(AG)$. Suppose that $x \in Ker\ \delta_0$. Then $K \otimes_R P \cong K \otimes_R Q$. Now, since AG is an R-algebra finitely generated as an R-module, and the Cartan map $K_0((A/\mathbf{m})G) \rightarrow G_0((AG) \rightarrow G_0((A/\mathbf{m})G)$ is a monomorphism (see [187], p.132, corollary 1 to theorem 35), then by [213], p.12, theorem 1.10), $P \cong Q$, i.e., $x = 0$. So, $K_0(AG) \rightarrow K_0(DG)$ is injective.

Now, it is clear from the diagram (VI) that $SK_0(G, \mathcal{P}(A)) \cong SG_0(AG) \cong$ Coker $\bar{\gamma}_1$. So, by considering the exact sequence associated with the left-hand square, namely, $\beta_0 \gamma_1 = \bar{\gamma}_1$, we have Ker $\beta_0 \rightarrow$ Coker $\gamma_1 \rightarrow$ Coker $\bar{\gamma}_1 \rightarrow$ Coker $\beta_0 \rightarrow 0$. Now Coker $\gamma_1 \cong$ Ker $\delta_0 = 0$. Hence Coker $\bar{\gamma}_1 \cong$ Coker $\beta_0 \cong$ Ker $\bar{\delta}_0 = SK_0(G, \mathcal{P}(A))$.

(ii) Since Γ is regular, we show that $SG_0(\Gamma) = 0$. Now DG is a separable F-algebra. So, if $x = [M] - [N] \in G_0(\Gamma)$, and $F \otimes_R M \cong F \otimes_R N$, then $M \cong N$ (see [213], p.101, theorem 5.27). So $G_0(\Gamma) \rightarrow G_0(DG)$ is injective.

The last statement follows by considering the exact sequence associated with the right-hand square of diagram (VII).   ▯

Theorem 7.7.4 *Let G be a finite p-group of order p^s, R the ring of integers in a p-adic field F, and A a maximal R-order in a central division F-algebra D, then $SK_0(G, \mathcal{P}(A))$ is a finite cyclic p-group of order $\leq p^s$.*

PROOF   Note that $A/\mathbf{m} = k$ is a finite field of characteristic p. So, $(A/\mathbf{m})G$ is Artinian, and $G_0(kG)$ is freely generated by the simple kG-modules (see [213]). But the only simple kG-module is k, and so, $G_0(kG) \cong \mathbb{Z}$. Moreover, $K_0(kG)$ is freely generated by the indecomposable projective kG-modules, namely, the projective envelopes of the simple kG-modules which, in this case, is the only envelope of the simple module k. So, the Cartan map is multiplication by p^s, and so, Coker μ_0 is isomorphic to Z/p^s, and hence

Coker $\beta_0 \cong SK_0(G, P(A))$ as a homomorphic image of $\mathbb{Z}/p^s$ is a finite cyclic group of order $p^r, r \leq s$. $\qquad\qquad$ ▯

$(7.7)^C$ **Higher K-theory of** $\mathcal{P}(A)_G$, A **a maximal order in division algebras over number fields**

The aim of this subsection is to obtain global results. So, let F be an algebraic number field (i.e., F is any finite extension of Q) and R the ring of integers of F (i.e., R is the integral closure of $\mathbb{Z}$ in F). If A is a maximal R-order in a central division algebra D over F, the AG is an R-order in DG.

We now have the following.

Theorem 7.7.5 *Let R be the ring of integers in an algebraic number field F, A a maximal R-order in a central division algebra D over F, G a finite group, p a prime ideal of R, $A_{\mathbf{p}} = A \otimes_R R_{\mathbf{p}}$. Then for all $n \geq 1$, we have:*

(i) $K_{2n-1}(G, \mathcal{P}(A_p))$ *is finitely generated.*

(ii) $K_n(G, \mathcal{P}(A))$ *is finitely generated.*

(iii) $SK_{2n-1}(G, \mathcal{P}(A))$ *is finite, and* $SK_{2n}(G, \mathcal{P}(A)) = 0$.

PROOF Since $[G, \mathcal{P}(A)] = \mathcal{P}_A(G)$, we identify $K_n(G, \mathcal{P}(A))$ with $G_n(AG)$. Also, AG is an R-order in DG. Hence the above results follow from earlier results on G_n of R-orders (see, e.g., 7.1.13). $\qquad$ ▯

7.7.5 The next result is the global version of the local result stated in theorem 7.7.3 on the Cartan maps. For each prime ideal p of R, let m_p be the unique maximal two-sided ideal of A lying above p, i.e., such that $m_p \supseteq p_p A$, $A_{\mathbf{p}} = R_{\mathbf{p}} \otimes_R A$, $\hat{A}_{\mathbf{p}} = \hat{R}_{\mathbf{p}} \otimes_R A$, where $R_{\mathbf{p}}$ is the localization of R at $\mathbf{p}$, and $\hat{R}_{\mathbf{p}}$ is the completion of R at $\mathbf{p}$. It is well known that $A/m_{\mathbf{p}} \cong A_{\mathbf{p}}/m_p A_{\mathbf{p}} \cong A_{\mathbf{p}}/rad\ A_{\mathbf{p}} \cong \hat{A}_{\mathbf{p}}/rad\ \hat{A}_{\mathbf{p}}$ (see [171], theorem 22.4 and its proof), and so, $A/m_{\mathbf{p}}$ is a finite simple algebra of characteristic p, where p is the rational prime lying below $\mathbf{p}$. Moreover, we have the following commutative diagram:

$$
\begin{array}{ccccccc}
\cdots \longrightarrow & K_{n+1}(DG) & \xrightarrow{\gamma_{n+1}} & K_n(\mathcal{H}_S(AG)) & \xrightarrow{\rho_n} & SK_n(AG) & \longrightarrow 0 \\
& \downarrow & & \downarrow{\scriptstyle \beta_n} & & \downarrow{\scriptstyle \alpha_n} & \qquad (I)\\
\cdots \longrightarrow & G_{n+1}(DG) & \longrightarrow & \underset{p}{\oplus}\, G_n((A/m_{\mathbf{p}})G) & \xrightarrow{\bar{\rho}_n} & SK_n(AG) & \longrightarrow 0
\end{array}
$$

and the following global version of diagram (IV) of 2.3, i.e.,

$$0 \longrightarrow \mathrm{Coker}(\delta_{n+1}) \xrightarrow{\gamma_1} K_n(\mathcal{H}_S(AG)) \xrightarrow{\rho_n} SK_n(AG) \longrightarrow 0$$

$$\downarrow \qquad\qquad \downarrow \qquad\qquad \downarrow \alpha_n \qquad\qquad (\mathrm{II})$$

$$0 \longrightarrow \mathrm{Coker}(\bar{\delta}_{n+1}) \longrightarrow \bigoplus_p G_n((A/\mathbf{m_p})G) \xrightarrow{\bar{\rho}_n} SK_n(AG) \longrightarrow 0.$$

Theorem 7.7.6 *Let R be the ring of integers in a number field F, A a maximal R-order in a central division algebra D over F, and G a finite group. Then, for each $n \geq 1$, the Cartan homomorphisms $K_n(AG) \xrightarrow{\chi_n} G_n(AG)$ induce a surjection $SK_n(AG) \xrightarrow{\alpha_n} SG_n(AG)$.*

For $n = 0$, $K_0(AG) \xrightarrow{\chi_0} G_0(AG)$ induces a homomorphism $SK_0(AG) \xrightarrow{\alpha_0} SG_0(AG)$ whose cokernel is a p-torsion group.

PROOF   Let $k_\mathbf{p} = \hat{A}_\mathbf{p}/rad\, \hat{A}_\mathbf{p}$, which we have seen to be isomorphic to A/m_p. By using an argument similar to that in the proof of theorem 7.7.2, we see that $\oplus_p \mathrm{Coker}(K_n(k_\mathbf{p}G) \to G_n(k_\mathbf{p}G))$ maps onto $\mathrm{Coker}\,\beta_n$ for all $n \geq 1$. Now, as in theorem 7.7.2, where each $(K_n(k_\mathbf{p}G) \to G_n(k_\mathbf{p}G))$ is surjective for $n \geq 1$, we have that $\mathrm{Coker}\,\beta_n = 0$ for all $n \geq 1$, and by applying the Snake lemma to diagram (II), we obtain that $SK_n(AG) \to (SG_n(AG))$ is surjective for $n \geq 1$. When $n = 0$, each $\mathrm{Coker}\,(K_0(k_\mathbf{p}G) \to G_0(k_\mathbf{p}G))$ is a finite p-group, and so, $\mathrm{Coker}\,(\beta_0)$ is a p-torsion group. So, $SK_0(AG) \to SG_0(AG)$ has a cokernel that is a p-torsion group.   ⬜

Remark 7.7.3 It follows from [213], p.112, corollary (Jacobinski) that $K_0(AG), G_0(AG) \cong K_0(G, \mathcal{P}(A))$ are finitely generated Abelian groups since R satisfies Jordan - Zassenhaus theorem (see [213], p.43), and AG is an R-order in the separable F-algebra DG. We conclude this subsection with the following theorem.

Theorem 7.7.7 *Let R be the ring of integers in a number field F, A a maximal R-order in a central division algebra D over F, and G a finite group. Then,*

(i) $SK_0(AG) = Ker(K_0(AG) \xrightarrow{\delta} K_0(DG))$ *is a finite group.*

(ii) $SK_0(G, \mathcal{P}(A)) \cong SG_0(AG) = Ker(G_0(AG) \to G_0(DG))$ *is a finite group.*

PROOF

(i) Let δ be the canonical map $K_0(AG) \to K_0(DG)$. We know that generators of $K_0(AG)$ have the form $[P] - [Q]$, where $P, Q \in \mathcal{P}(AG)$. Suppose $\delta([P] - [Q]) = 0$ in $G_0(DG) \cong K_0(DG)$, then $[K \otimes_R P] =$

$[K \otimes_R Q]$ and $K \otimes_R P \cong K \otimes_R Q$. Now, since there exists Q' such that $Q \oplus Q' = T$, where T is a finitely generated free AG-module, $[P] - [Q] = [P \oplus Q] - [Q \oplus Q'] = [P \oplus Q] - [T]$. So, Ker δ is generated by all $[P] - [T]$ where T is free and $F \otimes_R P \cong F \otimes_R T$. Since $(|G|, \text{char} F) = 1$, we can show that $P = T' \oplus I$, where $T = T' \oplus AG$, T a free AG-module, and I an ideal of AG. (Note that a slight modification of the proof of theorem 3.3 of [213] works where one puts $RG = AG$, etc., since the major theorem used in the proof, namely, Roiters' theorem, holds for an arbitrary R-order like AG since one needs only consider AG as an R-algebra finitely generated as an R-module.) So, $[P] - [T] = I - (AG)$, where I is an ideal such that $K \otimes_R I = DG$. By the Jordan - Zassenhaus theorem, there exists only a finite number of such I. So Ker δ is finite.

(ii) Consider the following diagram:

$$\ldots \longrightarrow K_1(DG) \longrightarrow K_0(\mathcal{H}_S(AG)) \longrightarrow SK_0(AG) \longrightarrow 0$$
$$\downarrow_{\imath\theta} \qquad\qquad \downarrow \qquad\qquad \downarrow_{\alpha_0}$$
$$\ldots \longrightarrow G_1(DG) \longrightarrow \underset{p}{\oplus} \; G_0(A\pi/\mathbf{m}_p G) \longrightarrow SG_0(AG) \longrightarrow 0$$

where the rows are exact, and θ is an isomorphism. We know by theorem 7.7.6 that Coker α_0 is a p-torsion group. Now, by (i), $SK_0(AG)$ is finite, and so, $Im(\alpha_0)$ is a finite subgroup of $SG_0(AG)$. Since $SG_0(AG)/(Im \; \alpha_0)$ is p-torsion, it means that $SG_0(AG)$ is torsion. Now, we know from remark 7.7.3 that $SG_0(AG)$ is a finitely generated Abelian group. So $SG_0(AG)$ is finite. $\square$

Exercises

7.1 Let Λ be any $\mathbb{Z}$-order or $\hat{\mathbb{Z}}_p$-order where p is any rational prime. Show that for all $n \geq 1$,

(i) $K_{2n}(\Lambda/p^s\Lambda)$ is a finite p-group.

(ii) $K_{2n-1}(\Lambda/p^s\Lambda)(\frac{1}{p}) \simeq K_{2n-1}((\Lambda/p^s\Lambda)/I)(\frac{1}{p})$ where $I = rad(\Lambda/p^s\Lambda)$.

7.2 Prove lemma 7.1.4 in the text.

7.3 Let Σ be a semi-simple algebra over a number field F, $WK_n(\Sigma)$ the wild kernel of Σ, and $W'K_n(D)$ the pseudo-wild kernel of Σ. (See definition 7.1.1.). Show that $WK_n(\Sigma)/W'K_n(\Sigma)$ is a finite 2-group with 8-rank 0 if $n \equiv 0, 4, 6 \pmod{8}$.

7.4 Let R be the ring of integers in a number field F, Λ any R-order in a semi-simple F-algebra Σ. Show that for all $n \geq 1$, $G_n(\Lambda_p)$ is a finitely generated Abelian group for all prime ideals p of R (where Λ_p is Λ localized at p).

7.5 Let R be the ring of integers in a number field F, p a rational prime. Show that $G_n(RG)(p) \simeq G_n(R(\frac{1}{p})G)(p)$ for all $n \geq 1$.

7.6 Let D be a central simple F-algebra of division s^2 over F. Show that the inclusion $i : F \subset D$ induces isomorphisms

$$K_n(i) \otimes \ \text{id} : K_n(F) \underset{\mathbb{Z}}{\otimes} \mathbb{Z}(\frac{1}{s}) \approx K_n(D) \underset{\mathbb{Z}}{\otimes} \mathbb{Z}(\frac{1}{s}).$$

7.7 Let R be the ring of integers in a number field F, Γ a maximal R-order in a semi-simple F-algebra Σ. If for each prime ideal p of $R, SK_{2n-1}(\Gamma_p) = 0$, show that Σ is unramified over its center (Here Γ_p is Γ localized at p).

7.8 Let R be a Dedekind domain with field of fractions F, Λ a Noetherian R-algebra that is R-torsion free (e.g., Λ an R-order). Write $\Sigma = F \otimes_R \Lambda, \bar{K}_n(\Lambda) := \text{Coker}(K_n(\Lambda) \to K_n(\Sigma))$ for all $n \in \mathbb{Z}$. Show that for all integers n, there exists an exact sequence

$$0 \to \bar{K}_{n+1}(\Lambda) \to \underset{p}{\oplus} \bar{K}_{n+1}(\hat{\Lambda}_p) \to \bar{K}_n(\Lambda) \to \underset{p}{\oplus} K_n(\hat{\Lambda}_p) \to 0$$

whose p ranges over all maximal ideals of R.

7.9 Let R be the ring of integers in a p-adic field F, Λ any R-order in a semi-simple F-algebra Σ, Γ a maximal R-order containing Λ. Show that for all $n \geq 2$,

(i) The canonical map $K_n(\Gamma) \to K_n(\Sigma)$ has finite kernel and cokernel.

(ii) The canonical map $G_n(\Lambda) \to G_n(\Sigma)$ has finite kernel and cokernel.

(iii) $\alpha_n : G_n(\Gamma) \to G_n(\Lambda)$ has finite kernel and cokernel where α_n is the map induced by the functor $\mathcal{M}(\Gamma) \to \mathcal{M}(\Lambda)$ given by restriction of scalars.

7.10 Let F be a p-adic field, R the ring of integers of F, Γ a maximal order in a central division algebra over F, $\bar{\Gamma}$ the residue class field of Γ. Show that for all $n \geq 1$,
$$K_n(\Gamma) \otimes \hat{\mathbb{Z}}_q \simeq K_n(\bar{\Gamma}) \otimes \hat{\mathbb{Z}}_q, q \neq p.$$

7.11 Let F be a number field and D a central division algebra over F with $[D : F] = m^2$, l an odd rational prime. Show that

(i) If l does not divide m, then $div(K_3(D)(l)) = WK_3(D)(l)$.

(ii) If $F = Q$ and l does not divide m, then div $(K_n(D))(l) \subset WK_n(D)(l)$ for all $n \geq 1$.

7.12 Prove that the conjecture of Hambleton, Taylor, and Williams is true for groups of square free order.

7.13 Let G be a finite group. Prove that for all odd $d > 0$, there is an isomorphism $Hmg^d(\hat{\mathcal{G}}) \xrightarrow{\sim} \underset{\chi \epsilon C_0}{\Pi} H^0(Q(\chi), Q/\mathbb{Z}(d))$ where C_0 is the set of representatives of non-trivial $\mathbb{Q}$-irreducible characters of G.

Chapter 8

Mod-m and profinite higher K-theory of exact categories, orders, and groupings

Let ℓ be a rational prime, $\mathcal{C}$ an exact category. In this chapter we define and study for all $n \geq 0$, the profinite higher K-theory of $\mathcal{C}$, that is $K_n^{pr}(\mathcal{C}, \hat{\mathbb{Z}}_\ell) := [M_{\ell^\infty}^{n+1}, BQ(\mathcal{C})]$, as well as $K_n(\mathcal{C}, \hat{\mathbb{Z}}_\ell) := \varprojlim_s [M_{\ell^\infty}^{n+1}, BQ(\mathcal{C})]$, where $M_{\ell^\infty}^{n+1} := \varinjlim_s M_{\ell^s}^{n+1}$, and $M_{\ell^s}^{n+1}$ is the $(n+1)$-dimensional mod-ℓ^S Moore space. We study connections between $K_n^{pr}(\mathcal{C}, \hat{\mathbb{Z}}_\ell)$ and $K_n(\mathcal{C}, \hat{\mathbb{Z}}_\ell)$ and prove several ℓ-completeness results involving these and associated groups, including the cases where $\mathcal{C} = \mathcal{M}(\Lambda)$ (resp. $\mathcal{P}(\Lambda)$) is the category of finitely generated (resp. finitely generated projective) modules over orders Λ in semi-simple algebras over number fields and p-adic fields. We also define and study continuous K-theory $K_n^c(\Lambda)(n \geq 1)$ of orders Λ in p-adic semi-simple algebras and show some connections between the profinite and continuous K-theory of Λ. The results in this chapter are due to A.O. Kuku (see [117]).

8.1 Mod-m K-theory of exact categories, rings, and orders

8.1.1 Let X be an H-space, m, n positive integers, M_m^n an n-dimensional mod-m Moore space, that is, the space obtained from S^{n-1} by attaching an n-cell via a map of degree m (see [158] or examples 5.2.2(viii) and (ix)). We shall write $\pi_n(X, \mathbb{Z}/m)$ for $[M_m^n, X]$ for $n \geq 2$. Note that $\pi_n(X, \mathbb{Z}/m)$ is a group for $n \geq 2$, and that $\pi_n(X, \mathbb{Z}/m)$ is an Abelian group for $n \geq 3$. For $n = 1$, if $\pi_1(X)$ is Abelian, we define $\pi_1(X, \mathbb{Z}/m)$ as $\pi_1(X) \otimes \mathbb{Z}/m$ (see [237]).

8.1.2 The cofibration sequence

$$S^{n-1} \xrightarrow{m} S^{n-1} \xrightarrow{\bar{\beta}} M_m^n \xrightarrow{\bar{\rho}} S^n \xrightarrow{m} S_n$$

yields an exact sequence

$$\pi_n(X) \xrightarrow{m} \pi_n(X) \xrightarrow{\rho} \pi_n(X, \mathbb{Z}/m) \xrightarrow{\beta} \pi_{n-1}(X) \xrightarrow{m} \pi_{n-1}(X) \qquad \text{(I)}$$

where the map 'm' means multiplication by m, ρ is the mod-m reduction map, and β is the Bockstein map (see [158]). We then obtain from (I) the following short exact sequence for all $n \geq 2$:

$$0 \longrightarrow \pi_n(X)/m \longrightarrow \pi_n(X, \mathbb{Z}/m) \longrightarrow \pi_{n-1}(X)[m] \longrightarrow 0$$

where

$$\pi_{n-1}(X)[m] = \{x \in \pi_{n-1}(X) \mid mx = 0\}.$$

Examples 8.1.1   (i) If $\mathcal{C}$ is an exact category, we shall write $K_n(\mathcal{C}, \mathbb{Z}/m)$ for $\pi_{n+1}(BQ\mathcal{C}, \mathbb{Z}/m), n \geq 1$, and write $K_0(\mathcal{C}, \mathbb{Z}/m)$ for $K_0(\mathcal{C}) \otimes \mathbb{Z}/m$.

(ii) If in(i), $\mathcal{C} = \mathcal{P}(A)$, the category of finitely generated projective modules over any ring A with identity, we shall write $K_n(A, \mathbb{Z}/m)$ for $K_n(\mathcal{P}(A), \mathbb{Z}/m)$. Note that $K_n(A, \mathbb{Z}/m)$ can also be written for $\pi_n(BGL(A)^+, \mathbb{Z}/m)$, $n \geq 1$.

(iii) If Y is a scheme, let $\mathcal{P}(Y)$ be the category of locally free sheaves of O_{Y^-} modules of finite rank. If we write $K_n(Y, \mathbb{Z}/m)$ for $K_n(\mathcal{P}(Y), \mathbb{Z}/m)$, then when Y is affine, that is, $Y = Spec(A)$ for some commutative ring A with identity, we recover $K_n(A, \mathbb{Z}/m) = K_n(\mathcal{P}(A), \mathbb{Z}/m)$ in the commutative case.

(iv) Let A be a Noetherian ring (non-commutative), $\mathcal{M}(A)$ the category of finitely generated A-modules. We shall write $G_n(A, \mathbb{Z}/m)$ for $K_n(\mathcal{M}(A), \mathbb{Z}/m)$.

(v) If Y is a Noetherian scheme, and $\mathcal{M}(Y)$ the category of coherent sheaves of O_Y-modules, we write $G_n(Y, \mathbb{Z}/m)$ for $\pi_n(\mathcal{M}(Y), \mathbb{Z}/m)$, and when $Y = Spec(A)$, for some commutative Noetherian ring A, we recover $G_n(A, \mathbb{Z}/m) = K_n(\mathcal{M}(A), \mathbb{Z}/m)$ as in (iv).

Remarks 8.1.1   (i) If $\mathcal{C}$ is an exact category, and $m \not\equiv 2(4)$, then the sequence

$$0 \longrightarrow K_n(\mathcal{C})/m \longrightarrow K_n(\mathcal{C}, \mathbb{Z}/m) \longrightarrow K_{n-1}(\mathcal{C})[m] \longrightarrow 0$$

splits (not naturally), and so, $K_n(\mathcal{C}, \mathbb{Z}/m)$ is a $\mathbb{Z}/m$-module (see [236]). If $m \equiv 2(4)$ then $K_n(\mathcal{C}, \mathbb{Z}/m)$ is a $\mathbb{Z}/2m$-module (see [6]).

(ii) For our applications, we shall be interested in the case $m = \ell^s$ where ℓ is a rational prime and s is some positive integer.

(iii) Note that for any Noetherian ring A (not necessarily commutative), the inclusion map $\mathcal{P}(A) \longrightarrow \mathcal{M}(A)$ induces Cartan maps $K_n(A) \longrightarrow G_n(A)$ and also $K_n(A, \mathbb{Z}/\ell^s) \longrightarrow G_n(A, \mathbb{Z}/\ell^s)$.

We now prove the following theorem.

Theorem 8.1.1 *Let p be a rational prime, F a p-adic field (i.e., any finite extension of $\hat{Q}_p$, R the ring of integers of F, Γ a maximal R-order in a semi-simple F-algebra Σ, $\mathbf{m}$ the maximal ideal of Γ, ℓ a prime such that $\ell \neq p$. Then, for all $n \geq 1$,*

(i) $K_n(\Gamma, \mathbb{Z}/\ell^s) \simeq K_n(\Gamma/\mathbf{m}, \mathbb{Z}/\ell^s)$.

(ii) $K_{2n}(\Gamma)$ *is ℓ-divisible.*

(iii) $K_{2n-1}(\Gamma)[\ell^s] \cong K_{2n-1}(\Gamma/\mathbf{m})[\ell^s]$.

(iv) There exists an exact sequence

$$0 \longrightarrow K_{2n+1}(\Gamma)/\ell^s \longrightarrow K_{2n+1}(\Gamma/\mathbf{m})/\ell^s \longrightarrow K_{2n}(\Gamma)[\ell^s] \longrightarrow 0.$$

PROOF

(i) By the Wedderburn theorem, $\Sigma = \Pi_{i=1}^r M_{n_i}(D_i)$ where D_i is a division algebra over F and $\Gamma = \Pi_{i=1}^r M_{n_i}(\Gamma_i)$ where Γ_i is the maximal order in D_i. Moreover, $\Gamma/\mathbf{m} = \Pi_{i=1}^r M_{n_i}(\Gamma_i/\mathbf{m_i})$ where $\Gamma_i/\mathbf{m_i}$ is a finite division ring, that is, a finite field. Now, by the rigidity result of Suslin and Yufryakov (see [204]), $K_n(\Gamma_i, \mathbb{Z}/\ell^S) \simeq K_n(\Gamma_i/\mathbf{m_i}, \mathbb{Z}/\ell^s)$ for all $n \geq 1$. Hence

$$K_n(\Gamma, \mathbb{Z}/\ell^s) \simeq \Pi_{i=1}^r K_n(\Gamma_i, \mathbb{Z}/\ell^s)$$
$$\simeq \Pi_{i=1}^r K_n(\Gamma_i/\mathbf{m_i}, \mathbb{Z}/\ell^s) \simeq K_n(\Gamma/\mathbf{m}, \mathbb{Z}/\ell^s).$$

(ii) From the following commutative diagram

$$
\begin{array}{ccccccccc}
0 & \longrightarrow & K_{2n}(\Gamma)/\ell^s & \longrightarrow & K_{2n}(\Gamma, \mathbb{Z}/\ell^s) & \longrightarrow & K_{2n-1}(\Gamma)[\ell^s] & \longrightarrow & 0 \\
& & \downarrow \alpha & & \downarrow \beta & & \downarrow \gamma & & \\
0 & \longrightarrow & K_{2n}(\Gamma/\mathbf{m})/\ell^s & \longrightarrow & K_{2n}(\Gamma/\mathbf{m}, \mathbb{Z}/\ell^s) & \longrightarrow & K_{2n-1}(\Gamma/\mathbf{m})[\ell^s] & \longrightarrow & 0 \\
\end{array}
$$
$$(\text{I})$$

we observe that $\Gamma/\mathbf{m}$ is a finite semi-simple ring, and so, $K_n(\Gamma/\mathbf{m}) = 0$ since $\Gamma/\mathbf{m}$ is a direct product of matrix algebras over finite fields. So, by applying the Snake lemma to diagram (I), and using (i) (i.e., β is an isomorphism), we obtain $K_{2n}(\Gamma) \simeq \ell^s K_{2n}(\Gamma)$ proving (ii).

(iii) Applying the Snake lemma to diagram (I) also yields $\text{Ker}\gamma = 0 = \text{Coker}\gamma$, proving (iii).

(iv) Applying the Snake lemma to the following diagram

$$0 \longrightarrow K_{2n+1}(\Gamma)/\ell^s \longrightarrow K_{2n+1}(\Gamma, \mathbb{Z}/\ell^s) \longrightarrow K_{2n}(\Gamma)[\ell^s] \longrightarrow 0$$
$$\downarrow \alpha' \qquad\qquad\quad \downarrow \beta' \qquad\qquad\quad \downarrow \gamma'$$
$$0 \longrightarrow K_{2n+1}(\Gamma/\mathbf{m})/\ell^s \longrightarrow K_{2n+1}(\Gamma/\mathbf{m}, \mathbb{Z}/\ell^s) \longrightarrow K_{2n}(\Gamma/\mathbf{m})[\ell^s] \longrightarrow 0$$

we observe that $Ker\alpha' = 0, Ker\gamma' = Coker\alpha'$. But $Ker\gamma' = K_{2n-1}(\Gamma)[\ell^s]$. Hence we have the result.

$\square$

Definition 8.1.1 *Let R be a Dedekind domain with quotient field F, Λ any R-order in a semi-simple F-algebra Σ. We define*

$$SK_n(\Gamma, \mathbb{Z}/\ell^s) := Ker(K_n(\Gamma, \mathbb{Z}/\ell^s) \longrightarrow K_n(\Sigma, \mathbb{Z}/\ell^s)).$$
$$SG_n(\Gamma, \mathbb{Z}/\ell^s) := Ker(G_n(\Gamma, \mathbb{Z}/\ell^s) \longrightarrow G_n(\Sigma, \mathbb{Z}/\ell^s)).$$

The following result is due to Laubenbacher and Webb (see 7.1.15 for a proof).

Theorem 8.1.2 *Let R be a Dedekind domain with quotient field F, Λ any R-order in a semi-simple F-algebra. Assume that,*

(i) $SG_1(\Lambda) = 0$.

(ii) $G_n(\Lambda)$ is finitely generated for all $n \geq 1$.

(iii) $R/\mathbf{p}$ is finite for all primes $\mathbf{p}$ of R.

(iv) If ζ is an ℓ^sth root of 1 for any rational prime ℓ and positive integer s, $\tilde{R}$ the integral closure of R in $F(\zeta)$, then $SG_1(\tilde{R} \otimes_R \Lambda) = 0$.

Then, (a) $SG_n(\Lambda, \mathbb{Z}/\ell^s) = 0$ for all odd $n \geq 1$ and rational primes ℓ.
   (b) $SG_n(\Lambda) = 0$ for all $n \geq 1$.

PROOF   See 7.1.15. $\square$

REMARK 8.1   If F is an algebraic number field (i.e., any finite extension of Q, R the ring of integers of F, G any finite group, then $\Lambda = RG$ satisfies the hypothesis of 8.1.7, and so, $SG_n(RG) = 0$. $\square$

Theorem 8.1.3 *Let R be the ring of integers in a number field F, Σ a semi-simple F-algebra, and Λ any R-order satisfying the hypothesis of theorem 8.1.2, ℓ a rational prime. Then for all $n \geq 1$ we have,*

(i) $SG_n(\hat{\Lambda}_p)$ for all prime ideals $\mathbf{p}$ of R.

(ii) (a) $SG_{2n}(\Lambda, \mathbb{Z}/\ell^s) = 0$,
   (b) $SG_n(\hat{\Lambda}_p, \mathbb{Z}/\ell^s) = 0$ *for all prime ideals* **p** *of R.*

(iii) (a) $SK_n(\Lambda) \simeq Ker(K_n(\Lambda) \longrightarrow G_n(\Lambda))$,
   (b) $SK_n(\hat{\Lambda}_p) \simeq Ker(K_n(\hat{\Lambda}_p) \longrightarrow G_n(\hat{\Lambda}_p))$ *for all prime ideals* **p** *of R.*

(iv) (a) $SK_n(\Lambda, \mathbb{Z}/\ell^s) \simeq Ker(K_n(\Lambda, \mathbb{Z}/\ell^s) \longrightarrow G_n(\Lambda, \mathbb{Z}/\ell^s))$,
   (b) $SK_n(\hat{\Lambda}_p, \mathbb{Z}/\ell^s) \simeq Ker(K_n(\hat{\Lambda}_p, \mathbb{Z}/\ell^S) \longrightarrow G_n(\hat{\Lambda}_p, \mathbb{Z}/\ell^s))$ *for all prime ideals* **p** *of R.*

PROOF

(i) First note that for almost all prime ideals **p** for R, $\hat{\Lambda}_p$ is a maximal $\hat{R}_p$-order in a split semi-simple $\hat{F}_p$-algebra $\hat{\Lambda}_p$. So, by remarks 7.1.5, $SG_{2n-1}(\hat{\Lambda}_p) = 0$ for almost all **p**. Also, by theorem 7.1.13(v), $SG_{2n}(\hat{\Lambda}_p) = 0$ for all **p**. Hence, for all $n \geq 1$, $SG_n(\hat{\Lambda}_p) = 0$ for almost all **p**. Now suppose that there are r non-maximal orders $\hat{\Lambda}_{p_1}, \hat{\Lambda}_{p_2}, \ldots \hat{\Lambda}_{p_r}$. Then, by applying the Snake lemma to the commutative diagram

$$
\begin{array}{ccccccccc}
0 \to & \text{Coker}(G_{n+1}(\Lambda) \to G_{n+1}(\Sigma)) & \to & K_n(\mathcal{M}_S(\Lambda)) & \to & SG_n(\Lambda) & \to 0 \\
& \downarrow{\scriptstyle \alpha_n} & & \downarrow{\scriptstyle \gamma_n} & & \downarrow & \\
0 \to & \bigoplus_{\mathbf{p}}(\text{Coker}(G_{n+1}(\hat{\Lambda}_p)) \to G_{n+1}(\hat{\Sigma}_p)) & \to & \bigoplus_{\mathbf{p}} G_n(\hat{\Lambda}_p/\mathbf{p}\hat{\Lambda}_p) & \to & \bigoplus_{i=1}^{r} SG_n(\hat{\Lambda}_{p_i}) & \to 0
\end{array}
$$

(I)

where $S = R - 0$, $\mathcal{M}_S(\Lambda)$ is the category of finitely generated S-torsion Λ-modules, and γ_n is an isomorphism (see remark 7.1.5), we obtain

$$0 \longrightarrow \text{Coker}\,\alpha_n \longrightarrow SG_n(\Lambda) \longrightarrow \bigoplus_{i=1}^{r} SG_n(\hat{\Lambda}_{p_i}) \longrightarrow 0.$$

But $SG_n(\Lambda) = 0$ by theorem 8.1.2. Hence, each $SG_n(\hat{\Lambda}_{p_i}) = 0$ for $i = 1, \ldots r$. So we have shown that $SG_n(\hat{\Lambda}_p) = 0$ for all prime ideals **p** of R.

(ii)

(a) Apply the Snake lemma to the commutative diagram

$$
\begin{array}{ccccccccc}
0 \longrightarrow & G_{2n}(\Lambda)/\ell^s & \longrightarrow & G_{2n}(\Lambda, \mathbb{Z}/\ell^s) & \longrightarrow & G_{2n-1}(\Lambda)[\ell^s] & \longrightarrow 0 \\
& \downarrow & & \downarrow & & \downarrow & \\
0 \longrightarrow & G_{2n}(\Sigma)/\ell^s & \longrightarrow & G_{2n}(\Sigma, \mathbb{Z}/\ell^s) & \longrightarrow & G_{2n-1}(\Sigma)[\ell^s] & \longrightarrow 0
\end{array}
$$

(II)

and using the fact that $SG_{2n}(\Lambda) = 0$ for all $n \geq 1$ (see theorem 7.1.32) we have that $SG_{2n}(\Lambda, \mathbb{Z}/\ell^s) = 0$.

(b) The argument is similar to that of (i) above using a commutative diagram of mod-ℓ^S localization sequence for Λ and $\hat{\Lambda}_{\mathbf{p}}$ analogous diagram (I) and obtaining a surjective map $SG_n(\Lambda, \mathbb{Z}/\ell^S) \to \oplus_{i=1}^r SG_n(\hat{\Lambda}_{\mathbf{p}_i}, \mathbb{Z}/\ell^s)$ for all $n \geq 1$ to conclude that each $SG_n(\hat{\Lambda}_{\mathbf{p}_i} = 0)$ using (ii)(a) and theorem 8.1.2(a). Details are left to the reader.

(iii) (a) Consider the commutative diagram

$$
\begin{array}{ccccccc}
0 & \longrightarrow & SK_n(\Lambda) & \longrightarrow & K_n(\Lambda) & \longrightarrow & K_n(\Sigma) \\
 & & \downarrow{\scriptstyle \alpha_n} & & \downarrow{\scriptstyle \beta_n} & & \downarrow{\scriptstyle \gamma_n} \\
0 & \longrightarrow & SG_n(\Lambda) & \longrightarrow & G_n(\Lambda) & \longrightarrow & G_n(\Sigma)
\end{array}
\qquad \text{(III)}
$$

where the rows are exact and γ_n is an isomorphism, and use the fact that $SG_n(\Lambda) = 0$ for all $n \geq 1$. Details are left to the reader.

(b) Proof similar to (a) by using similar diagram to III with respect to Λ replaced by $\hat{\Lambda}_{\mathbf{p}}$ using the fact that $SG_n(\hat{\Lambda}_{\mathbf{p}}) = 0$.

(iv) (a) Consider the following commutative diagram

$$
\begin{array}{ccccccc}
0 \longrightarrow & SK_n(\Lambda, \mathbb{Z}/\ell^s) & \xrightarrow{\rho_n} & K_n(\Lambda, \mathbb{Z}/\ell^s) & \xrightarrow{\eta_n} & K_n(\Sigma, \mathbb{Z}/\ell^s) & \longrightarrow \cdots \\
 & \downarrow{\scriptstyle \alpha_n} & & \downarrow{\scriptstyle \beta_n} & & \downarrow{\scriptstyle \gamma_n} & \\
0 \longrightarrow & SG_n(\Lambda, \mathbb{Z}/\ell^s) & \xrightarrow{\rho_n} & G_n(\Lambda, \mathbb{Z}/\ell^s) & \xrightarrow{\eta_n} & G_n(\Sigma, \mathbb{Z}/\ell^s) & \longrightarrow \cdots
\end{array}
$$
$$\text{(IV)}$$

where the rows are exact and γ_n is an isomorphism and use the fact that $SG_n(\Lambda, \mathbb{Z}/\ell^s) = 0$. Details are left to the reader.

(b) Proof similar to (a) using similar diagram to IV but replacing Λ by $\hat{\Lambda}_{\mathbf{p}}$, Σ by $\hat{\Sigma}_{\mathbf{p}}$, and using the fact that $SG_n(\hat{\Lambda}_{\mathbf{p}}, \mathbb{Z}/\ell^s) = 0$.

$\square$

Remarks 8.1.2 Let R be the ring of integers in a number field, and Λ any R-order in a semi-simple F-algebra Σ. Then the two exact sequences below are split

$$0 \to K_n(\Lambda)/\ell^s) \to K_n(\Lambda, \mathbb{Z}/\ell^s) \to K_{n-1}(\Lambda)[\ell^s] \to 0.$$
$$0 \to G_n(\Lambda)/\ell^s \to G_n(\Lambda, \mathbb{Z}/\ell^s) \to G_{n-1}(\Lambda)[\ell^s] \to 0.$$

This follows from [161], 7.3, and the fact that $K_n(\Lambda)$, $G_n(\Lambda)$ are finitely generated for all $n \geq 1$ (see theorem 7.1.11 and theorem 7.1.13).

8.2 Profinite K-theory of exact categories, rings and orders

Definition 8.2.1 *Let $\mathcal{C}$ be an exact category, ℓ a rational prime, s a positive integer, n non-negative integer, $M_{\ell^s}^{n+1}$ the $(n+1)$-dimensional mod-ℓ^s-Moore space and $M_{\ell^\infty}^{n+1} := \varinjlim M_{\ell^s}^{n+1}$. We define the profinite K-theory of $\mathcal{C}$ by*

$$K_n^{pr}(\mathcal{C}, \hat{\mathbb{Z}}_\ell) := [M_{\ell^\infty}^{n+1}, BQ\mathcal{C}],$$

and also write $K_n(\mathcal{C}, \hat{\mathbb{Z}}_\ell)$ for $\varprojlim K_n(\mathcal{C}, \mathbb{Z}/\ell^s)$.
If A is any ring with identity (not necessarily commutative), we write $K_n^{pr}(A, \hat{\mathbb{Z}}_\ell)$ for $[M_{\ell^\infty}^{n+1}; BQ\mathcal{P}(A)]$ for all $n \geq 0$ and call this profinite K-theory of A.

If A is a Noetherian ring (not necessarily commutative), we write

$$G_n^{pr}(A, \hat{\mathbb{Z}}_\ell) := [M_{\ell^\infty}^{n+1}, BQ\mathcal{M}(A)]$$

for all $n \geq 0$ and call this profinite G-theory of A. We shall also write $K_n(A, \hat{\mathbb{Z}}^\ell)$ for $\varprojlim_s K_n(A, \mathbb{Z}/\ell^s)$ and $G_n(A, \hat{\mathbb{Z}}^\ell)$ for $\varprojlim_s G_n(A, \mathbb{Z}/\ell^s)$.

Remarks 8.2.1 (i) Note that $K_n^{pr}(\mathcal{C}, \hat{\mathbb{Z}}_\ell)$ and hence $K_n^{pr}(A, \hat{\mathbb{Z}}_\ell)$, $G_n^{pr}(A, \hat{\mathbb{Z}}_\ell)$ are $\hat{\mathbb{Z}}_\ell$-modules. So are $K_n(\mathcal{C}, \hat{\mathbb{Z}}_\ell)$ and hence $K_n(A, \hat{\mathbb{Z}}_\ell)$, $G_n(A, \hat{\mathbb{Z}}_\ell)$ where we identify $\hat{\mathbb{Z}}_\ell$ with $[M_{\ell^\infty}^{n+1}, M_{\ell^\infty}^{n+1}]$ (see [13]).

(ii) If X is a scheme and $\mathcal{P}(X)$ is the category of locally free sheaves of O_X-modules of finite rank, and we restrict to the affine case $X = Spec(A)$, where A is a commutative ring with identity, we obtain the theory $K_n^{cts}(X, \hat{\mathbb{Z}}_\ell)$ treated by Banaszak and Zelewski in [13]. So, our theory in this article is a generalization of that in [13] to non-commutative rings. Note that the case of Noetherian schemes X whose affine specialization to $X = Spec(A)$ (A-commutative) could yield theories $G_n^{cts}(X, \hat{\mathbb{Z}}_\ell)$ that was not considered in [13], where

$$G_n^c(X, \hat{\mathbb{Z}}_\ell) := [M_{\ell^\infty}^{n+1}, BQ\mathcal{M}(X)],$$

and $\mathcal{M}(X)$ is the category of coherent sheaves of O_X-modules.

Lemma 8.2.1 *Let $\mathcal{C}$ an exact category, ℓ a rational prime. Then for all $n \geq 1$, there exists an exact sequence*

$$0 \longrightarrow \varprojlim_s{}^1 K_{n+1}(\mathcal{C}, \mathbb{Z}/\ell^s) \longrightarrow K_n^{pr}(\mathcal{C}, \hat{\mathbb{Z}}_\ell) \longrightarrow K_n(\mathcal{C}, \hat{\mathbb{Z}}_\ell) \longrightarrow 0$$

PROOF   Follows from the fact that $[-; BQ\mathcal{C}]$ is a cohomology theory (see [152]).   □

Remarks 8.2.2 It follows from 8.2.3 that

(i) If A is any ring with identity, then there exists an exact sequence

$$0 \longrightarrow \varprojlim_s {}^1 K_{n+1}(A, \mathbb{Z}/\ell^s) \longrightarrow K_n^{pr}(A, \hat{\mathbb{Z}}_\ell) \longrightarrow K_n(A, \hat{\mathbb{Z}}_\ell) \longrightarrow 0.$$

(ii) If A is Noetherian (not necessarily commutative), then we have an exact sequence

$$0 \longrightarrow \varprojlim_s {}^1 G_{n+1}(A, \mathbb{Z}/l^s) \longrightarrow G_n^{pr}(A, \hat{\mathbb{Z}}_\ell) \longrightarrow G_n(A, \hat{\mathbb{Z}}_\ell) \longrightarrow 0$$

Definition 8.2.2 *Let R be a Dedekind domain with quotient field F, Λ any R-order in a semi-simple F-algebra Σ, ℓ a rational prime. We define:*

$$SK_n^{pr}(\Lambda, \hat{\mathbb{Z}}_\ell) := Ker(K_n^{pr}(\Lambda, \hat{\mathbb{Z}}_\ell) \longrightarrow K_n^{pr}(\Sigma, \hat{\mathbb{Z}}_\ell)).$$
$$SG_n^{pr}(\Lambda, \hat{\mathbb{Z}}_\ell) := Ker(G_n^{pr}(\Lambda, \hat{\mathbb{Z}}_\ell) \longrightarrow G_n^{pr}(\Sigma, \hat{\mathbb{Z}}_\ell)).$$
$$SK_n(\Lambda, \hat{\mathbb{Z}}_\ell) := Ker(K_n(\Lambda, \hat{\mathbb{Z}}_\ell) \longrightarrow K_n(\Sigma, \hat{\mathbb{Z}}_\ell)).$$
$$SG_n(\Lambda, \hat{\mathbb{Z}}_\ell) := Ker(G_n(\Lambda, \hat{\mathbb{Z}}_\ell) \longrightarrow G_n(\Sigma, \hat{\mathbb{Z}}_\ell)).$$

Remarks 8.2.3   (i) With notation as in definition 8.2.2 and R, F, Λ, Σ as in theorem 8.1.3, **p** any prime ideal of R, it follows from arguments similar to those in the proof of theorem 8.1.3(ii) that $SG_n(\hat{\Lambda}_{\mathbf{p}}, \hat{\mathbb{Z}}_\ell) = 0$ and that $SG_n^{pr}(\hat{\Lambda}_{\mathbf{p}}, \hat{\mathbb{Z}}_\ell) = 0$ for all prime ideals **p** of R. Also, one can see easily, by using theorem 8.1.2, that $SG_n^{pr}(\Lambda, \hat{\mathbb{Z}}_\ell) = 0 = SG_n(\Lambda, \hat{\mathbb{Z}}_\ell)$.

(ii) Furthermore, by arguments similar to those in the proof of theorem 8.1.3(iii) and (iv), and the use of (i) above, one can deduce that

$$SK_n^{pr}(\Lambda, \hat{\mathbb{Z}}_\ell) := Ker(K_n^{pr}(\Lambda, \hat{\mathbb{Z}}_\ell) \longrightarrow G_n^{pr}(\Lambda, \hat{\mathbb{Z}}_\ell)),$$
$$SK_n(\Lambda, \hat{\mathbb{Z}}_\ell) := Ker(K_n(\Lambda, \hat{\mathbb{Z}}_\ell) \longrightarrow G_n(\Lambda, \hat{\mathbb{Z}}_\ell)),$$
$$SK_n^{pr}(\hat{\Lambda}_{\mathbf{p}}, \hat{\mathbb{Z}}_\ell) := Ker(K_n^{pr}(\hat{\Lambda}_{\mathbf{p}}, \hat{\mathbb{Z}}_\ell) \longrightarrow G_n^{pr}(\hat{\Lambda}_{\mathbf{p}}, \hat{\mathbb{Z}}_\ell)),$$
$$SK_n(\hat{\Lambda}_{\mathbf{p}}, \hat{\mathbb{Z}}_\ell) := Ker(K_n(\hat{\Lambda}_{\mathbf{p}}, \hat{\mathbb{Z}}_\ell) \longrightarrow G_n(\hat{\Lambda}_{\mathbf{p}}, \hat{\mathbb{Z}}_\ell)),$$

for any prime ideal **p** of R.

Theorem 8.2.1 *Let $\mathcal{C}$ be an exact category, ℓ a prime such that $K_n(\mathcal{C})_\ell$ contains no non-trivial divisible subgroups for each $n \geq 1$. Then there exists isomorphisms*

(i) $K_n(\mathcal{C})[\ell^s] \overset{\beta}{\simeq} K_n^{pr}(\mathcal{C}, \hat{\mathbb{Z}}_\ell)[\ell^s]$ *for all $n \geq 1$.*

(ii) $K_n(\mathcal{C})/\ell^s \overset{\alpha}{\simeq} K_n^{pr}(\mathcal{C}, \hat{\mathbb{Z}}_\ell)/\ell^s$ *for all* $n \geq 2$.

PROOF The canonical map $M_{\ell\infty}^{n+1} \to S^{n+1}$ induces a map $[S^{n+1}, BQ\mathcal{C}] \to [M_{\ell\infty}^{n+1}, BQ\mathcal{C}]$, that is, $K_n(\mathcal{C}) \to K_n^{pr}(\mathcal{C}, \hat{\mathbb{Z}}_\ell)$, and hence maps $K_n(\mathcal{C})/\ell^s \to K_n^{pr}(\mathcal{C}, \hat{\mathbb{Z}}_\ell)/\ell^s$ as well as $K_n(\mathcal{C})[\ell^s] \to K_n^{pr}(\mathcal{C}, \hat{\mathbb{Z}}_\ell)[\ell^s]$.

Moreover, as $s \to \infty$, the cofibration sequence $M_{\ell^s}^n \to M_{\ell^{s+t}}^n \to M_{\ell^t}^n$ induces the cofibration sequence $M_{\ell^s}^n \to M_{\ell\infty}^n \to M_{\ell\infty}^n$. Now, since for any exact category $\mathcal{C}$, $[; BQ\mathcal{C}]$ is a cohomology theory, we have a long exact sequence

$$\ldots K_{n+1}^{pr}(\mathcal{C}, \hat{\mathbb{Z}}_\ell) \overset{\ell^s}{\longrightarrow} K_{n+1}^{pr}(\mathcal{C}, \hat{\mathbb{Z}}_\ell) \longrightarrow K_{n+1}(\mathcal{C}, \mathbb{Z}/\ell^s) \longrightarrow K_n^{pr}(\mathcal{C}, \hat{\mathbb{Z}}_\ell) \longrightarrow \ldots \tag{I}$$

and hence an exact sequence

$$0 \longrightarrow K_{n+1}^{pr}(\mathcal{C}, \hat{\mathbb{Z}}_\ell)/\ell^s \longrightarrow K_{n+1}(\mathcal{C}, \mathbb{Z}/\ell^s) \longrightarrow K_n^{pr}(\mathcal{C}, \hat{\mathbb{Z}}_\ell)[\ell^s] \longrightarrow 0 \tag{II}$$

Now, by applying the Snake lemma to the commutative diagram

$$
\begin{array}{ccccccccc}
0 & \longrightarrow & K_{n+1}(\mathcal{C})/\ell^s & \longrightarrow & K_{n+1}(\mathcal{C}, \mathbb{Z}/\ell^s) & \longrightarrow & K_n(\mathcal{C})[\ell^s] & \longrightarrow & 0 \\
& & \downarrow & & \| & & \downarrow & & \\
0 & \longrightarrow & K_{n+1}^{pr}(\mathcal{C}, \hat{\mathbb{Z}}_\ell)/\ell^s & \longrightarrow & K_{n+1}(\mathcal{C}, \mathbb{Z}/\ell^s) & \longrightarrow & K_n^{pr}(\mathcal{C}, \hat{\mathbb{Z}}_\ell)[\ell^s] & \longrightarrow & 0
\end{array}
\tag{III}
$$

we see that $\alpha : K_{n+1}(\mathcal{C})/\ell^s \longrightarrow K_{n+1}^{pr}(\mathcal{C}, \hat{\mathbb{Z}}_\ell)/\ell^s$ is injective and $K_n(\mathcal{C})[\ell^s] \overset{\beta}{\to} K_n^{pr}(\mathcal{C}, \hat{\mathbb{Z}}_\ell)[\ell^s]$ is surjective and hence that $K_n(\mathcal{C})_\ell \overset{\beta^n}{\to} K_n^{pr}(\mathcal{C}, \hat{\mathbb{Z}}_\ell)_\ell$ is surjective (IV).

Also from the exact sequence associated to the composite $\delta = \eta\beta^*$ in the diagram V below

$$
\begin{array}{ccc}
K_n(\mathcal{C})_\ell & \overset{\beta^*}{\longrightarrow} & K_n^{pr}(\mathcal{C}, \hat{\mathbb{Z}}_\ell)_\ell \\
& \searrow{\scriptstyle \delta} \qquad \swarrow{\scriptstyle \eta} & \\
& K_n(\mathcal{C}, \hat{\mathbb{Z}}_\ell) &
\end{array}
\tag{V}
$$

we have $0 \to Ker\ \beta^* \to Ker\ \delta = divK_n(\mathcal{C})_\ell$ (see [13]), that is, $\text{Ker}(K_n(\mathcal{C})_\ell \to K_n^{pr}(\mathcal{C}, \hat{\mathbb{Z}}_\ell)_\ell) \subseteq \text{div}\ K_n(\mathcal{C})_\ell = 0$. So $K_n(\mathcal{C})_\ell \overset{\beta^*}{\to} K_n^{pr}(\mathcal{C}, \hat{\mathbb{Z}}_\ell)_\ell$ is injective (VI).

(IV) and (VI) imply that $K_n(\mathcal{C})_\ell \overset{\beta^*}{\simeq} K_n^{pr}(\mathcal{C}, \hat{\mathbb{Z}}_\ell)_\ell$ is an isomorphism. Hence $(K_n(\mathcal{C})_\ell[\ell^s] \simeq K_n^{pr}(\mathcal{C}, \hat{\mathbb{Z}}_\ell)[\ell^s]$ is an isomorphism. It now follows from diagram III above that $K_{n+1}(\mathcal{C})/\ell^s \simeq K_{n+1}^{pr}(\mathcal{C}, \hat{\mathbb{Z}}_\ell)/\ell^s$ is also an isomorphism. $\square$

Corollary 8.2.1 Let $\mathcal{C}$ be an exact category, ℓ a rational prime such that for all $n \geq 2$, $K_n(\mathcal{C})_\ell$ has no non-zero divisible subgroups. Let $\varphi : K_n(\mathcal{C}) \to$

$K_n^{pr}(\mathcal{C}, \hat{\mathbb{Z}}_\ell)$ be the map $[S^{n+1}; BQ\mathcal{C}] \to [M_{\ell\infty}^{n+1}; BQ\mathcal{C}]$ induced by the canonical map $M_{\ell\infty}^{n+1} \to S^{n+1}$. Then $\text{Ker}\varphi$ and $\text{Coker}\varphi$ are uniquely ℓ-divisible.

PROOF   Follows from theorem 8.2.1 and lemma 2 of [13].   ▯

Corollary 8.2.2 Let R be the ring of integers in a number field F, Λ any R-order in a semi-simple F-algebra Σ, ℓ a rational prime. Then the maps $\varphi: K_n(\Lambda) \to K_n^{pr}(\Lambda, \hat{\mathbb{Z}}_\ell)$ and $\varphi: G_n(\Lambda) \to G_n^{pr}(\mathcal{C}, \hat{\mathbb{Z}}_\ell)$ (defined in the proof of theorem 8.2.1, respectively, for $\mathcal{C} = \mathcal{P}(\Lambda), \mathcal{C} = \mathcal{M}(\Lambda)$) are injective.

PROOF   By theorem 8.2.1, $\text{Ker}(K_n(\Lambda \xrightarrow{\varphi} K_n^{pr}(\Lambda, \hat{\mathbb{Z}}_\ell))$ and $\text{Ker}(G_n(\Lambda \xrightarrow{\varphi'} G_n^{pr}(\Lambda, \hat{\mathbb{Z}}_\ell))$ are uniquely ℓ-divisible. But $K_n(\Lambda), G_n(\Lambda)$ are finitely generated (see theorem 7.1.11 and theorem 7.1.13), and so, $div\ Ker\varphi = div\ Ker\varphi' = 0$ as subgroups of $K_n(\Lambda), G_n(\Lambda)$. Hence, $Ker\varphi' = 0 = Ker\varphi$, as required.   ▯

Definition 8.2.3 *Let ℓ be a rational prime, G an Abelian group, and $\{G, \ell\} := \cdots \to^\ell G \to^\ell G \to^\ell \ldots$ an inverse system. It is well known that $\varprojlim(G, \ell)$ and $\varprojlim^1(G, \ell)$ are uniquely ℓ-divisible (see [91]).*

G is said to be weakly ℓ-complete if $\varprojlim(G, \ell) = 0 = \varprojlim^1(G, \ell)$.

An Abelian group G is said to be ℓ-complete if $G \simeq \varprojlim_s(G/\ell^s)$. We shall sometimes write $\hat{G}$ for $\varprojlim_s(G/\ell^s)$. Since for any G and ℓ we always have an exact sequence

$$0 \longrightarrow \varprojlim \ell^s G \longrightarrow G \longrightarrow \hat{G} \longrightarrow \varprojlim^1 \ell^s G \longrightarrow 0, \qquad (I)$$

it follows that G is ℓ-complete iff $\varprojlim \ell^s G = 0$ and $\varprojlim^1 \ell^s G = 0$.

Remarks 8.2.4   (i) For any Abelian group G and a rational prime ℓ, we have an exact sequence of inverse systems $0 \longrightarrow \{G[\ell^s]\} \longrightarrow \{G\} \longrightarrow \{\ell^s G\} \longrightarrow 0$ with associated $\lim - \lim^1$ exact sequence

$$0 \longrightarrow \varprojlim_s G[\ell^s] \longrightarrow \varprojlim(G, \ell) \longrightarrow \varprojlim_s(\ell^s G) \longrightarrow \varprojlim_s^1 G[\ell^s]$$
$$\longrightarrow \varprojlim_s^1(G, \ell) \longrightarrow \varprojlim_s^1(\ell^s G) \longrightarrow 0 \qquad (II)$$

So, if G is weakly ℓ-complete, then we have from (I) and (II) that

$$\varprojlim_s G[\ell^s] = 0 \text{ (III)} \varprojlim_s^1 G[\ell^s] = Ker(G \to \hat{G}) \text{ (IV)} \varprojlim_s^1(\ell^s G)$$
$$= Coker(G \to \hat{G}) = 0 \text{(V)}.$$

Note that $\varprojlim_s(\ell^s G) = \cap \ell^s G = \ell - div G$ (V) (see [91]).

(ii) The weakly ℓ-complete groups form a full subcategory of the category $\mathcal{A}b$ of Abelian groups and the category of weakly ℓ-complete groups is the smallest Abelian subcategory of $\mathcal{A}b$ containing the ℓ-complete groups (see [91]).

(iii) If two groups in an exact sequence $0 \longrightarrow G' \longrightarrow G \longrightarrow G'' \longrightarrow 0$ are weakly ℓ-complete, then so is the third (see [91], 4.8 for a proof).

(iv) If (G_n) is an inverse system of Abelian groups such that each G_n has a finite ℓ-power exponent, then $\varprojlim_s^1 G_n$ is weakly ℓ-complete (see [91] for a proof).

(v) An Abelian group G is ℓ-complete if $G = \varprojlim G_n$ where each G_n is a $\hat{\mathbb{Z}}_\ell$-module with finite exponent, if all the groups in the exact sequence (II) above are zero (see [91] for proofs).

Theorem 8.2.2 *Let $\mathcal{C}$ be an exact category, ℓ a rational prime. Then for all $n \geq 1$,*

(i) $\varprojlim_s K_n^{pr}(\mathcal{C}, \hat{\mathbb{Z}}_\ell)[\ell^s] = 0$.

(ii) $\varprojlim_s^1 K_{n+1}(\mathcal{C}, \mathbb{Z}/\ell^s) = div K_n^{pr}(\mathcal{C}, \hat{\mathbb{Z}}_\ell)$.

PROOF

(i) From lemma 8.2.1 we have the exact sequence

$$0 \longrightarrow \varprojlim_s^1 K_{n+1}(\mathcal{C}, \mathbb{Z}/\ell^s) \longrightarrow K_n^{pr}(\mathcal{C}, \hat{\mathbb{Z}}_\ell) \longrightarrow K_n(\mathcal{C}, \hat{\mathbb{Z}}_\ell) \longrightarrow 0. \quad \text{(I)}$$

Now, $K_n(\mathcal{C}, \hat{\mathbb{Z}}_\ell) = \varprojlim K_n(\mathcal{C}, \mathbb{Z}/\ell^s)$ is ℓ-complete by remarks 8.2.4(v) and hence weakly ℓ-complete by remarks 8.2.4(ii). Also, $\varprojlim_s^s K_{n+1}(\mathcal{C}, \mathbb{Z}/\ell^s)$ is weakly ℓ-complete by remarks 8.2.4(iv). Hence $K_n^{pr}(\mathcal{C}, \hat{\mathbb{Z}}_\ell)$ is weakly ℓ-complete by remarks 8.2.4(iii) and the exact sequence (I) above. It then follows from remarks 8.2.4(i)(III) that $\varprojlim_s K_n^{pr}(\mathcal{C}, \hat{\mathbb{Z}}_\ell)[\ell^s] = 0$.

(ii) As already deduced in (II) in the proof of theorem 8.2.1, there exists an exact sequence

$$0 \longrightarrow K_n^{pr}(\mathcal{C}, \hat{\mathbb{Z}}_\ell)/\ell^s \longrightarrow K_n(\mathcal{C}, \mathbb{Z}/\ell^s) \longrightarrow K_{n-1}^{pr}(\mathcal{C}, \hat{\mathbb{Z}}_\ell)[\ell^s] \longrightarrow 0. \quad \text{(II)}$$

So, by (i) above, $\varprojlim_s K_n^{pr}(\mathcal{C}, \hat{\mathbb{Z}}_\ell)/\ell^s \simeq K_n(\mathcal{C}, \hat{\mathbb{Z}}_\ell)$ (III). Now, for any $\hat{\mathbb{Z}}_\ell$-module M, $div M = Ker(M \longrightarrow \varprojlim_s (M/\ell^s))$ (see [13]). So, since

$K_n^{pr}(\mathcal{C}, \hat{\mathbb{Z}}_\ell)$ is a $\hat{\mathbb{Z}}_\ell$-module (see remarks 8.2.1(i)), we have

$$
\begin{aligned}
div\, K_n^{pr}(\mathcal{C}, \hat{\mathbb{Z}}_\ell) &= (Ker K_n^{pr}(\mathcal{C}, \hat{\mathbb{Z}}_\ell) \longrightarrow \varprojlim_s (K_n^{pr}(\mathcal{C}, \hat{\mathbb{Z}}_\ell)/\ell^s)), \\
&= Ker(K_n^{pr}(\mathcal{C}, \hat{\mathbb{Z}}_\ell) \longrightarrow K_n(\mathcal{C}, \hat{\mathbb{Z}}_\ell)) \text{ from III above,} \\
&= \varprojlim{}^1 K_{n+1}(\mathcal{C}, \mathbb{Z}/\ell^s) \quad \text{by lemma 8.2.1.}
\end{aligned}
$$

$\square$

Remarks 8.2.5 Let R be the ring of integers in a number field F, Λ an R-order in a semi-simple F-algebra Σ. Then, theorem 8.2.1 applies to $\mathcal{C} = \mathcal{P}(\Lambda)$, $\mathcal{C} = \mathcal{M}(\Lambda)$ since for all $n \geq 1$, $K_n(\Lambda)$, $G_n(\Lambda)$ are finitely generated Abelian groups (see theorem 7.1.11 and theorem 7.1.13), and so, $K_n(\Lambda)_\ell$, $G_n(\Lambda)_\ell$ being finite groups, have no non-trivial divisible subgroups. So, for all $n \geq 1$,

$$
K_n(\Lambda)[\ell^s] \simeq K_n^{pr}(\Lambda, \hat{\mathbb{Z}}_\ell)[\ell^s], \qquad G_n(\Lambda)[\ell^s] \xrightarrow{\sim} G_n^{pr}(\Lambda, \hat{\mathbb{Z}}_\ell)[\ell^s],
$$

and for all $n \geq 2$,

$$
K_n(\Lambda)/\ell^s \simeq K_n^{pr}(\Lambda, \hat{\mathbb{Z}}_\ell)/\ell^s, \qquad \text{and} \qquad G_n(\Lambda)/\ell^s \simeq G_n^{pr}(\Lambda, \hat{\mathbb{Z}}_\ell)/\ell^s.
$$

Theorem 8.2.3 Let $\mathcal{C}$ be an exact category such that $K_n(\mathcal{C})$ is a finitely generated Abelian group for all $n \geq 1$. Let ℓ be a rational prime. Then $K_n^{pr}(\mathcal{C}, \hat{\mathbb{Z}}_\ell)$ is an ℓ-complete profinite Abelian group for all $n \geq 2$.

The proof of theorem 8.2.3 makes use of the following.

Lemma 8.2.2 Let $\mathcal{C}$ be an exact category, ℓ a rational prime. Then $\varprojlim_s (K_n^{pr}(\mathcal{C}, \hat{\mathbb{Z}}_\ell)/\ell^s) \simeq K_n(\mathcal{C}, \hat{\mathbb{Z}}_\ell)$ for all $n \geq 2$.

PROOF By taking inverse limits in the following exact sequence,

$$
0 \longrightarrow K_{n+1}^{pr}(\mathcal{C}, \hat{\mathbb{Z}}_\ell)/\ell^s \longrightarrow K_{n+1}(\mathcal{C}, \mathbb{Z}/\ell^s) \longrightarrow K_n^{pr}(\mathcal{C}, \hat{\mathbb{Z}}_\ell)[\ell^s] \longrightarrow 0 \quad (I)
$$

for all $n \geq 1$, and using the fact that $\varprojlim_s (K_n^{pr}(\mathcal{C}, \hat{\mathbb{Z}}_\ell)[\ell^s]) = 0$ (see therem 8.2.2), we have that

$$
\varprojlim_s (K_n^{pr}(\mathcal{C}, \hat{\mathbb{Z}}_\ell)/\ell^s) \simeq K_{n+1}(\mathcal{C}, \hat{\mathbb{Z}}_\ell). \quad (I)
$$

$\square$

Proof of theorem 8.2.3 First observe from the exact sequence

$$
0 \longrightarrow K_{n+1}(\mathcal{C})/\ell^s \longrightarrow K_{n+1}(\mathcal{C}, \mathbb{Z}/\ell^s) \longrightarrow K_n(\mathcal{C})[\ell^s] \longrightarrow 0
$$

that $K_n(\mathcal{C}, \mathbb{Z}/\ell^s)$ is finite for all $n \geq 2$. Also, from the exact sequence

$$0 \longrightarrow \varprojlim^1 K_{n+1}(\mathcal{C}, \mathbb{Z}/\ell^s) \longrightarrow K_n^{pr}(\mathcal{C}, \hat{\mathbb{Z}}_\ell) \longrightarrow K_n(\mathcal{C}, \hat{\mathbb{Z}}_\ell) \longrightarrow 0$$

and the fact that $\varprojlim^1 K_{n+1}(\mathcal{C}, \mathbb{Z}/\ell^s) = 0$ for all $n \geq 1$, we have that $K_n^{pr}(\mathcal{C}, \hat{\mathbb{Z}}_\ell) \simeq K_n(\mathcal{C}, \hat{\mathbb{Z}}_\ell)$ for all $n \geq 1$. Result now follows from lemma 8.2.2.

Corollary 8.2.3 Let $\mathcal{C}$ be an exact category such that $K_n(\mathcal{C})$ is finitely generated for all $n \geq 1$. Then for all $n \geq 2$, we have

$$K_n(\mathcal{C}) \otimes \hat{\mathbb{Z}}_\ell \simeq K_n^{pr}(\mathcal{C}, \hat{\mathbb{Z}}_\ell) \simeq K_n(\mathcal{C}, \hat{\mathbb{Z}}_\ell)$$

are ℓ-complete profinite Abelian groups.

PROOF   Since $K_n(\mathcal{C})$ is finitely generated, then $K_n(\mathcal{C})_\ell$ has no non-zero divisible subgroups, and so, by theorem 8.2.1, $K_n(\mathcal{C})/\ell^s \xrightarrow{\sim} K_n^{pr}(\mathcal{C}, \hat{\mathbb{Z}}_\ell)/\ell^s$. Taking inverse limits, we have $\varprojlim_s K_n(\mathcal{C})/\ell^s \xrightarrow{\sim} \varprojlim_s K_n^{pr}(\mathcal{C}, \hat{\mathbb{Z}}_\ell)/\ell^s \simeq K_n^{pr}(\mathcal{C}, \hat{\mathbb{Z}}_\ell)$ since by theorem 8.2.3, $K_n^{pr}(\mathcal{C}, \hat{\mathbb{Z}}_\ell)$ is ℓ-complete. So, $K_n(\mathcal{C}) \otimes \hat{\mathbb{Z}}_\ell \simeq K_n^{pr}(\mathcal{C}, \hat{\mathbb{Z}}_\ell)$. Also since $K_n(\mathcal{C}, \mathbb{Z}/l)$ is finite for all $n \geq 2$, we have that $\varprojlim_s^1 K_{n+1}(\mathcal{C}, \mathbb{Z}/l^s) = 0$ in the exact sequence

$$0 \longrightarrow \varprojlim_s^1 K_{n+1}(\mathcal{C}, \mathbb{Z}/l^s) \longrightarrow K_n^{pr}(\mathcal{C}, \hat{\mathbb{Z}}_\ell) \longrightarrow K_n(\mathcal{C}, \hat{\mathbb{Z}}_\ell) \longrightarrow 0$$

for all $n \geq 1$. So, for all $n \geq 2$, we have $K_n(\mathcal{C}) \otimes \hat{\mathbb{Z}}_\ell \simeq K_n^{pr}(\mathcal{C}, \hat{\mathbb{Z}}_\ell) \simeq K_n(\mathcal{C}, \hat{\mathbb{Z}}_\ell)$ as required. The groups are ℓ-complete by theorem 8.2.3.   □

Remarks 8.2.6   (i) If R is the ring of integers in a number field F and Λ is any R-order in a semi-simple F-algebra Σ, it follows from theorem 8.2.3 and corollary 8.2.3 that $K_n(\Lambda) \otimes \hat{\mathbb{Z}}_\ell \simeq K_n^{pr}(\Lambda, \hat{\mathbb{Z}}_\ell) \simeq K_n(\Lambda, \hat{\mathbb{Z}}_\ell)$ and $G_n(\Lambda) \otimes \hat{\mathbb{Z}}_\ell \simeq G_n^{pr}(\Lambda, \hat{\mathbb{Z}}_\ell) \simeq G_n(\Lambda, \hat{\mathbb{Z}}_\ell)$ are ℓ-complete profinite Abelian groups for all $n \geq 2$ since, by theorem 7.1.11 and theorem 7.1.13, $K_n(\Lambda), G_n(\Lambda)$ are finitely generated Abelian groups for all $n \geq 1$.

(ii) If in (i) R, F, Λ, Σ satisfy the hypothesis of theorem 8.1.2, then $SG_n(\Lambda) = 0$ for all $n \geq 1$, and so, we have an exact sequence

$$0 \longrightarrow G_n(\Lambda) \longrightarrow G_n(\Sigma) \longrightarrow \bigoplus_{\mathbf{p}} G_{2n-1}(\Lambda/\mathbf{p}\Lambda) \longrightarrow 0$$

for all $n \geq 2$. In particular, $G_{2n-1}(\Lambda) \simeq G_{2n-1}(\Sigma)$ since $G_{2n}(\Lambda/\mathbf{p}\Lambda) = 0$ because $\Lambda/\mathbf{p}\Lambda$ is a finite ring (see theorem 7.1.12). But $G_n(\Sigma) \simeq K_n(\Sigma)$ since Σ is regular. Hence $K_{2n-1}(\Sigma)$ is a finitely generated Abelian group for all $n \geq 2$, and so, by theorem 8.2.3 and corollary 8.2.3 we have $K_{2n-1}(\Sigma) \otimes \hat{\mathbb{Z}}_\ell \simeq K_{2n-1}^{pr}(\Sigma, \hat{\mathbb{Z}}_\ell) \simeq K_{2n-1}(\Sigma, \hat{\mathbb{Z}}_\ell)$, which is an ℓ-complete profinite Abelian group.

(iii) If A is any finite ring, it also follows from theorem 8.2.3 and corollary 8.2.3 that $K_n(A) \otimes \hat{\mathbb{Z}}_\ell \simeq K_n^{pr}(A, \hat{\mathbb{Z}}_\ell) \simeq K_n(A, \hat{\mathbb{Z}}_\ell)$ and $G_n(A) \otimes \hat{\mathbb{Z}}_\ell \simeq G_n^{pr}(A, \hat{\mathbb{Z}}_\ell) \simeq G_n(A, \hat{\mathbb{Z}}_\ell)$ are ℓ-complete profinite Abelian groups since, by theorem 7.1.12, $K_n(A), G_n(A)$ are finite groups.

Corollary 8.2.4 Let R be the ring of integers in a number field F, Λ an R-order in a semi-simple F-algebra Σ satisfying the hypothesis of theorem 8.1.2, (e.g. $\Lambda = RG$, G a finite group, $\Sigma = FG$). Let ℓ be a rational prime. Then $SK_n^{pr}(\Lambda, \hat{\mathbb{Z}}_\ell) \simeq SK_n(\Lambda, \hat{\mathbb{Z}}_\ell)$ is an ℓ-complete profinite Abelian group.

PROOF   It follows from theorem 8.2.3 and corollary 8.2.3 that $\varprojlim_s K_n^{pr}(\Lambda, \hat{\mathbb{Z}}_\ell)$

$/\ell^s \simeq K_n(\Lambda, \hat{\mathbb{Z}}_\ell) \simeq K_n^{pr}(\Lambda, \hat{\mathbb{Z}}_\ell)$, and $\varprojlim_s G_n^{pr}(\Lambda, \hat{\mathbb{Z}}_\ell)/\ell^s \simeq G_n(\Lambda, \hat{\mathbb{Z}}_\ell) \simeq$

$G_n^{pr}(\Lambda, \hat{\mathbb{Z}}_\ell)$.   Hence, in view of remarks 2.6 (ii), we now have $\varprojlim_s SK_n^{pr}(\Lambda, \hat{\mathbb{Z}}_\ell)/\ell^s \simeq SK_n(\Lambda, \hat{\mathbb{Z}}_\ell) \simeq SK_n^{pr}(\Lambda, \hat{\mathbb{Z}}_\ell)$ as required.   ⬜

8.3   Profinite K-theory of p-adic orders and semi-simple algebras

8.3.1 Let R be the ring of integers in a number field F, $\mathbf{p}$ a prime ideal of R, $p := char(R/\mathbf{p})$. Then, $\hat{F}_{\mathbf{p}}$ is a p-adic field (i.e., a finite extension of $\hat{Q}_p$) and $\hat{R}_{\mathbf{p}}$ is the ring of integers of $\hat{F}_{\mathbf{p}}$. If Λ is any R-order in a semi-simple F-algebra Σ, then $\hat{\Lambda}_{\mathbf{p}}$ is an $\hat{R}_{\mathbf{p}}$-order in the semi-simple algebra $\hat{\Sigma}_{\mathbf{p}}$. Let ℓ be a rational prime such that $\ell \neq p$. Now, since we do not have finite generation results for $K_n(\hat{\Lambda}_{\mathbf{p}}), G_n(\hat{\Lambda}_{\mathbf{p}}), n \geq 1$, we cannot apply theorem 8.2.3 to conclude that $K_n^{pr}(\hat{\Lambda}_{\mathbf{p}}, \hat{\mathbb{Z}}_\ell)$ and $G_n^{pr}(\hat{\Lambda}_{\mathbf{p}}, \hat{\mathbb{Z}}_\ell)$ are ℓ-complete. However, we are able to show in this section that $G_n^{pr}(\hat{\Lambda}_{\mathbf{p}}, \hat{\mathbb{Z}}_\ell)$ is ℓ-complete by first showing that $K_n^{pr}(\hat{\Sigma}_{\mathbf{p}}, \hat{\mathbb{Z}}_\ell) = G_n^{pr}(\hat{\Sigma}_{\mathbf{p}}, \hat{\mathbb{Z}}_\ell)$ is ℓ-complete and that $K_n^{pr}(\hat{\Gamma}_{\mathbf{p}}, \hat{\mathbb{Z}}_\ell)$ is ℓ-complete for any maximal $\hat{R}_{\mathbf{p}}$-order $\hat{\Gamma}_{\mathbf{p}}$ in $\hat{\Sigma}_{\mathbf{p}}$. We also show in this section that if R, F, Λ, Σ satisfy the hypothesis of theorem 8.1.2 (e.g., when $\Lambda = RG$, G a finite group), then $SK_n^{pr}(\hat{\Lambda}_{\mathbf{p}}, \hat{\mathbb{Z}}_\ell)$ is ℓ-complete.

Theorem 8.3.1 *Let p be a rational prime, F a p-adic field (i.e., F is any finite extension of $\hat{Q}_p$), R the ring of integers of F, Γ a maximal R-order in a semi-simple F-algebra Σ, ℓ a rational prime such that $\ell \neq p$. Then, for all $n \geq 2$,*

(i) $K_n^{pr}(\Sigma, \hat{\mathbb{Z}}_\ell) \simeq K_n(\Sigma, \hat{\mathbb{Z}}_\ell)$ is an ℓ-complete profinite Abelian group.

(ii) $K_n^{pr}(\Gamma, \hat{\mathbb{Z}}_\ell) \simeq K_n(\Gamma, \hat{\mathbb{Z}}_\ell)$ is an ℓ-complete profinite Abelian group.

Remarks 8.3.1 Since in theorem 8.3.1, $\Sigma = \prod_{i=1}^{r} M_{n_i}(D_i), \Gamma = \prod_{i=1}^{r} M_{n_i}(\Gamma_i)$, say, where Γ_i is a maximal order in some division algebra D_i over F, it suffices to prove the following result in order to prove theorem 8.3.1.

Theorem 8.3.2 *Let Γ be a maximal order in a central division algebra D over a p-adic field F. Assume that ℓ is a rational prime such that $\ell \neq p$. Then, for all $n \geq 2$,*

(i) $K_n^{pr}(D, \hat{\mathbb{Z}}_\ell) \simeq K_n(D, \hat{\mathbb{Z}}_\ell)$ is an ℓ-complete profinite Abelian group.

(ii) $K_n^{pr}(\Gamma, \hat{\mathbb{Z}}_\ell) \simeq K_n(\Gamma, \hat{\mathbb{Z}}_\ell)$ is an ℓ-complete profinite Abelian group.

PROOF Let **m** be the unique maximal ideal of Γ. Consider the following localization sequence:

$$\cdots \longrightarrow K_n(\Gamma/\mathbf{m}, \mathbb{Z}/l^s) \longrightarrow K_n(\Gamma, \mathbb{Z}/l^s) \longrightarrow K_n(D, \mathbb{Z}/l^s)$$
$$\longrightarrow K_{n-1}(\Gamma/\mathbf{m}, \mathbb{Z}/l^s) \longrightarrow \cdots \qquad \text{(I)}$$

We know that $K_n(\Gamma, \mathbb{Z}/l^s) \simeq K_n(\Gamma/\mathbf{m}, \mathbb{Z}/l^s)$ is finite for all $n \geq 1$ (see [204], corollary 2 to theorem 2). Now, since the groups $K_n(\Gamma/\mathbf{m}, \mathbb{Z}/l^s), n \geq 1$ are finite groups with uniformly bounded orders (see [204]), so are the groups $K_n(D, \mathbb{Z}/l^s)$ and $K_n(\Gamma, \mathbb{Z}/l^s)$. So, in the exact sequences

$$0 \longrightarrow \varprojlim_s{}^1 K_{n+1}(D, \mathbb{Z}/l^s) \longrightarrow K_n^{pr}(D, \hat{\mathbb{Z}}_\ell) \longrightarrow K_n(D, \hat{\mathbb{Z}}_\ell) \longrightarrow 0 \qquad \text{(I)}$$

and

$$0 \longrightarrow \varprojlim_s{}^1 K_{n+1}(\Gamma, \mathbb{Z}/l^s) \longrightarrow K_n^{pr}(\Gamma, \hat{\mathbb{Z}}_\ell) \longrightarrow K_n(\Gamma, \hat{\mathbb{Z}}_\ell) \longrightarrow 0$$

we have $\varprojlim_s{}^1 K_{n+1}(D, \mathbb{Z}/l^s) = 0 = \varprojlim_s{}^1 K_{n+1}(\Gamma, \mathbb{Z}/l^s)$, and so,

$$K_n^{pr}(D, \hat{\mathbb{Z}}_\ell) \simeq K_n(D, \hat{\mathbb{Z}}_\ell); K_n^{pr}(\Gamma, \hat{\mathbb{Z}}_\ell) \simeq K_n(\Gamma, \hat{\mathbb{Z}}_\ell). \qquad \text{(II)}$$

This proves part of (i) and (ii).
To show that $K_n^{pr}(D, \hat{\mathbb{Z}}_\ell)$ and $K_n^{pr}(\Gamma, \hat{\mathbb{Z}}_\ell)$ are ℓ-complete, it suffices to show that

$$\varprojlim_s K_n^{pr}(D, \hat{\mathbb{Z}}_\ell)/\ell^s \simeq K_n(D, \hat{\mathbb{Z}}_\ell)(\simeq (K_n^{pr}(D, \hat{\mathbb{Z}}_\ell)),$$

and

$$\varprojlim_s K_n^{pr}(\Gamma, \hat{\mathbb{Z}}_\ell)/\ell^s \simeq K_n(\Gamma, \hat{\mathbb{Z}}_\ell)(\simeq (K_n^{pr}(\Gamma, \hat{\mathbb{Z}}_\ell)).$$

But these follow from the exact sequence

$$0 \longrightarrow \varprojlim_s K_n^{pr}(D, \hat{\mathbb{Z}}_\ell)/\ell^s \longrightarrow K_n(D, \hat{\mathbb{Z}}_\ell) \longrightarrow \varprojlim_s K_{n-1}^{pr}(D, \hat{\mathbb{Z}}_\ell)[\ell^s] \longrightarrow 0$$

and

$$0 \longrightarrow \varprojlim_s K_n^{pr}(\Gamma, \hat{\mathbb{Z}}_\ell)/\ell^s \longrightarrow K_n(\Gamma, \hat{\mathbb{Z}}_\ell) \longrightarrow \varprojlim_s K_{n-1}^{pr}(\Gamma, \hat{\mathbb{Z}}_\ell)[\ell^s] \longrightarrow 0$$

where $\varprojlim_s K_{n-1}^{pr}(D, \hat{\mathbb{Z}}_\ell)[\ell^s] = 0 = \varprojlim K_{n-1}^{pr}(\Gamma, \hat{\mathbb{Z}}_\ell)[\ell^s]$ by theorem 8.2.2(i).
Finally, observe that $K_n(D, \hat{\mathbb{Z}}_\ell) = \varprojlim K_n(D, \mathbb{Z}/l^s)$ and $K_n(\Gamma, \hat{\mathbb{Z}}_\ell) = \varprojlim K_n(\Gamma, \mathbb{Z}/l^s)$ are profinite since $K_n(D, \mathbb{Z}/l^s)$ and $K_n(\Gamma, \mathbb{Z}/l^s)$ are finite.
□

Theorem 8.3.3 *Let R be the ring of integers in a p-adic field F, ℓ a rational prime such that $\ell \neq p$, Λ an R-order in a semi-simple F-algebra Σ. Then, for all $n \geq 2$, $G_n^{pr}(\Lambda, \hat{\mathbb{Z}}_\ell)$ is an ℓ-complete profinite Abelian group.*

PROOF   First we show that for all $n \geq 2$, $G_n(\Lambda, \mathbb{Z}/l^s)$ is a finite Abelian group. Observe that $G_n(\Sigma, \mathbb{Z}/l^s)$ is finite by the proof of theorem 8.3.2 since $G_n(\Sigma, \mathbb{Z}/l^s) \simeq \overset{r}{\underset{i=1}{\otimes}} G_n(D_i, \mathbb{Z}/l^s)$ for some division algebra D_i over F and each $G_n(D_i, \mathbb{Z}/l^s)$ is finite.
That $G_n(\Lambda, \mathbb{Z}/l^s)$ is finite would follow from the following localization sequence:

$$\ldots \longrightarrow G_n(\Lambda/p\Lambda, \mathbb{Z}/l^s) \longrightarrow G_n(\Lambda, \mathbb{Z}/l^s) \longrightarrow G_n(\Sigma, \mathbb{Z}/l^s) \longrightarrow \ldots,$$

since $G_n(\Lambda/p\Lambda, \mathbb{Z}/l^s)$ is finite. Note that the finiteness of $G_n(\Lambda/p\Lambda, \mathbb{Z}/l^s)$ can be seen from the exact sequence

$$0 \longrightarrow G_n(\Lambda/p\Lambda)/\ell^s \longrightarrow G_n(\Lambda/p\Lambda, \mathbb{Z}/l^s) \longrightarrow G_{n-1}(\Lambda/p\Lambda)[\ell^s] \longrightarrow 0,$$

using the fact that $G_n(\Lambda/p\Lambda)$ is finite (G_n of a finite ring is finite) (see theorem 7.1.12). It follows from the exact sequence

$$0 \longrightarrow \varprojlim_s {}^1 G_{n+1}(\Lambda, \mathbb{Z}/l^s) \longrightarrow G_n^{pr}(\Lambda, \hat{\mathbb{Z}}_\ell) \longrightarrow G_n(\Lambda, \hat{\mathbb{Z}}_\ell) \longrightarrow 0$$

that $G_n^{pr}(\Lambda, \hat{\mathbb{Z}}_\ell) \simeq G_n(\Lambda, \hat{\mathbb{Z}}_\ell)$ since $\varprojlim_s {}^1 G_{n+1}(\Lambda, \mathbb{Z}/l^s) = 0$. Also, we have $\varprojlim_s G_{n+1}^{pr}(\Lambda, \hat{\mathbb{Z}}_\ell)/\ell^s \simeq G_{n+1}(\Lambda, \hat{\mathbb{Z}}_\ell)$ from the exact sequence

$$0 \longrightarrow \varprojlim_s G_{n+1}^{pr}(\Lambda, \hat{\mathbb{Z}}_\ell)/\ell^s \longrightarrow G_{n+1}(\Lambda, \hat{\mathbb{Z}}_\ell) \longrightarrow \varprojlim_s G_n^{pr}(\Lambda, \hat{\mathbb{Z}}_\ell)[\ell^s] \longrightarrow 0,$$

since $\varprojlim_s G_n^{pr}(\Lambda, \hat{\mathbb{Z}}_\ell)[\ell^s] = 0$.   □

Theorem 8.3.4 *Let R be the ring of integers in s p-adic field F, Λ an R-order in a semi-simple F-algebra Σ, ℓ a rational prime such the $\ell \neq p$. Then, for all $n \geq 1$,*

(i) $G_n(\Lambda)_\ell$ are finite groups.

(ii) $K_n(\Sigma)_\ell$ are finite groups.

(iii) Kernel and cokernel of $G_n(\Lambda) \to G_n^{pr}(\Lambda, \hat{\mathbb{Z}}_\ell)$ are uniquely ℓ-divisible.

(iv) Kernel and cokernel of $K_n(\Sigma) \to K_n^{pr}(\Sigma, \hat{\mathbb{Z}}_\ell)$ are uniquely ℓ-divisible.

PROOF (iii) and (iv) would follow from corollary 8.2.1 once we prove (i) and (ii) since, then, $G_n(\Lambda)_\ell$ and $K_n(\Sigma)_\ell$ would have no non-zero divisible subgroups.

(i) To prove that $K_n(\Sigma)_\ell$ is finite, it suffices to prove that $K_n(D)_\ell$ is finite where D is a central division algebra over some p-adic field F. Now, in the exact sequence

$$0 \longrightarrow K_{n+1}(D)/\ell^s \longrightarrow K_{n+1}(D, \mathbb{Z}/l^s) \longrightarrow K_n(D)[\ell^s] \longrightarrow 0,$$

we know from the proof of theorem 8.3.2 that for all $n \geq 1$, $K_{n+1}(D, \mathbb{Z}/l^s)$ is a finite group. Hence $K_n(D)[\ell^s]$ is a finite group having uniformly bounded orders. But $K_n(D)[\ell^{s-1}] \subset K_n(D)[\ell^s]$, for all s. Hence, the orders of the groups $K_n(D)[\ell^s]$ are the same for some $s \geq s_0$. But $K_n(D)_\ell = \bigcup_{s=1}^\infty K_n(D)[\ell^s]$. Hence $K_n(D)_\ell$ is finite.

(ii) To show that $G_n(\Lambda)_\ell$ is finite for all $n \geq 1$, we consider the exact sequence

$$0 \longrightarrow G_{n+1}(\Lambda)/\ell^s \longrightarrow G_{n+1}(\Lambda, \mathbb{Z}/l^s) \longrightarrow G_n(\Lambda)[\ell^s] \longrightarrow 0$$

and conclude that $G_n(\Lambda)[\ell^s]$ are finite groups with uniformly bounded orders since this is true for $G_{n+1}(\Lambda, \mathbb{Z}/l^s)$. Hence, $G_n(\Lambda)_\ell = \bigcup_{s=1}^\infty G_n(\Lambda)[\ell^s]$ is finite.

$\square$

Remarks Note that in the global case, $G_n(\Lambda)_\ell$ is known to be finite since $G_n(\Lambda)$ is finitely generated.

Theorem 8.3.5 *Let R be the ring of integers in a number field F, Λ an R-order in a semi-simple F-algebra Σ satisfying the hypothesis of theorem 8.1.2, $\mathbf{p}$ any prime ideal of R, and ℓ a rational prime such that $\ell = char(R/\mathbf{p})$. Then,*

(i) *$K_n(\hat{\Lambda}_\mathbf{p})_\ell$ is a finite group.*

(ii) *The map $\varphi : K_n(\hat{\Lambda}_\mathbf{p})_\ell \to K_n^{pr}(\hat{\Lambda}_\mathbf{p}, \hat{\mathbb{Z}}_\ell)_\ell$ is an isomorphism.*

PROOF

(i) By theorem 8.1.3(iii)(b), $SK_n(\hat{\Lambda}_{\mathbf{p}}) = Ker(K_n(\hat{\Lambda}_{\mathbf{p}}) \rightarrow G_n(\hat{\Lambda}_{\mathbf{p}}))$. We first show that $SK_n(\hat{\Lambda}_{\mathbf{p}})$ is finite or zero. Now, $SK_n(\Lambda)$ is finite for all $n \geq 1$ (see theorem 7.1.11). Moreover, it is well known that $\hat{\Sigma}_{\mathbf{p}}$ splits for almost all prime ideals $\mathbf{p}$ of R and that $\hat{\Lambda}_{\mathbf{p}}$ is a maximal $\hat{R}_{\mathbf{p}}$-order in a split semi-simple algebra $\hat{\Sigma}_{\mathbf{p}}$ for almost all $\mathbf{p}$. Hence $SK_n(\hat{\Lambda}_{\mathbf{p}}) = 0$ for all $n \geq 1$ (see theorem 7.1.9). Now, suppose $\hat{\Lambda}_{\mathbf{p}_1}, \hat{\Lambda}_{\mathbf{p}_2}, \dots \hat{\Lambda}_{\mathbf{p}_m}$ are the non-maximal orders. Also, it was shown in theorem 7.1.10 that there exists a surjection $SK_n(\Lambda) \rightarrow \overset{m}{\underset{i=1}{\otimes}} SK_n(\hat{\Lambda}_{\mathbf{p}})$. Hence each $SK_n(\hat{\Lambda}_{\mathbf{p}})$ is finite for $i = 1 \dots m$. So, we have shown that for all prime ideals $\mathbf{p}$ of R, $SK_n(\hat{\Lambda}_{\mathbf{p}})$ is finite or zero. Hence $SK_n(\hat{\Lambda}_{\mathbf{p}})_\ell = Ker(K_n(\hat{\Lambda}_{\mathbf{p}})_\ell \rightarrow G_n(\hat{\Lambda}_{\mathbf{p}})_\ell))$ is finite or zero. Now, by theorem 8.3.4(i), $G_n(\hat{\Lambda}_{\mathbf{p}})_\ell$ is finite. Hence $K_n(\hat{\Lambda}_{\mathbf{p}})_\ell$ is finite.

(ii) Now, by (i), $K_n(\hat{\Lambda}_{\mathbf{p}})_\ell$ is finite and hence has no non-zero divisible subgroups, and so, by 2.8, $Ker(K_n(\hat{\Lambda}_{\mathbf{p}}) \overset{\varphi}{\rightarrow} K_n^{pr}(\hat{\Lambda}_{\mathbf{p}}, \hat{\mathbb{Z}}_\ell))$ is uniquely ℓ-divisible. But $div\, K_n(\hat{\Lambda}_{\mathbf{p}})_\ell = 0$, since $K_n(\hat{\Lambda}_{\mathbf{p}})_\ell$ is finite. Consider the commutative diagram

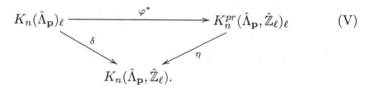

$$ (V) $$

So, we have

$$ Ker(K_n(\hat{\Lambda}_{\mathbf{p}})_\ell \longrightarrow K_n^{pr}(\hat{\Lambda}_{\mathbf{p}}, \hat{\mathbb{Z}}_\ell)_\ell) \subseteq Ker\delta = div\, K_n(\hat{\Lambda}_{\mathbf{p}})_\ell = 0. $$

So, $K_n(\hat{\Lambda}_{\mathbf{p}})_\ell \longrightarrow K_n^{pr}(\hat{\Lambda}_{\mathbf{p}}, \hat{\mathbb{Z}}_\ell)_\ell$ is injective. Also, $K_n(\hat{\Lambda}_{\mathbf{p}})_\ell \longrightarrow K_n^{pr}(\hat{\Lambda}_{\mathbf{p}}, \hat{\mathbb{Z}}_\ell)_\ell$ is surjective since each $K_n(\hat{\Lambda}_{\mathbf{p}})[\ell^s] \longrightarrow K_n^{pr}(\hat{\Lambda}_{\mathbf{p}}, \hat{\mathbb{Z}}_\ell)[\ell^s]$ is surjective by theorem 8.2.1(i). Hence $K_n(\hat{\Lambda}_{\mathbf{p}})_\ell \simeq K_n^{pr}(\hat{\Lambda}_{\mathbf{p}}, \hat{\mathbb{Z}}_\ell)_\ell$.

$\square$

Our next aim is to prove the local analogue of corollary 8.2.4. As we said earlier, we cannot exploit theorem 8.2.3 and corollary 8.2.3 since we do not have finite generation results for $K_n(\hat{\Lambda}_{\mathbf{p}}), G_n(\hat{\Lambda}_{\mathbf{p}}), n \geq 1$. However, we exploit the fact that $SK_n(\hat{\Lambda}_{\mathbf{p}})$ and $SG_n(\hat{\Lambda}_{\mathbf{p}})$ are finite to prove ℓ-completeness of $SK_n^{pr}(\hat{\Lambda}_{\mathbf{p}}, \hat{\mathbb{Z}}_\ell)$ when Λ, Σ, R, F satisfy the hypothesis of theorem 8.1.2.

Theorem 8.3.6 *Let R be the ring of integers in a number field F, Λ an R-order in a semi-simple F-algebra Σ satisfying the hypothesis of theorem 8.1.2 (e.g., $\Lambda = RG$, G any finite group). Let $\mathbf{p}$ be a prime ideal of R, ℓ a rational prime such that $\ell \neq char(R/\mathbf{p})$. Then for all $n \geq 1$,*

(i) $SK_n^{pr}(\hat{\Lambda}_\mathbf{p}, \hat{\mathbb{Z}}_\ell)$ *is isomorphic to a subgroup of* $SK_n(\hat{\Lambda}_\mathbf{p}, \hat{\mathbb{Z}}_\ell)$.

(ii) $\varprojlim\limits_{s} SK_n^{pr}(\hat{\Lambda}_\mathbf{p}, \hat{\mathbb{Z}}_\ell)/\ell^s \simeq SK_n(\hat{\Lambda}_\mathbf{p}, \hat{\mathbb{Z}}_\ell)$.

(iii) $SK_n^{pr}(\hat{\Lambda}_\mathbf{p}, \hat{\mathbb{Z}}_\ell)$ *is an ℓ-complete profinite Abelian group.*

PROOF

(i) First observe from theorem 8.1.3(iv)(b) that $SK_n(\hat{\Lambda}_\mathbf{p}, \mathbb{Z}/\ell^s)$
= $\mathrm{Ker}(K_n(\hat{\Lambda}_\mathbf{p}, \mathbb{Z}/\ell^s) \to G_n(\hat{\Lambda}_\mathbf{p}, \mathbb{Z}/\ell^s))$ and from remarks 8.2.3(ii) that

$$SK_n^{pr}(\hat{\Lambda}_\mathbf{p}, \hat{\mathbb{Z}}_\ell) = \mathrm{Ker}(K_n^{pr}(\hat{\Lambda}_\mathbf{p}, \hat{\mathbb{Z}}_\ell) \to G_n^{pr}(\hat{\Lambda}_\mathbf{p}, \hat{\mathbb{Z}}_\ell))$$

and

$$SK_n(\hat{\Lambda}_\mathbf{p}, \hat{\mathbb{Z}}_\ell) = \mathrm{Ker}(K_n(\hat{\Lambda}_\mathbf{p}, \hat{\mathbb{Z}}_\ell) \to G_n(\hat{\Lambda}_\mathbf{p}, \hat{\mathbb{Z}}_\ell)).$$

Now, by applying the Snake lemma to the following commutative diagram

$$0 \longrightarrow \varprojlim\limits_{s}{}^1 K_{n+1}(\hat{\Lambda}_\mathbf{p}, \mathbb{Z}/\ell^s) \longrightarrow K_n^{pr}(\hat{\Lambda}_\mathbf{p}, \hat{\mathbb{Z}}_\ell) \longrightarrow K_n(\hat{\Lambda}_\mathbf{p}, \hat{\mathbb{Z}}_\ell) \longrightarrow 0$$
$$\downarrow \qquad\qquad \downarrow \qquad\qquad \downarrow \qquad\qquad (I)$$
$$0 \longrightarrow \varprojlim\limits_{s}{}^1 G_{n+1}(\hat{\Lambda}_\mathbf{p}, \mathbb{Z}/\ell^s) \longrightarrow G_n^{pr}(\hat{\Lambda}_\mathbf{p}, \hat{\mathbb{Z}}_\ell) \longrightarrow G_n(\hat{\Lambda}_\mathbf{p}, \hat{\mathbb{Z}}_\ell) \longrightarrow 0$$

where the rows are exact by remarks 8.2.2(i),(ii), we obtain a sequence

$$0 \longrightarrow \varprojlim\limits_{s}{}^1 SK_{n+1}(\hat{\Lambda}_\mathbf{p}, \mathbb{Z}/\ell^s) \longrightarrow SK_n^{pr}(\hat{\Lambda}_\mathbf{p}, \hat{\mathbb{Z}}_\ell) \longrightarrow SK_n(\hat{\Lambda}_\mathbf{p}, \hat{\mathbb{Z}}_\ell)$$

$$\longrightarrow \varprojlim\limits_{s}{}^1 (\mathrm{Coker}(K_{n+1}(\hat{\Lambda}_\mathbf{p}, \mathbb{Z}/\ell^s)) \longrightarrow G_{n+1}(\hat{\Lambda}_\mathbf{p}, \hat{\mathbb{Z}}/\ell^s)) \longrightarrow \ldots$$

So, to prove (i), it suffices to show that for all $n \geq 1$, $\varprojlim\limits_{s}{}^1 SK_n(\hat{\Lambda}_\mathbf{p}, \mathbb{Z}/\ell^s) = 0$.

To do this, it suffices to show that $SK_n(\hat{\Lambda}_\mathbf{p}, \mathbb{Z}/\ell^s)$ is finite since $\varprojlim\limits_{s}{}^1 G_s = 0$ for any inverse system $\{G_s\}$ of finite groups G_s. This is what we set out to do now.

That $SK_n(\hat{\Lambda}_\mathbf{p}, \mathbb{Z}/\ell^s)$ is finite follows from applying the Snake lemma to the following commutative diagram

$$0 \to K_n(\hat{\Lambda}_\mathbf{p})/\ell^s \to K_n(\hat{\Lambda}_\mathbf{p}, \hat{\mathbb{Z}}/\ell^s) \to K_{n-1}(\hat{\Lambda}_\mathbf{p})[\ell^s] \to 0$$
$$\downarrow \qquad\qquad \downarrow \qquad\qquad \downarrow \qquad\qquad (II)$$
$$0 \to G_n(\hat{\Lambda}_\mathbf{p})/\ell^s \to G_n(\hat{\Lambda}_\mathbf{p}, \hat{\mathbb{Z}}/\ell^s) \to G_{n-1}(\hat{\Lambda}_\mathbf{p})[\ell^s] \to 0.$$

and using the fact that $SK_n(\hat{\Lambda}_\mathbf{p}) = \mathrm{Ker}(K_n(\hat{\Lambda}_\mathbf{p}) \to G_n(\hat{\Lambda}_\mathbf{p}))$ is finite for all $n \geq 1$ (see theorem 8.1.3(iii)(b) and proof of theorem 8.3.5).

(ii) Consider the following commutative diagram:

$$0 \to K_n^{pr}(\hat{\Lambda}_{\mathbf{p}}, \hat{\mathbb{Z}}_\ell)/\ell^s) \to K_{n+1}(\hat{\Lambda}_{\mathbf{p}}, \hat{\mathbb{Z}}/\ell^s) \to K_n^{pr}(\hat{\Lambda}_{\mathbf{p}}, \hat{\mathbb{Z}}_\ell)[\ell^s] \to 0$$

$$\downarrow \qquad\qquad \downarrow \qquad\qquad \downarrow$$

$$0 \to G_n^{pr}(\hat{\Lambda}_{\mathbf{p}}, \hat{\mathbb{Z}}_\ell)/\ell^s) \to G_{n+1}(\hat{\Lambda}_{\mathbf{p}}, \hat{\mathbb{Z}}/\ell^s) \to G_n^{pr}(\hat{\Lambda}_{\mathbf{p}}, \hat{\mathbb{Z}}_\ell)[\ell^s] \to 0.$$
$$\text{(III)}$$

Now, by taking $\varprojlim_s$ of the rows and using the fact that $\varprojlim_s K_n^{pr}(\hat{\Lambda}_{\mathbf{p}}, \hat{\mathbb{Z}}_\ell)[\ell^s] = 0 = \varprojlim_s G_n^{pr}(\hat{\Lambda}_{\mathbf{p}}, \hat{\mathbb{Z}}_\ell)[\ell^s]$ (see theorem 8.2.2(i) with $\mathcal{C} = \mathcal{P}(\hat{\Lambda}_{\mathbf{p}}), \mathcal{C} = \mathcal{M}(\hat{\Lambda}_{\mathbf{p}})$), we have the result from the commutative diagram

$$\varprojlim_s K_n^{pr}(\hat{\Lambda}_{\mathbf{p}}, \hat{\mathbb{Z}}_\ell)/\ell^s \simeq K_n(\hat{\Lambda}_{\mathbf{p}}, \hat{\mathbb{Z}}_\ell)$$

$$\downarrow \qquad\qquad\qquad \downarrow$$

$$\varprojlim_s G_n^{pr}(\hat{\Lambda}_{\mathbf{p}}, \hat{\mathbb{Z}}_\ell)/\ell^s \simeq G_n(\hat{\Lambda}_{\mathbf{p}}, \hat{\mathbb{Z}}_\ell).$$

(iii) From (i) and (ii) above, we have an exact sequence

$$0 \to SK_n^{pr}(\hat{\Lambda}_{\mathbf{p}}, \hat{\mathbb{Z}}_\ell) \overset{\varphi}{\to} SK_n(\hat{\Lambda}_{\mathbf{p}}, \hat{\mathbb{Z}}_\ell) \simeq \varprojlim_s SK_n(\hat{\Lambda}_{\mathbf{p}}, \hat{\mathbb{Z}}_\ell)/\ell^s. \qquad \text{(IV)}$$

But we know from the proof of theorem 8.2.12 that $K_n^{pr}(\hat{\Lambda}_{\mathbf{p}}, \hat{\mathbb{Z}}_\ell)$ is weakly ℓ-complete. So, $SK_n^{pr}(\hat{\Lambda}_{\mathbf{p}}, \hat{\mathbb{Z}}_\ell)$ is also weakly ℓ-complete as a subgroup of $K_n^{pr}(\hat{\Lambda}_{\mathbf{p}}, \hat{\mathbb{Z}}_\ell)$. But, for any weakly ℓ-complete Abelian group G, $\mathrm{Coker}(G \overset{\varphi}{\to} \varprojlim_s G/\ell^s) = 0$ (see remarks 8.2.4(i)(V)). So, $\mathrm{Coker}\varphi = 0$ in (IV) above. Hence $SK_n^{pr}(\hat{\Lambda}_{\mathbf{p}}, \hat{\mathbb{Z}}_\ell) \simeq \varprojlim_s SK_n^{pr}(\hat{\Lambda}_{\mathbf{p}}, \hat{\mathbb{Z}}_\ell)/\ell^s$.

□

Remarks 8.3.2 It is conjectured that $K_n^{pr}(\hat{\Lambda}_{\mathbf{p}}, \hat{\mathbb{Z}}_\ell)$ is ℓ-complete even though the author is not yet able to prove it. It follows from earlier results that $K_n^{pr}(\hat{\Lambda}_{\mathbf{p}}, \hat{\mathbb{Z}}_\ell)$ is weakly ℓ-complete.

8.4 Continuous K-theory of p-adic orders

8.4.1 Let F be a p-adic field, R the ring of integers of F, Λ any R-order in a semi-simple F-algebra Σ. We present in this section a definition of continuous K-theory $K_n^c(\Lambda)$ of Λ and examine some of the properties.

Definition 8.4.1 $K_n^c(\Lambda) := \varprojlim_s K_n(\Lambda/p^s\Lambda)$ *for all* $n \geq 1$.

Theorem 8.4.1   *(i) $K_n^c(\Lambda)$ is a profinite Abelian group for all $n \geq 1$.*

(ii) $K_{2n}^c(\Lambda)$ is a pro-p-group for $n \geq 1$.

PROOF

(i) Since $\Lambda/p^s\Lambda$ is a finite ring, the proof follows from the fact that K_n of a finite ring is finite (see theorem 7.1.12).

(ii) $\Lambda/p^s\Lambda$ is a $\mathbb{Z}/p^s$-algebra, and so, $I = rad(\Lambda/p^s\Lambda)$ is a nilpotent ideal in the finite ring $\Lambda/p^s\Lambda$. So, by [236], 5.4 we have, for all $n \geq 1$ that $K_n(\Lambda/p^s\Lambda, I)$ is a p-group. By tensoring with $\mathbb{Z}\left(\frac{1}{p}\right)$ the exact sequence

$$\cdots \longrightarrow K_n(\Lambda/p^s\Lambda, I) \longrightarrow K_n(\Lambda/p^s\Lambda) \longrightarrow K_n(\Lambda/p^s\Lambda)/I$$
$$\longrightarrow K_n(\Lambda/p^s\Lambda, I) \longrightarrow \cdots,$$

we have that $K_n((\Lambda/p^s\Lambda)/I)(1/p) \simeq K_n(\Lambda/p^s\Lambda)(1/p)$. But $(\Lambda/p^s\Lambda)/I$ is a finite semi-simple ring and hence a direct product of matrix algebras over fields. Hence $K_{2n}((\Lambda/p^s\Lambda)/I)(1/p) \simeq K_{2n}(\Lambda/p^s\Lambda)(1/p) = 0$ by Quillen's results. Hence $K_{2n}(\Lambda/p^s\Lambda)$ is a finite p-group, and so, $K_{2n}^c(\Lambda) := \varprojlim_s K_{2n}(\Lambda/p^s\Lambda)$ is a pro-p-group.

$\square$

8.4.2 Let R be the ring of integers in a number field or a p-adic field F, Λ any R-order in a semi-simple F-algebra Σ, $\mathbf{q}$ a two-sided ideal of Λ of finite index. Let $\overline{GL(\Lambda/\mathbf{q})}$ denote the image of $GL(\Lambda)$ under the canonical map $GL(\Lambda) \to GL(\Lambda/\mathbf{q})$. Suppose that $K(\Lambda, \mathbf{q})$ denotes the connected component of the homotopy fiber of the map $BGL^+(\Lambda) \to BGL^+(\Lambda/\mathbf{q})$. Then we have a fibration

$$K(\Lambda, \mathbf{q}) \to BGL^+(\Lambda) \to BGL^+(\Lambda/\mathbf{q})$$

and hence a long exact sequence

$$\dots K_{n+1}(\Lambda/\mathbf{q}) \longrightarrow K_n(\Lambda, \mathbf{q}) \xrightarrow{\delta_n} K_n(\Lambda) \xrightarrow{\alpha_n} K_n(\Lambda/\mathbf{q}) \dots$$

where $K_n(\Lambda, \mathbf{q}) := \pi_n(K(\Lambda, \mathbf{q}))$.

Definition 8.4.2 *Let R be the ring of integers in a p-adic field F, Λ any R-order in a (p-adic) semi-simple F-algebra Σ.*

Define

$$D_n(\Lambda) := \varprojlim_s \operatorname{Coker}(K_n(\Lambda, p^s\Lambda) \xrightarrow{\delta_n} K_n(\Lambda)),$$

$$D_n(\Sigma) := \varprojlim_s \operatorname{Coker}(K_n(\Lambda, p^s\Lambda) \xrightarrow{\eta_n} K_n(\Sigma)),$$

$$R_n(\Lambda) := \varprojlim_s \operatorname{Ker}(K_n(\Lambda, p^s\Lambda) \xrightarrow{\delta_n} K_n(\Lambda)),$$

$$R_n(\Sigma) := \varprojlim_s \operatorname{Ker}(K_n(\Lambda, p^s\Lambda) \xrightarrow{\eta_n} K_n(\Sigma)),$$

where η_n is the composite $K_n(\Lambda, p^s\Lambda) \to K_n(\Lambda) \to K_n(\Sigma)$.

Remarks 8.4.1 In [159] R. Oliver defines

$$K_2^c(\Lambda) := \varprojlim_s \operatorname{Coker}(K_2(\Lambda, p^s\Lambda) \to K_2(\Lambda))$$

where his own $K_2(\Lambda, p^s\Lambda)$ is not $\pi_2(K(\Lambda, p^s\Lambda))$ defined in 8.4.2 above. In the context of [159], $K_2(\Lambda, p^s\Lambda)$ was defined as $\operatorname{Ker}(K_2(\Lambda) \to K_2(\Lambda/p^s\Lambda))$. However, he shows in [159] theorem 3.6 that $K_2^c(\Lambda)$ as defined in [159] is isomorphic to $\varprojlim_k K_2(\Lambda/p^k\Lambda)$. So, in a sense, our definition is a generalization of that of Oliver in [159]. Note also that our definition is closely related to $K_n^{top}(R)$ of Wagoner [223] where R is the ring of integers in a p-adic field. In that article, Wagoner defined $K_n^{top}(R) := \varprojlim_s K_n(R/\underline{p}^sR)$ where $\mathbf{p}$ is a maximal ideal above p.

Theorem 8.4.2 *Let R be the ring of integers in a p-adic field F, Λ any R-order in a semi-simple F-algebra Σ. Then for all $n \geq 1$,*

(i) There exists an exact sequence

$$0 \to D_n(\Lambda) \to K_n^c(\Lambda) \to R_{n-1}(\Lambda) \to 0.$$

(ii) If Γ is a maximal R-order in Σ, we also have an exact sequence

$$0 \to D_{2n}(\Gamma) \to D_{2n}(\Sigma) \to K_{2n-1}(\Gamma/rad\Gamma) \to SK_{2n-1}(\Gamma) \to 0.$$

PROOF

(i) From the exact sequence

$$\cdots \longrightarrow K_{n+1}(\Lambda/p^s\Lambda) \longrightarrow K_n(\Lambda, p^s\Lambda)$$
$$\longrightarrow K_n(\Lambda) \longrightarrow K_n(\Lambda/p^s\Lambda) \longrightarrow K_{n-1}(\Lambda, p^s\Lambda) \longrightarrow \cdots,$$

we obtain

$$0 \longrightarrow \operatorname{Coker}(K_n(\Lambda, p^s\Lambda) \longrightarrow K_n(\Lambda)) \longrightarrow K_n(\Lambda/p^s\Lambda)$$
$$\longrightarrow \operatorname{Ker}(K_{n-1}(\Lambda, p^s\Lambda) \longrightarrow K_{n-1}(\Lambda)) \longrightarrow 0.$$

Now, by taking inverse limits, we have the result.

(ii) First, we have the following localization sequence

$$\ldots K_{2n}(\Gamma/rad\Gamma) \longrightarrow K_{2n}(\Gamma) \longrightarrow K_{2n}(\Sigma) \longrightarrow K_{2n-1}(\Gamma/rad\Gamma) \qquad \text{(I)}$$
$$\longrightarrow SK_{2n-1}(\Gamma) \longrightarrow 0.$$

Now, $(\Gamma/rad\Gamma)$ is a finite semi-simple ring that is a direct product of matrix algebras over finite fields, and so, we have $K_{2n}(\Gamma/rad\Gamma) = 0$. Hence $0 \to K_{2n}(\Gamma) \to K_{2n}(\Sigma)$ is exact. So,

$$0 \longrightarrow \frac{K_{2n}(\Gamma)}{Im(K_{2n}(\Gamma, p^s\Gamma))} \longrightarrow \frac{K_{2n}(\Sigma)}{Im(K_{2n}(\Gamma, p^s\Gamma))}$$

is also exact.

Taking inverse limits, we have $0 \to D_{2n}(\Gamma) \to D_{2n}(\Sigma)$ is exact.

So, we have the required sequence

$$0 \to D_{2n}(\Gamma) \to D_{2n}(\Sigma) \to K_{2n-1}(\Gamma/rad\Gamma) \to SK_{2n-1}(\Gamma) \to 0.$$

       □

Theorem 8.4.3 *In the notation of definition 8.4.2,*

(a) $R_n(\Lambda), D_n(\Lambda)$ *are profinite groups for all* $n \geq 1$.

(b) $R_n(\Sigma), D_{2n}(\Sigma)$ *are profinite groups for all* $n \geq 1$.

PROOF

(a) Consider the exact sequence

$$\cdots \longrightarrow K_{n+1}(\Lambda/p^s\Lambda) \longrightarrow K_n(\Lambda, p^s\Lambda) \longrightarrow K_n(\Lambda) \longrightarrow K_n(\Lambda/p^s\Lambda)$$
$$\longrightarrow K_{n-1}(\Lambda, p^s\Lambda) \longrightarrow \cdots .$$

Now, since $K_n(\Lambda/p^s\Lambda)$ is finite for all $n \geq 1$ (see theorem 7.1.12), it follows that $\text{Ker}(K_n(\Lambda, p^s\Lambda) \to K_n(\Lambda))$ is finite as the image of the finite group $K_{n+1}(\Lambda, p^s\Lambda)$. Hence $R_n(\Lambda)$ is profinite. Also, $\text{Coker}(K_n(\Lambda, p^s\Lambda) \to K_n(\Lambda))$ is finite as the image of $K_n(\Lambda) \to K_n(\Lambda/p^s\Lambda)$. Hence $D_n(\Lambda)$ is profinite.

(b) Consider the commutative diagram

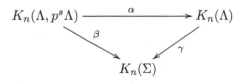

and the associated exact sequence

$$\cdots \to \mathrm{Ker}\alpha \to \mathrm{Ker}\beta \to SK_n(\Lambda) \to \mathrm{Coker}\alpha \to \mathrm{Coker}\beta \to \mathrm{Coker}\gamma \to 0.$$

Now, $\mathrm{Ker}\alpha$ has been shown to be finite in the proof of (a) above, and $SK_n(\Lambda)$ is finite (theorem 7.1.11(ii)). Hence $\mathrm{Ker}\beta$ is finite. Hence $R_n(\Sigma)$ is profinite.

Since $K_{2n-1}(\Gamma/rad\Gamma)$ is finite, it follows from the sequence

$$0 \longrightarrow D_n(\Gamma) \longrightarrow D_{2n}(\Sigma) \longrightarrow k_{2n-1}(\Gamma/rad\Gamma) \longrightarrow \cdots$$

that $D_{2n}(\Gamma)$ has finite index in $D_{2n}(\Sigma)$. Now, since $D_{2n}(\Gamma)$ is profinite, so is $D_{2n}(\Sigma)$.

□

Theorem 8.4.4 *Let R be the ring of integers in a number field F, Λ any R-order in a semi-simple F-algebra Σ. Then for all $n \geq 1$, we have the following:*

(i) For almost all prime ideals $\mathbf{p}$ in R,

$$R_n(\hat{\Lambda}_{\mathbf{p}}) \simeq R_n(\hat{\Sigma}_{\mathbf{p}})$$

$$0 \to D_n(\hat{\Lambda}_{\mathbf{p}}) \to D_n(\hat{\Sigma}_{\mathbf{p}}) \to \frac{K_n(\hat{\Sigma}_{\mathbf{p}})}{Im(K_n(\hat{\Lambda}_{\mathbf{p}}))} \to 0.$$

Hence $\frac{K_n(\hat{\Sigma}_{\mathbf{p}})}{ImK_n(\hat{\Lambda}_{\mathbf{p}})} \simeq \frac{D_n(\hat{\Sigma}_{\mathbf{p}})}{D_n(\hat{\Lambda}_{\mathbf{p}})}$ for almost all prime ideals $\mathbf{p}$ in R.

(ii) $\frac{D_{2r}(\hat{\Sigma}_{\mathbf{p}})}{D_{2r}(\hat{\Lambda}_{\mathbf{p}})} \simeq K_{2r-1}(\hat{\Lambda}_{\mathbf{p}}/rad\hat{\Lambda}_{\mathbf{p}})$ for almost all $\mathbf{p}$.   In this situation $\left| \left(\frac{D_{2r}(\hat{\Sigma}_{\mathbf{p}})}{D_{2r}(\hat{\Lambda}_{\mathbf{p}})} \right) \right| \equiv -1 \mod p$ for some rational prime p lying below $\mathbf{p}$.

PROOF   It is well known that for almost all $\mathbf{p}$, $\hat{\Lambda}_{\mathbf{p}}$ is a maximal order in the semi-simple $\hat{F}_{\mathbf{p}}$-algebra $\hat{\Sigma}_{\mathbf{p}}$ that splits. Also, for such $\mathbf{p}$, $SK_n(\hat{\Lambda}_{\underline{p}}) = 0$ for all $n \geq 1$ (see theorem 7.1.9). Note that for each $\mathbf{p}$, there is a rational prime p below $\mathbf{p}$.

Now, from the commutative diagram

$$K_n(\hat{\Lambda}_{\mathbf{p}}, p^s \hat{\Lambda}_{\mathbf{p}})$$

$$\searrow \alpha$$

$$\downarrow \beta \qquad K_n(\hat{\Sigma}_{\mathbf{p}}),$$

$$\nearrow \gamma$$

$$K_n(\hat{\Lambda}_{\mathbf{p}})$$

we have an exact sequence

$$0 \to \text{Ker}(K_n(\hat{\Lambda}_\mathbf{p}, p^s\hat{\Lambda}_\mathbf{p}) \to K_n(\Lambda)) \to \text{Ker}(K_n(\hat{\Lambda}_\mathbf{p}, p^s\hat{\Lambda}_\mathbf{p}) \to K_n(\hat{\Sigma}_\mathbf{p}))$$
$$\to SK_n(\hat{\Lambda}_\mathbf{p}) \to \text{Coker}(K_n(\hat{\Lambda}_\mathbf{p}, p^s\hat{\Lambda}_\mathbf{p}) \to K_n(\hat{\Lambda}_\mathbf{p}))$$
$$\to \text{Coker}(K_n(\hat{\Lambda}_\mathbf{p}, p^s\hat{\Lambda}_\mathbf{p}) \to K_n(\hat{\Sigma}_\mathbf{p})) \to \frac{K_n(\hat{\Sigma}_\mathbf{p})}{Im(K_n(\hat{\Lambda}_\mathbf{p}))} \to 0.$$

By taking inverse limits $\varprojlim\limits_{k}$ and observing that $SK_n(\hat{\Lambda}_\mathbf{p}) = 0$ for all $n \geq 1$, for such $\mathbf{p}$, we have (i).

(i) From theorem 8.4.2(i), we have an exact sequence

$$0 \to D_{2n}(\hat{\Lambda}_\mathbf{p}) \to D_{2n}(\hat{\Sigma}_\mathbf{p}) \to K_{2n-1}(\hat{\Lambda}_\mathbf{p}/rad\hat{\Lambda}_\mathbf{p}) \to SK_{2n-1}(\hat{\Lambda}_\mathbf{p}) \to 0$$

for almost all $\mathbf{p}$. Since $SK_{2n-1}(\hat{\Lambda}_\mathbf{p}) = 0$ for almost all $\mathbf{p}$, we have $D_{2n}(\hat{\Sigma}_\mathbf{p})/D_{2n}(\hat{\Lambda}_\mathbf{p}) \simeq K_{2n-1}(\hat{\Lambda}_\mathbf{p}/rad\hat{\Lambda}_\mathbf{p})$ as required.

Now, $\hat{\Lambda}_\mathbf{p}/rad\hat{\Lambda}_\mathbf{p}$ is a finite semi-simple ring and hence a product of matrix algebras over finite fields. Hence, the result follows by applying Quillen's result on $K_{2n-1}(F)$ where F is a finite field of order p^s, say for some rational prime p lying below $\mathbf{p}$.

We close this section with a connection between profinite and continuous K-theory of p-adic orders. $\quad\square$

Theorem 8.4.5 *Let F be a p-adic field with ring of integers R, Λ an R-order in a semi-simple F-algebra Σ, ℓ a rational prime such that $\ell \neq p$. Then, for all $n \geq 2$,*

$$\varprojlim_{t} K_n^{pr}(\Lambda/p^t\Lambda, \hat{\mathbb{Z}}_\ell) \simeq K_n^c(\Lambda) \times \hat{\mathbb{Z}}_\ell.$$

PROOF   Since $\Lambda/p^t\Lambda$ is a finite ring, then for all $n \geq 1$, $K_n(\Lambda/p^t\Lambda)$ is a finite group (see theorem 7.1.12). We apply corollary 8.2.3 to get

$$K_n(\Lambda/p^t\Lambda) \otimes \hat{\mathbb{Z}}_\ell \simeq K_n^{pr}(\Lambda/p^t\Lambda, \hat{\mathbb{Z}}_\ell).$$

Taking inverse limits, we have

$$K_n^c(\Lambda) \otimes \hat{\mathbb{Z}}_\ell \simeq \varprojlim_{t} K_n^{pr}(\Lambda/p^t\Lambda, \hat{\mathbb{Z}}_\ell).$$

$\quad\square$

Exercises

8.1 Let F be a number field, G a finite group $D_n(FG)$, the subgroup of divisible elements in $K_n(FG)$. Show that

(a) $D_{2n-1}(FG) = 0$ for $n \geq 1$.

(b) $D_{2n}(FG)$ is a finite group for all $n \geq 1$.

8.2 Let R be the ring of integers in a number field F, Λ any R-order in a semi-simple F-algebra Σ, ℓ an odd rational prime. Show that the exact sequence

$$0 \to K_n(\Lambda)/\ell^2 \to K_n(\Lambda, \mathbb{Z}/\ell^s) \to K_{n-1}(\Lambda)[\ell^s] \to 0$$

is split.

8.3 Let $\{G_n\}$ be an inverse system of Abelian groups such that each G_n has a finite ℓ-power exponent. Show that $\underleftarrow{\lim}{}^1 \, G_n$ is weakly ℓ-complete.

8.4 Let $\mathcal{C}$ be an exact category, ℓ a rational prime. Show that there exists an exact $\hat{\mathbb{Z}}_\ell$ sequence

$$0 \to K_n^{pr}(\mathcal{C}, \hat{\mathbb{Z}}_\ell) \otimes \hat{\mathbb{Q}}_\ell/\hat{\mathbb{Z}}_\ell \to K_{n+1}(\mathcal{C}, \hat{\mathbb{Q}}_\ell/\hat{\mathbb{Z}}_\ell) \to K_n^{pr}(\mathcal{C}, \hat{\mathbb{Z}}_\ell)(\ell) \to 0.$$

Part III

Mackey Functors, Equivariant Higher Algebraic K-Theory, and Equivariant Homology Theories

Chapter 9

Mackey, Green, and Burnside functors

In this chapter, we present Mackey, Green, and Burnside functors and their bases in a way that will prepare us for applications in equivariant K and homology theories.

Many of the results in this chapter were originally due to A. Dress see [47, 48, 50].

9.1 Mackey functors

Definition 9.1.1 *Let $\mathcal{D}$ be an Abelian category and let $\mathcal{B}$ be a category with finite sums, final object, and finite pull-backs (and hence finite products). Then the pair of functors $(M_*, M^*) : \mathcal{B} \to \mathcal{D}$ is called a Mackey functor if*

(i) $M_* : \mathcal{B} \to \mathcal{D}$ *is covariant,* $M^* : \mathcal{B} \to \mathcal{D}$ *is contravariant, and* $M_*(X) = M^*(X) := M(X)$ *for all* $X \in ob\mathcal{C}$.

(ii) For any pull-back diagram

$$
\begin{array}{ccc}
A' & \xrightarrow{p_2} & A_2 \\
\downarrow{p_1} & & \downarrow{f_2} \\
A_1 & \xrightarrow{f_1} & A
\end{array}
\quad in\ \mathcal{C}, \quad the\ diagram \quad
\begin{array}{ccc}
M(A') & \xrightarrow{p_{2*}} & M(A_2) \\
\uparrow{p_1^*} & & \uparrow{f_2^*} \\
M(A_1) & \xrightarrow{f_{1*}} & M(A)
\end{array}
\quad commutes.
$$

(iii) M^* *transforms finite coproducts in $\mathcal{B}$ into finite products in $\mathcal{D}$, i.e., the embeddings $X_i \hookrightarrow \coprod\limits_{i=1}^{n} X_i$ induce an isomorphism*

$$
M(X_1 \coprod X_2 \coprod \cdots \coprod X_n) \simeq M(X_1) \times \cdots \times M(X_n).
$$

Remarks 9.1.1 Note that (ii) above is an axiomatization of the Mackey subgroup theorem in classical representation theory (see [39]). As a first step towards seeing this connection, one can show that if $\mathcal{B} = GSet$, $A_1 = G/H$, $A_2 = G/H'$, then the orbit space of $G/H \times G/H'$ can be identified with the set $D(H, H') = \{HgH' : g \in G\}$ of double cosets of H and H' in G. This identification is also crucial for the connection between Mackey functors and Green's G-functors discussed in Remark (9.1.2 (ii)).

Examples 9.1.1   (i) Let G be a finite group and B a $\mathbb{Z}G$-module. For any G-set S, let $\mathcal{H}_G(S, B)$ denote the set of G-maps $f : S \to B$. Then, $\mathcal{H}_G(S, B)$ is an Abelian group where addition is defined in $\mathcal{H}_G(S, B)$ by $(f_1 + f_2)(s) = f_1(s) + f_2(s)$ $(f_1, f_2 \in \mathcal{H}_G(S, B))$. Now, if $h : S \to T$ is a $G - map$, we define $h^* : \mathcal{H}_G(T, B) \to \mathcal{H}_G(S, B)$ by $f \mapsto fh$, and $h_* : \mathcal{H}_G(S, B) \to \mathcal{H}_G(T, B)$ by $g \mapsto h_*g : t \to \sum_{s \in h^{-1}(t)} g(s)$ $(t \in T)$. Then, $\mathcal{H}_G(-, B) : GSet \to \mathbb{Z} - \mathcal{M}od$ given by $S \mapsto \mathcal{H}_G(S, B)$, $h \mapsto (h_*, h^*)$ satisfies property (i) of definition 9.1.1.
Now suppose that

$$
\begin{array}{ccc}
S_1 \underset{S}{\times} S_2 & \xrightarrow{p_2} & S_2 \\
\downarrow{\scriptstyle p_i} & & \downarrow{\scriptstyle f_2} \\
S_1 & \xrightarrow{f_1} & S
\end{array}
$$

is a pull-back square in $GSet$ and $g \in \mathcal{H}_G(S_1, B)$ where B is a $\mathbb{Z}G$-module.
We have to show that $f_2^* f_{1*}g(s) = p_{2*}p_1^* g(s)$ for all $s \in S_1$. Now,

$$f_2^* f_{1*}g(s) = f_{1*}g(f_2(s)) = \sum_{x \in f_1^{-1}(f_2(s))} g(x)$$

and

$$p_{2*}p_1^* g(s) = \sum_{(x,y) \in p_2^{-1}(s)} p_1^* g(x, y) = \sum_{(x,y) \in p_2^{-1}(s)} g(p_1(x, y))$$

$$= \sum_{(x,y) \in p_2^{-1}(s)} g(x).$$

Moreover, $(x, y) \in p_2^{-1}(s) \subseteq S_1 \times S_2$ if and only if $y = p_2(x, y) = s$ and $f_1(x) = f_2(s)$, and so the two sums coincide. That $\mathcal{H}_G(-, B)$ takes sums into products can be checked easily. Details are left to the reader.

(ii) Let B be a $\mathbb{Z}G$-module. We define a bifunctor $B = (B^*, B_*) : GSet \to \mathbb{Z} - \mathcal{M}od$ as follows:
for $S \in GSet$, $B(S)$ is the set of all set-theoretic maps from S to B as $\mathbb{Z}G$-module where for $f_1, f_2 : S \to B$ one puts $(f_1 + f_2)s = f_1s + f_2s$, and the G-action $G \times B(S) \to B(S) : (g, f) \mapsto gf$ is defined by $(gf)(s) = g\dot{f}(g^{-1}s)$ for any $s \in S$. Note that $\mathcal{H}_G(S, B)$ is just the subgroup of G-invariant elements in $B(S)$. Moreover, given a G-map $\varphi : S \to T$, then one has $\varphi^* = B^*(\varphi) : B(T) \to B(S)$ given by $f \mapsto f\dot{\varphi}$ and $\varphi_* = B_*(\varphi) :$

$B(S) \to B(T)$ given by $f \mapsto \varphi_* f$ where $\varphi_* f(t) = \sum_{s \in \varphi^{-1}(t)} f(s)$, which are now $\mathbb{Z}G$-homomorphisms. As above, it can be easily checked that $B = (B_*, B^*)$ is a Mackey functor.

Now, if $F : \mathbb{Z}G\text{-}\mathcal{M}od \to \mathbb{Z}\text{-}\mathcal{M}od$ is any additive functor (e.g., $B \mapsto B^G$) where $B^G = \{x \in B | gx = x \forall g \in G\}$ or, $(B \mapsto B_G = B/I_G B$ where $I_G = \{\sum_{g \in G} n_g g \in \mathbb{Z}G | \sum n_g = 0\}$ is the argumentation ideal in $\mathbb{Z}G$), then $F(-, B) : GSet \to \mathbb{Z}G - \mathcal{M}od \to \mathbb{Z} - \mathcal{M}od$ is a Mackey functor where

$$(F(\varphi, B)_*, F(\varphi, B)^*) = (F(B_*(\varphi)), F(B^*(\varphi)))$$

if φ is a $GSet$ morphism.

(iii) Let $\mathcal{C}$ be an exact category; G a finite, profinite, or compact Lie group. Then for all $n \geq 0$, the functor $K_n^G(-, \mathcal{C}) : GSet \to Ab$ is a Mackey functor. We shall consider these functors in some detail under equivariant higher K-theory in chapters 10 and 12.

(iv) Let $(\mathcal{C}, \perp)$ be a symmetric monoidal category. Then $K_o^G(-, \mathcal{C})$ is a Mackey functor.

PROOF Let $\varphi : S \to T$ be a map in $GSet$. Then φ defines a functor $\underline{\varphi} : \underline{S} \to \underline{T} : s \mapsto \varphi(s); (g, s) \mapsto (g, \varphi(s))$, and φ induces a functor $\varphi_* : [\underline{T}, \mathcal{C}] \to [\underline{S}, \mathcal{C}]$ given by $\zeta \to \zeta \circ \varphi$ such that $(\zeta \perp \eta) \cdot \underline{\varphi} = \zeta \cdot \underline{\varphi} \perp \eta \cdot \underline{\varphi}$. So, $\underline{\varphi}^*$ induces a homomorphism $[\underline{T}, \mathcal{C}]^+ \to [\underline{S}, \mathcal{C}]^+$, and so, we have a composite map $[\underline{T}, \mathcal{C}]^+ \to [\underline{S}, \mathcal{C}]^+ \to K_o^G(S, \mathcal{C})$. The universal property of the map $[\underline{T}, \mathcal{C}]^+ \to K_o^G(T, \mathcal{C})$ implies that we have a homomorphism $K_o^G(T, \mathcal{C}) \to K_o^G(S, \mathcal{C})$. Define $\underline{\varphi}_* : [\underline{S}, \mathcal{C}] \to [\underline{T}, \mathcal{C}]$ by $\zeta \mapsto \varphi_*(\zeta)$ where

$$\varphi_*(\zeta)_t = \underset{s \in \varphi^{-1}(t)}{\perp} \zeta_s \quad \text{and}$$

$$(\underline{\varphi}_* \zeta)_{(g,t)} : (\underline{\varphi}_* \zeta)_t = \underset{s \in \varphi^{-1}(t)}{\perp} \zeta_s \xrightarrow{\underset{s \in \varphi^{-1}(t)}{\perp} \zeta_{gs}} \underset{s \in \varphi^{-1}(t)}{\perp} \zeta_{gs}$$

$$= \underset{s \in \varphi^{-1}(gt)}{\perp} \zeta_s = (\underline{\varphi}_* \zeta)_{gt}.$$

It can be easily checked that $\underline{\varphi}_*(\zeta \perp \eta) \simeq \underline{\varphi}_*(\zeta) \perp \underline{\varphi}_*(\eta)$, and so, it induces a homomorphism $\underline{\varphi}_* : K_o^G(S, \mathcal{C}) \to K_o^G(T, \mathcal{C})$. So $K_o^G(-, \mathcal{C})$ is a bifunctor. We leave the checking of other properties of Mackey functors as an exercise for the reader.

If in addition to $\perp$, our category $\mathcal{C}$ in (iv) possesses a further associative composition o, which is naturally distributive with respect to $\perp$ and has a "unit", i.e., an element $X \in ob\mathcal{C}$, such that the functor $\mathcal{C} \to \mathcal{C} : Y \mapsto Y \circ X$ as well as $\mathcal{C} \to \mathcal{C} : Y \mapsto Y \circ Y$ are naturally equivalent to the identity $\mathcal{C} \to \mathcal{C} : Y \mapsto Y$, then o induces a multiplicative structure on $[\underline{S}, \mathcal{C}]$, which makes

$K_o^G(S, \mathcal{C})$ a ring with an identity, and $K_o^G(-, \mathcal{C}) : GSet \to \mathbb{Z} - \mathcal{M}od$ a Green functor. Note that $K_o^G(*, FSet) \simeq \Omega(G)$. Also, $K_o^G(G/G, \mathbb{C}) \simeq char(G)$ since $[G/G, \mathcal{P}(\mathbb{C})] \simeq \mathcal{P}(\mathbb{C})_G$. More generally, $K_o^G(*, \mathcal{C}) \simeq K(\mathcal{C}_G)$ since $[*, \mathcal{C}] \simeq \mathcal{C}_G$.

Remarks 9.1.2 (i) One can show that if $\mathcal{B}$ is small, then the Mackey functors from $\mathcal{B}$ to $\mathcal{D}$ form an Abelian category, and so, one can do homological algebra in the category of Mackey functors (see [48, 68, 111]).

(ii) One can relate Mackey functors with G-functors, as defined by J.A. Green (see [64]). Let G be a finite group and let δG denote the subgroup category whose objects are the various subgroups of G with $\delta G(H_1, H_2) = \{(g, H_1, H_2) : g \in G, gH_1g^{-1} \subseteq H_2\}$ and composition of $(g, H_1, H_2) \in \delta G(H_1, H_2)$ and $(h, H_2, H_3) \in \delta G(H_2, H_3)$ defined by $(h, H_2, H_3) \circ (g, H_1, H_2) = (hg, H_1, H_3)$ so that $(e, H, H) \in \delta G(H, H)$ is the identity where $H \leq G$ and $e \in G$ is the trivial element.

There is a canonical functor C, the *coset functor* from δG into $GSet$ G/H, and with each morphism $(g, H_1, H_2) \in \delta G(H_1, H_2))$ the G-map $\psi_{g^{-1}} : G/H_1 \to G/H_2 : xH_1 \to xg^{-1}H_2$. If $M : GSet \to \mathcal{D}$ is a Mackey functor, then the composition $\check{M} = M \circ C : \delta G \to \mathcal{D}$ is a bifunctor from δG into $\mathcal{D}$, which has the following properties: (G1) if $g \in H \leq G$, then $\check{M}^*(g, H, H) : \check{M}(H) \to \check{M}(H)$ and $\check{M}_*(g, H, H) : \check{M}(H) \to \check{M}(H)$ are the identity; (G2) if $H_1, H_2 < H < G$ and if $H = \cup_{i=1}^k H_1 g_i H_2$ is the double coset decomposition of H with respect to H_1 and H_2, then the composition of

$$\check{M}_*(1, H_1, H) : \check{M}(H_1) \to \check{M}(H)$$

and

$$\check{M}^*(1, H_2, H) : \check{M}(H) \to \check{M}(H_2)$$

coincides with the sum of the compositions of the various maps:

$$\check{M}^*(1, H_1 \cap g_i H_2 g_i^{-1}, H_1) : \check{M}(H_1) \to \check{M}(H_1 \cap g_i H_2 g_i^{-1})$$

and

$$\check{M}_*(g_i^{-1}, H_1 \cap g_i H_2 g_i^{-1}, H_2) : \check{M}(H_1 \cap g_i H_2 g_i^{-1}) \to \check{M}(H_2)$$

for $i = 1, \ldots, k$. The last statement follows from the fact that there is a pull-back diagram

$$
\begin{array}{ccc}
\cup_{i=1}^k G/(H_1 \cap g_i H_2 g_i^{-1}) & \xrightarrow{\Phi} & G/H_2 \\
\Psi \downarrow & & \downarrow \psi \\
G/H_1 & \xrightarrow{\phi} & G/H
\end{array}
$$

with

$$\varphi : G/H_1 \to G/H : xH_1 \to xH, \psi : G/H_2 \to G/H : xH_2 \to xH,$$

$$\Psi_{|G/(H_1 \cap g_i H_2 g_i^{-1})} G/(H_1 \cap g_i H_2 g_i^{-1}) \to G/H_1 : x(H_1 \cap g_i H_2 g_i^{-1}) \mapsto xH_1$$

$$\Phi_{|G/(H_1 \cap g_i H_2 g_i^{-1})} G/(H_1 \cap g_i H_2 g_i^{-1}) \to G/H_2 : x(G_1 \cap g_i H_2 g_i^{-1}) \mapsto xH_2$$

to which definition (9.1.1(ii)) has to be applied. In [64]), Green considered G-functors from δG into Abelian categories $\mathcal{D}$, which satisfy (**G1**) and (**G2**), and called them G-functors. It can be shown that the above construction of G-functors $\check{M} : \delta G \to \mathcal{D}$ from Mackey functors $M : GSet \to \mathcal{D}$ yields a one-one correspondence (up to isomorphism) between G- and Mackey-functors, the inverse construction being given by associating to each G-functor $\check{F} : \delta G \to \mathcal{D}$ the Mackey functors $F : GSet \to \mathcal{D}$, which maps any G-set S onto the G-invariant elements in $\oplus_{s \in S} F(G_s)$ (or in case $\mathcal{D}$ is something more abstract than a category of modules, the category theoretic equivalent object in $\mathcal{D}$) and associating to any G-map $\varphi : S \to T$ the induced morphisms

$$F_*(\varphi) : F(S) = (\oplus_{s \in S} \check{F}(G_s))^G \to F(T) = (\oplus_{t \in T} \check{F}(G_t))^G$$

and

$$F^*(\varphi) F_*(T) = (\oplus_{t \in T} \check{F}(G_t))^G \to F(S) = (\oplus_{s \in S} \check{F}(G_s))^G,$$

which come from morphisms

$$\check{F}_*(1, G_s, G_{\varphi(s)}) : \check{F}(G_s) \to \check{F}(G_{\varphi(s)})$$

and

$$\check{F}^*(1, G_s, G_{\varphi(s)}) : \check{F}(G_{\varphi(s)}) \to \check{F}(G_s)$$

$g \in G$ acts on

$$\oplus_{s \in S} x_s \in \oplus_{s \in S} \check{F}(G_s) \quad \text{via} \quad g(\oplus_{s \in S} x_s) = \oplus_{s \in S} \check{F}_*(g, G_{g^{-1}s}, G_s)(x_{g^{-1}s}).$$

Since Mackey functors are definable on arbitrary categories with finite pull-backs and sums, whereas G-functors are defined only on the various subgroup categories, we will mainly stick to Mackey functors but use this relation to identify Mackey functors with their associated G-functors, which sometimes seem more familiar to representation theorists.

In particular, for any Mackey functor $M : GSet \to \mathcal{D}$, we will identify $\check{M} : \bar{\delta}G \to \mathcal{D}$ with M and thus sometimes write M(H) instead of M(G/H).

(iii) When $\mathcal{B} = GSet$ in definition 9.1.1, $\mathcal{D} = \mathcal{A}b$ or $R - \mathcal{M}od$, then any Mackey functors $\mathcal{B} \to \mathcal{D}$ are completely determined by their behavior on the orbit category or (G) which is a full subcategory of $GSet$ of the form G/H $(H \leq G)$ (see [48, 50, 111]).

If G is a compact Lie group and $\mathcal{B}$ is the category of G-spaces of the G-homotopy type of finite G-CW complexes, the behavior of Mackey functors $\mathcal{B} \to \mathcal{A}b$ is determined by the behavior on the category $\mathcal{A}(G)$ of homogeneous spaces where $ob\mathcal{A}(G) = \{G/H : H$, a closed subgroup of $G\}$. And $mor_{\mathcal{A}(G)}(G/H, G/K)$ is the free Abelian group on the e-quivalence class of diagrams $G/H \leftarrow G/L \to G/K$ (see [218]).

(iv) Let $\mathcal{B}$ be a category with finite coproducts, final object, and finite pull-backs (and hence with initial objects and products).

Let $\overline{\mathcal{B}}$ be a category such that $ob(\overline{\mathcal{B}}) = ob\mathcal{B}$, $mor_{\mathcal{B}}(X,Y) =$ free Abelian group on the equivalence classes of diagrams $X \xleftarrow{\alpha} V \xrightarrow{\beta} Y(I)$ where $X \xleftarrow{\alpha} V \xrightarrow{\beta} Y$ is equivalent to $X \xleftarrow{\alpha'} V' \xrightarrow{\beta'} Y$ if and only if there exists an isomorphism $i : V \to V'$ such that $\alpha'i = \alpha$, $\beta'i = \beta$. Then, the Mackey functor $\overline{M} : \mathcal{B} \to \mathbb{Z} - \mathcal{M}od$ as defined in (9.1.1) is equivalent to an additive functor $\overline{M} : \overline{\mathcal{B}} \to \mathbb{Z} - \mathcal{M}od$ (see [134]). Note that the situation applies notably to $\mathcal{B} = GSet$ (G finite) or $\mathcal{A}(G)$ (G compact Lie group).

(v) **Note:** If we denote by $\overline{\mathcal{B}}_R$ (R is a commutative ring with identity) a category such that $ob(\overline{\mathcal{B}}_R) = ob(\mathcal{B})$, $mor_{\overline{\mathcal{B}}_R}(X,Y) =$ free R-module generated by equivalence classes of diagram (I), then a Mackey functor $M : \mathcal{B} \to R - \mathcal{M}od$ can be equivalently defined as an additive functor $\overline{M}, \overline{\mathcal{B}}_R \to R - \mathcal{M}od$.

Definition 9.1.2 *For any category* $\mathcal{C}$, *define a pre-ordering* $<$ *of* $ob(\mathcal{C})$ *by* $A < B$ *if* $mor_{\mathcal{C}}(A, B) \neq \emptyset$ *and an equivalence relation* $\diamond$ *on* $ob(\mathcal{C})$ *by* $A\diamond B$ *if and only if* $A < B$ *and* $B < A$.

Suppose $\mathcal{C}$ has finite coproducts $\coprod$ and products $\times$. One can easily check that $X < X \coprod Y$, $Y < X \coprod Y$, and $X < Z, Y < Z \Rightarrow X \coprod Y < Z$. Also, $X \times Y < X$, $X \times Y < Y$, and $Z < X, Z < Y \Rightarrow Z < X \times Y$. Let $\mathcal{C}$ be a category with finite coproducts and products. A class $\mathcal{R}$ of $\mathcal{C}$-objects is said to be r-closed (resp. l-closed) if $\mathcal{R} \neq \emptyset$ and if $X < Y, Y \in \mathcal{R} \Rightarrow X \in \mathcal{R}$ and $X, Y \in \mathcal{R} \Rightarrow X \coprod Y \in \mathcal{R}$ (resp. $X < Y, X \in \mathcal{R} \Rightarrow Y \in \mathcal{R}$ and $X, Y \in \mathcal{R} \Rightarrow X \times Y \in \mathcal{R}$) hold.

So, for any $\mathcal{C}$-object X, the class $\mathcal{R}_r(X) = \{Y : Y < X\}$ is r-closed and X is maximal in $\mathcal{R}_r$ with respect to $<$, while the class $\mathcal{R}_l(X) = \{Y : X < Y\}$ is l-closed and X is minimal in $\mathcal{R}_l(X)$.

Moreover, $\mathcal{R}_r(X) = \mathcal{R}_r(Y) \iff X \diamond Y \iff \mathcal{R}_l(X) = \mathcal{R}_l(Y)$.

Conversely, if $\mathcal{R}$ is r-closed in $\mathcal{C}$ and if $X \in \mathcal{R}$ is maximal/minimal in $\mathcal{R}$ with respect to $<$, then $\mathcal{R} = \mathcal{R}_r(X)/\mathcal{R}_l(X)$ since, sure enough, $\mathcal{R}_r(X) \subseteq$

$\mathcal{R}/\mathcal{R}_l(X) \subseteq \mathcal{R}$ whereas, vice versa, $Y \in \mathcal{R}$ implies $X \coprod Y \in \mathcal{R}/X \times X \in \mathcal{R}$, and thus using $X < X \coprod Y$ and the maximality of $X/X > X \times Y$ and the minimality of X- the equivalence of X and $X \coprod Y/X \times Y$, we get

$$Y \in \mathcal{R}_r(X \coprod Y) = \mathcal{R}_r(X)/Y \in \mathcal{R}_l(X \times Y) = \mathcal{R}_l(X).$$

In particular, if C contains only finitely many equivalence classes with respect to the equivalence relation $\diamondsuit$, or if, more generally, any class $\mathcal{R} \subseteq C$ contains at least one maximal/minimal object with respect to $<$, then any r-closed /l-closed class $\mathcal{R} \subseteq C$ is of the form $\mathcal{R}_r(X)/\mathcal{R}_l(X)$ for some $X \in \mathcal{R}$. In order to apply the foregoing to *GSet*, we first give the following.

Definition 9.1.3 *For any* $S \in GSet$, *define* $\mathcal{U}(S) = \{H < G : S^H \neq \emptyset\}$. *Note that* $\mathcal{U}(S)$ *is subconjugately closed, i.e., contains with any* $H \leq G$ *also any* $H' \overset{G}{\leq} H$.

Theorem 9.1.1 *(i) If S, T are GSets, then* $S < T$ *if and only if* $\mathcal{U}(S) \subseteq \mathcal{U}(T)$. *Hence* $S \diamondsuit T$ *if and only if* $\mathcal{U}(S) = \mathcal{U}(T)$, *and there are only finitely many equivalence classes with respect to* $\diamondsuit$ *in GSet.*

(ii) Any r-closed (resp. l-closed) class of G-sets is of the form $\mathcal{R}_r(S)$ *(resp. $\mathcal{R}_l(S)$) for some G-set S.*

PROOF We only have to prove the first part of (i) since (ii) and the second part of (i) follow easily from the above remarks.
Let $S < T$, and let $\varphi : S \to T$ be a G-map. Since $\varphi(S^H) \subseteq T^H$, the assumption $S^H \neq \emptyset$ implies $T^H \neq \emptyset$. Thus $\mathcal{U}(S) \subseteq \mathcal{U}(T)$. On the other hand, if $\mathcal{U}(S) \subseteq \mathcal{U}(T)$, let $S = \cup_{i=1}^n (G/H_i)$ for some $H_i \leq G$. Then, $H_i \in \mathcal{U}(S) \subseteq \mathcal{U}(T)$, and so, $G/H_i < T$. So, $S = \cup_{i=1}^n (G/H_i) < T$ by definition (9.1.2). □

Remarks 9.1.3 Let $\mathcal{U}$ be a collection of subgroups of finite group G, and let $S(\mathcal{U}) = \cup_{H \in \mathcal{U}} G/H$. Then, $\overline{\mathcal{U}}(S(\mathcal{U})) = \overline{\mathcal{U}}$ where $\overline{\mathcal{U}} = \{H \leq G : \exists H' \in \mathcal{U}$ *with* $H \leq H'\}$ is the subconjugate closure of $\mathcal{U}$ since $(\cup_{H \in \mathcal{U}} G/H)^G \neq \emptyset \iff \exists H' \in \mathcal{U}$ with $(G/H')^H \neq \emptyset$ iff there exists $H' \in \mathcal{U}$ with $H \leq H'$. In particular, $\overline{\mathcal{U}} = \mathcal{U}(S(\mathcal{U})) = \mathcal{U}$ if and only if $\mathcal{U}$ is subconjugately closed.
Now define a relation $\sim$ on sets of subgroup of G by $\mathcal{A} \sim \mathcal{B}$ if and only if $\overline{\mathcal{A}} = \overline{\mathcal{B}}$. Then $\sim$ is an equivalence relation. The following facts now emerge:

(i) For any G-set T, one has $T \diamondsuit S\mathcal{U}(T)$ since $\mathcal{U}(S(\mathcal{U}(T))) = \overline{\mathcal{U}(T)}$.

(ii) There exists a one-one correspondence between $\diamondsuit$-equivalence classes of G-sets, r-closed classes of G-sets, ℓ-closed classes of G-sets, subconjugately closed sets of subgroups, and $\sim$-equivalence classes of sets of subgroups.

Definition 9.1.4 *Let $\mathcal{B}$ be as in definition 9.1.1, $M : \mathcal{B} \to \mathcal{D}$ a Mackey functor, X a $\mathcal{B}$-object. Define $M_X : \mathcal{B} \to R - \mathcal{M}od$ by $M_X(Y) = M(X \times Y)$. Then M_X is also a Mackey functor where $M_X^*(f) = M^*(id_X \times f)$ and $M_{*X}(f) = M_*(id_X \times f)$.*

The projection map $pr = pr_2 : X \times Y \to Y$ defines natural transformations of bifunctors $\theta^X = \theta_M^X : M \to M_X$ where $\theta^X(Y) = pr^* : M(Y) \to M(X \times Y)$, and $\theta_X = \theta_X^M : M_X \to M$ where $\theta_X(Y) = pr_* : M(X \times Y) \to M(Y)$. The functor M is said to be X-projective (resp. X-injective) if θ_X (resp. θ^X) is split surjective (resp. injective) as a natural transformations of bifunctors, i.e., there exists a natural transformation $\varphi : M \to M_X$ (resp. $\varphi' : M_X \to M$) such that $\theta_X \varphi = id_M$ (resp. $\varphi' \theta^X = id_M$), in which case $\varphi : M \to M_X$ (resp. $\varphi' : M_X \to M$) is split injective (resp. surjective).

A sequence $M' \xrightarrow{\varphi'} M \xrightarrow{\varphi''} M''$ of functors in $[\mathcal{B}, \mathcal{D}]$ is said to be X-split if $M_X' \xrightarrow{\varphi_X'} M_X \xrightarrow{\varphi_X''} M_X''$ splits where

$$\varphi_X'(Y) = \varphi'(X \times Y) : M_X'(Y) = M'(X \times Y) \to M(X \times Y) = M_X(Y)$$

and φ_X'' defined accordingly, i.e., if there exist natural transformations $\psi' : M_X \to M_X'$ and $\psi'' : M_X'' \to M_X$ with $\varphi_X' \psi' + \psi'' \varphi_X'' = id_{M_X}$.

Proposition 9.1.1 (a) For any Mackey functor $M : \mathcal{B} \to \mathcal{D}$, the sequences

$$0 \longrightarrow M \xrightarrow{\theta^X} M_X$$

and

$$M_X \xrightarrow{\theta_X} M \longrightarrow 0$$

are X-split, and M_X is X-projective as well as X-injective.

(b) For any Mackey functor $M : \mathcal{B} \to \mathcal{D}$, the following statements are equivalent:

(i) M is X-projective.

(ii) M is X-injective.

(iii) M is isomorphic to a direct summand of M_X.

(iv) M is a direct summand of an X-injective Mackey functor (i.e., there exists an X-injective functor M' and a split-injective natural transformation $M \to M'$ or equivalently a split-surjective natural transformation $M' \to M$),

(v) M is a direct summand of an X-projective Mackey functor.

(vi) For any X-split sequence $N' \xrightarrow{\varphi} N'' \to 0$ of Mackey functors and any natural transformation $\psi'' : M \to N''$, there exists a natural transformation $\psi' : M \to N'$ with $\psi'' = \varphi \circ \psi'$.

(vii) For any X-split sequence $0 \to N' \overset{\psi}{\to} N''$ of Mackey functors and any natural transformation $\psi' : N' \to M$, there exists a natural transformation $\psi'' : N'' \to M$ with $\psi' = \psi'' \circ \psi$.

(c) The class $\mathcal{B}_M$ of objects $X \in \mathcal{B}$ such that M is X-projective is ℓ-closed in the sense of definition 9.1.2. Thus, if $\mathcal{B}$ contains only finitely many $\Diamond$-equivalence classes, then there exists for any Mackey functor M an object $X \in \mathcal{B}$ unique up to $\Diamond$-equivalence such that M is Y-projective if and only if $X < Y$. Such an object X will be called a *vertex* or a *vertex-object* for M. In particular, if $\mathcal{B} = GSet$, then for any Mackey functor M defined on $GSet$, there exists a unique subconjugately closed class of subgroups $\mathcal{U} = \mathcal{U}_M$ such that M is Y-projective if and only if $Y^H \neq \emptyset$ for all $H \in \mathcal{U}$. $\mathcal{U}$ will be called the class of vertex-subgroups of M.

PROOF

(a) For any product $X_1 \times \cdots \times X_n$ of n objects in $\mathcal{B}$, let us denote by $pr_i = pr_i(X_1, \ldots, X_n)$ the ith projection morphism $X_1 \times \cdots \times X_n \to X_i$. For any object Y, let us identify the morphisms $Y \overset{\varphi}{\to} X_1 \times \cdots \times X_n$ of Y into $X_1 \times \cdots \times X_n$ with the family of morphisms $(\varphi_1 \times \cdots \times \varphi_n) = (pr_1 \circ \varphi, \ldots pr_n \circ \varphi)$ of Y into the X_i: $\varphi = \varphi_1 \times \cdots \times \varphi_n = (\varphi_1, \ldots, \varphi_n)$. In particular, if $Y = Y_1 \times \cdots \times Y_m$ is itself a product, and if for any $i = 1, \ldots, n$, an index $j(i) \in \{1, \ldots, m\}$ and a morphism $\psi_i : Y_{j(i)} \to X_i$ are given, then $\psi_1 pr_{j(1)}(Y_1, \ldots, Y_m) \times \cdots \times \psi_n pr_{j(n)}(Y_1, \ldots, Y_m)$ denotes the (unique) morphism $\psi : Y_1 \times \cdots \times Y_m \to X_1 \times \cdots \times X_n$, for which all the diagrams

$$
\begin{array}{ccc}
Y_1 \times \cdots \times Y_m & \overset{\psi}{\longrightarrow} & X_1 \times \cdots \times X_n \\
\downarrow{\scriptstyle pr_{j(i)}} & & \downarrow{\scriptstyle pr_i} \\
Y_{j(i)} & \overset{\psi_i}{\longrightarrow} & X_i
\end{array}
$$

$(i = 1, \ldots, n)$ commute.
Thus the natural transformation $\theta^X = \theta_M^X : M \to M_X$ is given by the family of morphisms

$$\theta^X(Y) : M(Y) \overset{M^*(pr_2(X,Y))}{\longrightarrow} M_X(Y) = M(X \times Y), \quad \text{and}$$

the induced natural transformation

$$(\theta^X)_X : M_X \to (M_X)_X$$

is given by

$$(\theta^X)_X(Y) : M_X(Y)$$
$$= M(X \times Y) \xrightarrow{M^*(pr_1(X,X,Y) \times pr_3(X,X,Y))} M(X \times (X \times Y))$$
$$= (M_X)_X(Y), \quad \text{while}$$

the natural transformation

$$\theta^X_{M_X} : M_X \to (M_X)_X \quad \text{is given by}$$

$$\theta^X_{M_X}(Y) : M_X(Y)$$
$$= M(X \times Y) \xrightarrow{M^*(pr_2(X,X,Y) \times pr_3(X,X,Y))} M(X \times (X \times Y))$$
$$= (M_X)_X(Y),$$

and a left inverse of both, $(\theta^X_M)_X$ and $\theta^X_{M_X}$, is given by

$$\alpha : (M_X)_X \to M_X : (M_X)_X(Y) =$$
$$= M(X \times X \times Y) \xrightarrow{M^*(pr_1(X,Y) \times pr_1(X,Y) \times pr_2(X,Y))} M(X \times Y)$$
$$= M_X(Y),$$

since for both $i = 1$ and $i = 2$ one has

$$M^*(pr_1(X, Y) \times pr_1(X, Y) \times pr_2(X, Y)) \circ M^*(pr_i(X, X, Y) \times pr_3(X, X, Y))$$
$$= M^*(pr_i(X, X, Y) \times pr_3(X, X, Y)) \circ pr_1(X, Y) \times pr_1(X, Y) \times pr_2(X, Y))$$
$$= M^*(pr_1(X, Y) \times pr_2(X, Y)) = M^*(Id_{X \times Y}) = Id_{M(X \times Y)}.$$

Similarly, the same morphism $X \times Y \xrightarrow{pr_2(X,X,Y) \times pr_3(X,X,Y)} X \times X \times Y$ gives rise to a right inverse of $(\theta^M_X)_X : (M_X)_X \to M_X$ and of $(\theta^{M_X}_X)_X :$ $(M_X)_X \to M_X$. Thus $0 \to M \xrightarrow{\theta^M_X} M_X$ and $M_X \xrightarrow{\theta^X_M} M \to 0$ are X-split, and M_X is X-projective and X-injective for all Mackey functors M.

(b) The implications "(i) $\Rightarrow$ (iii)" and "(ii) $\Rightarrow$ (iii)" follow directly from the definitions, "(iii) $\Rightarrow$ (iv)" and "(iii) $\Rightarrow$ (v)" follow from (a), "(iv) $\Rightarrow$ (ii)" (resp. "(v) $\Rightarrow$ (i)") follow from the fact that for any natural transformation $\tau : M \to M'$ (resp. $\tau' : M' \to M$), and we have a commutative diagram

$$\begin{array}{ccc} M & \xrightarrow{\tau} & M' \\ \theta^X_M \downarrow & & \downarrow \theta^X_{M'} \\ M_X & \xrightarrow{\tau_X} & M'_X \end{array}$$

(resp.)

$$M'_X \xrightarrow{\tau'_X} M_X$$

$$\theta_X^{M'} \downarrow \qquad \downarrow \theta_X^M$$

$$M' \xrightarrow{\tau'} M$$

so that the existence of left inverse $\tau' : M' \to M$ of τ and left inverse $\alpha : M'_X \to M'$ of $\theta_{M'}^X$ (resp. a right inverse $\tau : M \to M'$ of τ' and a right inverse $\beta : M' \to M'_X$ of $\theta_X^{M'}$) imply the existence of the left inverse

$$\tau' \circ \alpha \circ \tau_X : M_X \to M \quad \text{of} \quad \theta_M^X : M \to M_X$$

since

$$\tau' \circ \alpha \circ \tau_X \circ \theta_M^X = \tau' \circ \alpha \circ \theta_{M'}^X \circ \tau = \tau' \circ \tau = Id_M :$$

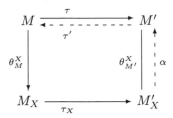

(resp. the existence of right inverse $\tau'_X \circ \beta \circ \tau : M \to M_X$ of $\theta_X^M :$ $M_X \to M$, since $\theta_X^M \circ \tau'_X \circ \beta \circ \tau = \tau' \circ \theta_X^{M'} \circ \beta \circ \tau = \tau' \circ \tau = Id_M$:

$$M'_X \xrightarrow{\tau'_X} M_X$$

$$\beta \uparrow \quad \theta_X^{M'} \downarrow \qquad \qquad \downarrow \theta_X^M$$

$$M' \xleftarrow{\tau'} M$$

Applying (vi) to the X-split sequence $M_X \xrightarrow{\theta_X} M \to 0$ and the identity $M \to M$ shows that (vi) implies (i). Similarly, applying (vii) to the X-split sequence $0 \to M \xrightarrow{\theta^X} M_X$ and the identity $M \to M$ shows (vii) implies (ii). Finally, if $N' \xrightarrow{\varphi} N'' \to 0$ is X-split, then $\varphi_X : N'_X \to N''_X$ has a right inverse $\alpha : N''_X \to N'_X$. So, if $\psi'' : M \to N''$ is a natural transformation, and if $\theta_X^M : M_X \to M$ has a right inverse $\beta : M \to M_X$, then, considering the commutative diagram

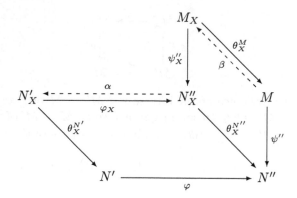

we see that $\psi' : M \to M_X \xrightarrow{\psi_X''} N_X'' \xrightarrow{\alpha} N_X' \xrightarrow{\theta_X^{N'}} N'$ is a natural transformation with $\varphi \circ \psi' = \psi''$, i.e., (i) implies (vi). And similarly, if $0 \to N' \xrightarrow{\varphi} N''$ is X-split, then $\varphi_X' : N_X' \to N_X''$ has left inverse $\alpha : N_X'' \to N_X'$. So, if $\psi' : N' \to M$ is a natural transformation and if $\theta_M^X : M \to M_X$ has a left inverse $\beta : M_X \to M$, then considering the commutative diagram

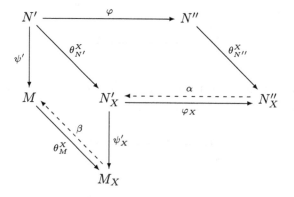

we see that $\psi'' : N'' \xrightarrow{\theta_{N''}^X} N_X'' \xrightarrow{\alpha} N_X' \xrightarrow{\psi_X'} M_X \xrightarrow{\beta} M$ is a natural transformation with $\psi'' \circ \varphi = \psi\prime$, i.e., (ii) implies (vii).

(c) If $X < Y$, then any morphism $X \to Y$ induces a commutative diagram of natural transformations

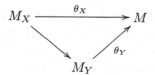

Thus, if $X \in \mathcal{B}_{M'}$, i.e., if $\theta_X : M_X \to M$ is split-surjective, so is $\theta_Y :$ $M_Y \to M$, i.e., $X \in \mathcal{C}_M$ and $X < Y$ implies $Y \in \mathcal{B}_M$.

If M is X- and Y-projective, then the right inverse $\alpha : M \to M_X$ of $\theta_X : M_X \to M$ induces a right inverse $\alpha_Y : M_Y \to (M_X)_Y = M_{X \times Y}$ of $(\theta_X)_Y : (M_X)_Y = M_{X \times Y \to M_Y}$, so M – being a direct sum in M_Y because of $Y \in \mathcal{B}_M$ – is also a direct summand in $M_{X \times Y}$ and therefore $X \times Y$-projective, i.e., $X, Y \in \mathcal{B}_M$ implies $X \times Y \in \mathcal{B}_M$.

The rest follows from definition 9.1.2.

$\square$

Remark 9.1.1 By applying proposition 9.1.1 to the Mackey functors $\mathcal{H}_G(-, B) : \mathcal{G}Set \to \mathbb{Z} - \mathcal{M}od$ as defined in 9.1.1, one easily obtains the various equivariant characterizations of relative projective (or injective) G-modules (see [39, 49, 64]) and their vertices (see [47, 64]).

9.2 Cohomology of Mackey functors

In this subsection, we will derive a more refined machinery using relative homological algebra to draw further consequences of X-projectivity, which in particular will allow us to compute $M(Y)$ as the difference kernel of the two maps $M(Y \times X) \rightrightarrows M(Y \times X \times X)$ induced by the two projections $pr_1 \times pr_i : Y \times X \times X \to Y \times X$, $(i = 2, 3)$, whenever M is an X-projective Mackey functor.

Later on we will show how X-projectivity can be established for various Mackey functor M and appropriate objects X, so that our homological results can be put in use.

Definition 9.2.1 *For any object $X \in \mathcal{B}$, one has a simplicial Amitsur-complex*

$$\bullet \xleftarrow{\quad f_0 \quad} X \underset{f_1^\alpha}{\overset{f_0^1}{\rightleftarrows}} X^2 \rightleftarrows \cdots \quad \vdots \quad X^{i+1} \quad \vdots \quad X^{i+2} \quad \cdots$$

where $f_i^j : X^{i+1} \to X^i$ eliminates the $(j+1)^{st}$ factor in X^{i+1}, i.e., $f_i^j =$

$$pr_1 \times \ldots \times pr_j \times pr_{j+2} \times \ldots \times pr_{i+1}.$$

Thus, the natural transformations $\theta^X : M \to M_X$ and $\theta_X : M_X \to M$ can be considered as the beginning (or the end) of an infinite complex, namely, the "augmented" Amitsur-complexes

$$\mathcal{A}^*(M) = \mathcal{A}_X^*(M) : 0 \to M \xrightarrow{\theta^X} M_X \xrightarrow{d^1} M_{X^2} \to \cdots \to M_{X^2} \xrightarrow{d^i} M_{X^{i+1}} \to \ldots$$

or

$$\mathcal{A}_*(M) = \mathcal{A}_*^X(M) : 0 \leftarrow M \overset{\theta^X}{\leftarrow} M_X \overset{d_1}{\leftarrow} M_{X^2} \leftarrow \ldots \leftarrow M_{X^2} \overset{d_i}{\leftarrow} M_{X^{i+1}} \leftarrow \ldots$$

with $d^i = \sum_{j=0}^{i}(-1)^j f_j^{i*}$ and $d_i = \sum_{j=0}^{i}(-1)^j f_{j*}^{i}$ where by abuse of notation, f_j^{i*} denotes the natural transformation $f_j^{i*} : M_{X^{i+1}} \to M_{X^i}$ given by $f_j^i(Y) = pr_1 \times \ldots \times pr_j \times pr_{j+2} \times \ldots \times pr_{i+1} \times pr_{i+2}$:

$$M_{X^{i+1}}(Y) = M(\underbrace{X \times \ldots \times X}_{i+1 \quad times} \times Y) \to M_{X^i}(Y) = M(\underbrace{X \times \ldots \times X}_{i \quad times}$$

$(Y \in \mathcal{B})$ and $f_{j*}^i : M_{X^i} \to M_{X^{i+1}}$ is defined correspondingly. If we cut off the augmentation

$$0 \to M \overset{\theta^X}{\longrightarrow} M_X \ldots \quad \text{and}$$

$$0 \leftarrow M \overset{\theta^X}{\longleftarrow} M_X \ldots,$$

we get the non-augmented complexes

$$\mathcal{H}^*(M) = \mathcal{H}_X^*(M) : 0 \longrightarrow M \overset{\theta^X}{\longrightarrow} M_X \overset{d^1}{\longrightarrow} M_{X^2} \overset{d^2}{\longrightarrow} \ldots$$

$$\mathcal{H}_*(M) = \mathcal{H}_*^X(M) : 0 \overset{d_0}{\longleftarrow} M_X \overset{d_1}{\longleftarrow} M_{X^2} \overset{d_2}{\longleftarrow} \ldots$$

whose homology Mackey functors will be denoted by $\mathcal{H}^i M$ and $\mathcal{H}_i M$, or more precisely by $\mathcal{H}_X^i M$ and $\mathcal{H}_i^X M$:

$$\mathcal{H}^i M = Ker d^{i+1}/Im d^i.$$

$$\mathcal{H}_i M = Ker d_i/Im d_{i+1}.$$

Note that the augmentation defines homomorphisms $M \to \mathcal{H}^0 M, \mathcal{H}_0 M \to M$.

Definition 9.2.2 *One can "splice together" the two complexes $\mathcal{A}^*(M)$ and $\mathcal{A}_*(M)$ via the augmentation to get a doubly infinite complex*

$$\mathcal{H}^*(M) = \mathcal{H}_X^*(M) : \cdots \to M_{X^2} \overset{d^{-1}=d_1}{\longrightarrow} M_X \overset{M}{\overset{\theta_X \nearrow \quad \searrow \theta^X}{\longrightarrow}} M_X \overset{d^0}{\to} M_X \to M_{X^2} \to \cdots,$$

the "Tate - Amistur-complex", whose cohomology Mackey functors will be denoted by $\hat{\mathcal{H}}^i M = Ker d^{i+1}/Im d^i$, with $i \in \mathbb{Z}$ with d^0 now denoting the map $M_X \overset{\theta^X}{\to} M \overset{\theta_X}{\to} M_X$. Note that $\hat{\mathcal{H}}^i M = \mathcal{H}^i M$ for $i \geq 1$ and that $\hat{\mathcal{H}}^i M = \mathcal{H}_{-i-1} M$ for $i \leq -2$, whereas there are exact sequences

$$0 \to \hat{\mathcal{H}}^{-1} M \to \mathcal{H}_0 M$$

and

$$\mathcal{H}^0 M \to \hat{\mathcal{H}}^0 M \to 0.$$

Definition 9.2.3 *From standard techniques in homological algebra (see, e.g., [31, 86, 140, 238]), it is obvious that any natural transformations $\theta : M \to M'$ of Mackey functors induces natural transformations between the associated complexes and thus between the (co-)homology functors:*

$$\mathcal{H}^i \theta : \mathcal{H}^i M \to \mathcal{H}^i M',$$

$$\mathcal{H}_i \theta : \mathcal{H}_i M \to \mathcal{H}_i M',$$

$$\hat{\mathcal{H}}^i \theta : \mathcal{H}^i M \to \hat{\mathcal{H}}^i M'.$$

Moreover if we define a sequence $0 \to M' \xrightarrow{\theta'} M \xrightarrow{\theta''} M'' \to 0$ of Mackey functors to be X-exact if for any $Y < X$ the sequence

$$0 \to M'(Y) \xrightarrow{\theta'(Y)} M(Y) \xrightarrow{\theta''(Y)} M''(Y) \to 0$$

is exact, then for any such X-exact sequence, the associated sequences of complexes

$$0 \to \mathcal{H}^* M' \to \mathcal{H}^* M \to \mathcal{H}^* M'' \to 0$$

$$0 \to \mathcal{H}_* M' \to \mathcal{H}_* M \to \mathcal{H}_* M'' \to 0$$

and

$$0 \to \mathcal{H} M' \to \mathcal{H} M \to \mathcal{H} M'' \to 0$$

are exact and thus give rise to a long exact sequence

$$0 \to \mathcal{H}^0 M' \to \mathcal{H}^0 M \to \mathcal{H}^0 M'' \to \mathcal{H}^1 M' \to \dots \mathcal{H}^i M' \to \mathcal{H}^i M \to \mathcal{H}^i M'' \to$$

$$\to \mathcal{H}^{i+1} M'' \to \dots \to \mathcal{H}_i M \to \mathcal{H}_i M'' \to \mathcal{H}_{i-1} M \to \dots \to \mathcal{H}_1 M'' \to$$

$$\mathcal{H}_0 M' \to \mathcal{H}_0 M \to \mathcal{H}_0 M'' \to 0 \quad and \quad \dots \to \hat{\mathcal{H}}^i M \to \hat{\mathcal{H}}^i M'' \to \hat{\mathcal{H}}^{i+1} M' \to \dots$$

Theorem 9.2.1 *(a) The complexes $\mathcal{A}^* M$, $\mathcal{A}_* M$, and $\mathcal{H} M$ are X-split.*

(b) If M is X-projective, then $\mathcal{H}^i M = \mathcal{H}_i M = 0$ for $i \geq 1$, $\hat{\mathcal{H}}^i M = 0$ for $i \in \mathbb{Z}$, and $\mathcal{H}^0 M = \mathcal{H}_0 M = M$.

PROOF

(a) The complex $(\mathcal{A}^* M)_X = \mathcal{A}^*(M)_X$ looks basically like $\mathcal{A}^* M$, except for a dimension shift by -1 and the elimination of the maps $f_i^{i*} : M_{X^i} \to M_{X^{i+1}}$ in the alternating sum defining d_X^{i-1}. We claim that that map

$$(-1)^{i+1} h^{i*} : M_{X^{i+1}} \to M_{X^i}$$

defined by

$$h^i = pr_1 \times pr_2 \times \dots \times pr_i \times pr_i : X^i \to X^{i+1}$$

is a splitting of $(\mathcal{A}^* M)_X$, i.e., it satisfies

$$d_X^{i-1}\big((-1)^{i+1}h^{i*}\big) + \big((-1)^{i+2}h^{i+1*}\big)d_X^i = Id_{X^{i+1}}.$$

This follows from

$$d^{i-1}h^{i*} = \Big(\sum_{j=0}^{i-1}(-1)^j f_j^{i*}\Big)h^{i*} = \sum_{j=0}^{i-1}(-1)^j\big(h^i f_j^i\big)^*,$$

$$h^{i+1*}d_X^i = h^{i+1*}\Big(\sum_{j=0}^{i-1}(-1)^j f_j^{i+1*}\Big) = \sum_{j=0}^{i-1}(-1)^j\big(f_j^{i+1}h^{i+1}\big)^*,$$

$$h^i f_j^i = f_j^{i+1}h^{i+1} = pr_1 \times \ldots \times pr_j \times pr_{j+2} \times \ldots \times pr_{i+1} : X^{i+1} \to X^{i+1}$$

for $j = 0,\ldots,i-1$ and

$$f_i^{i+1}h^{i+1} = Id : X^{i+1} \to X^{i+1}.$$

The same argument applies to $(\mathcal{A}_* M)_X$ and – by splicing together these two splittings – to $(\mathcal{H}M)_X$.

(b) Without loss of generality, we assume $M = M_X$, in which case the statements follows from (a) and $\mathcal{A}^*(M_X) = (\mathcal{A}^* M)_X$, $\mathcal{A}_*(M_X) = (\mathcal{A}_* M)_X$, and $\mathcal{H}(M_X) = (\mathcal{H}M)_X$.

$\square$

Corollary 9.2.1 For any $X < Y$ we have $\mathcal{H}_X^i M(Y) = \mathcal{H}_i^X M(Y) = 0$ for $i \geq 1$, $\mathcal{H}_X^0 M(Y) = \mathcal{H}_0^X M(Y) = M(Y)$ and $\hat{\mathcal{H}}_X^i M(Y) = 0$ for $i \in \mathbb{Z}$.

PROOF   If $\varphi : Y \to X$ is a morphism, then $Id_Y \times \varphi : Y \to Y \times X$ is a right inverse of $pr_1 : Y \times X \to Y$, and so, for any Mackey functor N, the group $N(Y)$ is a direct summand of $N(Y \times X)$. So, without loss of generality, we may replace Y by $Y \times X$. But then

$$\mathcal{H}^i M(Y \times X) \cong \mathcal{H}^i M_X(Y) = 0$$

and

$$\mathcal{H}_i M(Y \times X) \cong \mathcal{H}_i M_X(Y) = 0$$

for $i \geq 1$,

$$\mathcal{H}^0 M(Y \times X) = \mathcal{H}^0 M_X(Y) = M_X(Y) = M(Y \times X),$$

$$\mathcal{H}_0 M(Y \times X) = \mathcal{H}_0 M_X(Y) = M_X(Y) = M(Y \times X)$$

and

$$\hat{\mathcal{H}}^i M(Y \times X) = \hat{\mathcal{H}}^i M_X(Y) = 0 \quad \text{for } i \in \mathbb{Z}.$$

$\square$

Corollary 9.2.2 If M is X-projective, then M is canonically isomorphic to the difference kernel of the natural transformations $M_X \rightrightarrows M_{X \times X}$ and to the difference cokernel of the two natural transformations $M_{X \times X} \to M_X$ associated with the two projections $X \times X \to X$.

Remarks 9.2.1 (i) If $R^X M$ denotes the cokernel of $\theta^X : M \to M_X$ and $R_X M$ denotes the kernel of $\theta_X : M_X \to M$, then the sequences

$$0 \to M \xrightarrow{\theta^X} M_X \to R^X M \to 0 \quad \text{and} \quad 0 \leftarrow M \xleftarrow{\theta_X} M_X \leftarrow R_X M \leftarrow 0$$

are X-split and therefore exact at any $Y < X$. So, the long exact sequences associated with them together with the triviality of $\mathcal{H}^i M_X$ give rise to canonical isomorphisms $\hat{\mathcal{H}}^{i-1}(R^X M) \simeq \hat{\mathcal{H}}^i M \simeq \hat{\mathcal{H}}^i(R_X M)$. Thus, to deal with the Tate - Amitsur cohomology of Mackey functors one uses the technique of dimension shifting as usual.

(ii) Using standard techniques of homological algebra (see [31, 86, 140, 238]) one can show that for any X-split complexes

$$0 \to M \to M^0 \xrightarrow{d^1} M^1 \xrightarrow{d^2} \ldots$$

or

$$0 \leftarrow M \leftarrow M_0 \xleftarrow{d_1} M_1 \xleftarrow{d_2} \ldots$$

with $M^0, M^1, \ldots$ and $M_0, M_1, \ldots$ being X-projective, one has canonical isomorphisms $\mathcal{H}^i M \cong Ker d^{i+1}/Im d^i$ and $\mathcal{H}_i M \cong Ker d_i/Im d_{i+1}$ (with $d^0 : 0 \to M^0$ and $d_0 : M^0 \to 0$). Thus our cohomology functors can also be defined by certain universal properties as derived functors (see [86, 140, 238]).

Theorem 9.2.2 *If $Y < X$ and if $\varphi : Y \to X$ is some morphism from Y into X, then the induced natural transformations*

$$\mathcal{H}_X^i M \xrightarrow{\varphi^*} \mathcal{H}_Y^i M$$

and

$$\mathcal{H}_i^X M \xrightarrow{\psi_*} \mathcal{H}_i^Y M$$

do not depend on the chosen morphism φ, and any morphism $\varphi : X \to X$ induces the identity on $\mathcal{H}_i^X M$ and $\mathcal{H}_X^i M$.

PROOF Let $x \in M(X^i \times Z)$ be in the kernel of $d_i = \sum\limits_{j=0}^{i} (-1)^j f_{j*}^i$: $M_{X^i}(Z) \to M_{X^{i+1}}(Z)$, then $\varphi, \psi : Y \to X$ be two $\mathcal{B}$-morphisms, let h_j^i : $Y^i \to X^{i+1}$ denote the morphism $pr_1 \times \varphi pr_2 \times \ldots \times \varphi pr_j \times \psi pr_j \times \ldots \times \psi pr_i$ $(j = 1, \ldots, i)$ and let $g_j^i : Y^i \to X^i$ denote the morphism

$$\varphi pr_1 \times \ldots \times \varphi pr_j \times \psi pr_{j+1} \times \ldots \times \psi pr_i \quad (j = 1, \ldots, i).$$

270   *A.O. Kuku*

Thus we have commutative diagrams:

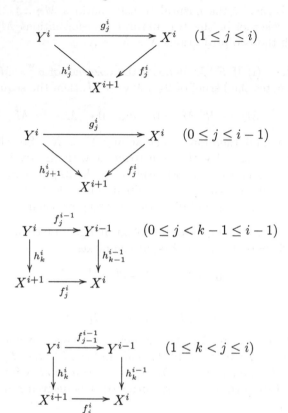

and

$$
\begin{array}{ccc}
Y^i & \xrightarrow{f_{j-1}^{i-1}} & Y^{i-1} \\
\downarrow{h_k^i} & & \downarrow{h_k^{i-1}} \\
X^{i+1} & \xrightarrow{f_j^i} & X^i
\end{array}
\qquad (1 \le k < j \le i)
$$

Now consider the double sum:

$$
\sum_{k=1}^{i}(-1)^k h_k^{i*}\Big(\sum_{j=0}^{i}(-1)^j f_j^{i*}(x)\Big).
$$

On the one hand, this sum is zero since $x \in Ker\big(\sum_{j=0}^{i}(-1)^j f_j^{i*}\big)$. On the other hand, we can compute it by splitting it into the four partial sums $k = j$, $k = j + 1$, $k < 1$, and $k > j + 1$. Using the commutativity of the above diagrams, our double sum becomes:

$$
\sum_{k=j=1}^{i}(-1)^{k+j}g_j^{i*}(x) + \sum_{\substack{j=0\\k=j+1}}^{i}(-1)^{k+j}g_j^{i*}(x) + \sum_{j=2}^{i}\sum_{k=1}^{j-1}(-1)^{k+j}f_{j-1}^{i-1*}(h_k^{i-1*}(x)) +
$$

$$
+\sum_{j=0}^{i-2}\sum_{k=j+2}^{i}(-1)^{k+j}f_j^{i-1*}(h_{k-1}^{i-1*}(x)) = \big(g_1^{i*}(x) + \cdots + g_i^{i*}(x)\big)
$$

$$-(g_0^{i*}(x) + \cdots + g_{i-1}^{i*}(x)) + \sum_{j=1}^{i-1}(-1)^{j+1} f_j^{i-1*}\left(\sum_{k=1}^{j}(-1)^k h_k^{i-1*}(x)\right)$$

$$+\sum_{j=0}^{i-2}(-1)^j f_j^{i-1*}\left(\sum_{k=j+1}^{i-1}(-1)^{k+1} h_k^{i-1*}(x)\right) =$$

$$= g_i^{i*}(x) - g_0^{i*}(x) + \sum_{j=0}^{i-1}(-1)^j f_j^{i-1*}\left(\sum_{k=1}^{i-1}(-1)^k h_k^{i-1*}(x)\right).$$

So, $g_i^{i*}(x) = (\varphi \times \ldots \times \varphi)^*(x)$ differs from $g_0^{i*}(x) = (\psi \times \ldots \times \psi)^*(x)$ by an element in the image of $d_{i-1} = \sum_{j=0}^{i-1}(-1)^j f_j^{i-1*}$, only, so both define the same element in $\mathcal{H}^{i-1}M(Z)$.

Similar reasoning shows that $(\varphi \times \ldots \times \varphi)_*$ and $(\psi \times \ldots \times \psi)_*(x)$ define the same map from $\mathcal{H}_{i-1}^Y M(Z)$ into $\mathcal{H}_{i-1}^X M(Z)$.

Note that, in general, $\varphi : Y \to X$ does not define a morphism $\varphi_* : \mathcal{H}_Y^* M \to \mathcal{H}_X^* M$, nor a morphism $\varphi^* : \mathcal{H}_*^X M \to \mathcal{H}_*^X M$. $\quad\Box$

Examples 9.2.1 (i) Let G be a finite group and B a $\mathbb{Z}G$-module, $\overline{B} = \mathcal{H}_G(-, B)$ the Mackey functor associated with B as defined in, e.g., 9.1.1(i). The (co-)homology groups $\mathcal{H}_X^i \overline{B}(*)$, $\mathcal{H}_i^X \overline{B}(*)$, $\hat{\mathcal{H}}_X^i \overline{B}(*)$ coincide with the standard (co-)homology groups $\mathcal{H}^i(G, B)$, $\mathcal{H}_i(G, B)$, and $\hat{\mathcal{H}}^i(G, B)$ of G with coefficients in B, when $X = G/e$.

More generally, one has, for $X = G/e$,

$$\mathcal{H}_X^i \overline{B}(G/H) \simeq \mathcal{H}^i(H, B),$$

$$\mathcal{H}_i^X \overline{B}(G/H) \simeq \mathcal{H}_i(H, B),$$

and

$$\hat{\mathcal{H}}_X^i \overline{B}(G/H) \simeq \hat{\mathcal{H}}^i(H, B)$$

for any $H \leq G$ and for $H' \leq H \leq G$ the maps between $\mathcal{H}^i \overline{B}(G/H)$, $\mathcal{H}_i \overline{B}(G/H)$, and $\hat{\mathcal{H}}^i \overline{B}(G/H)$ on the one side and $\mathcal{H}^i \overline{B}(G/H')$, $\mathcal{H}_i \overline{B}(G/H')$, and $\hat{\mathcal{H}}^i \overline{B}(G/H')$ on the other side induced by $G/H' \to G/H : xH' \mapsto xH$ to coincide with the standard restriction and corestriction maps between $\mathcal{H}^i(H, B)$, $\mathcal{H}_i(H, B)$, and $\hat{\mathcal{H}}^i(H, B)$ on the one side and $\mathcal{H}^i(H', B)$, $\mathcal{H}_i(H', B)$, and $\hat{\mathcal{H}}^i(H', B)$ on the other. Also, if $H \leq G$, with quotient group $G/H = \overline{G}$, then

$$\mathcal{H}_i^{G/H} \overline{B}(*) \simeq \mathcal{H}_i(\overline{G}, B); \quad \mathcal{H}_{G/H}^i \overline{B}(*) \simeq \mathcal{H}^i(\overline{G}, B^H),$$

$$\hat{\mathcal{H}}_{G/H}^i \overline{B}(*) \simeq \hat{\mathcal{H}}^i(G, B^H)$$

with $B^H = \{b \in B : gb = b \quad \text{for all} \quad g \in H\}$ considered as a G-module in the natural way.

(ii) Let $\mathcal{C} = GSet$ and $S = \cup_{H \in \mathcal{U}} G/H$ where $\mathcal{U}$ is the collection of subgroups of G. If $M : GSet \to \mathbb{Z} - \mathcal{M}od$ is a Mackey functor, then by the identification $M(G/H) \simeq M(H)$, discussed in remark 9.1.1, the map $M(*) \to M(S)$ associated with the unique map $S \to *$ coincides with $M(G) \to \coprod M(H)$. Now, 9.2.2 says that for M being S-projective, the sequence:

$$0 \to M(G) \simeq M(*) \to M(S) \simeq \prod_{H \in \mathcal{U}} M(H) \to M(S \times S)$$

$$\simeq \coprod_{\substack{HgH \subset G \\ H \in \mathcal{U}}} M(H \cap gH'g^{-1})$$

is exact, and so, $M(G)$ is isomorphic to the difference kernel or equalizer of the two maps $M(S) \to M(S \times S)$ induced by the projection maps $S \times S \to S$. This difference kernel in turn can be shown to be isomorphic to $\varprojlim_{H \in \overline{\mathcal{U}}} M(H) - \overline{\mathcal{U}}$ – the subconjugate closure of $\mathcal{U}$ – where by definition $\varprojlim_{\mathcal{U}} M(H)$ is the subgroup of all $(x_H) \in \coprod_{H \in \mathcal{U}} M(H)$ such that for any $H', H \in \mathcal{U}$ and any $g \in G$ with $gH'g^{-1} \subseteq H$, we have $M(\varphi)(x_H) = x_{H'}$ where $\varphi : H' \to H$, $h \mapsto ghg^{-1}$.

In other words, for any S-projective Mackey functor M, it is possible to compute $M(*) = M(G)$ in terms of the $M(H)$ with $H \in \mathcal{U}$.

9.3   Green functors, modules, algebras, and induction theorems

Throughout this section, $\mathcal{B}$ is a category with final object, finite pull-backs, and finite coproducts, and $\mathcal{D}$ is R-$\mathcal{M}od$ for some ring R with identity.

Definition 9.3.1 *Let* $L, M, N : \mathcal{B} \to \mathcal{D}$ *be three Mackey functors. A pairing* $\Gamma = <,>: L \times M \to N$ *is a family of R-bilinear maps* $<>_X = \Gamma_X : L(X) \times M(X) \to N(X)$ *(where* $X \in ob\mathcal{B}$*) such that for any $\mathcal{B}$-morphism* $f : X \to Y$, *the following hold:*

(i) $f^*(< a, b >_Y) = < f^*(a), f^*(b) >_X$ *(a $\in L(Y)$, $b \in M(Y)$).*

(ii) $f_*(< f^*(a), c >_X) = < a, f_*(c) >_Y$ *(a $\in L(Y)$, $c \in M(X)$).*

(iii) $f_*(< d, f^*(b) >_X) = < f_*(d), b >_Y$ *(d $\in L(X)$, $b \in M(Y)$).*

Here, f_* (resp. f^*) means the covariant (resp. contravariant) part of the relevant Mackey functor applied to f.
Note that (i) and (iii) above may be regarded as an axiomatization of the Frobenius reciprocity law (see [38, 39]) .

9.3.1 The following statements are more or less direct consequences of the above definition:

(a) Let $<,>: L \times M \to N$ be a pairing of Mackey functors $L, M, N : \mathcal{B} \to \mathcal{D}$, and let $f : X \to Y$ be a $\mathcal{B}$-morphism. For any Mackey functor $\mathcal{Q} : \mathcal{C} \to \mathcal{D}$, let $I_f \mathcal{Q} \subseteq \mathcal{Q}(Y)$ denote the image of $\mathcal{Q}(X) \to^{f_*} \mathcal{Q}(Y)$, and let $K_f \mathcal{Q} \subseteq \mathcal{Q}(Y)$ denote the kernel of $\mathcal{Q}(Y) \to^{f^*} \mathcal{Q}(X)$. Then:

(i) $< K_f L, M >_Y, \ < L, K_f M >_Y \subseteq K_f N,$

(ii) $< I_f L, M >_Y, \ < L, I_f M >_Y \subseteq I_f N,$

(iii) $< K_f L, I_f M >_Y, \ < I_f L, K_f M >_Y = 0.$

(b) Any pairing $L \times M \to N$ of Mackey functors induces a pairing $L \times M_X \to N_X$ ($X \in Ob\mathcal{B}$), and for any $\mathcal{B}$-morphism $g : Z \to X$ we have commutative diagrams:

$$
\begin{array}{ccc}
L \times M_X & \longrightarrow & N_X \\
\downarrow{\scriptstyle id \times g_*} & & \downarrow{\scriptstyle g^*} \\
L \times M_Z & \longrightarrow & N_Z
\end{array}
\qquad \text{and} \qquad
\begin{array}{ccc}
L \times M_Z & \longrightarrow & N_Z \\
\downarrow{\scriptstyle id \times g_*} & & \downarrow{\scriptstyle g_*} \\
L \times M_X & \longrightarrow & M_X
\end{array}
$$

In particular, any pairing $L \times M \to N$ of Mackey functors induces pairings $L \times R^X M \to R^X N$ and $L \times R_X M \to R_X N$, and thus, if $R^j M = R^j_X M$ is defined inductively by $R^0 M = 0$, $R^{j+1} M = R_X(R^j M)$ for $j \geq$), and $R^{j-1} M = R^X(R^j M)$ for ($j \leq 0$), it induces pairings $R^i L \times R^j M \to R^i(R^j N)$, $i, j \in \mathbb{Z}$.

(c) Any pairing $L \times M \to N$ of Mackey functors induces pairings $\mathcal{H}^0 L \times \mathcal{H}^0 M \to \mathcal{H}^0 N$ and $\hat{\mathcal{H}}^0 L \times \hat{\mathcal{H}}^0 M \to \hat{\mathcal{H}}^0 N$ (with $\mathcal{H}^0 = \mathcal{H}^0_X$ and $\hat{\mathcal{H}}^0 = \hat{\mathcal{H}}^0_X$ for some $X \in ob\mathcal{C}$). The commutativity of the diagram

$$
\begin{array}{ccc}
L(X \times Y) \times M(X \times Y) & \xrightarrow{<,>_{X \times Y}} & N(X \times Y) \\
\downarrow & & \downarrow{\scriptstyle N^*(pr_i \times pr_3)} \\
{\scriptstyle L^*(pr_i \times pr_3) \times M^*(pr_i \times pr_3)} & & \\
L(X \times X \times Y) \times M(X \times X \times Y) & \xrightarrow[<,>_{X \times X \times Y}]{} & N(X \times X \times Y)
\end{array}
$$

($i = 1, 2$) implies that $<,>_{X \times Y}$ maps $\mathcal{H}^0 L \times \mathcal{H}^0 M$ into $\mathcal{H}^0 N$, whereas for $a \in L(X \times Y)$ and $b \in H^0 M(Y) \subseteq M(X \times Y)$, the Mackey axiom, applied to the pull-back diagram

$$
\begin{array}{ccc}
X \times X \times Y & \xrightarrow{g = pr_2 \times pr_3} & X \times Y \\
\downarrow{\scriptstyle f = pr_1 \times pr_3} & & \downarrow{\scriptstyle pr_2} \\
X \times Y & \xrightarrow[pr_2]{} & Y
\end{array}
$$

together with the Frobenius reciprocity and $g^*(b) = f^*(b)$ imply

$$< pr_2(< pr_{2*}(a)), b > = < g_*(f^*(a)), b >$$
$$= g_* < f^*(a), g^*(b) >= g_* < f^*(a), f^*(b) >$$
$$= g_* f^* < a, b >= pr_2^* pr_{2*} < a, b >,$$

so the above pairing $L \times M \to N$ induces a pairing $\mathcal{H}^0 L \times \mathcal{H}^0 M \to \mathcal{H}^0 N$ which – by the same argument applied to $a \in M(X \times Y)$ and $b \in \mathcal{H}^0 L(Y)$ – even induces a pairing $\hat{\mathcal{H}}^0 L \times \hat{\mathcal{H}}^0 M \to \hat{\mathcal{H}}^0 N$.

(d) In consequence, any pairing $L \times M \to N$ of Mackey functors induces pairings

$$\hat{\mathcal{H}}^i L \times \mathcal{H}^j M = \hat{\mathcal{H}}^0 (R^{-i} L) \times \hat{\mathcal{H}}^0 (R^j M), \hat{\mathcal{H}}^0 (R^{-i} (R^{-j} N)) = \hat{\mathcal{H}}^i (R^j N)$$
$$= \hat{\mathcal{H}}^{i+j}(N).$$

A perhaps more direct way to construct a pairing $\mathcal{H}^i L \times \mathcal{H}^j M \to \mathcal{H}^{i+j} N$, $i, j \geq 0$ (which in case $i, j > 0$ coincides with the pairing $\hat{\mathcal{H}}^i L \times \hat{\mathcal{H}}^j M \to \mathcal{H}^{i+j} N$, defined above, is the following: to avoid notational confusion let us construct a pairing $\mathcal{H}^i_X L \times \mathcal{H}^j_X M \to \mathcal{H}^{i+j}_X N$ since the $\diamond$-equivalence of X and $X \times X$ defines a canonical isomorphism between $\mathcal{H}^{i+j}_X N$ and $\mathcal{H}^{i+j}_{X \times X} N$.

So, let $\varphi_i : (X \times Y)^{i+j} \to X^i$ denote the projection of $(X \times Y)^{i+j}$ onto the first i X-factors, and let $\psi_i : (X \times Y)^{i+j} \to Y^i$ denote the projection onto the last j Y-factors. Then our pairing $<, >: L \times N \to N$ together with the maps $\varphi_i : L_{X^i} \to L^{i+j}_{X \times Y}$ and $\psi_i : M_{Y^i} \to M^{i+j}_{X \times Y}$ induces a pairing $L_{X^i} \times M_{Y^i} \to N^{i+j}_{X \times Y}$, which maps

$$\mathrm{Ker}(L_{X^i} \to L_{X^{i+1}}) \times Ker(M_{Y^i} \to M_{Y^{i+1}}) \quad \text{into} \quad Ker(N^{i+j}_{X \times Y} \to N^{i+j+1}_{X \times Y})$$

and

$$Im(L_{X^{i-1}} \to L_{X^i}) \times Ker(M_{Y^j} \to M_{Y^{j+1}})$$

as well as

$$\mathrm{Ker}(L_{X^i} \to L_{X^{i+1}}) \times Im(M_{Y^{j-1}} \to M_{Y^j}) \quad \text{into} \quad Im(N^{i+j-1}_{X \times Y} \to N^{i+j}_{X \times Y}).$$

Thus, induces indeed a pairing

$$\mathcal{H}^i_X L \times \mathcal{H}^j_X M \to \mathcal{H}^{i+j}_{X \times Y} N.$$

Definition 9.3.2 *A Green functor $\mathcal{G} : \mathcal{B} \to R\text{-}\mathcal{M}od$ is a Mackey functor together with a pairing $\mathcal{G} \times \mathcal{G} \to \mathcal{G}$ such that for any $\mathcal{B}$-object X, the R-bilinear map $\mathcal{G}(X) \times \mathcal{G}(X) \to \mathcal{G}(X)$ makes $\mathcal{G}(X)$ into an R-algebra with a unit $1 \in \mathcal{G}(X)$ such that for any morphism $f : X \to Y$, one has $f^*(1_{\mathcal{G}(Y)}) = 1_{\mathcal{G}(X)}$.*

A left (resp. right) $\mathcal{G}$-module is a Mackey functor $M : \mathcal{B} \to R-\mathcal{M}od$ together with a pairing $\mathcal{G} \times M \to M$ (resp. $M \times \mathcal{G} \to M$) such that for any $\mathcal{B}$-object X, $M(X)$ becomes a left (resp. right) unitary $\mathcal{G}(X)$-module. We shall refer to left $\mathcal{G}$-modules simply as $\mathcal{G}$-modules and call M a Green module over $\mathcal{G}$.

Let $\mathcal{G}, \mathcal{G}' : \mathcal{B} \to R\text{-}\mathcal{M}od$ be Green functors. A (Green) homomorphism from $\mathcal{G}$ to $\mathcal{G}'$ is a natural transformation $\theta : \mathcal{G} \to \mathcal{G}'$ such that, for any $\mathcal{B}$-object X, the map $\theta_X : \mathcal{G}(X) \to \mathcal{G}'(X)$ can be used to define a $\mathcal{G}$-module structure on $\mathcal{G}'$ (i.e., $\hat{\theta} : \mathcal{G} \times \mathcal{G}' \to \mathcal{G}' : \hat{\theta}_X : \mathcal{G}(X) \times \mathcal{G}'(X) \to \mathcal{G}'(X)$:) $(x, x') \mapsto \theta_X(x)\dot{x}'$ such that the multiplication becomes $\mathcal{G}$-bilinear.

On the other hand, given a $\mathcal{G}$-module structure $\rho : \mathcal{G} \times \mathcal{G}' \to \mathcal{G}'$ such that the multiplication $\mathcal{G}' \times \mathcal{G}' \to \mathcal{G}'$ is $\mathcal{G}$-bilinear, then the map

$$\bar{\rho} : \mathcal{G} \to \mathcal{G}' : \bar{\rho}_X : \mathcal{G}(X) \to \mathcal{G}'(X) : x \mapsto \rho(x, 1_{\mathcal{G}'(X)}) = x \cdot 1_{\mathcal{G}'(X)}$$

is a (Green homomorphism from $\mathcal{G}$ to $\mathcal{G}'$. Moreover, $(\theta)^- = \theta$, $(\rho)\hat{} = \rho$, and so, we have a one-one correspondence between bilinear $\mathcal{G}$-module structures on $\mathcal{G}'$ and homomorphisms $\mathcal{G} \to \mathcal{G}'$. A Green functor $\mathcal{G}'$ together with a homomorphism $\theta : \mathcal{G} \to \mathcal{G}'$ is called a Green algebra over $\mathcal{G}$ or just a $\mathcal{G}$-algebra.

Examples and Remarks 9.3.1 (i) We mentioned in 9.1.1(iii) that for $\mathcal{C}$ any exact category, G finite, profinite, or compact Lie group, then for all $n \geq 0$, $K_n^G(-, \mathcal{C}) : GSet \to \mathcal{A}b$ is a Mackey functor. If $\mathcal{C}$ possesses a further associative composition $'\circ'$ such that $\mathcal{C}$ is distributive with respect to $'\oplus'$ and $'\circ'$, we also have that $K_0^G(-, \mathcal{C}) : GSet \to \underline{Ab}$ is a Green functor (see chapters 10 - 12). For finite and profinite group actions, we shall study in chapters 10 and 11 the situation when $\mathcal{C} = \mathcal{P}(R)$, R any commutative ring, $\mathcal{C} = M(R)$, R a commutative Noetherian ring. For G, a compact Lie group, we shall focus attention on $\mathcal{C} = \mathcal{P}(\mathbb{C})$, the category of finite-dimensional complex vector spaces.

(ii) Let G be a discrete group, G_1 a finite group, and $\varphi : G \to G_1$ a group epimorphism. Then φ induces a functor $\hat{\varphi} : G_1 Set \to GSet$ given by $S \mapsto S_{|G}$ where $S_{|G}$ is obtained from S by restriction of the action of G_1 on S to G via φ, i.e., for $s \in S$, $g \in G$, $gs = \varphi(g)s$. If $\alpha : S \to T$ is a G_1-map, then α is also a G-map $\alpha_{|G} = \varphi(\alpha) : S_{|G} \to T_{|G}$. Hence we have a canonical functor $\underline{S_{|G}} \xrightarrow{\delta} \underline{S}$ given by $S \to S$; $(g, s) \mapsto (\varphi(g), s)$. If $\mathcal{C}$ is an exact category, then δ induces an exact functor $[\underline{S}, \mathcal{C}] \to [\underline{S_{|G}}, \mathcal{C}]$, which associates to any $\zeta \in [\underline{S}, \mathcal{C}]$ a G-equivariant $\mathcal{C}$-bundle $\zeta_{|G}$ over $S_{|G}$. Note that $\zeta_{|G}$ has the same fibers as ζ with G-action defined by restriction of G_1-action to G via φ. Hence, we have a homomorphism $\delta : K_n^G(S, \mathcal{C}) \to K_n^G(S_{|G}, \mathcal{C})$ of K-groups. In particular, if $S = G_1/G_1$, then $S_{|G} = G/G$, and we have a homomorphism $K_n^G(G_1/G_1, \mathcal{C}) \to K_n^G(G/G, \mathcal{C})$.

(iii) Recall from (1.4.5) that if G is a finite group, S a G-set, $\underline{S}$ the translation category of S, then the Swan group $Sw^f((\underline{S})) := K_0[\underline{S}, \mathcal{F}((\mathbb{Z})]'$ has a ring structure as follows: for $M, N \in [\underline{S}, \mathcal{F}((\mathbb{Z})]$ $M \otimes_{\mathbb{Z}} N \in [\underline{S}, \mathcal{F}((\mathbb{Z})]$, and so, $\otimes_{\mathbb{Z}}$ induces a pairing $Sw^f((S)) \times Sw^f((S)) \to Sw^f((S))$, which turns $Sw^f((S))$ into a commutative ring with identity. One can easily check that $\overline{Sw}^f : GSet \to \mathcal{A}b$ is a Green functor. If $\varphi : G \to G_1$ is a group epimorphism where G_1 is finite and G is discrete, then

$$Sw^f : GSet \to \mathcal{A}b, \quad S \mapsto Sw^f(S_{|\overline{G}})$$

is a Green functor where $S_{|G}$ is a finite G-set explained in (ii) above. This set up will be used in chapter 14 in the context of equivariant homology theories to extend induction results to discrete groups.

(iv) One can prove that if $\mathcal{G} : GSet \to \mathbb{Z}\text{-}\mathcal{M}od$ is a Green functor such that $\mathcal{G}(S) \to \mathcal{G}(G/G)$ is surjective, and if $\tau : M \to N$ is a morphism of G-modules such that $\tau_S : M_S \to N_S$ is an isomorphism, then $\tau : M \to N$ is an isomorphism.

Lemma 9.3.1 *Let* $\mathcal{G} : \mathcal{B} \to \mathbb{Z}\text{-}\mathcal{M}od$ *be a Green functor, M a $\mathcal{G}$-module, $f : X \to Y$ a $\mathcal{B}$-morphism. Then in the notation of 9.3.1 we have:*

(i) $K_f M$ *and* $I_f M$ *are* $\mathcal{G}(Y)$*-submodules of* $M(Y)$*. In particular,* $I_f(\mathcal{G}) = f_*(\mathcal{G}(X))$ *is a two-sided ideal in* $\mathcal{G}(Y)$*.*

(ii) *If* $f_* : \mathcal{G}(X) \to \mathcal{G}(Y)$ *is surjective, then* $M(X) \to M(Y)$ *is split surjective.*

(iii) $K_f \mathcal{G} \cdot M(Y) \subseteq K_f M$, $I_f \mathcal{G} \cdot M(Y) \subseteq I_f M$, $K_f \mathcal{G} \cdot I_f M = I_f \mathcal{G} \cdot K_f M = 0$.

(iv) *If* $g : Z \to Y$ *is any other* $\mathcal{B}$*-map such that* $n \cdot 1_{\mathcal{G}(Y)} \in K_f \mathcal{G} + I_g \mathcal{G}$ *for some natural number n, then* $K_f M + I_g M \supseteq n \cdot M(Y)$, *and* $n \cdot (K_g M \cap I_f M) = 0$.

PROOF

(i) Follows directly from 9.3.1 (a) by putting $L = \mathcal{G}$, $M = N$. The second statement follows by putting $L = M = N = \mathcal{G}$.

(ii) Define $g : M(Y) \to M(X)$ by $a \mapsto fr^*(a)$ where $\dot{a} \in \mathcal{G}(X)$ is such that $f_*(r) = 1_{\mathcal{G}(Y)}$. Then, $f_*(g(a)) = f_*(rf^*(a)) = f_*(r)a = 1a = a$ as required.

(iii) The three statements follow directly from 9.3.1 (a)(i), (ii), and (iii), respectively, by putting $L = \mathcal{G}$, $N = M$.

(iv) $n \cdot M(Y) = n \cdot 1_{\mathcal{G}(Y)} \cdot M(Y) \subseteq (K_f \mathcal{G} + I_g \mathcal{G})M(Y) = K_f \mathcal{G} M(Y) + I_g \mathcal{G} M(Y) \subseteq K_f M + I_g M$.
Also $n \cdot (K_g M \cap I_f M) = n \cdot 1_{\mathcal{G}(Y)}(K_g M \cap I_f M) \subseteq (K_f \mathcal{G} + I_g \mathcal{G})(K_g M \cap I_f M) \subseteq K_f \mathcal{G} \cdot I_f M + I_g \mathcal{G} \cdot K_g M = 0$.

☐

Remarks 9.3.1 Assume, as in lemma 9.3.1(ii), that $f_* : \mathcal{G}(X) \to \mathcal{G}(Y)$ is surjective. Then we have the following facts :

(i) If $M(X) = 0$, then $M(Y) = 0$.

(ii) If $\theta : M \to N$ is a natural transformation of $\mathcal{G}$-modules, and $\theta_X : M(X) \to N(X)$ is surjective (resp. split-surjective, injective, split-injective, or bijective), then so is $\theta_Y : M(Y) \to N(Y)$.

(iii) If $M' \to M \to M''$ is a sequence of G-modules and $M'(X) \to M(X) \to M''(X)$ is exact (resp. split-exact) then so is $M'(Y) \to M(Y) \to M''(Y)$.

Theorem 9.3.1 *Let* $\mathcal{G} : \mathcal{B} \to R\text{-}\mathcal{M}od$ *be a Green functor,* X *a* $\mathcal{B}$-*object,* $f : X \to *$ *the unique* $\mathcal{B}$-*morphism from* X *to* $*$. *Then, the following assertions are equivalent:*

(i) $f_* : \mathcal{G}(X) \to \mathcal{G}(*)$ *is surjective.*

(ii) $\mathcal{G}$ *is* X-*projective.*

(iii) *Any* $\mathcal{G}$-*module* M *is* X-*projective.*

PROOF $(iii) \to (ii)$: since $\mathcal{G}$ is a $\mathcal{G}$-module,
$(ii) \to (i)$: since by the definition of X-projectivity, $\mathcal{G}_X(*) \to \mathcal{G}(*)$, i.e., $\mathcal{G}(X) \to \mathcal{G}(*)$ is split-surjective,
$(i) \to (iii)$: choose $a \in \mathcal{G}(X)$ with $f_*(a) = 1$, and define a natural transformation $\eta : M_X \to M$ by $\eta(Y) : M(X \times Y) \to M(Y) : b \mapsto q_*(p^*a b)$ where $p = pr_1 : X \times Y \to X$ and $q = pr_2 : X \times Y \to Y$ are the two projections. Then, η is a natural transformation of Mackey functors. Moreover, η is left inverse to $\theta^X : M \to M_X$ since, if $c \in M(Y)$, we have $\eta \theta^X(Y)(c) = q_* < p^*a, q^*c > = < q_*p^*a, c >$ by 9.3.1 (iii), and by considering the pull-back diagram

$$
\begin{array}{ccc}
X \times Y & \xrightarrow{\ q\ } & Y \\
{\scriptstyle p}\downarrow & & \downarrow{\scriptstyle g} \\
X & \xrightarrow{\ f\ } & *
\end{array}
$$

we have $q_*p^*a = g^*f_*a = g^*1 = 1$. So, $\eta \theta^X(X)(c) = c$ as required. ☐

Remarks 9.3.2 In chapter 14 we shall see that an important criteria for the G-homology theories $\mathcal{H}_n^G : G\mathcal{S}et \to \mathcal{A}b$ is that it is isomorphic to some Mackey functor $G\mathcal{S}et \to \mathcal{A}b$.

Corollary 9.3.1 Let $\mathcal{G} : \mathcal{B} \to R\text{-}\mathcal{M}od$ be a Green functor, M a $\mathcal{G}$-module, and X a $\mathcal{B}$-object such that $\mathcal{G}(X) \to \mathcal{G}(*)$ is surjective. then $\hat{\mathcal{H}}_X^n(M) = 0$ for all $n \in \mathbb{Z}$, and the (augmented) Amitsur complexes

$$0 \to M \to M_X \to M_{X^2} \to \ldots$$

and

$$0 \leftarrow M \leftarrow M_X \leftarrow M_{X^2} \leftarrow \ldots$$

are split exact.

Remark 9.3.1 Let $\mathcal{G} : \mathcal{B} \to R\text{-}\mathcal{M}od$ be a Green functor, X,Y $\mathcal{B}$-objects. Then, $\mathcal{G}(X) \to \mathcal{G}(*)$ and $\mathcal{G}(Y) \to \mathcal{G}(*)$ are surjective if and only if $\mathcal{G}(X \times Y) \to \mathcal{G}(*)$ is surjective.

Remark 9.3.2 If $\mathcal{G}$ is a Green functor and M a $\mathcal{G}$-module, then the pairings defined in 9.3.1 and 9.3.2 induce pairings $\mathcal{H}^i\mathcal{G} \times \mathcal{H}^j M \to \mathcal{H}^{i+j} M$ $(i, j \geq 0)$ and $\hat{\mathcal{H}}^i\mathcal{G} \times \hat{\mathcal{H}}^j M \to \hat{\mathcal{H}}^{i+j} M$ $(i, j \in \mathbb{Z})$. In particular, putting $i = j = 0$ and $M = \mathcal{G}$, we see that $\mathcal{H}^0\mathcal{G}$ and $\hat{\mathcal{H}}^0\mathcal{G}$ are Green functors and that $\mathcal{H}^i M$ $(j \geq 0)$ and $\hat{\mathcal{H}}^j M$ $(j \in \mathbb{Z})$ are $\mathcal{H}^0\mathcal{G}$ and $\hat{\mathcal{H}}^0\mathcal{G}$-modules, respectively. Thus, if for $f : X \to *$ and $g : Y \to *$, one has

$$n \cdot 1_{\mathcal{G}(*)} \in K_f\mathcal{G} + I_g\mathcal{G}, \quad \text{therefore} \quad n \cdot 1_{\mathcal{H}_X^0\mathcal{G}(*)} \in I_g\mathcal{H}_X^0\mathcal{G}.$$

It follows that n annihilates $K_g\hat{\mathcal{H}}_X\mathcal{G}$ for all $i \in \mathbb{Z}$. In particular, in case $f = g$ and thus $K_g\hat{\mathcal{H}}_X^i = \hat{\mathcal{H}}_X^i\mathcal{G}$, the assumption $n \cdot 1_{G(*)} \in K_f\mathcal{G} + I_g\mathcal{G}$ implies $n \cdot \hat{\mathcal{H}}^i\mathcal{G} = 0$ for all $i \in \mathbb{Z}$.

9.4 Based category and the Burnside functor

$(9.4)^A$ Burnside ring of a based category

Definition 9.4.1 *Let $\mathcal{B}$ be a category with finite coproducts $\coprod$ and an initial object $\emptyset$. A $\mathcal{B}$-object X is said to be indecomposable if and only if $X = X_1 \coprod X_2 \Rightarrow X_1 = \emptyset$ or $X_2 = \emptyset$. For example, if $\mathcal{B} = GSet$, S is indecomposable if and only if S is simple and any simple G-set is isomorphic to G/H for some subgroup H of G.*

9.4.1 A category $\mathcal{B}$ is said to be based if

(i) $\mathcal{B}$ has a final object $*$, finite coproducts, and pull-backs. (Note that this implies also the existence in $\mathcal{B}$ of an initial object $\emptyset$ and products $\coprod$).

(ii) The two squares in a commutative diagram

$$
\begin{array}{ccccc}
Y_1 & \xrightarrow{j_1} & Z & \xrightarrow{j_2} & Y_2 \\
\downarrow & & \downarrow & & \downarrow \\
X_1 & \xrightarrow{q_1} & X_1 \coprod X_2 & \xrightarrow{q_2} & X_2
\end{array}
$$

are pull-backs if and only if the upper horizontal arrows in the diagram represent Z as a coproduct (sum) of Y_1 and Y_2.

(iii) There exists only a finite number of isomorphism classes of indecomposable objects.

(iv) If X,Y are indecomposable $\mathcal{B}$-objects, then $\mathcal{B}(X,Y)$ is finite and $End_{\mathcal{B}}(X) = Aut_{\mathcal{B}}(X)$, i.e., $\mathcal{B}$ is a finite EI category.
Any set I of representatives of the isomorphism classes of indecomposable $\mathcal{B}$-objects is called a basis $\mathcal{B}$. Since by (iv), for $X, X' \in I$, $X < X' < X$ implies $X \simeq X'$, it means that I contains precisely one object out of any $\Diamond$-equivalence class of indecomposable objects. We shall sometimes write $(\mathcal{B}, I)$ to denote a based category with basis I.

Remarks Since there are only a finite number of $\Diamond$-equivalence classes in $\mathcal{B}$, any Mackey functor $M : \mathcal{B} \to R\text{-}\mathcal{M}od$ has a vertex or defect object X. Moreover, the $\Diamond$-equivalence class of X is uniquely determined by the finite set $\mathcal{D}(M) = \{Z \in I : Z < X\}$, which is called the defect set of M.

Examples 9.4.1 (i) *GSet* is a based category with basis $I = \{G/H : H \leq G\}'$, the prime indicating that one has to take one subgroup H out of any conjugacy class of subgroup.

(ii) If $\mathcal{B}$ is a based category with basis I, and if X is a $\mathcal{B}$-object, then $\mathcal{B}/X$ is also based with basis $I/X = \{\varphi : Z \to X | Z \in I\}$, modulo isomorphisms in $\mathcal{B}/X$. Here, for $(Z, \varphi : Z \to X)$, $(Z\prime, \varphi' : Z' \to X)$, two objects of $\mathcal{B}/X$, their sum is $(Z \coprod Z', \varphi \coprod \varphi', Z \coprod Z' \to X)$ and their products is $\delta : Z \coprod Z' \to X$ where δ is the diagonal map in the pull-back diagram

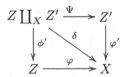

Hence $GSet/X$ is based for any $X \in GSet$.

(iii) If $\mathcal{B}$ is a based category and G is a finite group, then $\mathcal{B}_G$ is also based. Hence, the fact that $GSet = FSet_G$ is based can be derived from the fact that $FSet$ is based.

(iv) If $\mathcal{B}, \mathcal{B}'$ are based categories, so is $\mathcal{B} \times \mathcal{B}'$.

9.4.2 Note that if $\mathcal{B}$ is based category, property 9.4.1(ii) ensures that for $X, Y, Z \in \mathcal{B}$, the natural map $(Z \times X) \coprod (Z \times Y) \to Z \times (X \prod Y)$ is an isomorphism (put $Y_1 = Z \times X, Y_2 = Z \times Y, Z = Z(X \prod Y)$ in the commutative diagram of (ii). Hence, it can easily be checked that the isomorphism classes of $\mathcal{B}$-objects form a semi-ring $\mathcal{B}^+ = \Omega^+(\mathcal{B})$ with respect to sums and products where φ represents $0 \in \Omega^+(\mathcal{B})$ and $*$ represents 1 in $\Omega^+(\mathcal{B})$. We denote the associated Grothendieck ring $K(\mathcal{B}^+)$ by $\Omega(\mathcal{B})$.

If X is a fixed $\mathcal{B}$-object, then as already observed in 9.4.1(ii), $\mathcal{B}/X$ is a based category and the isomorphism classes of objects in $\mathcal{B}/X$ also form a semi-ring $(\mathcal{B}/X)^+ = \Omega^+(X)$ with respect to sums and products defined in 9.4.1(ii), and we denote the associated Grothendieck ring $K((\mathcal{B}/X)^+)$ by $\Omega(X)$. If $*$ is the final object of $\mathcal{B}$, then $\Omega(*) \simeq \Omega(\mathcal{B})$.

If $\mathcal{B} = GSet$, then we shall write $\Omega(G)$ for $\Omega(GSet)$ and call this the Burnside ring G. If $S \in ob(GSet)$, we shall write $\Omega(S)$ for $K(GSet/S)$. Note that $\Omega G \simeq \Omega(GSet/*)$ and that $\Omega(S) = K_0^G(S, FSet)$ since $[\underline{S}, FSet] \simeq GSet/S$.

Theorem 9.4.1 $\Omega : \mathcal{B} \to \mathbb{Z}\text{-}\mathcal{M}od$ *is a Green functor.*

PROOF   Let $\varphi : X \to X'$ be a $\mathcal{B}$-morphism. We want to define additive maps $\Omega_*(\varphi) = \varphi_* : \Omega(X) \to \Omega(X')$ and $\Omega^*(\varphi) = \varphi^* : \Omega(X') \to \Omega(X)$ in such a way that Ω becomes a Mackey functor. To do this, it suffices to define additive maps $\varphi_* : (\mathcal{B}/X)^+ \to (\mathcal{B}/X')^+$ and $\varphi^* : (\mathcal{B}/X')^+ \to (\mathcal{B}/X)^+$. Now, for $(Y, \beta) \in (\mathcal{B}/X)^+$, define $\varphi_*(Y, \beta) = (Y, \varphi\beta : Y \to X') \in (\mathcal{B}/X')^+$, and for $(Y, \beta) \in (\mathcal{B}/X')^+$, define $\varphi^*(Y, \beta) = (Y \underset{X'}{\times} X, \overline{\beta} : Y \underset{X'}{\times} X \to X) \in (\mathcal{B}/X)^+$ where $\overline{\beta} =: \beta|_\varphi$ arises from the pull-back diagram

$$
\begin{array}{ccc}
Y \underset{X'}{\times} X & \xrightarrow{\overline{\varphi}} & Y \\
{\scriptstyle \overline{\beta}=\beta|_\varphi} \downarrow & & \downarrow {\scriptstyle \beta} \\
X & \xrightarrow{\varphi} & X'
\end{array}
$$

It can be easily verified that both maps are additive and extend to Ω, and so, Ω is a bifunctor. To prove that Ω satisfies definition (9.1.1(ii)) (i.e., the Mackey property), we observe from general category theory that in a diagram

$$
\begin{array}{ccccc}
\overline{Y} & \xrightarrow{\alpha|_{\overline{\Psi}}} & X_1 \underset{X}{\times} X_2 & \xrightarrow{\overline{\varphi}} & X_2 \\
\downarrow & & {\scriptstyle \overline{\Psi}} \downarrow & & \downarrow {\scriptstyle \Psi} \\
\overline{Y} & \xrightarrow{\alpha} & X_1 & \xrightarrow{\varphi} & X
\end{array}
$$

with the second square a pull-back, the first square is a pull-back if and only if the rectangle is a pull-back.

So, if $(Y, \alpha) \in \Omega^+(X_1)$, then

$$\Psi^* \varphi_*(\alpha) = \Psi^*(\varphi\alpha) = (\varphi\alpha)|_\Psi = \overline{\varphi}(\alpha|_{\overline{\Psi}}) = \overline{\varphi}_* \overline{\Psi}^*(\alpha).$$

So Ω satisfies the Mackey property.
Also by 9.4.1(iii), the map $q_i^* : \Omega^+(X_1 \coprod X_2) \to \Omega^+(X_i)$ induces a bijection

$$q_1^* \times q_2^* : \Omega^+(X_1 \coprod X_2) \to \Omega^+(X_1) \times \Omega^+(X_2),$$

the inverse map being given by

$$(Y_1, j_1 : Y_1 \to X_1) \times (Y_2, j_2 : Y_2 \to X_2 \to (Y_1 \coprod Y_2; j_1 \coprod j_2; Y_1 \coprod Y_2 \to X_1 \coprod X_2),$$

i.e., Ω satisfies definition 9.1.1(ii) (the additivity condition). So Ω is a Mackey functor.
Finally, the ring structure on $\Omega(X)$ defines the required map $\Omega \times \Omega \to \Omega$, and so, Ω is a Green functor. $\qquad\square$

$(9.4)^B$ Universality of the Burnside functor

Remarks and definition 9.4.1 Note that if R is a commutative ring with identity, then $R \otimes \Omega : \mathcal{B} \to R\text{-}\mathcal{M}od$ defined by $(R \otimes \Omega)(X) = R \otimes \Omega(X)$ is also a Green functor. We call $R \otimes \Omega : \mathcal{B} \to R\text{-}\mathcal{M}od$ the Burnside functor associated with the based category $\mathcal{B}$. We shall prove the universality of the Burnside functor in theorem 9.4.2 below.

Lemma 9.4.1 *Let $\mathcal{B}$ be a based category and let $M : \mathcal{B} \to R\text{-}\mathcal{M}od$ be a Mackey functor. Then M_* transforms finite sums into finite sums.*

PROOF Since M^* transforms finite sums into finite products, we have $M(\emptyset) = 0$. This is because $\emptyset$ can be considered as the sum of $\emptyset$ and $\emptyset$ with identities as the canonical maps of summands into the sum. Then, definition 9.1.1(iii) implies that $M(\emptyset) \xrightarrow{id^* \times id^*} M(\emptyset) \times M(\emptyset)$ is an isomorphism, and this holds if and only if $M(\emptyset) = 0$. Also since $\mathcal{B}$ satisfies (i) and (ii) of 9.4.1,

the squares $\begin{array}{ccc} X & \xrightarrow{id} & X \\ {\scriptstyle id}\downarrow & & \downarrow \\ X & \longrightarrow & X \coprod Y \end{array}$ and $\begin{array}{ccc} X & \longleftarrow & \emptyset \\ \downarrow & & \downarrow \\ X \coprod Y & \longleftarrow & Y \end{array}$ are pull-backs, and by applying

M to the squares we get a diagram:

$$
\begin{array}{ccc}
M(X) & & M(Y) \\
\quad\searrow M_* & & M_*\nearrow \\
\downarrow id & M(X\coprod Y) & \downarrow id \\
\quad\nearrow M^* & & M^*\searrow \\
M(X) & & M(Y)
\end{array}
$$

where the composite diagonal maps are zero. Since

$$ M^* \times M^* : M(X \coprod Y) \to M(X) \times M(Y) $$

is an isomorphism, then

$$ M_* \oplus M_* : M(X) \oplus M(Y) \to M(X \coprod Y) $$

is also an isomorphism.   ☐

Theorem 9.4.2 *Let $\mathcal{B}$ be a based category. Then any Mackey functor $M :$ $\mathcal{B} \to \mathbb{Z}$-Mod is in a natural way, an Ω module, and any Green functor $\mathcal{G} :$ $\mathcal{B} \to \mathbb{Z}$-Mod is an Ω-algebra.*

Note: The proof of theorem 9.4.2 will depend on the following lemmas 9.4.2, 9.4.3, 9.4.4.

Lemma 9.4.2 *If $\varphi : X \to Y$ is a $\mathcal{B}$-map (i.e., object of $\mathcal{B}/Y$), and $a \in M(Y)$, define $<\varphi, a>$ in $M(Y)$ by $<\varphi, a> = \varphi_*(\varphi^*(a))$. Then we have, for two $\mathcal{B}$-maps $\varphi : X' \to Y$, $\varphi' : X' \to Y$:*

(i) $<\varphi \coprod \varphi', a> = <\varphi, a> + <\varphi', a>$ *where $\varphi \coprod \varphi' : X \coprod X' \to Y$ is the sum of φ and φ' in $\mathcal{B}/Y$.*

(ii) $<\varphi, a + a'> = <\varphi, a> + <\varphi, a'>$ *for all a, a' in $M(Y)$.*

(iii) $<\varphi \underset{Y}{\times} \varphi', a> = <\varphi, <\varphi', a>>=<\varphi', <\varphi, a>>$.

(iv) $<idY, a>= a$.

PROOF

(iv) Let $i : X \to X \coprod X'$, $j : X' \to X \coprod X'$ be the canonical imbedding so that $\varphi = (\varphi \coprod \varphi')i$ and $\varphi' = (\varphi \coprod \varphi')j$ and hence

$$
\begin{aligned}
< \varphi, a > &+ < \varphi, a' > \\
&= \varphi_*(\varphi^*(a)) + \varphi'_* \varphi'^*(a) \\
&= (\varphi \coprod \varphi')_* i_* i^* (\varphi \coprod \varphi')^* a + (\varphi \coprod \varphi')_* j_* j^* (\varphi \coprod \varphi')^* a \\
&= (\varphi \coprod \varphi')_* (i_* i^* + j_* j^*)(\varphi \coprod \varphi')^* a \\
&= (\varphi \coprod \varphi')_* (\varphi \coprod \varphi')^* a \quad \text{by lemma 9.4.1} \\
&= < \varphi \coprod \varphi', a >
\end{aligned}
$$

as required.

(ii) $< \varphi, a + a' > = \varphi_*(\varphi^*(a + a')) = \varphi_* \varphi^* a + \varphi_* \varphi^* a' = < \varphi, a > + < \varphi, a' >$.

(iii) Using 9.1.1(ii) for the diagram

we have $< \underset{Y}{\varphi \times} \varphi', a > = (\varphi \underset{Y}{\times} \varphi')_* (\varphi \underset{Y}{\times} \varphi')^* a (\varphi \overline{\varphi}')_* (\varphi' \overline{\varphi})^* a = \varphi_*(\overline{\varphi}'_* \overline{\varphi}^*)\varphi'^* a = \varphi_*(\varphi^* \varphi_*)\varphi'^* a = \varphi_* \varphi^* < \varphi', a > = < \varphi, < \varphi', a >>$. Also since $\varphi \underset{Y}{\times} \varphi' \simeq \varphi' \underset{Y}{\times} \varphi$, we have $< \varphi \underset{Y}{\times} \varphi', a > = < \varphi', < \varphi, a >>$.

(iv) Is trivial.

$\square$

Remark 9.4.1 By lemma 9.4.2(i), the pairing $<,>: \Omega^+(Y) \times M(Y) \to M(Y) : (\varphi, x) \mapsto < \varphi, x >$ is bilinear and thus extends to a bilinear map $\Omega(Y) \times M(Y) \to M(Y)$.

Lemma 9.4.3 *Let $\mathcal{B}$ be a based category, $\alpha : X' \to X$ a $\mathcal{B}$-morphism, $M : \mathcal{B} \to \mathbb{Z}$-Mod a Mackey functor, $\Omega : \mathcal{B} \to \mathbb{Z}$-Mod the Burnside functor, $<,>_X : \Omega(X) \times M(X) \to M(X)$ defined as in 9.4.2 and examples 9.4.1. Then,*

(i) For $b \in \Omega(X)$, $a \in M(X)$, we have $\alpha^*(< b, a >) = < \alpha^* b, \alpha^* a >$.

(ii) $< b, \alpha^*(a) > = \alpha^*(< \alpha^* b, a >)$. where $a \in M(X')$, $b \in \Omega(X)$,

(iii) For $b \in \Omega(X')$, $a \in M(X)$, we have $< \alpha^* b, a > = \alpha^*(< b, \alpha^* a >)$.

PROOF   It suffices to prove the above formula for elements in $\Omega^+(X)$, i.e. for maps $\varphi : Y \to Y$ in $\mathcal{B}/X$.

(i) and (ii). Let $\varphi : Y \to X$ represent an element in $\Omega^+(X)$ and apply definition 9.1.1(ii) to the diagram

$$
\begin{array}{ccc}
X' \underset{X}{\times} Y & \xrightarrow{\ \overline{\alpha}\ } & Y \\
\Big\downarrow{\overline{\varphi}} & & \Big\downarrow{\varphi} \\
X' & \xrightarrow[\ \alpha\ ]{} & X
\end{array}
$$

Then we have for $a \in M(X)$,

$$
\begin{aligned}
< \alpha^*\varphi, a > &= \alpha^*\varphi_*\varphi^*a = \overline{\varphi}_*\overline{\alpha}^*\varphi^*a = \overline{\varphi}_*\overline{\varphi}^*\alpha^*a = < \overline{\varphi}, \alpha^*a > \\
&= < \alpha^*(\varphi), \alpha^*(a) >.
\end{aligned}
$$

Also, for $a \in M(X')$, $< \varphi, \alpha_*a > = \varphi_*\varphi^*\alpha_*a = \varphi_*\overline{\alpha}_*\overline{\varphi}^*a = \alpha_* < \overline{\varphi}, x >= \alpha_* < \alpha^*\varphi, x >$.

(iii) If $\varphi' : Y' \to X' \in \Omega^+(X')$, $a \in M(X)$, then

$$
\begin{aligned}
< \alpha^*(\varphi'), a > &= < \alpha\varphi', a > = \alpha_*\varphi'_*\varphi'^*\alpha_*a \\
&= \alpha_* < \varphi', \alpha^*a >.
\end{aligned}
$$

□

Lemma 9.4.4 *Let $\mathcal{B}$ be a based category, $M : \mathcal{B} \to \mathbb{Z}$-$\mathcal{M}$od a Mackey functor that is an Ω-module. Then, any pairing $<,>: M_1 \times M_2 \to M_3$ is Ω-bilinear.*

PROOF   If X is a $\mathcal{B}$-object, $\varphi : Y \to X$ a $\mathcal{B}$-morphism, $a_i \in M_i(X)$, we show that $\ll \varphi, a_1 >, a_2 > \ = \ < a_1, < \varphi, a_2 \gg = < \varphi, < a_1, a_2 \gg$.

Now, $< \varphi, a_1 > \ = \varphi_*\varphi^*$, and so, we have

$$
\begin{aligned}
\ll \varphi, a_1 >, a_2 > \ &= \ < \varphi_*\varphi^*a_1, a_2 > \ = \varphi_* < \varphi^*a_1, \varphi^*a_2 >_Y \ = \varphi_*\varphi^* < a_1, a_2 > \\
&= \ < \varphi, < a_1, a_2 \gg.
\end{aligned}
$$

Similarly, $< a_1, < \varphi, a_2 \gg = < \varphi, < a_1, a_2 \gg$.   □

PROOF of Theorem 9.4.2   Define $<,>_X : \Omega(X) \times M(X) \to M(X)$ by first defining $<,>_X : \Omega^+(X) \times M(Y) \to M(X) :< \varphi, a >_X = \varphi_*\varphi^*(a)$ where $a \in M(X), \varphi : Y \to X$ in $\mathcal{B}/X$ and extend to $\Omega \times M \to M$. Lemma 9.4.3 shows that $<,>$ is indeed a pairing of bifunctors. We only have to show that for $c, d \in \Omega(X)$, $a, b \in M(X)$:

$$
< c + d, a > \ = \ < c, a > + < d, a >.
$$

$$
< c, a + b > \ = \ < c, a > + < c, b >.
$$

$$< c, < d, a \gg = \ll c, d >, a > \quad \text{and} < id_X, a > = a.$$

But without loss of generality we can take $c = \varphi : Y \to X$, $d = \Psi : Z \to X$ in Ω^+, $c + d = \varphi \coprod \Psi$, $< \varphi, \Psi > = \varphi \underset{X}{\times} \Psi : Y \underset{X}{\times} Z \to X$, and the result follows from lemma 9.4.2.

The last statement follows from lemma 9.4.4.   ⬜

Remark 9.4.2 Note that if in 9.4.2 we have a Mackey functor $M : \mathcal{B} \to R\text{-}\mathcal{M}od$, then M is an $\Omega^R = R \otimes \Omega$-module, and any Green functor $\mathcal{G} : \mathcal{B} \to R\text{-}\mathcal{M}od$ is an Ω^R-algebra.

$(9.4)^C$ Arithmetic structure of $\Omega(\mathcal{B})$, $\mathcal{B}$ a based category

9.4.3 Let $\mathcal{B}$ be a based category with basis I. In 9.4.2, we defined the Burnside ring $\Omega(\mathcal{B})$ as the Grothendick ring associated with the semi-ring $\Omega^+(\mathcal{B})$ and showed that $\Omega : \mathcal{B} \to \mathbb{Z}\text{-}\mathcal{M}od$ is a Green functor while any Mackey functor $M : \mathcal{B} \to \mathbb{Z}\text{-}\mathcal{M}od$ is, in a canonical way, an Ω-module. The aim of this subsection is to study in some detail the arithmetic structure of $\Omega(\mathcal{B})$. We shall adopt the following notation: If X is a $\mathcal{B}$-object and n a positive integer, we write nX for $X \coprod X \coprod \cdots \coprod X$ (n summands) and we write $o\dot{X}$ for $\emptyset$. Also if $X_1, ; X_k$ are $\mathcal{B}$-objects, we write $\sum\limits_{i=1}^{k} X_i$ for $X_1 \coprod X_2 \coprod \cdots \coprod X_k$. So any $\mathcal{B}$-object has the form $X = \sum n_i X_i$ where $X_i \in I$.

9.4.4 Let $(\mathcal{B}, I)$ be a based category, and $T \in I$. We define $\varphi_T : \Omega^+(\mathcal{B}) \to \mathbb{Z}$ by $\varphi_T(X) = | \mathcal{B}(T, X) |$, the number of elements in the set $\mathcal{B}(T, X)$. This definition satisfies the following properties.

Lemma 9.4.5 *Let* X, X' *be* $\mathcal{B}$-*objects,* $T, T' \in I$, *then*

(i) $\varphi_T(X \times X') = \varphi_T(X) \cdot \varphi_T(X')$.

(ii) $\varphi_T(X \coprod X') = \varphi_T(X) + \varphi_T(X')$.

(iii) $\varphi_T(X) \neq 0$ *if and only if* $T < X$.

(iv) $\varphi_T(X) = \varphi_{T'}(X) \; \forall X \in \mathcal{B}$ *if and only if* $T \simeq T'$.

PROOF

(i) Follows from the definition of products in categories.

(ii) Follows from property (ii) of based category.

(iii) Follows from the definitions of "<" and φ_T.

(iv) $\varphi_T(X) = \varphi_{T'}(X)$ for all $X \in \mathcal{B}$ implies $T \diamond T'$. Now, if $\varphi : T \to T'$ and $\psi : T' \to T$ two $\mathcal{B}$-morphism, then $\varphi\psi \in \mathcal{B}(T',T') = Aut_{\mathcal{B}}(T')$, and $\psi\varphi \in \mathcal{B}(T,T) = Aut_{\mathcal{B}}(T)$. Hence φ and ψ are isomorphism. So, $T \simeq T'$.

$\square$

Remark 9.4.3 (i) and (ii) of lemma 9.4.5 imply that $\varphi_T(\cdot) : \Omega^+(\mathcal{B}) \to \mathbb{Z}$ induces a ring homomorphism $\varphi_T : \Omega(\mathcal{B}) \to \mathbb{Z}$, so we have also a ring homomorphism

$$\prod_{T \in I} \varphi_T : \Omega(\mathcal{B}) \to \prod_{T \in I} \mathbb{Z} =: \widetilde{\Omega}(\mathcal{B}).$$

The next lemma generalizes a theorem of Burnside.

Lemma 9.4.6 *Let $(\mathcal{B}, I)$ be a based category, and let $X \cong \sum_{T \in I} n_T T$ and $X' \cong \sum_{T \in I} n'_T T$ be two objects in $\mathcal{B}$. Then, the following statements are equivalent:*

(i) $X = X'$.

(ii) $\varphi_T(X) = \varphi_T(X')$ for all $T \in I$.

(iii) $n_T = n_{T'}$ for all $T \in I$.

In particular, $\mathcal{B}$ satisfies the Krull - Schmidt Theorem.

PROOF   "(i) $\Rightarrow$ (ii)" and "(iii) $\Rightarrow$ (i)" are trivial. Now, suppose $\varphi_T(X) = \varphi_T(X')$ for all $T \in I$ and let $I' = \{T \in I : n_T \neq n'_T\}$. We have to prove that $I' = \emptyset$. Otherwise there exists $Y \in I'$, which is maximal with respect to $<$, and for this Y we have

$$O = \varphi_Y(X) - \varphi_Y(X') = \sum_T (n_T - n'_T)\varphi_Y(T) = (n_Y - n'_Y)\varphi_Y(Y)$$

since all other summands vanish, either because $n_T = n'_T$ or because in case $n_T \neq n'_T$ and thus $T \in I'$, but $Y \not< T$ and therefore $\varphi_Y(T) = 0$. But $\varphi_Y(Y) \neq 0$ and therefore $n_Y = n'_Y$, contradiction. $\square$

Corollary 9.4.1 If X, Y, Z are $\mathcal{B}$-objects, then

$$X \coprod Z \cong Y \coprod Z \text{ implies } X \cong Y.$$

PROOF   By lemma 9.4.6, $(X \coprod Z) \cong Y \coprod Z \Rightarrow \varphi_T(X \coprod Z) \cong \varphi_T(Y \coprod Z)$ $\forall T \in I \Rightarrow \varphi_T(X) + \varphi_T(Z) = \varphi_T(Y) + \varphi_T(Z)$ (by lemma 9.4.5) $\Rightarrow \varphi_T(X) = \varphi_T(Y) \Rightarrow X \cong Y$ (by 9.4.6). $\square$

Theorem 9.4.3 *Let $(\mathcal{B}, I)$ be a based category. Then,*

(i) $\Omega^+(\mathcal{B})$ *(resp. $\Omega(\mathcal{B})$) is a free Abelian semi-group (resp. group) with basis I, and the canonical map $\Omega^+(\mathcal{B}) \to \Omega(\mathcal{B})$ is injective.*

(ii) *The map $\varphi = \prod \varphi_T : \Omega(\mathcal{B}) \to \prod_{T \in I} \mathbb{Z} = \widetilde{\Omega}(\mathcal{B})$ is injective with finite cokernel.*

PROOF

(i) Follows from lemma 9.4.6 and corollary 9.4.1. Note that if we write $[X]$ for the image of X under the map $\Omega^+(\mathcal{B}) \to \Omega(\mathcal{B})$, then the last statement says that "$[X] = [Y] \Rightarrow X = Y$", i.e., "$X \coprod Z \cong Y \coprod Z$ for some $Z \in \mathcal{B}$ $\Rightarrow X \cong Y$", which is just corollary 9.4.1.

(ii) If $a \in \Omega(\mathcal{B})$, then $a = [X] - [X']$ for some $X, X' \in \mathcal{B}$. Now, suppose $\varphi_T(a) = 0$ for all $T \in I$, then $\varphi_T(X) = \varphi_T(X')$ for all $T \in I$, and so, by Lemma 9.4.6, $X \cong X'$, i.e., $a = 0$. So, $\prod \varphi_T$ is injective.
Now by (i), $\Omega(\mathcal{B})$ has $\mathbb{Z}$-rank $|I| = rank_{\mathbb{Z}}\widetilde{\Omega}(\mathcal{B})$, and since $\prod \varphi_T$ is injective, then $Im(\prod \varphi_T)$ and $\widetilde{\Omega}(\mathcal{B})$ have the same rank $|I|$, and so, the cokernel of $Im(\prod \varphi_T)$ is finite.

$\Box$

Remark 9.4.4 (i) Since $\varphi = \prod_{T \in I} \varphi_T : \Omega(\mathcal{B}) \to \prod_{T \in T} \mathbb{Z} = \widetilde{\Omega}(\mathcal{B})$ is injective, we can identify $\Omega(\mathcal{B})$ with its image in $\widetilde{\Omega}(\mathcal{B})$. Since $\widetilde{\Omega}(\mathcal{B})/\Omega(\mathcal{B})$ is finite, it has a well-defined exponent $\|\mathcal{B}\|$. We call $\|\mathcal{B}\|$ the Artin index of $\mathcal{B}$. So, $n\widetilde{\Omega}(\mathcal{B}) \subset \Omega(\mathcal{B})$ if and only if $\|\mathcal{B}\|$ divides n.

(ii) Note that an element $x \in \Omega(\mathcal{B})$ is a non-zero divisor if and only if $\varphi(x)$ has no zero component. Hence $\widetilde{\Omega}(\mathcal{B}) \otimes \mathbb{Q}$ is the total quotient ring of $\Omega(\mathcal{B})$. If $y \in \Omega(\mathcal{B}) \otimes \mathbb{Q}$ is integral over $\Omega(\mathcal{B})$, then the components of y are integral over $\mathbb{Z}$ and hence are in $\mathbb{Z}$. Conversely, $\prod \mathbb{Z}$ is integral over $\varphi(\Omega(\mathcal{B}))$ (since, for example, $\prod \mathbb{Z}$ is generated by idempotent elements that are integral over any subring). So, φ is the inclusion of $\Omega(\mathcal{B})$ into the integral closure of $\Omega(\mathcal{B})$ in its total quotient ring.

Lemma 9.4.7 *Let $(\mathcal{B}, I)$ be a based category, X a $\mathcal{B}$-object, $T \in I$, then $T \times X = \varphi_T(X) \cdot T + \sum_{\substack{Y \in I \\ Y \lneq T}} n_Y Y$ where the n_Y are some non-negative integers* in $\mathbb{Z}$.

PROOF We know from theorem 9.4.3 that we can write $T \times X = \sum_{Y \in I} n_Y Y$ where $n_Y \in \mathbb{Z}$ and $n_Y \geq 0$. Now, $\varphi_Z(T \times X) = \varphi_Z(T) \cdot \varphi_Z(X)$

(by lemma 9.4.5) $= 0$ if $Z \not< T$ (by lemma 9.4.5). So, $0 = \varphi_Z \left(\sum_{Y \in I} n_Y Y \right) \geq$
$n_Z \varphi_Z(Z) \geq n_Z \geq 0$. i.e., $n_Z = 0$ for $Z \not< T$. Hence,

$$T \times X = \sum_{Y < T} n_Y Y \tag{9.1}$$

where $n_Y \in \mathbb{Z}$, $n_Y \geq 0$. Applying φ_T on both sides of I, we have $\varphi_T(T) \cdot \varphi_T(X) = n_T \varphi_T(T)$, which implies that $\varphi_T(X) = n_T$ since $\varphi_T \neq 0$. Hence $T \times X = \varphi_T(X) \cdot T + \sum_{\substack{Y \in I \\ Y \lneq}} n_Y Y$ as required. ⬜

Theorem 9.4.4 *Let $(\mathcal{B}, I)$ be a based category, R an integral domain, $\psi : \Omega(\mathcal{B}) \to R$ any ring homomorphism, and $I_\psi = \{X \in I : \psi X \neq 0\}$. Then, there exists exactly one element $T \in I_\psi$ that is minimal with respect to $<$ in I_ψ. Moreover, $\psi(X) = \varphi_T(X) \cdot 1_R$ for all $X \in \Omega(\mathcal{B})$ and this minimal T in I_ψ.*

PROOF   Let $T, T' \in I_\psi$ be minimal, and let $T \times T' = \sum_{X \in I} n_X X$. Since $\psi(T \times T') = \psi(T) \cdot \psi(T') \neq 0$ in R, there exists $X \in I$ with $\psi(n_X X) = n_X \psi(X) \neq 0$, that is, $X \in I_\psi$ and $X < T$, $X < T'$ since $\varphi_X(T) \cdot \varphi_X(T') = \varphi_X(T \times T') \geq n_X \varphi_X(X) > 0$. Hence, by the minimality of T and T', we have $T = X = T'$. So, the minimal T in I_ψ is unique.
Now, using this T in 9.4.7, we have

$$\psi(T \times X) = \psi(T) \cdot \psi(X) = \varphi_T(X) \cdot T + \sum_{\substack{X \in I \\ Y \lneq T}} n_X X = \varphi_T(X) \cdot \psi(T)$$

since one has $\psi(X) \lneq 0$ for $X \lneq T$ by the minimality of T. Since R is an integral domain, we can divide both sides by $\psi(T) \neq 0$ and we then have $\psi(X) = \varphi_T(X) \cdot 1_R$ as required. ⬜

Corollary 9.4.2 *Let $(\mathcal{B},I)$ be a based category, $T \in I$, p a characteristic (i.e., $p = 0$ or p a prime), $\underline{p}(T,p)$ the prime ideal $\{x \in \Omega(\mathcal{B}) : \varphi_T(x) \equiv 0(p)\}$. Then, any prime ideal $\underline{p}$ in $\Omega(\mathcal{B})$ is of the form $\underline{p}(T,p)$ for some $T \in I$ and some p.*

PROOF   Consider the natural map $\psi : \Omega(\mathcal{B}) \to \Omega(\mathcal{B})/\underline{p}$. Then $\psi(x) = \varphi_T(x) \cdot 1_R$ for some $T \in I$ by theorem 9.4.4.
So,

$$\underline{p} = \{x \in \Omega(\mathcal{B}) : \psi(x) = 0\} = \{x \in \Omega(\mathcal{B}) : \varphi_T(x) \cdot 1_R = 0\}$$
$$= \{x \in \Omega(\mathcal{B}) : \varphi_T(x) \equiv 0(p)\} = \underline{p}(T,p),$$

where $p = $ characteristic of $\Omega(\mathcal{B})/\underline{p}$. ⬜

Remarks 9.4.1 (i) Let $\mathbb{F}_p$ be a prime field of characteristic p. Note that

$$\underline{p} = \underline{p}(T, p) = Ker(\Omega(\mathcal{B}) \xrightarrow{\varphi_T} \mathbb{Z} \to \mathbb{F}_p).$$

It follows from corollary 9.4.2 that given any prime ideal $\underline{p}$ in $\Omega(\mathcal{B})$, the quotient ring $\Omega(\mathcal{B})/\underline{p}$ is a prime ring (i.e., either isomorphic to $\mathbb{Z}$ or $\mathbb{F}_p$, $p \neq 0$). In particular, if R is an integral domain, then two homomorphisms $\psi, \psi' : \Omega(\mathcal{B}) \to R$ coincide if and only if they have the same kernel.

(ii) Note that if $\psi : \Omega(\mathcal{B}) \to \Omega(\mathcal{B})/\underline{p} = R$ where $\underline{p}$ is a prime ideal of $\Omega(\mathcal{B})$, then $I_\psi = I - (I \cap \underline{p})$. We shall sometimes denote the unique minimal $T \in I_\psi$ by $T_{\underline{p}}$.

(iii) If we define a relation "$\overset{p}{\sim}$" on $\mathcal{B}$-objects by $T \overset{p}{\sim} T'$ if $\underline{p}(T, p) = \underline{p}(T', p)$, then "$\overset{p}{\sim}$" is an equivalence relation.

$(9.4)^D$ Arithmetic structure of $\Omega(G)$, G a finite group

Henceforth, we restrict our study of Burnside rings to the case $\Omega(\mathcal{B})$ where $\mathcal{B} = GSet$, i.e., to $\Omega(G)$. First, we have the following definition:

Definition 9.4.2 *Let G be a finite group, $H \leq G$. Define $\varphi_H : GSet \to \mathbb{Z}$ by $\varphi_H(S) = |S^H|$, the number of elements in S^H where*

$$S^H = \{s \in S : gs = s \forall g \in H\}.$$

The following lemma is analogous to lemma 9.4.5 and also gives the connection between $\varphi_T(S)$ and $\varphi_H(S)$.

Lemma 9.4.8 *Let S, T be G-sets, $H, H' \leq G$. Then:*

(i) $\varphi_H(S) \simeq \varphi_T(S)$ if and only if $T \simeq G/H$.

(ii) $\varphi_H(S_1 \cup S_2) = \varphi_H(S_1) + \varphi_H(S_2)$, where $S_1, S_2 \in GSet$.

(iii) $\varphi_H(S_1 \times S_2) = \varphi_H(S_1) \cdot \varphi_H(S_2)$, where $S_1, S_2 \in GSet$.

(iv) $\varphi_H(S) \neq 0$ if and only if $H \in \mathcal{U}(S)$, i.e., if and only if $S^H \neq \emptyset$.

(v) If $H, H' \leq G$, then $\varphi_H(S) \leq \varphi_{H'}(S)$ for all GSets S if and only if $H' \underset{\sim}{\leq} H$. In particular, $\varphi_H(S) = \varphi_{H'}(S)$ for all GSets S if and only if $H' \overset{G}{\underset{\sim}{\,}} H$.

(vi) *For any two simple GSets S, T with $S \simeq G/H$, we have $\varphi_S(S) = |Aut_G(S)| = (N_G(H) : H)$, and $\varphi_S(S)$ divides $\varphi_T(S)$.*

PROOF

(i) Follows from the fact that the map $GSet(G/H, S) \to S$ given by $\varphi \mapsto \varphi(eH)$ induces a bijection $GSet(G/H, S) \simeq S^H$.

(ii) and (iii) are left as easy exercises.

(iv) Follows from the definition of φ, $<$, and $\mathcal{U}(S)$.

(v) If $\varphi_H(S) \leq \varphi_{H'}(S)$ for all G-set S, then in particular $\varphi_{H'}(G/H) \neq 0$ since $\varphi_H(G/H) > 0$, and so, $H' \lesssim H$. On the other hand, if $H' \leq gHg^{-1} = K$, then $\varphi_H(S) = |S^H| = |S^K| \leq |S^{H'}| = \varphi_{H'}(S)$.

(vi) Left as an exercise.

$\square$

Remarks 9.4.2 (i) It follows from lemma 9.4.8(ii) and (iii) that we have a ring homomorphism $\varphi_H : \Omega(G) \to \mathbb{Z}$. Since I consists of only a finite number of non-isomorphic G-sets, each isomorphic to some G/H, for some $H \leq G$, (since $G/H \lozenge G/H'$ if and only if H and H' are conjugate), it means that we also have a homomorphism $w = \prod'_{H \leq G} \varphi_H : \Omega(G) \to \prod'_{H \leq G} \mathbb{Z} = \widetilde{\Omega}(G)$ where $\prod'$ is taken over representatives of conjugacy classes of subgroups of G.

(ii) In view of lemma 9.4.8(i), one can restate the Burnside theorem as follows:
If S, S' are G-sets, then $S \simeq S'$ if and only if $\varphi_H(S) = \varphi_H(S')$ for all subgroups H of G. It then follows that if
$$w = \prod'_{H \leq G} \varphi_H : \Omega(G) \to \prod'_{H \leq G} \mathbb{Z}, \text{ then } S \simeq S' \text{ if and only if}$$
$w(S) = w(S')$.
Hence w is injective.

(iii) It follows from theorem 9.4.4 that any homomorphism $\psi : \Omega(G) \to R$ (R is an integral domain) factors through some $\varphi_H : \Omega(G) \to \mathbb{Z}$ ($H \leq G$) and the unique homomorphism $\mathbb{Z} \to R$ given by $n \mapsto n \cdot 1_R$.

(iv) If $H \leq G$, p a characteristic, (i.e., $p = 0$ or prime), and if $\underline{p} = \{x \in \Omega(G) : \varphi_H x \equiv 0(p)\}$, then any prime ideal $\underline{p} \in \Omega(G)$ is of the form $\underline{p}(H, p)$ for some $H \leq G$ and some rational prime p (see corollary 9.4.1).

(v) If the number of conjugacy classes of G is k, then $\widetilde{\Omega}(G) = \prod \mathbb{Z}$ (k products of $\mathbb{Z}$) is a free Abelian group of rank k, $\Omega(G)$ is also a free Abelian group of the same rank by theorem 9.4.3, and the cokernel of w is finite.

Our main aim in this subsection is to compute more explicitly the cokernel of w.

Lemma 9.4.9 *Consider* $\Omega(G)$ *and* $\widetilde{\Omega}(G)$ *as subrings of*

$$\prod_{T \in I} \mathbb{Q} = \mathbb{Q} \otimes \Omega(G) = \{(x_T)_{T \in I} : x_T \in \mathbb{Q}\}, \quad i.e.,$$

$$\widetilde{\Omega}(G) = \{(x_T)_{T \in I} \in \prod \mathbb{Q} : x_T \in \mathbb{Z}\} \quad and$$

$$\Omega(G) = \sum_{S \in I} \mathbb{Z}(\varphi_T S)_{T \in I} \subseteq \prod_{T \in I} \mathbb{Q}.$$

Then, $J' = \left\{ \frac{1}{|Aut_G(S)|} S = \left(\frac{\varphi_T(S)}{\varphi_S(S)} \right)_{T \in I} \middle| S \in I \right\}$ *is a basis for* $\widetilde{\Omega}(G)$.

PROOF By lemma 9.4.8, we know that $|Aut_G(S)| = \varphi_S(S)$ divides $\varphi_T(S)$, and so, $\left(\frac{\varphi_T(S)}{\varphi_S(S)} \right)_{T \in I}$ is an element of $\widetilde{\Omega}(G)$ for any $S \in I$. Now, compare the set J' with the canonical basis

$$J = \left\{ i_S = (\delta_S^T)_{T \in I} : S \in I \right\} \quad \text{of} \quad \widetilde{\Omega}(G), \quad \text{where} \quad \delta_S^T = \begin{cases} 0 \text{ if } & S \neq T \\ 1 \text{ if } & S = T \end{cases}$$

Since $|I| = |J| = |J'|$, it is enough to show that $i_S \in J$ is an integral linear combination of elements of J'. This is done by induction with respect to $<$. If $S = G/\epsilon$, then:

$\varphi_T \left(\frac{1}{|Aut_G(G/\epsilon)|} G/\epsilon \right) = \delta_{G/\epsilon}^T$, and so, $i_S = \frac{1}{|Aut_G(S)|} S \in J'$. For arbitrary S, we have $\varphi_S \left(\frac{1}{|Aut_G(S)|} S \right) = 1$ and $\varphi_T \left(\frac{1}{|Aut_G(S)|} S \right) = 0$ for $T \not< S$, and so,

$$\frac{1}{|Aut_G(S)|} S = i_S + \sum_{\substack{T \in I \\ T \lneq S}} n_{T,S} i_T \quad \text{with} \quad n_{T,S} = \frac{\varphi_T(S)}{\varphi_S(S)} \in \mathbb{Z}.$$

Now, by induction hypothesis, any i_T with $T \lneq S$ is an integral linear combination of the elements of J'. So, the same is true of i_S. $\square$

Remark 9.4.5 One alternative way of visualizing the proof of the above lemma is the following:
If we order $i = \{S_I = G/\epsilon, S_2, \cdots\}$ in such a way that $S_i < S_j \Rightarrow i \leq j$, then the matrix $(A_{ij})_{i,j} = \left(\frac{\varphi_{S_i}(S_j)}{\varphi_{S_j}(S_i)} \right)_{i,j}$, which transforms J into J', is integral and triangular with $1's$ on the diagonal and hence unimodular. So, J' is a basis of $\widetilde{\Omega}(\pi)$ as well as J.

Corollary 9.4.3 Let $H_1 = 1, H_2, \cdots H_k$ be a complete set of representatives of conjugacy classes of subgroups of G. Then

$$(\widetilde{\Omega}(G) : \Omega(G)) = \prod_{i=1}^{k} |\overline{N}(H_i)|.$$

PROOF   Without loss of generality, assuming that $i \leq j$ implies $|H_i| \leq |H_j|$. The set $\{G/H_i : i = 1, \cdots, k\}$ forms a basis for $\Omega(G)$. Now, consider the $k \times k$ matrix $M = (\varphi_{H_j}(G/H_i))$. Note that $\varphi_{H_j}(G/H_i) = 0$ unless $H_j \overset{G}{\leq} H_i$. So, M is triangular and $(\widetilde{\Omega}(G) : \Omega(G)) = det(M) = \prod_{i=1}^{k} \varphi_{H_i}(G/H_i) = \prod_{i=1}^{k} \overline{N}(H_i)$.   □

Theorem 9.4.5 $\{n \in \mathbb{Z} : n\widetilde{\Omega}(G) \subset \Omega(G)\} = |G|\mathbb{Z}$.

PROOF   For $n \in \mathbb{Z}$, $n\widetilde{\Omega}(G) \subseteq \Omega(G)$ if and only if $|Aut_G(S)|$ divides n for all simple G-sets S if and only if $(N_G(H) : H)$ divides n for all subgroups $H \leq G$ if and only if $|G|$ divides n.   □

Lemma 9.4.10 Let G be a finite group, $H \leq G$, S a G-set. Then S^H is a $N_G(H)/H$-set.

PROOF   Put $\overline{N}(H) = N_G(H)/H$. Define $\rho : \overline{N}(H) \times S^H \to S^H$ by $(\overline{g}, s) \mapsto gs$, where $\overline{g} = g \cdot H \in \overline{N}(H)$, $g \in N_G(H)$, $s \in S^H$. Then S^H is a well-defined $\overline{N}(H)$-set since one has $gs = gxs$, $(gh)s = g(hs)$ for all $x \in H$, $g, h \in N_G(H)$, $s \in S^H$, and $gs \in S^H$ (for, if $h \in H$, then $h(gs) = g(g^{-1}hg)s = gs$).   □

Remark 9.4.6 In view of lemma 9.4.10, we now define a map $\Psi_H : \Omega(H) \to \Omega(\overline{N}(H))$ by $S \mapsto S^H$. It is easy to see that Ψ_H is a well-defined ring homomorphism and that we have a commutative diagram

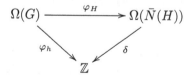

where $\delta(X) = |X|$, the number of elements of X.

Lemma 9.4.11 [Burnside's Lemma] Let S be a G-set.   Then, $\sum_{g \in G} \varphi_{<g>}(S) = \sum_{g \in G} |S^g|$ equals $|G|$ times the number of G-orbits in S.

PROOF   Let $S = S_1 \cup S_2 \cup \cdots \cup S_k$ be a decomposition of S into simple G-sets. Note that for each $s \in S_i$, one has an isomorphism $S_i \simeq G/G_s$, i.e.,

$|G_s| = \frac{|G|}{|S_i|}$. But

$$\sum_{g \in G} \varphi_{<g>}(S) = |\{(g,s) : g \in G, s \in S, gs = s\}| =$$

$$= \sum_{s \in S} |G_s| = \sum_{i=1}^{k} \sum_{s \in S_i} |G_s| = \sum_{i=1}^{k} \Big(\sum_{s \in S_i} \frac{|G|}{|S_i|} \Big) = \sum_{i=1}^{k} \Big(\sum_{i=1}^{k} (|G|/|S_i|)|S_i| \Big) = k|G|.$$

$\square$

Lemma 9.4.12 *Let G be a finite group. Then $\Omega(G)$ is contained in the kernel of the additive map $\Omega(G) \xrightarrow{\eta_G} \mathbb{Z}/|G|\mathbb{Z}$ given by $x = (x_{<g>})_{g \in G} \mapsto \sum_{g \in G} x_{<g>} \bmod |G|$.*

PROOF Let $x = (x_H)_{H \leq G}$ be an element of $\Omega(G) \subseteq \widetilde{\Omega}(G)$. If $x = S - T$, then we have $\sum x_{<g>} = \sum \varphi_{<g>}(S) - \sum \varphi_{<g>}(T) \equiv 0 \pmod{|G|}$. $\square$

Remarks 9.4.3 Let $H \leq L \leq G$ and let $\overline{L} = L/H < \overline{N}(H)$. Then, for any $x \in \Omega(G)$, $\varphi_{\overline{L}}(\varphi_H(x)) = \varphi_{\overline{L}}(x)$ since, if $x = S$, we have $(S^H)^{\overline{L}} = S^L$. Hence the ring homomorphism $\Phi : \Omega(G) \to \Omega(\overline{N}(H))$ can be extended to a ring homomorphism $\Phi_H : \widetilde{\Omega}(G) \to \widetilde{\Omega}(\overline{N}(H))$ such that the diagram

$$\begin{array}{ccc} \Omega(G) & \xrightarrow{\Phi_H} & \Omega(\bar{N}(H)) \\ \downarrow & & \downarrow \\ \widetilde{\Omega}(G) & \xrightarrow[\widetilde{\Phi}_H]{} & \widetilde{\Omega}(\bar{N}(H)) \end{array}$$

commutes, where $\widetilde{\Phi}_H : \widetilde{\Omega}(G) \to \widetilde{\Omega}(\overline{N}(H))$ is given by $(x_L)'_{L \leq G} \mapsto (y_{\overline{L}})'_{\overline{L} \leq \overline{N}(H)}$ with $y_{\overline{L}} = x_{\overline{L}}$ if $L = \cup_{gH \in \overline{L}} gH$.

Theorem 9.4.6 *If $\eta = \prod'_{H \leq G}(\eta_{\overline{N}(H)} \circ \Phi_H) : \widetilde{\Omega}(G) \to = \prod'_{H \leq G}(\mathbb{Z}/|\overline{N}(H)|\mathbb{Z})$, then,*

(i) The sequence

$$0 \to \Omega(G) \xrightarrow{w} \widetilde{\Omega}(G) \xrightarrow{\eta} \prod'_{H \leq G}(\mathbb{Z}/|\overline{N}(H)|\mathbb{Z}) \to 0$$

   is exact.

(ii) The sequence

$$0 \to \Omega(G) \to \widetilde{\Omega}(G) \xrightarrow{\eta_1} \prod'_{H \leq G} \prod'_{p||G|} (\mathbb{Z}/(|\overline{N}(H)|_p\mathbb{Z})) \to 0$$

is also exact where $\eta_1 = \prod'_{H \leq G} \prod_{p \mid \mid G} \eta_{\overline{N}(H)_p} \circ \widetilde{\Phi}^p_H$ *with* $\overline{N}(H)_p$ *denoting a Sylow-p-subgroup of* $\overline{N}(H)$ *and* $\widetilde{\Phi}^p_H$ *the composition*

$$\widetilde{\Omega}(G) \xrightarrow{\widetilde{\Phi}_H} \widetilde{\Omega}(\overline{N}(H)) \xrightarrow{res} \widetilde{\Omega}(\overline{N}^p_H).$$

PROOF   Since $\Omega(G)$, identified with its image, is contained in the kernel of η by lemma 9.4.12 and remark 9.4.3, and since the index of $\Omega(G)$ in $\widetilde{\Omega}(G)$ is equal to the order $\prod'_{H \leq G} \mathbb{Z}/|\overline{N}(H)|\mathbb{Z}$ by remark 9.4.3, it is enough to show that η is surjective.

To do this, we decompose η into the decomposition of the two maps $\rho :$ $\widetilde{\Omega}(G) \to \widetilde{\Omega}(G)$ and $\rho' : \widetilde{\Omega}(G) \to \prod'_{H \leq G} \mathbb{Z}/|\overline{N}(H)|\mathbb{Z}$, which are defined as follows:

$$\rho' : \widetilde{\Omega}(G) = \prod_{H \leq G}' \mathbb{Z} \longrightarrow \prod_{H \leq G}' \mathbb{Z}/|\overline{N}(H)|\mathbb{Z}$$

is just the obvious canonical surjection that maps each $(x_H)'_{H \leq G} \in \widetilde{\Omega}(G)$ onto $(x_H + |\overline{N}(H)|\mathbb{Z})'_{H \leq G}$, and $\rho : \widetilde{\Omega}(G) \to \widetilde{\Omega}(G)$ is defined by $(x_H)'_{H \leq G} \mapsto$ $(x'_H)'_{H \leq G}$ with $x'_H = \frac{1}{|H|} \sum_{g \in N(H)} x_{<H,g>}$. Obviously, $\eta = \rho' \circ \rho$, and so, we only have to show that ρ is an isomorphism. However, using the canonical basis of primitive idempotents in $\widetilde{\Omega}(G)$ (denoted by i_S), in the proof of lemma 9.4.9 where $S = \{G/H : H \leq G\}$ for computing the matrix of ρ, one sees easily this matrix has the form

$$\begin{pmatrix} 1 & * & * & \cdots & \cdots & * \\ 0 & 1 & * & & & * \\ 0 & 0 & 1 & * & & * \\ 0 & 0 & \ddots & \ddots & \ddots & \vdots \\ \vdots & & & 0 & 1 & * \\ 0 & 0 & \cdots & \cdots & 0 & 1 \end{pmatrix}.$$

i.e., it is upper triangular with $1's$ on the diagonal, if the basis is ordered appropriately. So, ρ is indeed an isomorphism.

(ii) is proved analogously. $\square$

9.4.5   Before closing this section, we shall show that a finite group G is solvable if and only if the prime ideal spectrum of $\Omega(G)$ is connected, i.e., if and only if 0 and 1 are the only idempotents in $\Omega(G)$. This will be done in 9.4.8.

Definition 9.4.3 *Let* G *be a finite group,* $H \leq G$. *We denote by* H^p *the (well-defined) smallest normal subgroup of* H *such that* H/H^P *is a p-group, and we denote by* H_p *the pre-image of any Sylow-subgroup of* $\overline{N_G(H^p)} =$

$N_G(H^p)/H^p$. If H is any finite group, define H^s as the minimal normal subgroup of H such that H/H^s is solvable.

Recall from remark 9.4.2(iv) that if p is a characteristic and G a finite group, then any prime ideal $\underline{p}$ of $\Omega(G)$ has the form $\underline{p} = \underline{p}(H,p) = \{x \in \Omega(G) : \varphi_H(x) \equiv 0(p)\}$ where H is some appropriate subgroup of G (see corollary 9.4.2).

Theorem 9.4.7 *Let* H, H' *be subgroups of a finite group* G, p *a rational prime. Then the following conditions are equivalent*

(i) $\underline{p}(H,p) = \underline{p}(H',p)$.

(ii) $H_p \overset{G}{\sim} H'_p$.

(iii) $H^p \overset{G}{\sim} H'^p$.

(iv) $\varphi_H \equiv \varphi_{H'} \pmod{p}$.

PROOF $(ii) \Rightarrow (iii)$ follows from $(H_p)^p = H^p$.
$(iii) \Rightarrow (iv)$: We first show that if $H \le H' \le G$, and $(H' : H) = p^k$, a power of p, then $\varphi_H \equiv \varphi_{H'} \pmod{p}$, i.e., $\varphi_H(S) \equiv \varphi_{H'}(S) \pmod{p}$for all $S \in GSet$, Now, $\varphi_H(S) = |S^H|$, and since $H \lhd H'$, S^H is H'-invariant (since if $s \in S^H$, $g \in H'$, then $h(gs) = (g'g^{-1}hg)s = gs$ for all $h \in H$). Moreover, H'/H acts on S^H, leaving $S^{H'}$ pointwise invariant, whereas $S^H - S^{H'}$ is a disjoint union of non-trivial transitive H'/H-sets, all of which have a length $\ne 1$ and dividing $|H'/H| = p^k$. So, $p \mid |S^H - S^{H'}|$ i.e., $\varphi_H(S) = |S^H| \equiv |S^{H'}| = \varphi_{H'}(S) \pmod{p}$. Now, let H, H' be two arbitrary subgroups of G. Then, since $(H : H^p)$ is a power of p, $\varphi_H(S) = \varphi_{H'}(S) \pmod{p}$ for all $S \in GSet$. Similarly, $\varphi_{H'}(S) \equiv \varphi_{H'^p}(S) \pmod{p}$ for all $S \in GSet$ (see lemma 9.4.8). Hence (iii) $\Rightarrow$ (iv).
$(iv) \Rightarrow (i)$ is trivial since

$$\underline{p}(H,p) = \{x = S - T \in \Omega(G) : \varphi_H(S) \equiv \varphi_H(T) \pmod{p}\}.$$

$(i) \Rightarrow (ii)$: Let $L = H_p$, and $L' = H'$ and $\bar{p} = \underline{p}(H,p) = \underline{p}(L,p) = \underline{p}(L',p)$. Since $H^p \equiv (H_p)^p$, the subgroup H^P is a characteristic subgroup of L, and so, we have $N_G(L) \subseteq N_G(H^p)$. Since $p \nmid (N_G(H^p) : L)$ by definition of $L = H_p$, we get *a fortiori*, $p \nmid (N_G(L) : L) = \varphi_L(G/L)$.
Hence, G/L is, up to isomorphism, the unique minimal G-set $T = T_{\bar{p}}$ with $T \in \bar{p}$ (see theorem 9.4.4). Since the same holds for G/L', we have $G/L \cong G/L'$, and so, $L \overset{G}{\sim} L'$. ⬚

Remark 9.4.7 Note that the above proof shows, in particular, that any two Sylow-p-subgroups must be conjugate. To see this take $H = \epsilon$, the trivial subgroup.

Theorem 9.4.8 *Two prime ideals $\underline{p}(H,p)$ and $\underline{p}(H',p)$ are the same connected components of $Spec(\Omega(G))$ if and only if $H \overset{G}{\sim} H'$. Hence the connected components of $Spec(\Omega(G))$ are in one-one correspondence with the class of conjugate perfect subgroups H of G (i.e., subgroups H with $[H,H] = H$).*

PROOF   It suffices to prove the first statement. Let A be a Noetherian ring. For any prime ideal $p \in Spec(A)$, let $\bar{p} = \{\underline{a} : \underline{a} \in Spec(A), p \subset \underline{a}\}$ be the closure of p in $Spec(A)$. Then, two prime ideals p and q are in the same connected component of $Spec(A)$ if and only if there exists a series of minimal ideals $\underline{p}_1, \cdots, \underline{p}_n$ with $\underline{p} \in \underline{p}_1$, $\underline{q} \in \underline{p}_n$, $\underline{p}_i \cap \underline{p}_{i+1} \neq \emptyset$ $(i = 1, 2, \cdots, n-1)$. But for $A = \Omega(G)$, we have $\bar{p}(H,0) \cap \bar{p}(H',0) \neq \emptyset$ if and only if $H^p \overset{G}{\sim} H'^p$ for some p, which implies that $H^p = (H^p)^s \overset{\pi}{\sim} (H'^p)^s = H'^s$.

Hence, if $\underline{p}(H,p)$ and $\underline{p}(H',q)$ are the same connected component of $Spec(\Omega(G))$, we have $H^s \overset{G}{\sim} H'^s$.

On the other hand, $\underline{p}(H,p)$ and $\underline{p}(H^s,0)$ are always in the same connected component since we can find a series of normal subgroups of H such that $H^s = H^{(n)} \lhd H^{(n-1)} \lhd \cdots \lhd H^{(1)} \lhd H^{(0)} = H$ with $H^{(i-1)}/H^{(i)}$ a q_i-group for some prime q_i $(i = 1, \cdots, n)$, which implies $\underline{p}(H,p) \in \bar{p}(H^{(0)},0)$, $\bar{p}(H^{(i-1)},0) \cap \bar{p}(H^{(i)},0) \neq \emptyset$ for $i = 1, \cdots, n$ and $\underline{p}(H^s,0) \in \bar{p}(H^{(n)},0)$.   ▯

Remark 9.4.8 Theorems 9.4.7 and 9.4.8 can also be deduced easily from theorem 9.4.6 (ii) and (i), respectively. We indicate how theorem 9.4.6 (i) implies theorem 9.4.8: Two prime ideals $\underline{p}(H,p)$ and $\underline{p}(H',q)$ are in the same connected component of $\Omega(G)$ if and only if for any idempotent $e \in \Omega(G)$ one has "$e - 1 \in \underline{p}(H,p) \Leftrightarrow e - 1 \in \underline{p}(H',q)$". But an idempotent in $\Omega(G)$ is an idempotent in $\widetilde{\Omega}(G)$, which is contained in the kernel of η, and an idempotent in $\widetilde{\Omega}(G)$ is always of the form $e = e_{\mathcal{U}} = (\delta_H^{\mathcal{U}})_{H \leq G} \in \widetilde{\Omega}(G) = \prod'_{H \leq G} \mathbb{Z}$ with
$$\delta_H^{\mathcal{U}} = \begin{array}{l} 1 \text{ if } H \in \mathcal{U} \\ 0 \text{ if } H \notin \mathcal{U} \end{array} \text{ for some set } \mathcal{U} \text{ of conjugacy classes of subgroups of } G.$$
But any such $e_{\mathcal{U}}$ is in the kernel of η if and only if for any $H \in \mathcal{U}$ one has $< H, g > \in \mathcal{U}$ for any $g \in N_G(H)$ and thus $e_{\mathcal{U}}(\Omega(G))$, if and only if $\mathcal{U}$ is a disjoint union of $\overset{s}{\sim}$-equivalent subgroups of G, if $\overset{s}{\sim}$-equivalence is defined by "$H \overset{s}{\sim} H' \Leftrightarrow H^s \overset{G}{\sim} H'^s$". Thus $\underline{p}(H,p)$ and $\underline{p}(H',q)$ are in the same connected component of $\Omega(G)$ if and only if H and H' are $\overset{s}{\sim}$-equivalent, q.e.d.

9.5   Induction Theorems for Mackey and Green functors

In this section, we obtain general induction theorems for Mackey functors, and we shall see that these results are in fact generalizations of Artin's induction Theorem.

Definition 9.5.1 *Let $\mathcal{B}$ be a based category, M a Mackey functor from $\mathcal{B}$ into $\mathbb{Z}$-Mod, $f : X \to Y$ a $\mathcal{B}$-morphism. Recall from lemma 9.3.1 that*

$$K_f(M) = Ker(f^* : M(Y) \to M(X)) \text{ and } I_f(M) = Im(f_* : M(X) \to M(Y)).$$

Some properties of $K_f(M)$ and $I_f(M)$ have been given in lemma 9.3.1. If $Y = *$, the final object of $\mathcal{B}$, and f is now the unique morphism $f_X : X \to *$, we write $K_X(M)$ and $I_X(M)$ for $K_f(M)$ and $I_f(M)$, respectively. Note that if X, Z are two $\mathcal{B}$-objects with $X < Z$, we have a commutative triangle

$$
\begin{array}{c}
X \\
\downarrow \qquad \searrow {f_X} \\
\qquad \qquad \star \\
\qquad \nearrow {f_Z} \\
Z
\end{array}
$$

and so, we have $I_X(M) \subseteq I_Z(M)$ and $K_Z(M) \subseteq K_X(M)$. Hence, if $X \diamond Z$, then $K_Z(M) = K_X(M)$, and $I_X(M) = I_Z(M)$. So, $K_X(M)$ and $I_X(M)$ depend only on the $\diamond$-equivalence class of X.
Now, let $\mathcal{B} = GSet$. Recall from theorem 9.1.1 and definition 9.1.3 that for any two G-sets S and T, one has $S \diamond T$ if and only if $\mathcal{U}(S) = \mathcal{U}(T)$ where $\mathcal{U}(S) = \{H \leq G : S^H \neq \emptyset\}$. If $\mathcal{U}$ is any collection of subgroups of G, define $S(\mathcal{U}) \in GSet$ by $S(\mathcal{U}) = \dot\bigcup_{H \in \mathcal{U}} G/H$. Denote $K_{S(\mathcal{U})}(M)$ by $K_{\mathcal{U}}(M)$, $I_{S(\mathcal{U})}(M)$ by $I_{\mathcal{U}}(M)$, and write $K_H(M)$, $I_H(M)$ if $\mathcal{U}$ contains exactly one subgroup H only. Clearly, if $\overline{\mathcal{U}}$ is the subconjugate closure of $\mathcal{U}$ (as defined in 9.1.3), then $K_{\mathcal{U}}(M) = K_{\overline{\mathcal{U}}}(M)$, and $I_{\mathcal{U}}(M) = I_{\overline{\mathcal{U}}}(M)$. Also, we have $K_{S \dot\cup T}(M) = K_S(M) \cap K_T(M)$ and $I_{S \dot\cup T}(M) = I_S(M) + I_T(M)$.
In particular, $K_{\mathcal{U}}(M) = \cap_{H \in \mathcal{U}} K_H(M)$, $I_{\mathcal{U}}(M) = \sum_{H \in \mathcal{U}} I_H(M)$.
Suppose $\mathcal{P}$ is a (possibly empty) set of prime numbers and $\mathcal{P}'$ its complement in the set of all prime numbers. So, any natural number n can be uniquely written as the product of its $\mathcal{P}$-part $n_{\mathcal{P}}$ and its $\mathcal{P}'$-party $n_{\mathcal{P}'}$ where if $n = \prod p^{\alpha_p}$, then $n_{\mathcal{P}} = \prod_{p \in \mathcal{P}} p^{\alpha_p}$, and $n'_{\mathcal{P}} = \prod_{p \notin \mathcal{P}} p^{\alpha_p}$. If $\mathcal{U}$ is a set of subgroups of G, define $h_{\mathcal{P}}\mathcal{U} = \overline{\mathcal{U}} \cup J$ where

$$J = \{H \leq G : \exists p \in \mathcal{P} \quad with \quad H^p \in \overline{\mathcal{U}}\}.$$

Put $h\mathcal{U} = h_{\mathcal{P}}\mathcal{U}$ if $\mathcal{P}$ is the set of all primes, i.e., $\mathcal{P}' = \emptyset$. Then $h_{\mathcal{P}}\mathcal{U}$ contains for any $H \in \overline{\mathcal{U}}$ all $H' \leq G$ such that $H^p \overset{G}{\sim} H'^p$ (i.e., such that $\underline{p}(H, p) = \underline{p}(H', p)$), (see theorem 9.4.7) for some $p \in \mathcal{P}$.

If T is a G-set, define $T_{\mathcal{P}} = S(h_{\mathcal{P}}(\mathcal{U}(T))) = \dot{\cup}_{H \in h_{\mathcal{P}}(\mathcal{U}(T))} G/H$. Then, $T < T_{\mathcal{P}}$, $T \lozenge S$ implies that $T_{\mathcal{P}} = S_{\mathcal{P}}$ and $h_{\mathcal{P}}(\mathcal{U}(T)) = \mathcal{U}(T_{\mathcal{P}})$.

The following lemma will be useful in proving the general induction theorem 9.5.1 for Mackey functors $M : GSet \to \mathbb{Z}\text{-}\mathcal{M}od$ as well as its generalization to Mackey functors $M : \mathcal{B} \to \mathbb{Z}\text{-}\mathcal{M}od$ where $\mathcal{B}$ is a based category (see theorem 9.5.2).

Lemma 9.5.1 *Let $\mathcal{B}$ be a based category, Z a $\mathcal{B}$-object, $\Omega : \mathcal{B} \to \mathbb{Z}\text{-}\mathcal{M}od$ the Burnside functor, and $*$ the final object of $\mathcal{B}$, with J a basis of $\mathcal{B}$. Then,*

(i) $K_Z(\Omega) = \{x \in \Omega(*) : \varphi_T(x) = 0 \quad \text{for all} \quad T \in J \quad T < Z\}.$

(ii) $I_Z(\Omega) = \{x \in \Omega(*) : \varphi_T(x) = 0 \quad \text{for all} \quad T \in J \quad T \not< Z\}.$

PROOF

(i) Let $x = X - X' \in \Omega(*)$, $X, X' \in GSet$. Then $x \in K_Z(\Omega)$ if and only if the projections $P_Z : X \times Z \to Z$ and $P'_Z : X' \times Z \to Z$ are isomorphic in the based category $\mathcal{B}/Z$, i.e., if and only if $\varphi_{T \to Z}(X \times Z \to Z) = \varphi_{T \to Z}(X' \times Z \to Z)$ for all objects $(T, \alpha : T \to Z)$ in J/Z. Now, $\varphi_{T \to Z}(X \times Z \to Z)$ is by definition the number of elements in the set

$$\{\psi : T \to X \times Z : \begin{array}{ccc} & T \longrightarrow X \times Z & \\ \alpha \searrow & & \swarrow P_Z \\ & Z & \end{array} \quad \text{commutes}\}$$

i.e., in the set of all $\mathcal{B}/Z$-morphism from $(T, \alpha : T \to Z)$ to $(X \times Z, P_Z : X \times Z \to Z)$. Moreover, a map $\psi : T \to X \times Z$ is nothing else but a pair (ψ_1, ψ_2) of maps $\psi_1 : T \to X$, $\Psi_2 : T \to Z$, and so,

$$\varphi_{T \to Z}(X \times Z \to Z) = |\{(\psi_1, \psi_2) : T \to X \times Z | \psi_2 = \alpha\}|$$
$$= |\{\psi_1 : T \to X\}| = \varphi_T(X).$$

So, $x = X - X' \in K_Z(\Omega)$ if and only if $\varphi_T(X) = \varphi_T(X')$ for all $T < Z$, $T \in J$, i.e., if and only if $\varphi_T(x) = 0$ for all $T \in J$ as required.

(ii) It follows from the definition of $I_Z(\Omega)$ that $I_Z(\Omega) = \sum_{T \in J, T < Z} \mathbb{Z} \cdot T$. So,

$$I_Z(\Omega) = \{x = \sum_{T \in J} x_T T \in \Omega(*) | x_T = 0 \text{ for all } T \in J, T \not< Z\}$$

i.e., $\{x \in \Omega(*) | \varphi_T(x) = 0 \text{ for } T \in J, T \not< Z\}.$

Vice-versa, if $x = \sum_{T \in J} x_T T \in \Omega(\mathcal{B})$ and $\varphi_T(x) = 0$ for all $T \in J$, with $T \not< Z$, we have $x_T = 0$ for all such T, since otherwise, choosing a maximal element T_0 with respect to $<$ in $\{T \in J | T \not< Z, \text{ and} x_T \neq 0\}$, we get

$$\varphi_{T_0}(x) = x_{T_0} \varphi_{T_0}(T_0) \neq 0.$$

▯

We now state the general induction theorem for Mackey functors $GSet \to \mathbb{Z}$-$\mathcal{M}od$. We shall later generalize this in theorem 9.5.2 to Mackey functors from based category to $\mathbb{Z}$-$\mathcal{M}od$.

Theorem 9.5.1 *Let G be a finite group, $M : GSet \to \mathbb{Z}$-$\mathcal{M}od$ a Mackey functor, $\mathcal{P}$ a set of primes, and $*$ the final object in $GSet$. Then, for any G-set S, one has*

(i) $|G|'_{\mathcal{P}} M(*) \subseteq K_S(M) + I_{S_{\mathcal{P}}}(M)$.

(ii) $|G|_{\mathcal{P}}(I_S(M) \cap K_{S_{\mathcal{P}}}(M)) = 0$.
 In particular, if $\mathcal{U}$ is any set of subgroups of G, we have

$$|G|'_{\mathcal{P}} M(*) \subseteq K_{\mathcal{U}}(M) + I_{h_{\mathcal{P}}\mathcal{U}}(M) \text{ and}$$

$$|G|'_{\mathcal{P}}(I_{\mathcal{U}}(M) \cap K_{h_{\mathcal{P}}\mathcal{U}}(M)) = 0.$$

PROOF   Since it has been shown in theorem 9.4.2 that the Burnside functor $\Omega : GSet \to \mathbb{Z}$-$\mathcal{M}od$ is a Green functor and that any Mackey functor $M : GSet \to \mathbb{Z}$-$\mathcal{M}od$ is an Ω-module, the theorem would follow from lemma 9.3.1(iv) once we show that $|G|'_{\mathcal{P}} 1_{\Omega(G)} \in K_S(\Omega) + I_{S_{\mathcal{P}}}(\Omega)$.
Now let J be the basis of $GSet$, $T \in J$. Define $e^{\mathcal{P}}_S, f^{\mathcal{P}}_S \in \widetilde{\Omega}(G)$ by

$$\varphi_T(e^{\mathcal{P}}_S) = \begin{cases} 1 \text{ if } & T < S_{\mathcal{P}} \\ 0 \text{ if } & T \not< S_{\mathcal{P}} \end{cases} \qquad \varphi_T(f^{\mathcal{P}}_S) = \begin{cases} 0 \text{ if } & T < S_{\mathcal{P}} \\ 1 \text{ if } & T \not< S_{\mathcal{P}}. \end{cases}$$

Then $(e^{\mathcal{P}}_S + f^{\mathcal{P}}_S) = 1_{\widetilde{\Omega}(G)} = 1_{\Omega(G)}$. Now, $|G|'_{\mathcal{P}} 1_{\Omega(G)} = |G|'_{\mathcal{P}} e^{\mathcal{P}}_S + |G|'_{\mathcal{P}} f^{\mathcal{P}}_S$, and $|G|'_{\mathcal{P}} e^{\mathcal{P}}_S, |G|'_{\mathcal{P}} f^{\mathcal{P}}_S \in \Omega(G)$ by theorem 9.4.6(ii). (The reader is requested to check that both elements are the kernel of η_1 by small, direct computation). But $\varphi_T \cdot (|G|'_{\mathcal{P}} e^{\mathcal{P}}_S) = 0$ for $T \not< S_{\mathcal{P}}$. So, by lemma 9.5.1, $|G|'_{\mathcal{P}} e^{\mathcal{P}}_S \in I_{S_{\mathcal{P}}}(\Omega)$. Also, $\varphi_T(|G|'_{\mathcal{P}} f^{\mathcal{P}}_S) = 0$ for $T < S_{\mathcal{P}}$. So, by lemma 9.5.1, $|G|'_{\mathcal{P}} e^{\mathcal{P}}_S \in K_S(\Omega)$. So, $|G|'_{\mathcal{P}} 1_{\Omega(G)} = |G|'_{\mathcal{P}} e^{\mathcal{P}}_S + |G|'_{\mathcal{P}} f^{\mathcal{P}}_S \in I_{S_{\mathcal{P}}}(\Omega) + K_S(\Omega)$ as required.   ▯

Remark 9.5.1   (i) Putting $\mathcal{P} = \emptyset$ in the above theorem, we get $S \diamond S_\emptyset$, $|G|'_{\mathcal{P}} = |G|$, and so, the theorem yields $|G| M(*) \subseteq K_S(M) + I_S(M)$. This latter form of our induction theorem yields the Artin induction theorem if we put $M = K^G_0(-, \mathcal{P}(\mathbb{C}))$ and $S = S(\mathcal{U}) = \cdot \cup G/H$ where $\mathcal{U} = \{H \le G | H \text{ cyclic}\}$ see [38]. This is because $K^G_0(G/H, \mathcal{P}(\mathbb{C}))$ is isomorphic to the generalized character ring $char(H)$ of H and the map

$$K^G_0(*, \mathcal{P}(\mathbb{C})) \simeq char(G) \to \prod_{H \in \mathcal{U}} K_G(G/H, \mathcal{P}(\mathbb{C}))$$

is injective, i.e., $K_S(K_0^G(-, \mathcal{P}(\mathbb{C})) = 0$. So,

$$|G| \cdot K_0^G(*, \mathcal{P}(\mathbb{C})) \subseteq Im(K_0^G(S, \mathcal{P}(\mathbb{C})) \to K_0^G(*, \mathcal{P}(\mathbb{C}))).$$

i.e.,

$$|G|char(G) \subseteq \sum_{H \in \mathcal{U}} Im(char(H) \to char(G)).$$

(ii) If $\mathcal{P}$ contains all the prime divisor of $|G|$, we have

$$M(*) \subseteq K_S(M) + I_{S_{\mathcal{P}}}(M).$$

In the special case $M = K_0^G(-, \mathcal{P}(\mathbb{C}))$ considered above, this is an induction theorem of the type proved by R.Brauer and others in the 40's (see [38, 39]).

(iii) Let $\mathcal{B}$ be a based category. It was shown in theorem 9.4.3 that $\Omega(\mathcal{B})$ has finite index in $\widetilde{\Omega}(\mathcal{B}) = \prod_{T \in J} \mathbb{Z}$, and we defined, in remarks 9.4.4, the Artin index $\|\mathcal{B}\|$ of $\mathcal{B}$ as the exponent of $\widetilde{\Omega}(\mathcal{B})/\Omega(\mathcal{B})$. It follows from theorem 9.4.5 that $\|G\mathcal{S}et\| = |G|$ for any finite group G.

Lemma 9.5.2 *Let* $\Omega : \mathcal{B} \to \mathbb{Z}$-$\mathcal{M}od$ *be the Burnside functor. Then* $\|\mathcal{B}\| \cdot 1_{\Omega(\mathcal{B})} \in K_X(\Omega) + I_X(\Omega)$ *for any* $\mathcal{B}$-*object* X.

PROOF The proof is similar to theorem 9.5.1 in case $\mathcal{P} = \emptyset$ where we use $\|\mathcal{B}\|$ instead of $|G| = |G|'_{\mathcal{P}}$ to conclude that $\|\mathcal{B}\|e_S^\varphi, \|\mathcal{B}\|f_S^\varphi \in \Omega(\mathcal{B})$, and then apply theorem 9.5.1 to conclude finally that $\|\mathcal{B}\| \cdot 1_{\Omega(\mathcal{B})} \in I_X(\Omega) + K_X(\Omega)$. ▯

Theorem 9.5.1 can now be generalized as follows.

Theorem 9.5.2 *Let* $\mathcal{B}$ *be a based category,* $M : \mathcal{B} \to R$-$\mathcal{M}od$ *a Mackey functor. Then* $\|\mathcal{B}\|$ *annihilates all cohomology groups* $\hat{\mathcal{H}}_X M(*)$ *(see 9.2.2). In particular,*

(i) $\|\mathcal{B}\| \cdot M(*) \subseteq K_X(M) + I_X(M)$.

(ii) $\|\mathcal{B}\| \cdot (K_X(M) \cap I_X(M)) = 0$.

PROOF Since the canonical map $M(*) \to \hat{\mathcal{H}}_X^0 M(*)$ has kernel equal to the right-hand side of (i), (i) follows from $\|\mathcal{B}\|\hat{\mathcal{H}}_X^0(M)(*) = 0$. Now, by theorem 9.4.2 it suffices to show that $\|\mathcal{B}\|\hat{\mathcal{H}}_X^0 \Omega(*) = 0$, which follows from $\|\mathcal{B}\|1_{\Omega(*)} \in K_X(\Omega) + I_X(\Omega)$, i.e., from lemma 9.5.2. (ii) follows also immediately from lemma 9.5.2. ▯

Remarks 9.5.1   (i) Theorem 9.5.2, apart from generalizing Artin's induction theorem, also generalizes the fact that $|G|$ annihilates all cohomology groups $\hat{\mathcal{H}}^n(G, B)$ where B is $\mathbb{Z}G$-module.

(ii) If $\|\mathcal{B}\|1_R$ is invertible in a ring R, then (i) and (ii) of remark 9.5.2 imply that $M = \mathrm{Ker}(M \to M_X) \oplus Im(M_X \to M)$, and in particular that $M(*) \to M(X)$ is injective if and only if $M(X) \to M(*)$ is surjective.

Corollary 9.5.1 Let $\|\mathcal{B}\| \cdot R = R$, $\mathcal{G} : \mathcal{B} \to R\text{-}\mathcal{M}od$ a Green functor and M a $\mathcal{G}$-module such that $M(*)$ is a faithful $\mathcal{G}(*)$-module. Then the following statements are equivalent:

(i) M is X-projective.

(ii) $M(X) \twoheadrightarrow M(*)$ is surjective.

(iii) $M(*) \hookrightarrow M(X)$ is injective.

(iv) $\mathcal{G}(*) \hookrightarrow \mathcal{G}(X)$ is injective.

(v) $\mathcal{G}(X) \twoheadrightarrow \mathcal{G}(*)$ is surjective.

(vi) $\mathcal{G}$ is X-projective.

PROOF   The implications $(i) \Rightarrow (ii) \Rightarrow (iii) \Rightarrow (iv) \Rightarrow (v) \Rightarrow (vi) \Rightarrow (i)$ follow easily form remarks 9.5.1 (ii) and proposition 9.1.1.   ▯

Corollary 9.5.2 Assume $\|\mathcal{B}\| \cdot R = R$ and let $\Omega^R : \mathcal{B} \to R\text{-}\mathcal{M}od : X \mapsto R \otimes_{\mathbb{Z}} \Omega(X)$ be the Burnside functor tensored with R. Then,

$$\Omega^R / Ker(\Omega^R \to \Omega^R_X) \simeq Im(\Omega^R_X \to \Omega^R)$$

is X-projective for any $\mathcal{B}$-object X.

PROOF   Put $M = \Omega^R_X$, $\mathcal{G} = Im(\Omega^R \to \Omega^R_X)$ in remarks 9.5.1(ii) and and corollary 9.5.1   ▯

Corollary 9.5.3 If $\|\mathcal{B}\| \cdot R = R$ and $M : \mathcal{B} \to R\text{-}\mathcal{M}od$ is a Mackey functor, then the following statements are equivalent:

(i) M is X-projective.

(ii) $M(X \times Y) \twoheadrightarrow M(Y)$ is surjective for all $\mathcal{B}$-objects Y.

(iii) $M(Y) \hookrightarrow M(X \times Y)$ is injective for all $\mathcal{B}$-objects Y.

In particular, any subfunctor and any quotient functor of an X-projective Mackey functor $M : \mathcal{B} \to R\text{-}\mathcal{M}od$ is X-projective.

PROOF

"$(i) \Rightarrow (ii)$" follows from the definition of X-projectivity (see definition 9.1.4)

"$(ii) \Rightarrow (iii)$" follows from the remarks in remarks 9.5.1(ii)

"$(iii) \Rightarrow (i)$" It follows from (iii) that the Ω^R-module M is a module over $\Omega^R/Ker(\Omega^R \to \Omega^R)$, which is an X-projective Green functor by corollary 9.5.2. So, M is X-projective by corollary 9.5.1.

The last statement follows since (iii) holds for any subfunctor of M and (ii) holds for any quotient functor of M if it holds for M.   ▯

Remarks 9.5.2 It follows from above that if $\|\mathcal{B}\| \cdot R = R$ and $M : \mathcal{B} \to R\text{-}\mathcal{M}od$ is a Mackey functor, then $Im(M \to M_X)$ and $\mathcal{H}_X^0(M)$ are X-projective as subfunctors of M_X, and $Im(M_X \to M)$ and $\mathcal{H}_X^0(M)$ are X-projective as quotient of M_X. Also, a Green functor $\mathcal{G} : \mathcal{B} \to \mathbb{R}\text{-}\mathcal{M}od$ is X-projective if and only if the image of Ω^R in $\mathcal{G}$ is X-projective.

Corollary 9.5.4 Assume $\|\mathcal{B}\| \cdot R = R$ and $M : \mathcal{B} \to \mathbb{R}\text{-}\mathcal{M}od$ is a Mackey functor and J a basis of $\mathcal{B}$. Define

$$M^Z = Im(M_Z \to M) \cap (\cap_{\substack{Z' \in J \\ Z' \neq Z, Z' < Z}} Ker(M \to M_Z))$$

for any $Z \in J$. Then, $M = \oplus_{Z \in J} M^Z$. Moreover, M^Z can be characterized as the largest Z-projective subfunctor of M, all of whose Z'-projective subfunctors are zero for $Z' \leq Z$ $(Z, Z' \in J)$.
A Green functor $\mathcal{G} : \mathcal{B} \to R\text{-}\mathcal{M}od$ is also a direct product $\mathcal{G} = \prod_{Z \in J} \mathcal{G}^Z$ of Green functors $\mathcal{G}^Z$.

PROOF   From definition of $\mathcal{B}$ and the fact that $\|\mathcal{B}\| \cdot R = R$, we have:

$$\Omega^R(\mathcal{B}) = R \otimes \Omega(\mathcal{B}) \simeq R \otimes \widetilde{\Omega}(\mathcal{B}) = \prod_{Z \in J} R.$$

So, we have a set $e_Z(Z \in J)$ of pairwise orthogonal idempotents of $\Omega^R(\mathcal{B}) = \Omega^R(*)$ with $\sum e_Z = 1$. The statement then follows from $M^Z(Y) = e_{Z|Y}(Y)$ for any $\mathcal{B}$-object Y (i.e., $M^Z = e_Z \cdot M$).   ▯

9.6   Defect Basis of Mackey and Green functors

9.6.1 Let $\mathcal{B}$ be a based category, and $M : \mathcal{B} \to R\text{-}\mathcal{M}od$ a Mackey functor. Recall from proposition 9.1.1 that a vertex object or a vertex of M is a $\mathcal{B}$-object X, unique up to $\diamond$-equivalence such that M is Y-projective for some

$\mathcal{B}$-object Y if and only if $X < Y$. Also, induction theory could be roughly understood as one way of computing vertex objects for M.

If M is a $\mathcal{G}$-module for some Green functor $\mathcal{G} : \mathcal{B} \to \mathbb{R}$-$\mathcal{M}$od, we saw in theorem 9.3.1 that M is Y-projective if $\mathcal{G}$ is Y-projective, which in turn is equivalent to the surjectivity of $\mathcal{G}(Y) \to \mathcal{G}(*)$. So, one can derive induction theorems for M (i.e., prove that M is Y-projective) by constructing a Green functor $\mathcal{G}$ which acts unitarily on M, and then proving that $\mathcal{G}(Y) \to \mathcal{G}(*)$ is surjective. A vertex object for a Mackey functor $M : GSet \to \mathbb{R}$-$\mathcal{M}$od is also called a defect object for M or just a defect set.

Our aim in this section is to study defect objects S for the case $\mathcal{B} = GSet$, and in particular the collection $\mathcal{U}(S)$ of subgroups of G, which is, of course, uniquely determined by M (see proposition 9.1.1(c)). Since this collection $\mathcal{U}(S)$ of subgroups of G does not depend on the particular defect object S chosen but only on M, we shall denote it by $\mathcal{D}_M$. This collection $\mathcal{D}_M$ of subgroups of G, called the class of vertex subgroups of M in proposition 9.1.1(c), will also be called a defect basis of M.

Remarks 9.6.1 (i) Note that proposition 9.1.1(c) is a generalization to Mackey functors of a result due to J.A. Green, that defect sets do exist for Green functors. Moreover, if M is any Mackey functor $GSets \to \mathbb{Z}$-$\mathcal{M}$od, proposition 9.1.1(c) then reduces the problem of determining all G-sets S for which M is S-projective to that of determining the defect basis of M.

(ii) If $M : GSets \to \mathbb{Z}$-$\mathcal{M}$od is a Mackey functor and X a G-set, let

$$M|_X : GSets/X \xrightarrow{f} GSet \xrightarrow{M} \mathbb{Z}-\mathcal{M}od$$

denote the composition of M with the forgetful functor $GSets/X \to GSets : (Y \to X) \mapsto Y$. Then, it follows from prooposition 9.1.1(c) that there exists a G-map $\alpha : S \to X$ such that for any G-map $\beta : T \to X$, $M|_X$ is β-projective if and only if there exists a morphism from α to β in $GSets/X$. Such a map $\alpha : S \to X$ is called a defect map with respect to X, and it is uniquely determined by M up to $\diamond$-equivalence in $GSets/X$.

X is said to be without defect with respect to M if the identity map $X \to X$ is a defect map over X, i.e., if for $\beta : T \to X$ a G-map, $M|_X$ is β-projective only if there exists a G-map $\beta' : X \to T$ such that $\beta\beta' = id_X$.

In some sense a Mackey functor is determined by its behavior on $GSets$ without defect, and the theory of defects may be considered as a way of reducing the study of Mackey functor in arbitrary G-sets to the case of G-sets without defect.

From now on, we shall concentrate on the defect sets of Green functors rather than Mackey functors in general.

Theorem 9.6.1 *Let $\mathcal{B}$ be a based category, $\mathcal{G} : \mathcal{B} \to \mathbb{Z}$-$\mathcal{M}od$ a Green functor, X a $\mathcal{B}$-object, $n \in \mathbb{N}$ a fixed natural number, $\theta : \Omega \to \mathcal{G}$ the unique homomorphism from Ω into $\mathcal{G}$. Consider the following statements*

(i) $n \cdot 1_{\mathcal{G}(*)} \in I_X(\mathcal{G})$.

(ii) $n \cdot \mathcal{G}(*) \subseteq I_X(\mathcal{G})$.

(iii) $n \cdot K_X(\mathcal{G}) = 0$.

(iv) $\theta(n \cdot K_X(\mathcal{G})) = 0$.

(v) $\|\mathcal{B}\| \cdot n \cdot 1_{\mathcal{G}(*)} \in n \cdot I_X(\mathcal{G}) \subseteq I_X(\mathcal{G})$.

Then, (i) $\Leftrightarrow$ (ii) $\Rightarrow$ (iii) $\Rightarrow$ (iv) $\Rightarrow$ (v); in particular, if $\|\mathcal{B}\| \cdot \mathcal{G}(*) = \mathcal{G}(*)$, then they are all equivalent.

PROOF

(i) $\Leftrightarrow$ (ii) is trivial since $I_X(\mathcal{G})$ is an ideal of $\mathcal{G}(*)$.

(ii) $\Rightarrow$ (iii) $n \cdot K_X(\mathcal{G}) = n \cdot \mathcal{G}(x) \cdot K_X(\mathcal{G}) \subseteq I_X(\mathcal{G}) \cdot K_X(\mathcal{G}) = 0$.

(iii) $\Rightarrow$ (iv) Let $\eta_X : X \to *$ be the unique map from $X \to *$, and $x \in K_X(\Omega)$. Then, in the commutative diagram

$$
\begin{array}{ccc}
\Omega(*) & \xrightarrow{\ \theta_* \ } & \mathcal{G}(*) \\
\downarrow{\scriptstyle \Omega^*(\eta_X)} & & \downarrow{\scriptstyle \mathcal{G}^*(\eta_X)} \\
\Omega(X) & \xrightarrow{\ \theta_X \ } & \mathcal{G}(X)
\end{array}
$$

we have $\Omega^*(\eta_X)(x) = 0 \Rightarrow \theta_X \Omega^*(\eta_X)(x) = 0 \Rightarrow \mathcal{G}^*(\eta_X)\theta(x) = 0$. Now, by (iii), $n\theta_*(x) = 0$. So, $\theta_*(n \cdot x) = n\theta_*(x) = 0$.

(iv) $\Rightarrow$ (v) By theorem 9.5.2, we have that $\|\mathcal{B}\| \cdot 1_{\Omega(*)} \in K_X(\Omega) + I_X(\Omega)$. Hence

$$
\begin{aligned}
\|\mathcal{B}\| \cdot n \cdot 1_{\Omega(*)} &\in \theta(n \cdot K_X(\Omega)) + \theta(n \cdot I_X(\Omega)) \\
&= 0 + n \cdot (I_X(\Omega)) \subseteq n \cdot I_X(G) \subseteq I_X(G).
\end{aligned}
$$

$\Box$

Corollary 9.6.1 *If $\mathcal{B} = GSet$ and $|G| \cdot \mathcal{G}(*) = \mathcal{G}(*)$, then for any $GSet$ S one has $\mathcal{D}_\mathcal{G} \subseteq \mathcal{U}(S)$ if and only if $K_S(\mathcal{G}) = 0$.*

PROOF Put $\mathcal{B} = GSet$, $n = 1$ in Theorem 9.6.1. $\Box$

Corollary 9.6.2 Let $\mathcal{G}' \subseteq \mathcal{G}$ be a sub-Green functor of $\mathcal{G}$ and assume $|G| \cdot \mathcal{G}'(*) = \mathcal{G}'(*)$. Then $\mathcal{D}_{\mathcal{G}} = \mathcal{D}_{\mathcal{G}'}$.

PROOF   Left as an exercise.   ▯

Corollary 9.6.3 Let $\mathcal{B}$ be a based category, $\mathcal{G} : \mathcal{B} \to \mathbb{Z}\text{-}\mathcal{M}od$ a Green functor. If $\mathcal{G}(*)$ is torsion free, $X \in ob(\mathcal{B})$, and $n \cdot 1_{\mathcal{G}(X)} \in I_X(\mathcal{G})$, then $(n, \|\mathcal{B}\|) \cdot 1_{\mathcal{G}(*)} \in I_X(\mathcal{G})$.

PROOF   Left as an exercise.   ▯

Notations 9.6.1 If $\mathcal{G} : GSet \to \mathbb{Z}\text{-}\mathcal{M}od$ is a Green functor and A a commutative ring with identity, then $A \otimes_{\mathbb{Z}} \mathcal{G} : GSet \to A\text{-}mod$ is also a Green functor, which we shall sometimes denote by $\mathcal{G}^A$. We also write $\mathcal{D}_{\mathcal{G}}^A$ for the defect basis of $A \otimes_{\mathbb{Z}} \mathcal{G}$. We shall write $\mathcal{D}_{\mathcal{G}}^{\mathcal{P}}$ for $\mathcal{D}_{\mathcal{G}}^A$ in the special case $A \subseteq \mathbb{Q}$,

$$\mathcal{P} = \mathcal{P}_A = \{p : p \text{ a prime with } \quad pA \neq A\}, \quad \text{so that}$$

$$A = \mathbb{Z}_{\mathcal{P}} = \mathbb{Z}[\frac{1}{q} : q \notin \mathcal{P}] = \mathbb{Z}[\frac{1}{q} : q \in \mathcal{P}']$$

where $\mathcal{P}'$ is the complement of $\mathcal{P}$ in the set of all primes.

Lemma 9.6.1 Let $\mathcal{G} : GSet \to \mathbb{Z}\text{-}\mathcal{M}od$ be a Green functor, $\mathcal{P}, \mathcal{P}_1, \mathcal{P}_2, \cdots$ sets or primes. Then,

(a) For a G-set S, we have $\mathcal{D}_{\mathcal{G}}^{\mathcal{P}} \subseteq \mathcal{U}(S)$ if and only if there exists a $\mathcal{P}'$-number $n \in \mathbb{N}$ such that $n \cdot 1_{\mathcal{G}(*)} \in I_S(\mathcal{G})$.

(b) If $\mathcal{P} \subseteq \mathcal{P}_1$, then $\mathcal{D}_{\mathcal{G}}^{\mathcal{P}} \subseteq \mathcal{D}_{\mathcal{G}}^{\mathcal{P}_1}$. In particular, $\mathcal{D}_{\mathcal{G}}^{\mathcal{P}_1 \cap \mathcal{P}_2} \subseteq \mathcal{D}_{\mathcal{G}}^{\mathcal{P}_1} \cap \mathcal{D}_{\mathcal{G}}^{\mathcal{P}_2}$.

(c) $\mathcal{D}_{\mathcal{G}}^{\mathcal{P}_1 \cup \mathcal{P}_2} = \mathcal{D}_{\mathcal{G}}^{\mathcal{P}_1} \cup \mathcal{D}_{\mathcal{G}}^{\mathcal{P}_2}$.

(d) If $\mathcal{P}_1 = \{p \in \mathcal{P} : p\mathcal{G}(*) \neq \mathcal{G}(*)\} = \mathcal{P} \cap \mathcal{P}_{\mathcal{G}(*)}$, then $\mathcal{D}_{\mathcal{G}}^{\mathcal{P}} = \mathcal{D}_{\mathcal{G}}^{\mathcal{P}_1}$.

(e) If $\mathcal{G}(*)$ is torsion free, or more generally, if all torsion elements in $\mathcal{G}(*)$ are nilpotent, and $\mathcal{P}_2 = \{p \in \mathcal{P} : p \quad divides \quad |G|\}$, then $\mathcal{D}_{\mathcal{G}}^{\mathcal{P}} = \mathcal{D}_{\mathcal{G}}^{\mathcal{P}_2}$.

PROOF

(a) The result follows from the fact that tensoring with $\mathbb{Z}_{\mathcal{P}}$ is the same as localization with respect to $\{n \cdot 1_{\mathcal{G}(*)} : n \quad a \quad \mathcal{P}'\text{-number}\}$.

(b) Since $\mathbb{Z}_{\mathcal{P}_1} \subseteq \mathbb{Z}_{\mathcal{P}}$, we have a morphism $\mathbb{Z}_{\mathcal{P}_1} \otimes \mathcal{G} \to \mathbb{Z}_{\mathcal{P}} \otimes \mathcal{G}$, and so, $\mathcal{D}_{\mathcal{G}}^{\mathcal{P}} \subseteq \mathcal{D}_{\mathcal{G}}^{\mathcal{P}_1}$.

(c) It follows from (b) that $\mathcal{D}_{\mathcal{G}}^{\mathcal{P}_1} \cup \mathcal{D}_{\mathcal{G}}^{\mathcal{P}_2} \subseteq \mathcal{D}_{\mathcal{G}}^{\mathcal{P}_1 \cup \mathcal{P}_2}$. Now, let S_i $(i = 1, 2)$ be defect sets for $\mathbb{Z}_{\mathcal{P}_i} \otimes \mathcal{G}$, i.e., $\mathcal{D}_{\mathcal{G}}^{\mathcal{P}_i} = \mathcal{U}(S_i)$ $(i = 1, 2)$. Then, there exists a natural $\mathcal{P}_i'$-number n_i such that $n_i \cdot 1_{\mathcal{G}(*)} \in I_{S_i}(\mathcal{G})$, and so,

$$(n_1, n_2) \cdot 1_{\mathcal{G}(*)} \in I_{S_1}(\mathcal{G}) + I_{S_2}(\mathcal{G}) = I_{S_1 \cdot \cup S_2}(\mathcal{G}).$$

Moreover, since (n_1, n_2) is a $(\mathcal{P}_1 \cup \mathcal{P}_2)'$-number,

$$\mathcal{D}_{\mathcal{G}}^{\mathcal{P}_1 \cup \mathcal{P}_2} \subseteq \mathcal{U}(S_1 \dot\cup S_2) = \mathcal{D}_{\mathcal{G}}^{\mathcal{P}_1} \cup \mathcal{D}_{\mathcal{G}}^{\mathcal{P}_2} = \mathcal{U}(S_1) \cup \mathcal{U}(S_2).$$

(d) Since $\mathcal{P}_1 \subseteq \mathcal{P}$, it follows from (b) that $\mathcal{D}_{\mathcal{G}}^{\mathcal{P}_1} \subseteq \mathcal{D}_{\mathcal{G}'}^{\mathcal{P}}$. On the other hand, let S be the defect set for $\mathbb{Z}_{\mathcal{P}_1} \otimes \mathcal{G}$, i.e., $\mathcal{U}(S) = \mathcal{D}_{\mathcal{G}}^{\mathcal{P}_1}$. Then, there exists a natural $\mathcal{P}_1'$-number n such that $n \cdot 1_{\mathcal{G}(*)} \in I_S(\mathcal{G})$. But $n = n_{\mathcal{P}} \cdot n_{\mathcal{P}}'$, and $n_{\mathcal{P}}$ is a $\mathcal{P} \cap \mathcal{P}_1' \subseteq \mathcal{P}_{\mathcal{G}(*)}'$-number, i.e., $n_{\mathcal{G}}(*) = \mathcal{G}(*)$, and so, $n_{\mathcal{P}}' \cdot 1_{\mathcal{G}(*)} \in n_{\mathcal{P}}'$ since

$$n_{\mathcal{P}} \mathcal{G}(*) = n\mathcal{G}(*) \subseteq I_S(\mathcal{G}). \quad \text{So,} \quad \mathcal{D}_{\mathcal{G}}^{\mathcal{P}} \subseteq \mathcal{U}(S)' = \mathcal{D}_{\mathcal{G}}^{\mathcal{P}_1}.$$

(e) It follows again (b) that $\mathcal{D}_{\mathcal{G}}^{\mathcal{P}_2} \subseteq \mathcal{D}_{\mathcal{G}}^{\mathcal{P}}$. So, let S be a defect set for $\mathbb{Z}_{\mathcal{P}_2} \otimes \mathcal{G}$. Then, there exists a natural $\mathcal{P}_2'$-number n such that $n \cdot 1_{\mathcal{G}(*)} \in I_S(\mathcal{G})$. But then, $(n, |G|) \cdot 1_{\mathcal{G}(*)} \in I_S(\mathcal{G})$ by corollary 9.6.2, and $(n, |G|)$ is a $\mathcal{P}'$-number since $\mathcal{P} \subseteq \mathcal{P}_2 \cup \{p \in \mathcal{P} : p \text{ does not devide } |G|\}$. So $\mathcal{D}_{\mathcal{G}}^{\mathcal{P}} \subseteq \mathcal{U}(S) = \mathcal{D}_{\mathcal{G}}^{\mathcal{P}_2}$.

$$\square$$

Our next aim is to study the relations between the defect sets $\mathcal{D}_{\mathcal{G}}^{\mathcal{P}}$ for various $\mathcal{P}$, using theorem 9.5.1.

Lemma 9.6.2 *Let $\mathcal{G} : GSet \to \mathbb{Z}\text{-}\mathcal{M}od$ be a Green functor and S a G-set such that any element of $K_S(\mathcal{G})$ is nilpotent. Then*

(i) *For any set $\mathcal{P}$ of primes, we have $\mathcal{D}_{\mathcal{G}}^{\mathcal{P}} \subseteq h_{\mathcal{P}}(\mathcal{U}(S))$, in particular, $\mathcal{D}_{\mathcal{G}} \subseteq h(\mathcal{U}(S))$.*

(ii) *$|G|^n K_S(\mathcal{G}) = 0$ for a certain power $|G|^n$ of $|G|$.*

PROOF

(i) If $\mathcal{P}$ is any set of primes, then by theorem 9.5.1 we have $|G|_{\mathcal{P}}' \cdot 1_{\mathcal{G}(*)} = x + y$ where $x \in K_S(\mathcal{G})$ and $y \in I_{S_{\mathcal{P}}}(\mathcal{G})$. Suppose that $x^n = 0$. Then,

$$|G|_{\mathcal{P}}'^n \cdot 1_{\mathcal{G}(*)} = (x+y)^n = x^n + y(nx^{n-1} + \cdots + y^{n-1}) = 0 + y \cdot z \in I_{S_{\mathcal{P}}}(\mathcal{G})$$

and so,

$$\mathcal{D}_{\mathcal{G}} \subseteq \mathcal{U}(S_{\mathcal{P}}) = h_{\mathcal{P}}(\mathcal{U}(S)) \quad \text{by lemma 9.6.1 (a).}$$

(ii) If $\mathcal{P}$ is empty, we get $|G|^n \cdot 1_{\mathcal{G}(*)} \in I_S(G)$ for some $n \in \mathbb{N}$, and so, $|G|^n K_S(\mathcal{G}) = 0$, by lemma 9.6.1.

$\square$

Theorem 9.6.2 *Let* $\mathcal{G} : \mathcal{G}Set \to \mathbb{Z}\text{-}\mathcal{M}od$ *be a Green functor such that all torsion elements in* $\mathcal{G}(*)$ *are nilpotent. Then, for any set* $\mathcal{P}$ *of primes, we have* $\mathcal{D}_{\mathcal{G}}^{\mathcal{P}} \subseteq h_{\mathcal{P}} \mathcal{D}_{\mathcal{G}}^{\mathbb{Q}}$, *in particular,* $\mathcal{D}_{\mathcal{G}} \subseteq h \mathcal{D}_{\mathcal{G}}^{\mathbb{Q}}$.

PROOF   Let S be a defect set of $\mathbb{Q} \otimes G$, i.e., $\mathcal{D}_{\mathcal{G}}^{\mathbb{Q}} = \mathcal{U}(S)$. Then, by lemma 9.6.1 (a) and theorem 9.6.1, we have $n \cdot K_S(\mathcal{G}) = 0$ for some $n \in \mathbb{N}$, and so, by hypothesis, $K_S(\mathcal{G})$ is nilpotent. Hence $\mathcal{D}_{\mathcal{G}}^{\mathcal{P}} \subseteq h_{\mathcal{P}} \mathcal{U}(S) = h_{\mathcal{P}} \mathcal{D}_{\mathcal{G}}^{\mathbb{Q}}$ by lemma 9.6.2.

$\square$

Remarks 9.6.2   (i) Note that the assumption in theorem 9.6.2 that all torsion elements in $\mathcal{G}(*)$ are nilpotent is equivalent to having $char(\mathcal{G}(*)/\underline{p}) = 0$ for any minimal prime $\underline{p}$ of $\mathcal{G}(*)$. A particular example is when $\mathcal{G}(*)$ is torsion free.

(ii) Let $\mathcal{G} : \mathcal{G}Set \to \mathbb{Z}\text{-}\mathcal{M}od$ be a Green functor and $\theta : \Omega \to \mathcal{G}$ be the canonical homomorphism. Then the image $\mathcal{G}' = \theta(\Omega)\mathcal{G}$ is a sub-Green functor of $\mathcal{G}$, and so, by lemma 9.6.2, $\mathcal{D}_{\mathcal{G}}^{\mathbb{Q}} = \mathcal{D}_{\mathcal{G}'}^{\mathbb{Q}}$. So, with the assumption in theorem 9.6.2 one can get useful upper bounds for $\mathcal{D}_{\mathcal{G}}$ by considering only the image of Ω in $\mathcal{G}$, tensored with $\mathbb{Q}$. This is one of the reasons why permutation representations are so important in the theory of induced representations.

(iii) Through theorem 9.6.2, the problem of determining the defect basis of a Green functor is partly reduced to studying those Green functors whose images are $\mathbb{Q}$-vector spaces, i.e., to Green functors: $\mathcal{G} : \mathcal{G}Set \to \mathbb{Q}\text{-}\mathcal{M}od$. This in turn is reduced by theorem 9.6.3 to the study of simple G-sets without defect. The next few results characterize sets without defects in a particular way.

Lemma 9.6.3 *Let* $\mathcal{R}$ *be an equivalence class of* G-sets *with respect to* $\diamond$. *Then one has*

(a) *If* $\alpha : X \to Y$ *is a* G-map *with* $X, Y \in \mathcal{R}$, *then* $\alpha(X)$, *considered as a* G-subset *of* Y, *is in* $\mathcal{R}$.

(b) *If* X_0 *is a* G-set *in* $\mathcal{R}$ *with a minimal number of elements, i.e.,* $|X_0| \le |Y|$ *for all* $Y \in \mathcal{R}$, *then*

(i) *Any* G-map $\alpha : X_0 \to Y$ *is injective and has a left inverse* $\alpha' : Y \to X_0$.

(ii) *Any G-map* $\beta : Y \rightarrow X_0$ *is surjective and has a right inverse* $\beta' : X_0 \rightarrow Y$.

(iii) *If* $X_1 \in \mathcal{R}$ *with* $|X_1| = |X_0|$, *then any G-map* $\alpha : X_0 \rightarrow X_1$ *is an isomorphism. In particular,* X_0 *is determined by* $\mathcal{R}$ *up to isomorphism.*

PROOF   Left as an exercise.   ⬜

Remarks and definition 9.6.1 (i) A G-object $X \in \mathcal{R}$ satisfying (b) of 9.6.3 is called a smallest G-set in $\mathcal{R}$. If Y is any G-set in $\mathcal{R}$, X_0 is called the smallest G-set with $X_0 \Diamond Y$.

(ii) If $I_{\mathcal{R}} = \{T \in I : T < Y$ for some $Y \in \mathcal{R}\}$, and $I'_{\mathcal{R}} = \{T \in I_{\mathcal{R}} :$ T is maximal in $I_{\mathcal{R}}$ with respect to $<\}$, then one can show $X_0 = \cup_{T \in I'_{\mathcal{R}}} T$ is a smallest G-set in $\mathcal{R}$.

Theorem 9.6.3   (a) *Two maps* $\alpha_1 : S_1 \rightarrow X_1$, $\alpha_2 : S_2 \rightarrow X_2$ *are defect maps if and only if* $\alpha_1 \dot{\cup} \alpha_1 : S_2 \dot{\cup} S_2 \rightarrow X_1 \dot{\cup} X_2$ *is a defect map over* $X_1 \dot{\cup} X_2$. *In particular,* $X_1 \dot{\cup} X_2$ *is without defect if and only if* X_1, X_2 *are without defects.*

(b) *If* M *is a Mackey functor from GSet into* $\mathbb{Z}$-$\mathcal{M}od$ *and* I *is a basis of GSet, and* $I_M = \{T \in I : T$ *without defect with respect to* $M\}$, *then* $S = \dot{\cup} T$ *is a defect set with respect to* M.

PROOF

(a) Follows from the definitions and lemma 9.6.3.

(b) Let X be a "smallest" defect set for M in the sense of lemma 9.6.3(b) and assume that $T \in I_M$. Since the X-projectivity of M implies that $M|_T$ is $(T \times X \xrightarrow{pr_1} T)$-projective, there must exist a map $\alpha : T \rightarrow T \times X$ with $T \xrightarrow{\alpha} T \times X \xrightarrow{pr_1} T$ being the identity of T. So , in particular, $T < X$, and so, $S = \underset{T \in I_M}{\dot{\cup}} T < X$.

Vice-versa, let X be the disjoint union of simple G-sets $T_1, T_2, \cdots, T_k$ so that $X = \overset{k}{\underset{i=1}{\dot{\cup}}} T_i$. We have to show that $T_i \in M$ for all $i = 1, 2 \cdots, k$. Suppose not. Then there would exist an index $i \in \{1, \cdots, k\}$ and a G-map $Y_i \xrightarrow{\beta_i} T_i$ without a right inverse, such that $M|_{T_i}$ is β_i-projective, which in turn would imply that M is

$$Y = T_1 \dot{\cup} \cdots \dot{\cup} T_{i-1} \dot{\cup} Y_i \dot{\cup} T_{i+1} \dot{\cup} \cdots \dot{\cup} T_p\text{-projective}$$

with $Y \lneq X$, a contradiction.

⬜

Remark 9.6.1 Although theorem 9.6.3(b) says that the disjoint union S of all non-isomorphic simple sets without defect (which by (a) is a set without defect) is a defect set, it does not follow that any $S' < S$ is without defect, i.e., $S' < S$, and S without defect does not necessarily imply that S' is without defect.

Theorem 9.6.4 *Let $\mathcal{G} : GSet \to \mathbb{Q}\text{-}\mathcal{M}od$ be a Green functor and S a simple G-set. Then*

(i) *S is without defect with respect to $\mathcal{G}$ if and only if there exists a linear map $\epsilon : \mathcal{G}(S) \to \mathbb{Q}$ such that*

$$
\begin{array}{ccc}
\Omega(S) & \xrightarrow{\theta_S} & \mathcal{G}(S) \\
\downarrow{\scriptstyle \epsilon_S} & & \downarrow{\scriptstyle \epsilon} \\
\mathbb{Z} & \hookrightarrow & \mathbb{Q}
\end{array}
$$

commutes, where the "augmentation" $\epsilon_S = \varphi_{id_S} : \Omega(S) \to \mathbb{Z}$ maps an object $\alpha : S' \to S$ in $GSet/S$ onto the number of simple components of S' isomorphic to S.

(ii) *S is not without defect if and only if there exists an element $x \in \Omega(S)$ with $\epsilon_S(x) \neq 0$, i.e., if and only if $\theta_S(x) = 0$ for one (all) element(s) $x \in \Omega(S)$ with $\epsilon_S(x) = \varphi_{id_S}(x) \neq 0$ but $\varphi_\alpha(x) = 0$ for all $\alpha \in I_S$, $\alpha \not\cong id_S$, where $I_S = \{\alpha : T \to S : T \in I\}$ and $\varphi_\alpha(\beta : S' \to S) = |\{\gamma \in GSet(T, S) : \beta\gamma = \alpha\}|$.*

PROOF

(i) Suppose that there exists a commutative diagram

$$
\begin{array}{ccc}
\Omega(S) & \xrightarrow{\theta_S} & \mathcal{G}(S) \\
\downarrow{\scriptstyle \epsilon_S} & & \downarrow{\scriptstyle \epsilon} \\
\mathbb{Z} & \hookrightarrow & \mathbb{Q}
\end{array}
$$

and that $\alpha : S' \to S$ is a G-map with $\alpha_* : \mathcal{G}(S') \to \mathcal{G}(S)$ surjective. We have to show that α is a right inverse. For this, it is enough to show that $S < S'$ since S being simple is a smallest $GSet$ in its equivalence class.

Since $S' < S$, we only have to show that $S < S'$. Now, by theorem 9.4.5, there exists an element $x \in \Omega(*) = \Omega(G)$ such that $\varphi_S(X) \neq 0$ and $\varphi_T(x) = 0$ for any simple G-set T with $T \not\cong S$. Also, by lemma 9.5.1 (i), for any G-set Y with $\eta_Y : Y \to *$ the unique map from Y to $*$, we have $(\eta_Y) * (x) \neq 0$ if and only if $S < Y$. So, we have to show that $(\eta_{S'}) * (X) \neq 0$. Now consider the commutative diagram

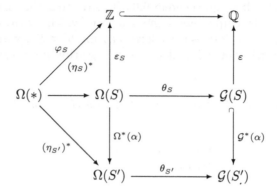

where the commutativity of the upper left hand triangle follows from the proof of lemma 9.5.1 (i) and the injectivity of $\mathcal{G}^*(\alpha)$ follows from corollary 9.2.2.

By considering the various images of x in this diagram, we get

$$\epsilon(\theta_S)(\eta)^*(x) = \varphi_S(x) \neq 0 \Rightarrow \theta_S((\eta_S)^*(x)) \neq 0 \Rightarrow 0 \neq \mathcal{G}^*(\alpha)\theta_S((\eta_S)^*(x))$$
$$= \theta_{S'}((\eta_{S'})^*(x)) \Rightarrow (\eta_{S'})^*(x) \neq 0.$$

Now, suppose that S is without defect. Then, in $GSet/S$ we have a unique maximal $\diamond$-equivalence class just below the final class that is represented, for instance, by the sum σ_S of all maps $\alpha : T \to S$ $(T \in I)$ such that

$$\alpha \neq id_S, \text{ i.e., } \sigma_S = \dot{\cup}_{\substack{\alpha \in I_S \\ \alpha \neq id_S}} \alpha : S^0 = \dot{\cup}_{\substack{T \in I_S \\ \alpha \neq id_S}} T \to S.$$

Note that by its definition σ_S has no section and moreover,

$$\Omega(S) = \sum_{\alpha \in I_S} \mathbb{Z}\alpha = \mathbb{Z} \cdot Id_S \oplus I_{\sigma_S}(\Omega), \quad Ker(\epsilon_S) = I_{\sigma_S}(\Omega).$$

Consider the following diagram:

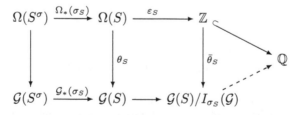

One can see easily that the exactness of the two rows and the surjectivity of ϵ_S implies the existence of $\overline{\theta}_S$. Since $1_{\mathcal{G}(S)} \in \theta_S(\Omega(S))$, but $1_{\mathcal{G}(S)} \notin$

$I_{\sigma_S}(\mathcal{G})$ (because S has no defect and σ_S has no section), the induced map $\mathbb{Z} \xrightarrow{\theta_S} \mathcal{G}(S)/I_{\sigma_S}(\mathcal{G})$ is non-zero and hence injective, since $\mathcal{G}(S)/I_{\sigma_S}(\mathcal{G})$ is a $\mathbb{Q}$-vector space. So, there exists the required map $\mathcal{G}(S)/I_{\sigma_S}(\mathcal{G}) \to \mathbb{Q}$, which makes the whole diagram commutative.

(ii) The proof of the first part follows from (i) above. Note that a map $\epsilon : \mathcal{G}(S) \to \mathbb{Q}$ with the properties in (i) exists if and only if $Ker(\theta_S : \Omega(S) \to \mathcal{G}(S)) \subset Ker(\epsilon_S : \Omega(S) \to \mathbb{Z})$. Now for $x \in \Omega(S)$, one has $\epsilon_S(x) \neq 0$) if and only if x has the form $x = y + n \cdot id_S$ where $y \in I_{\sigma_S}(\Omega)$ and $n = \epsilon_S(y + n \cdot id_S) = \epsilon_S(x) \neq 0$. Now, $\theta_S(x) = 0$ if and only if

$$0 = \theta_S(y + n \cdot id_S) = \theta_S(y) + n \cdot 1_{\mathcal{G}(S)} \Leftrightarrow n \cdot 1_{\mathcal{G}(S)}$$
$$= -\theta_S(y) \in I_{\sigma_S}(\mathcal{G}) \Leftrightarrow 1_{\mathcal{G}(S)} \in I_{\sigma_S}(\mathcal{G}) = \mathcal{G}(S) \Leftarrow S \text{ is not without defect}$$

For the second part let $y \in \Omega(S)$ be such that $\epsilon_S(y) \neq 0$, $\varphi_\alpha(y) = 0$ for all $\alpha \in I_S$, $\alpha \not\cong id_S$ (e.g., $y = (\eta_S)^*(x)$ with $x \in \Omega(*)$ as in the proof of (i)). If $\theta_S(y) = 0$, then, as has been shown above, S has a defect with respect to $\mathcal{G}$. On the other hand, if S has a defect with respect to $\mathcal{G}$, then there exists $x \in \Omega(S)$ with $\epsilon_S(x) \neq 0$ and $\theta_S(x) = 0$, and then,

$$\epsilon_S(x \cdot y) = \epsilon_S(x) \cdot \epsilon_S(y) \neq 0, \theta_S(x \cdot y) = \theta_S(x) \cdot \theta_S(y) = 0,$$
$$\varphi_\alpha(x \cdot y) = \varphi_\alpha(x) \cdot \varphi_\alpha(y) = 0$$

for all $\alpha \in I_S$, $\alpha \neq id_S$. So, there exists at least one such element (i.e., $x \cdot y$) in $\Omega(S)$, and since any other element z with $\varphi_\alpha(z) = 0$ for all $\alpha \in I_S$, $\alpha \neq id_S$ differs from $x \cdot y$ only by a scalar, we have $\theta_S(s) = 0$ for any such element.

⬜

Remark 9.6.2 (i) In theorem 9.6.4, the field $\mathbb{Q}$ in the diagram can be replaced by any field of characterstic zero, e.g., $\mathbb{R}, \mathbb{C}$ or $\hat{\mathbb{Q}}_p$ (see [51]).

(ii) Our next aim is to characterize those Green functors G for which any G-set S' with $S' < S$ for some S without defect is itself without defect. Although this is not necessarily the case for arbitrary Green functors, it is, however, true for all Green functors that occur naturally in integral representation theory, or more generally, equivariant algebraic K-theory. First, we prove a lemma 9.6.4 necessary for the proof of theorem 9.6.5 below.

Lemma 9.6.4 *Let $\beta : S' \to S$ be a G-map. Then the only ideal contained in $ker(\Omega_*(\beta) : \Omega(S') \to \Omega(S))$ is the zero ideal.*

PROOF Assume $0 \neq x \in \Omega(S')$ and $\Omega(S') \cdot x \in ker(\Omega_*(\beta))$. Since $x \neq 0$, there exists $(\alpha : T \to S') \in I_{S'}$ such that $\varphi_\alpha(x) \neq 0$. Similarly, from lemma 9.4.7 we have $x \cdot \alpha = \varphi_\alpha(x) + \sum_{\gamma \in I_S} n_\gamma \gamma$ where $n_\gamma \neq 0$ only for such $(\gamma : Y \to S') \in I_{S'}$ with $\gamma < \alpha$, $\gamma \neq \alpha$ in $G\mathcal{S}et/S'$. In particular, $|Y| \gneqq |T|$. But $\alpha \cdot x \in \Omega(S')x \subseteq Ker(\Omega_*(\beta))$ implies that $0 = \Omega_*(\beta)(\alpha \cdot x) = \varphi_\alpha(x)\beta\alpha + \sum_{\gamma \in I_{S'}} n_\gamma \beta\gamma$ in $\Omega(S)$, and since $\beta\alpha$, $\beta\gamma$ represents elements of the canonical basis I_S of $\Omega(S)$ and $\varphi_\alpha(x) \neq 0$, such an equation can hold only if $\beta\alpha = \beta\gamma \in \Omega(S)$, i.e., $(\beta\alpha : T \to S) \cong (\beta\gamma : Y \to S)$ in $G\mathcal{S}et$ for some $\gamma : Y \to S'$ with $n_\gamma \neq 0$ – a contradiction to $|Y| \neq |T|$. □

Theorem 9.6.5 *Let* $\mathcal{G} : G\mathcal{S}ets \to \mathbb{Q}\text{-}\mathcal{M}od$ *be a Green functor,* S *a* G*-set. Then, all* G*-sets* $S' < S$ *have no defect with respect to* $\mathcal{G}$ *if and only if the canonical map* $\theta_S : \Omega(S) \to \mathcal{G}(S)$ *is injective.*

PROOF   Suppose that any $S' < S$ has no defect with respect to $\mathcal{G}$ and $x \in Ker(T_S : \Omega(S) \to \mathcal{G}(S))$. We show that $x = 0$, i.e., $\varphi_\alpha(x) = 0$ for all $\alpha : T \to S$ in I_S. Now, by lemma 9.6.4 and our assumption, we have a commutative diagram

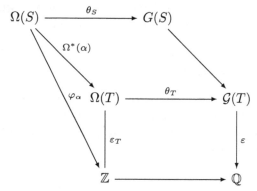

Note that $\varphi_\alpha = \epsilon_T \Omega^*(\alpha)$ since for $\beta : S' \to S$ we have

$$\varphi_\alpha(\beta) = |\{\gamma : T \to S' | \beta\gamma = \alpha\}| = |\{\gamma' : T \to S' \times_S T | pr_T \gamma' = id_T\}|$$
$$= \epsilon_T(pr_T) = \epsilon_T(\Omega^*(\alpha)(\beta))$$

where $pr_T = \Omega^*(\alpha)(\beta)$ is the projection of $S' \times_S T$ onto T. So, if $\theta_S(x) = 0$, then $\epsilon(\mathcal{G}^*(\alpha)\theta_S(x)) = \varphi_\alpha(x) = 0$.

To prove the converse, we first show that if $\mathcal{G} : G\mathcal{S}et \to \mathbb{Z}\text{-}\mathcal{M}od$ is a Green functor (not necessary $\mathbb{Q}$-vector spaces as images), then the injectivity of the canonical map $\theta_S : \Omega(S) \to \mathcal{G}(S)$ implies the injectivity of $\theta_{S'} : \Omega(S') \to \mathcal{G}(S')$ for any $S' < S$. To show this, choose a map $\beta : S' \to S$ and consider the diagram:

$$
\begin{array}{ccc}
\Omega(S') & \xrightarrow{\Omega_*(\beta)} & \Omega(S) \\
\downarrow{\scriptstyle \theta_{S'}} & & \downarrow{\scriptstyle \theta_S} \\
\mathcal{G}(S') & \xrightarrow{\mathcal{G}_*(\beta)} & \mathcal{G}(S)
\end{array}
$$

Since θ_S is injective, we have $Ker(\theta_{S'}) \subseteq Ker(\Omega_*\beta)$. But $Ker(\theta_{S'})$ is an ideal, so, by lemma 9.6.4, $Ker(\theta_{S'}) = 0$. Now, if $\mathcal{G} : GSet \to \mathbb{Q}\text{-}\mathcal{M}od$ is a Green functor, it follows easily from 9.6.4 that a G-set S' is without defect with respect to $\mathcal{G}$ if $\theta_{S'} : \Omega(S') \to \mathcal{G}(S')$ is injective. $\quad\Box$

9.7 Defect basis for K_0^G-functors

9.7.1 Let $(\mathcal{C}, \perp)$ be a symmetric monoidal category together with further associative composition 'o' such that $\mathcal{C}$ is distributive with respect to '$\perp$' and 'o', G a finite group. In example 1.3.2(iv) we defined a functor $K_0^G(-, \mathcal{C})$: $GSets \to \mathbb{Z}\text{-}\mathcal{M}od$ which, as we saw in example 9.1.1(iv), is a Green functor. If $\theta : H \to G$ is a group homomorphism, we defined in example (1.3.3) (v) a natural transformation of functors $K_0^G(-, \mathcal{C}) : GSets \to \mathbb{Z}\text{-}\mathcal{M}od$ to $K_0^H \circ \hat\theta$: $GSet \to HSet \to \mathbb{Z}\text{-}\mathcal{M}od$ where $\hat\theta$ is a functor $GSet \to HSet$ induced by restriction of the action of G on a G-set S to H via θ. These considerations can be formalized with the introduction of the concept of inversal family of Green functors as in definition 9.7.1 below. The aim of this section is to determine the defect basis for $K_0^G(-, \mathcal{C})$ for various categories $\mathcal{C}$. We shall derive these results for symmetric monoidal categories $(\mathcal{C}, \perp)$. If $(\mathcal{C}, \perp)$ possesses additional composition 'o' as above such that $\mathcal{C}$ is distributive with respect to '$\perp$', and 'o' we shall call $\mathcal{C}$ a distributive category and denote it by $(\mathcal{C}, \perp, \circ)$. We note, however, that most of our computations of defect basis will be on $\mathcal{C} = \mathcal{P}(R)$, which is also an exact category and so will apply also in chapter 10 on equivariant higher K-theory for exact categories.

Definition 9.7.1 *A universal family of Green functors is a family of Green functors $\mathcal{G}_G : GSet \to R\text{-}\mathcal{M}od$, one for each finite group G, together with natural transformations of Green functors $\eta_\theta : \mathcal{G}_G \to \mathcal{G}_H \cdot \hat\theta$, one for any group homomorphism $\theta : H \to G$ such that*

$$\eta_{id} = Id, \quad \eta_{\theta_1\theta_1} = (\eta_{\theta_2}\hat\theta_1) \circ \eta_{\theta_1} : \mathcal{G}_H \xrightarrow{\eta_{\theta_1}} \mathcal{G}_H \circ \hat\theta_1 \xrightarrow{\eta_{\theta_2}\hat\theta_1} \mathcal{G}_H \circ \hat\theta_2 \circ \hat\theta_1$$

for any $\theta_1 : H \to G, \theta_2 : H' \to H$ and

$$\mathcal{G}_G(G/H) \xrightarrow{\eta_\theta} \mathcal{G}_H(G/H|H) \to \mathcal{G}_H(H/H)$$

is an isomorphism for any embedding

$$\theta : H \hookrightarrow G.$$

Remark 9.7.1 The relevance of definition 9.7.1 lies in the fact that any small distributive category $\mathcal{C}$ defines a universal family of Green functors $K_0^G(-, \mathcal{C})$: $GSets \to \mathbb{Z}\text{-}\mathcal{M}od$ such that $K_0^G(G/G, \mathcal{C}) =: K(G/G, \mathcal{C})$ is the Grothendieck ring of G-objects in $\mathcal{C}$.

Theorem 9.7.1 *Let $\mathcal{G}$ be a universal family of Green functors with values in R-$\mathcal{M}$od. Let $\mathcal{D}'(\mathcal{G})$ be the class of all finite groups H such that H/H is a defect set of $\mathcal{G}_H$, i.e., such that $\sum_{H' \lneq H} \mathcal{G}_H(H/H') \to \mathcal{G}_H(H/H)$ is not surjective (resp. such that $\mathcal{G}_H(S) \to \mathcal{G}_H(*)$ is surjective if and only if $S^H \neq \emptyset$). Then,*

(i) $\mathcal{D}(\mathcal{G}_H) = \{H' \leq G | \exists H \in \mathcal{D}'(\mathcal{G})$ such that $H' \overset{G}{\leq} H < G\}$.

(ii) $\mathcal{D}'(\mathcal{G})$ is closed with respect to epimorphic images, i.e., if $\theta : H \twoheadrightarrow L$ is surjective and $H \in \mathcal{D}'(G)$, then $L \in \mathcal{D}'(\mathcal{G})$.

PROOF   We show that

(i) $\sum\limits_{\substack{H \leq G \\ H \in \mathcal{D}'(\mathcal{G})}} \mathcal{G}(H) = \sum\limits_{\substack{H < G \\ H \in \mathcal{D}'(\mathcal{G})}} \mathcal{G}_G(G/H) \to \mathcal{G}_G(G/G) = \mathcal{G}(G)$ is surjective

by induction with respect to $|G|$. If $|G| = 1$, or more generally, if $G \in \mathcal{D}'(G)$, then the required map is obviously surjective. Suppose $G \in \mathcal{D}'(G)$. Then by the definition of $\mathcal{D}'(G)$, the map $\sum_{\gamma \lneq \pi} \mathcal{G}(H) \twoheadrightarrow \mathcal{G}(G)$ is surjective, which implies by induction that

$$\sum\limits_{\substack{H \leq G \\ H \in \mathcal{D}'(\mathcal{G})}} \mathcal{G}(H) = \sum\limits_{\substack{H < G \\ H \in \mathcal{D}'(\mathcal{G})}} \mathcal{G}_G(G/H) \twoheadrightarrow \mathcal{G}_G(G/G) = \mathcal{G}(G)$$

is also surjective. On the other hand, if $\sum_{H \in \mathcal{D}} \mathcal{G}_G(G/H) \to \mathcal{G}_G(G/G)$ is surjective for some set $\mathcal{D}$ of subgroups of G, we have to show that for any $H \leq G$ such that $H \in \mathcal{D}'(\mathcal{G})$, there exists $L \in \mathcal{D}$ such that $H \overset{G}{\leq} L$, i.e., $(G/L)^H \neq \varphi$. However, by restricting the above formula to H via η_{i_H}, we get a commutative diagram

$$\begin{array}{ccc} \sum_{L \in \mathcal{D}} \mathcal{G}_G(G/L) = \mathcal{G}_G(\dot{\cup}_{L \in \mathcal{D}}(G/L)) & \longrightarrow & \mathcal{G}_G(G/G) \\ \downarrow & & \downarrow \\ \mathcal{G}(\dot{\cup}_{L \in \mathcal{D}}(G/L)|_H) & \longrightarrow & \mathcal{G}_G(H/H) \end{array} \quad .$$

Since η_{i_H} maps the unit e_G in $\mathcal{G}_G(G/G)$ onto the unit e_H in $\mathcal{G}_H(H/H)$ and the upper arrow is surjective, we see that e_H is contained in the image of the lower arrow, which image is an ideal, and so, the lower arrow is surjective too. Since $H \in \mathcal{D}'(G)$, this implies that $\dot{\cup}(G/L)^H \neq \varphi$ by the definition of $\mathcal{D}'(G)$.

(ii) Let $\theta : H \to G$ be a homomorphism and S a G-set. Consider the diagram

$$\begin{array}{ccc} \mathcal{G}_G(S) & \longrightarrow & \mathcal{G}_G(G/G) \\ \downarrow & & \downarrow \\ \mathcal{G}_H(S|_H) & \longrightarrow & \mathcal{G}_H(H/H) \end{array} \quad .$$

If $H \in \mathcal{D}'(G)$ and if $\mathcal{G}_G(S) \to \mathcal{G}_G(G/G)$ is surjective, then $\mathcal{G}_H(S|_H) \to \mathcal{G}_H(H/H)$ is also surjective, and so, $(S|_H)^H \neq \varphi$. But if $\theta : H \to G$ is surjective, one has $(S|_H)^H = S^H$. So, $H \in \mathcal{D}'(\mathcal{G})$ implies that $G \in \mathcal{D}'(\mathcal{G})$.

$\square$

Definition 9.7.2 *A universal family of Green functors $\mathcal{G}$ is said to be saturated if $\mathcal{D}'(\mathcal{G})$ is also closed with respect to subgroups. So, in this case, part (i) of theorem 9.7.1 could be written in the form*

$$\mathcal{D}(\mathcal{G}_G) = \{H \leq G | H \in \mathcal{D}'(\mathcal{G})\}.$$

Also, a universal family of Green functors is not necessarily saturated. The next result indicates that the K_0^G-functors in which we are interested are saturated.

Theorem 9.7.2 *Let $(\mathcal{C}, \bot, \theta)$ be a distributive category, and $K_0^G(-, \mathcal{C})$ the associated universal Green functor. Then $A \otimes K_0^G(-, \mathcal{C})$ is saturated for any ring A.*

PROOF If $\mathcal{G}$ is an universal Green functor, we write:

$$\overline{\mathcal{G}(G)} = \mathcal{G}(G)/Im \left(\sum_{H \leq G} \mathcal{G}(H) \to \mathcal{G}(G) \right)$$

so that $\overline{\mathcal{G}(G)} \neq 0$ if and only if $G \in \mathcal{D}'(\mathcal{G})$.
Now, if $\mathcal{G} = K_0^G(-, \mathcal{C})$, we have to show that $\overline{A \otimes \mathcal{G}(H)} = A \otimes \overline{\mathcal{G}(H)} = 0$ implies that $A \otimes \overline{\mathcal{G}(G)} = 0$ whenever $H \leq G$. To do this, it suffices to construct a ring homomorphism $\overline{\mathcal{G}(H)} \to \overline{\mathcal{G}(G)}$. Note that

$$\mathcal{G}(H) = K_0^G(H/H, \mathcal{C}) \simeq K^G(G/H, \mathcal{C}) = K([\underline{G/H}, \mathcal{C}]).$$

To the homomorphism $\varphi : G/H \to G/G$ are associated two functors

$$\varphi^* : [\underline{G/G}, \mathcal{C}] \to [\underline{G/H}, \mathcal{C}] \text{ and } \varphi_*^{\perp} : [\underline{G/H}, \mathcal{C}] \to: [\underline{G/G}, \mathcal{C}],$$

the latter using the composition in $\mathcal{C}$. (See 1.1.5.)
We also have a functor $\varphi_* : [\underline{G/H}, \mathcal{C}] \to [\underline{G/G}, \mathcal{C}]$, which associates to a G-equivariant $\mathcal{C}$-bundle ζ over G/H, the G-object $\theta(\zeta_K | K \in G/H)$. The last functor defines a θ-multiplicative map from isomorphism classes in $[\underline{G/H}, \mathcal{C}]$

onto isomorphism classes in $[G/G, \mathcal{C}]$, and so, we have a diagram

$$
\begin{array}{ccc}
K^+(H/H,\mathcal{C}) \simeq K^+(G/H,\mathcal{C}) & \longrightarrow & K^+(G/G,\mathcal{C}) \\
\downarrow & & \downarrow \\
K_0^H(H/H,\mathcal{C}) & & K_0^G(G/G,\mathcal{C}). \\
\downarrow & & \downarrow \\
\overline{K_0^G(H/H,\mathcal{C})} & \longrightarrow & \overline{K_0^G(G/G,\mathcal{C})}.
\end{array}
$$

We show that the lower dotted arrow exists as a ring homomorphism. This would follow from the following.   □

Lemma 9.7.1   *(a) If ζ_1, ζ_2 are two bundles over G/H, then $\theta(\perp (\zeta_1, \zeta_2)) = \perp (\theta(\zeta_1), \theta(\zeta_2))$ modulo $Im(\sum\limits_{H' \lneqq G} K_0^G(G/H',\mathcal{C}) \to K_0^G(G/G,\mathcal{C}))$, i.e.,*

$$
\theta(\perp (\zeta_1, \zeta_2)) = \perp (\theta(\zeta_1), \theta(\zeta_2)) \; in \; \overline{K(G,\mathcal{C})}.
$$

(b) If $\gamma = \varphi_*^{\perp}(\zeta)$ for some $\mathcal{C}$-bundle ζ over some GSet S with $S^H = \emptyset$ with respect to some G-map $\varphi : S \to G/H$, then

$$
\theta(\gamma) \in Im(\sum\limits_{H' \lneqq G} K_0^G(G/H',\mathcal{C}) \to K_0^G(G/G,\mathcal{C})), \quad i.e.,
$$
$$
\theta(\gamma) = 0 \quad in \quad \overline{K(G,\mathcal{C})}.
$$

PROOF

(a) First note that if η is a G-equivariant $\mathcal{C}$-bundle over T with $T^G = \varphi$, and $I = Ker(K(G,\mathcal{C}) \to \overline{K(G,\mathcal{C})}) = Im(\sum_{H' \lneqq \pi} K_0^G(G/H',\mathcal{C}) \to K_0^G(G/G,\mathcal{C}))$, then $\perp (\eta_t | t \in T) \in I$. Now,

$$
\begin{aligned}
\theta(\perp (\zeta_1, \zeta_2)) &= \theta(\perp (\zeta_{1K}, \zeta_{2K}) | K \in G/H) \\
&= \tfrac{\perp}{\alpha} \theta_K(\zeta_{\alpha(K),K} | K \in G/H | \alpha \in Hom(G/H, \{1,2\}))
\end{aligned}
$$

where $Hom(G/H, \{1,2\})$ is the G-set of all maps from G/H into $\{1,2\}$, identified with the set of all sections of projection $G/H \times \{1,2\} \to G/H$. We may consider $\theta(\zeta_{\alpha(K),K} | K \in G/H)_\alpha$ ($\alpha \in Hom(G/H, \{1,2\})$) as a G-equivariant $\mathcal{C}$-bundle over $Hom(G/H, \{1,2\})$. However, $Hom(G/H, \{1,2\})$ is a disjoint union of

$$
T_1 = Hom(G/H, \{1\}) \simeq G/G, \quad T_2 = Hom(G/H, \{2\}) \simeq G/G \quad and
$$
$$
T = \{\alpha \in Hom(G/H, \{1,2\}) | \alpha \; not \; constant\}. \quad So,
$$

$T^G = \varphi$ and $\perp_{\alpha \in T_i} (\theta_K(\zeta_{\alpha(K),K})) = \theta(\zeta_i)$ $(i = 1, 2)$. So,

$\tfrac{\perp}{\alpha} (\theta(\zeta_{\alpha(K),K} | K \in G/H) | \alpha \in Hom(G/H, \{1,2\})) \equiv \perp (\zeta_1, \zeta_2) \; mod \; I$ since by $T^G = \varphi$, $\perp (\cdots)$ applied to any bundle over T is contained in I.

(b) We have $\theta(\zeta) = \theta(\varphi_\alpha^\perp(\zeta)) = \perp (\theta(\zeta_{\delta(K)}|K \in G/H|\delta \in \Gamma))$ where Γ is the G-set of all sections $\delta : G/H \to S$ of $\varphi : S \to G/H$. Since $S^H = \varphi$, we have $\Gamma^G = \varphi$, and so, $\theta(\varphi_*^\perp(\zeta)) \in I$.

☐

Remarks 9.7.1 It is clear from theorem 9.7.2 that $\mathcal{D}(A \otimes K_0^G(-,\mathcal{C})) = \mathcal{D}'K_0^G(-,\mathcal{C}))$ and we write $\mathcal{D}_A(\mathcal{C})$ for this defect basis. So, in order to prove induction theorems for $A \otimes K_0^G(-,\mathcal{C}))$, we just have to compute $\mathcal{D}_A(\mathcal{C})$. This we now set out to do.

First we fix some notations. If $H \leq G$, R a commutative ring, and N an RH-module, we write $N^{H \to G}$ for the induced RG-module $RG \otimes_{RH} N$, i.e., the RG-module induced from the G-equivariant R-$\mathcal{M}od$-bundle $G \times_H N$ over G/H.

If S is a G-set, we write $R[S]$ for the associated permutation representation, i.e. the RG-module, which is induced from the trivial G-equivariant R-$\mathcal{M}od$-bundle $R \times S/S$ over S. So, $R[G/H] \simeq R^{H \to G}$ where $R[H/H] = R$ is the trivial $R[H]$-module.

Lemma 9.7.2 *Let $\mathcal{D}$ be a class of finite groups closed under epimorphic images and subgroups, p a prime. If $\mathbb{Z}_p \times \mathbb{Z}_p$ and any non-Abelian group of order $p \cdot q$ with $q \mid (p-1)$ (q another prime) is not contained in $\mathcal{D}$, then any group in $\mathcal{D}$ has a cyclic Sylow-p-subgroup and is p-nilpotent.*

PROOF   If $G \in \mathcal{D}$, and G_p a Sylow-p-subgroup of G, then any factor group of G_p is in $\mathcal{D}$. But $\mathbb{Z}_p \times \mathbb{Z}_p \notin \mathcal{D}$. So, G_p is cyclic.

Suppose G was not p-nilpotent. Then, by a popular transfer argument, there would exist some $g \in G$ such that $g \in N_G(G_p)$ but $g \notin C_G(G_p)$. Since the p-part of g is necessarily contained in $G_p \leq C_G(G_p)$, we may assume g to be p-regular and then assume as well that $g^q \in C_G(G_p)$ for some prime $q \neq p$. But then, with $G_p = <h>$, the group $<h,g> / <h^p,g^q>$ is non-Abelian of order pq with $q \mid (p-1)$, a contradiction to

$$G \in \mathcal{D} \Rightarrow <h,g> \in \mathcal{D} \Rightarrow <h,g> / <h^p,g^q> \in \mathcal{D}.$$

☐

Lemma 9.7.3 *Let R be a commutative ring with identity, p a prime such that $pR = R$. Then $\mathcal{D}_\mathcal{Q}(\mathcal{P}(R)) =: \mathcal{D}_\mathcal{Q}(R)$ contains neither $\mathbb{Z}_p \times \mathbb{Z}_p$ nor any non-Abelian group of order pq with $q \mid (p-1)$.*

PROOF   The proof follows from the following   ☐

Lemma 9.7.4 *(a) Under the hypothesis of lemma 9.7.3, let $pR = R$, $G = \mathbb{Z}_p \times \mathbb{Z}_p$, and $H_0, H_1, \cdots, H_p$ the $(p+1)$ subgroups of order p in G.*

Then, $\underbrace{R \oplus R \oplus \cdots \oplus R}_{p \text{ times}} \oplus R[G/e] \simeq \oplus_{i=0}^{p} R[G/H_i]$ where R here is the

trivial RG-module representing 1 in $K(G, R)$.

(b) Let $R = \mathbb{Z}(\frac{1}{p}, \zeta)$ where ζ is a primitive p-th root of unity, G the semi-direct product $\mathbb{Z}_p \rtimes A$ where $A = \mathrm{Aut}(\mathbb{Z}_p)$ is cyclic of order p-1.
Let $\widetilde{R}$ be R considered as a $\mathbb{Z}_p$-module with $z_i r = \zeta^i r$ (where $r \in R$, $i \in \mathbb{F}_p$) and the elements $z = z_i$ in $\mathbb{Z}_p$ indexed by the elements $i \in \mathbb{F}_p$ such that $z_i \cdot z_j = z_{i+j}$. Then, $R[G/A] \simeq R \oplus \widetilde{R}^{\mathbb{Z}_p \to G}$.

PROOF

(a) Define for any finite group G, G-set S, and commutative ring R, $I_R[S] = Ker(R[S] \to R)$ where $R[S] \to R$ is defined by $s \mapsto 1$ ($s \in S$). Then, $pR = R$ implies that $R[G/e] \simeq \overset{p}{\underset{i=0}{\oplus}} I[G/H_i]$. An explicit isomorphism is given by first restricting the canonical maps $R[G/e] \to R[G/H_i]$, $g \cdot e \mapsto gH_i$ to $I[\cdots]$ and taking their product, its inverse being the sum of the restriction to $I[\cdots]$ of the maps

$$R[G/H_i] \to R[G/e] : gH_i \mapsto \frac{1}{p} \sum_{x \in gH_i} x \cdot e.$$

(b) We index the elements in A by the elements in $\mathbb{F}_p^*$, $a = a_j$ ($j \in \mathbb{F}_p, j \neq 0$) such that $a_j(z_j) = z_{ij}$. Note that $R[G/A]$ has an r-basis $x_i = z_i A$ ($i \in \mathbb{F}_p$) such that $z_j x_i = x_{i+j}$, $a_j x_i = x_{ij}$. Consider $y_j = \sum_{i \in \mathbb{F}_p} \zeta^{-ij} x_i$ ($j \in \mathbb{F}_p$). Since the determinant

$$\begin{vmatrix} 1 & 1 & \cdots & 1 \\ 1 & \zeta & \cdots & \zeta^{p-1} \\ \vdots & \vdots & \ddots & \vdots \\ 1 & \zeta^{p-1} & \cdots & \zeta^{(p-1)^2} \end{vmatrix} = \prod_{0 \le i < j \le p-1} (\zeta^j - \zeta^i)$$

is invertible in $R (p = \prod_{i=1}^{p-1}(1 - \zeta^i)$ is a unit in $R)$, the set $\{y_j | j \in \mathbb{F}_p\}$ is also an R-basis for $R[G/A]$. But $z_t y_j = \zeta^{jt} y_j$, $a_t^{-1} y_j = y^{jt}$, and Ry_0 is a trivial RG-module, whereas the sub-R-modules Ry_j ($j \in \mathbb{F}_p$) are blocks of imprimitivity with $\mathbb{Z}_p$ being the stabilizer subgroup of the first (and – being normal – of any) block and $Ry_{1|\mathbb{Z}_p} \simeq \widetilde{R}$, and so,

$$R[G/A] \simeq \oplus_{j \in \mathbb{F}_p} Ry_j \simeq R \oplus \widetilde{R}^{\mathbb{Z}_p \to G}.$$

□

Theorem 9.7.3 *If $pR = R$ for p any prime, R a commutative ring with 1 (i.e. if R is a $\mathbb{Q}$-algebra), then $\mathcal{D}_{\mathbb{Q}}(R) \subseteq \{H | H \text{ cyclic}\}$. If any prime, except one, say ℓ, is invertible in R (e.g., R is a local ring with residue class characteristic ℓ), then $\mathcal{D}_{\mathbb{Q}}(R) \subseteq \{H | H \text{ cyclic mod } \ell\}$ where a group H is cyclic mod ℓ if the ℓ-Sylow subgroup H_ℓ is normal in H and $H | H_\ell$ is cyclic.*

PROOF If $pR = R$ for any p , then any group in $\mathcal{D}_{\mathbb{Q}}(R)$ is p-nilpotent and has a cyclic Sylow-p-subgroup for any p, and so, it is nilpotent with only a cyclic Sylow-subgroups, and hence cyclic.
If $pR = R$ for $p \neq \ell$, then any group H in $\mathcal{D}_{\mathbb{Q}}(R)$ has a normal p-complement for any $p \neq \ell$, and so, the intersection of all these normal p-complements, i.e., the Sylow-ℓ-subgroup H_ℓ of H is normal. Moreover, H/H_ℓ is p-nilpotent with cyclic Sylow-p-subgroup for any p dividing $|H/H_\ell|$, and so, by the above argument, it is cyclic. $\quad\square$

Lemma 9.7.5 *Let R be a Dedekind ring. Then $\mathcal{D}_A(R) = \cup_{\underline{m}} \mathcal{D}_A(R_{\underline{m}})$ where $\underline{m}$ runs through the maximal ideals in R, and A is a commutative ring with $1 \in A$.*

PROOF For any commutative ring A, we shall write $K_0^G(-, A) := K_0^G(-, \mathcal{P}(A))$.
Since $K_0^G(-, R_{\underline{m}})$ is a $K_0^G(-, R)$-algebra, we have $\mathcal{D}_A(R_{\underline{m}}) \subseteq \mathcal{D}_A(R)$. Now suppose that $G \in \mathcal{D}_A(R)$, but $G \notin \cup_{\underline{m}} \mathcal{D}_A(R_{\underline{m}})$. Then, for any $\underline{m}$, there exist elements $x_H \in \otimes K(H \lneqq G)$ such that

$$1_{A \otimes K(G, R_{\underline{m}})} = \sum_{H \lneqq G} x_H^{H \to G}$$

(where $x_H^{H \to G}$ is the image of $x_H \in \mathcal{G}(H)$) with respect to the induction map $\mathcal{G}(H) \to \mathcal{G}(G)$.
Since we have only finitely many $R_{\underline{m}} H$-modules and only a finite number of isomorphisms involved in the above equation, it can be realized already in a finite subextension of R in $R_{\underline{m}}$, and so, there exists an element $a_{\underline{m}}$ in $R - \underline{m}$ such that the above situation can be realized already over $R_{\{a_{\underline{m}}^n | n \in \mathbb{N}\}} = R_{a_{\underline{m}}}$. In particular, $G \notin \mathcal{D}_A(R_{\underline{m}})$. So it is enough to show that the set $I = \{a \in R | a = 0 \text{ or } G \notin \mathcal{D}_A(R_a)\}$ is an ideal of R – since $a_{\underline{m}} \in I$ would imply $I \not\subseteq \underline{m}$ for all $\underline{m}$ and so $I = R \ni 1$ and $G \notin \mathcal{D}_A(R)$ – a contradiction. So, assume $a, b \in I$. Then, without loss of generality, we may assume that $a + b \neq 0$ and even that $a + b = 1$ since $R_a \subseteq (R_{a+b})_{\frac{a}{a+b}}$, $R_b \subseteq (R_{a+b})_{\frac{a}{a+b}}$. Our result now follows from the next two lemmas. In 9.7.6 below, we shall write $K(G, A)$ for $K_0^G(G/G, \mathcal{P}(A))$ for any commutative ring A. $\quad\square$

Lemma 9.7.6 *Let $C \subset R$ be a multiplicatively closed subset of Dedekind ring R with $0 \in C$, and let R_C denote the associated ring of C-quotients of*

R with $0 \notin C$, and $i_C \subseteq A \otimes K(G, R)$ is the ideal generated by all $[L] - [N] \in A \otimes K(G, R)$ such that there exists $\varphi : L \to N, \psi : N \to L$ with $\varphi \circ \psi = c \cdot id_N, \psi \circ \varphi = c \cdot id_M$ for some $c \in C$. Then, the canonical map $A \otimes K(G, R) \to A \otimes K(G, R_C) : [L] \to [R_C \otimes L]$ induces an isomorphism $A \otimes K(G, R)/i_C \xrightarrow{\sim} A \otimes K(G, R_C)$.

PROOF   Clearly, i_C is the kernel of $A \otimes K(G, R) \to A \otimes K(G, R_C)$. We now construct an inverse of

$$A \otimes K(G, R)/i_C \longrightarrow A \otimes K(G, R_C).$$

If L' is a finitely generated R_C-projective $R_C G$-module, choose a finitely generated R-projective RG-module L such that $R_C \otimes L \simeq L'$, which is possible since R is a Dedekind ring. We then define the inverse $A \otimes K(G, R_C) \to A \otimes K(G, R)/i_C$ by $[L'] \to [L] + i_C$, which is well defined since it can easily be shown that $R_C \otimes L \simeq R_C \otimes N$ implies that $[L] - [N] \in i_C$.   ⊓

Lemma 9.7.7 Let $C_1, C_2 \subseteq R$ be multiplicative closed subsets of R and assume $c_1 R + c_2 R = R$ for $c_1 \in C_1$, $c_2 \in C_2$. Then, $i_{C_1} \cdot i_{C_2} = 0$.

PROOF   If $[L_\nu] - [N_\nu] \in i_{C_\nu}$ with maps $\varphi_\nu : L_\nu \to N_\nu$, $\psi_\nu : N_\nu \to L_\nu$, $\varphi_\nu \psi_\nu = c_\nu id_{N_\nu}$, $\psi_\nu \varphi_\nu = c_\nu id_{L_\nu} (c_\nu \in C_\nu)$, and $r_1 c_1 + R_2 c_2 = 1$, then we have an isomorphism from $(L_1 \otimes L_2) \oplus (N_1 \otimes N_2)$ into $(L_1 \otimes N_2) \oplus (N_1 \otimes L_2)$ given by the matrix

$$\begin{bmatrix} id_{L_1} \otimes \varphi_2 & \psi_1 \otimes r_1 id_{N_2} \\ \varphi_1 \otimes id_{L_2} & -r_2 id_{N_1} \otimes \psi_2 \end{bmatrix}$$

whose inverse is given by

$$\begin{bmatrix} r_2 id_{L_1} \otimes \psi_2 & \psi_1 \otimes r_1 id_{L_2} \\ \varphi_1 \otimes id_{N_2} & -id_{N_1} \otimes \varphi_2 \end{bmatrix}$$

So, $([L_1] - [N_1])([L_2] - [N_2]) = 0$.   ⊓

Conclusion of proof of 9.7.5 Using lemma 9.7.6, we have that there exists elements $x_H, y_H \in A \otimes K(H, R)$ $(H \lneq G)$ with

$$x = 1 - \sum_{H \lneq G} x_H^{H \to G} \in i_{\{a^n | n \in \mathbb{N}\}} = i^a$$

and

$$y = 1 - \sum_{H \lneq G} y_H^{H \to G} \in i_{\{b^n | n \in \mathbb{N}\}} = i^b.$$

By multiplying, we get $0 = x \cdot y = 1 - \sum_{H \lneqq G} z_H^{H \to G} \in i_a \cdot i_b$ for appropriate $z_H \in A \otimes K(H, R)$, and our claim that $G \notin \mathcal{D}_A(R_a)$, $G \notin \mathcal{D}_A(R_b)$ and $a+b = 1$ implies that $G \notin \mathcal{D}_A(R)$ follows from 9.7.7.

Remark 9.7.2 Lemma 9.7.5 holds even for arbitrary commutative rings R once we work with finitely presented R-modules (see [48]).

Theorem 9.7.4 *If R is a commutative ring with identity, then $\mathcal{D}_{\mathbb{Q}}(R) = \{H | H$ cyclic mod ℓ for some characteristic ℓ such that $\ell R \neq R\}=$*

$$\cup_{\ell R \neq R} \mathcal{C}_\ell \quad \text{where} \quad \mathcal{C}_\ell = \{H | H \text{ cyclic mod } \ell\}.$$

PROOF Let $R' = \mathbb{Z}[\frac{1}{p} | pR = R]$. Then R' is a Dedekind ring and R is an R'-algebra. So, $\mathcal{D}_{\mathbb{Q}}(R) \subseteq \mathcal{D}_{\mathbb{Q}}(R') \cup_m \mathcal{D}_{\mathbb{Q}}(R'_m)$. Moreover, $\mathcal{D}_{\mathbb{Q}}(R'_m) \subseteq \mathcal{C}_\ell$, if $\ell = char(R'/m)$, by theorem 9.7.3, and so, $\ell R' \neq R'$. So, $\ell R \neq R$ and hence $\mathcal{D}_{\mathbb{Q}} \subseteq \cup_{\ell R \neq R} \mathcal{C}_\ell$.
To show that $\mathcal{C}_\ell \subseteq \mathcal{D}_{\mathbb{Q}}(R)$ for $\ell R \neq R$, choose a maximal ideal m in R with $char(R/m) = \ell$ (resp. with arbitrary residue class characteristic if $\ell = 0$). In any case, we have $\mathcal{D}_{\mathbb{Q}}(R/m) \subseteq \mathcal{D}_{\mathbb{Q}}(R)$, and so, it is enough to show $\mathcal{C}_\ell \subseteq \mathcal{D}_{\mathbb{Q}}(R)$ whenever R is a field of characteristic ℓ. So, let G be cyclic mod ℓ, G_ℓ its Sylow-ℓ-subgroup (resp. e if $\ell = 0$), and $G = G_\ell < g >$ for some appropriate $g \in G$. We construct a non-zero linear map $K(G, R) \to \mathbb{C}$ which vanishes on $Im(\sum_{H \lneqq G}(K(H, R) \to K(G, R))$ (and thus proves $G \in \mathcal{D}_{\mathbb{Q}}(R)$) by associating to any RG-module N with direct decomposition $N \simeq \oplus_{j=1}^n N_j$ into indecomposable RG-module the sum $\sum' \chi_{N_j}(g)$ of the Brauer characters of g on those direct summands N_j that have vertex G_ℓ in the sense of proposition 9.1.1(c). (That is, are not a direct summand in any $L^{H \to G}$ with $H \lneqq G_\ell$, L any RH-module). This is well defined and additive by the Krull - Remark - Schmidt theorem. It is non-zero since the trivial RG-module R is mapped to 1 and it vanishes on any N that is induced from a proper subgroup H.
If $N = L^{H \to G}$ for some RH-module L that without loss of generality may be indecomposable, then either the vertex of L and thus the vertex of any indecomposable summand of L is properly contained in G_ℓ, and so, $0 = \sum' \chi_{N_i}(g)$, an empty sum, or $G_\ell \leq H$ and L is a direct summand of $L_1^{G \to H}$ for some indecomposable RG_ℓ-module L_1 with vertex G_ℓ, and then, any indecomposable summand N' of N restricted to G_ℓ is isomorphic to a direct sum of copies of G-conjugates of L_1 and so has vertex G_ℓ, too, in which case, we get $\sum' \chi_{N_i}(g) = \sum \chi_{N_i}(g) = \chi_L(g) = 0$ since $G_\ell \leq G$ implies $g \notin H$. $\Box$

Remarks and definition 9.7.1 A commutative ring R with identity is called a λ-ring if there exists a map

$$\lambda_t : R \longrightarrow R((t)) : a \mapsto \sum_{n=0}^{\infty} \lambda^n(a)t^n$$

of R into the ring of formal power series over R such that

(a) $\lambda_t(a) = 1 + at + \cdots$ for all $a \in R$.

(b) $\lambda_t(a + b) = \lambda_t(a)\lambda_t(b)$ for all $a, b \in R$.

Note that for any ring R, the exterior powers define a λ-ring structure on $K(G, R)$, and the next lemma that says that any torsion element in a λ-ring is nilpotent enables us to compute $\mathcal{D}_A(R)$ for arbitrary R via earlier results.

Lemma 9.7.8 *[G.Segal] Let R be an arbitrary λ-ring. Then any torsion element R is nilpotent.*

PROOF   Suppose $p^r x = 0$ for some p. Then,

$$1 = \lambda_t(p^r \cdot x) = \lambda_t(x)^{p^r} = (1 + xt + \cdots)^{p^r} = 1 + x^{p^r} \cdot t^{p^r} + \cdots \bmod (p).$$

So, $x^{p^r} = py$ for some $y \in R$ and hence $x^{(r+1)p^r} = x^{p^r} = (p \cdot x)x^{p^r - 1} \cdot y = 0$. Since any p-torsion element in R is nilpotent for any p, then any torsion element R is nilpotent.   □

Definition 9.7.3 *Let q, ℓ be primes, H a finite group. Then H is said to be a q-hyperelementary mod ℓ if there exists a normal series $e \lhd H_1 \lhd H_2 \lhd H$ with H_1 an ℓ-group, H_2/H_1 cyclic, and H/H_2 a q-group.*
A group H is said to be q-elementary mod ℓ if the Sylow-ℓ-subgroup H_ℓ of H is normal and H/H_ℓ is a direct product of a cyclic group and a q-group. Note that for $q = 0$ or $\ell = 0$, a group (resp. an ℓ-group) is always meant to be the trivial group.
By combining the above results with theorem 9.7.7, we have

Theorem 9.7.5 *Let A, R be commutative rings with identity. Then $\mathcal{D}_A(R) \subseteq \{H | H \ q\text{-hyperelementary mod} \ell \text{ for some } q \text{ with } qA \neq A \text{ and } \ell R \neq R\}$.*

Theorem 9.7.6 *If R contains a primitive p^{th} root ζ of 1 (i.e., R is a $\mathbb{Z}[\zeta]$-algebra) and $H \in \mathcal{D}_A(R)$, then there exists a normal series $e \trianglelefteq H_1 \trianglelefteq H_2 \trianglelefteq H$ such that H/H_2 acts trivially on the p-part of H_2/H_1.*

PROOF   Note that R is an R'-algebra where $R' = [\zeta, \frac{1}{r} | rR = R, \quad r \in \mathbb{N}]$ is a Dedekind ring. So, $H \in \mathcal{D}_A(R) \subseteq \mathcal{D}_A(R') = \underset{m}{\cup}\mathcal{D}_A(R'_m)$, and so, we may already assume R to be local Dedekind ring with residue class characteristic ℓ (possibly $\ell = 0$). So, H has a normal series $e \trianglelefteq H_1 \trianglelefteq H_2 \trianglelefteq H$ with H_1 and ℓ-group, ($H_1 = 0$ for $\ell = 0$), H_2/H_1 cyclic, and H/H_2 a q-group for some q with $qA \neq A$. If $\ell = p$ or $q = p$, we may assume that H_2/H_1 is p-regular, in which case our statement is trivial. If $\ell \neq p \neq q$, we use the fact that $\mathcal{D}_A(R)$ is closed with respect to subgroups and quotients, and so, if H/H_2 does not act trivially on the p-part of H_2/H_1, we may even assume H to be non-Abelian of order pq with $q \mid (p-1)$. But the isomorphism of lemma 9.7.4 (b) holds for any $\mathbb{Z}(\frac{1}{p}, \zeta)$-algebra, and so, especially for a local ring R of residue characteristic $\ell \neq p$, and by restriction, this isomorphism to $H = \mathbb{Z}_p \rtimes \mathbb{Z}_q \leq \mathbb{Z}_p \rtimes A$, $R[H\mathbb{Z}_p] \simeq \underbrace{R \oplus R^{\mathbb{Z}_p \to H} \oplus \cdots \oplus R^{\mathbb{Z}_p \to H}}_{(p-1) \text{ times}}$, and so, $\ell \in K(H, R)$ is induced from proper subgroups, i.e., $H \notin \mathcal{D}(R)$, and hence $H \notin \mathcal{D}_A(R)$ is a contradiction. □

Remark 9.7.3 It follows from theorem 9.7.6 that for any finite group G and a ring R, which contains a pth root of unity for any prime p dividing $|G|$, $A \otimes K_0^G(-, R)$ has a defect basis contained in $\mathcal{C}_A^R(G) := \{H \leq G \mid H \ q\text{-elementary mod } \ell \text{ for some characteristic } q \text{ with } qA \neq A \text{ and some characteristic } \ell \text{ with } \ell R \neq R\}$. More precisely, we have the following.

Theorem 9.7.7 *If G is a finite group, R a commutative ring with identity such that for any prime p dividing $|G|$, R contains a primitive p^{th} root of unity, then the defect basis of $A \otimes K_0^G(-, R) : GSet \to A\text{-}\mathcal{M}od$ is precisely $\mathcal{C}_A^R(G)$ for any commutative ring A with identity.*

PROOF We have to show that for any subgroup $H \in \mathcal{C}_A^R(G)$ of G, we have $A \otimes \overline{K(H, R)} \neq 0$. So, let H be q-elementary mod ℓ with $qA \neq A$ and $\ell R \neq R$. Without loss of generality, we may assume that A and R are algebraically closed fields of characteristic ℓ and q, respectively. It will be enough to construct a non-zero linear map $\chi : K(H, R) \to A$ that vanishes on $Im\left(\sum_{H' \lneq H} K(H', R) \to K(H, R)\right)$. So, let H_ℓ be the Sylow-ℓ-subgroup of H. By our assumption, we have $H_\ell \trianglelefteq H$ and $H/H_\ell \simeq (H_q \cdot H_\ell/H_\ell) \times (< g, H_\ell > /H_\ell)$ for some appropriate $g \in H$ of order n with $(n, q) = (n, \ell) = 1$. Choose a fixed isomorphism of the group of nth roots of unity in R onto the same group in A so that for any R-module N, we have a well-defined Brauer character $\chi_N(g)$ with values in A. Now define $\chi(N) = \sum' \chi_{N_i}(g)$ where $N = \oplus N_i$ is a decomposition of N into indecomposable RH-modules and the sum $\sum' \chi_{N_i}(g)$ is taken over all N_i with vertex H_ℓ. Note that χ is non-zero since it maps the trivial representation into 1, but it vanishes on any $N = \oplus N_i \simeq L^{H' \to H}$ if $H' \neq H$, since otherwise N must have a vertex H_ℓ, and in particular, $H_\ell \leq H'$, in which case all the N_i have vertex H_ℓ (as above since H_ℓ is normal in H). So, $\sum' \chi_{N_i}(g) = \chi_{N_i}(g) = 0$ unless also $g \in H'$, in which case $\chi_N(g) = (H : H')\chi_L(g)$ since H_q acts trivially on $< g >$. But then, again, $\chi_N(g) = 0$ since $(H : H')$ is a power of q and hence zero in A unless $H = H'$, which had been excluded. □

Exercises

9.1 Let $\mathcal{B}$ be a based category, p a prime ideal of $\Omega(\mathcal{B})$, and $p =$ characteristic of $\Omega(B)/\underline{p}$. Define a relation on $\mathcal{B}$-objects by $T \underset{p}{\sim} T'$ if $\underline{p}(T,p) = \underline{p}(T',p)$ (see remarks 9.4.1). Show that '$\sim$' is an equivalence relation.

9.2 In the notation of definition 9.4.2, show that for G-sets S_1, S_2,

(i) $\varphi_H(S_1 \cup S_2) = \varphi_H(S_1) + \varphi_H(S_2)$, and

(ii) $\varphi_H(S_1 \times S_2) = \varphi_H(S_1) \cdot \varphi_H(S_2)$ where $S_1, S_2 \in GSet$.

9.3 Let $\mathcal{B}$ be a based category, $\mathcal{G} : \mathcal{B} \to \mathbb{Z}\text{-}\mathcal{M}od$ a Green functor. If $\mathcal{G}(*)$ is torsion-free $X \in ob(\mathcal{B})$ and $n \cdot 1_{\mathcal{G}(X)} \in I_X(\mathcal{G})$, show that $(n, \|\mathcal{B}\|) \cdot 1_{\{\mathcal{G}(*)\}} \in I_X(\mathcal{G})$.

9.4 Let $\mathcal{G} : GSet \to \mathbb{Z}\text{-}\mathcal{M}od$ be a Green functor such that $\mathcal{G}(S) \to \mathcal{G}(G/G)$ is surjective. If $\tau : M \to N$ is a morphism of G-modules such that $\tau_S : M_S \to N_S$ is an isomorphism, show that $\tau : M \to N$ is also an isomorphism.

9.5 Let G be a finite group, B a $\mathbb{Z}G$-module. For any G-set S, let $\mathcal{H}_G(S, B)$ denote the set all G-maps $f : S \to B$. Show that $\mathcal{H}_G(-, B) : GSet \to \mathbb{Z}\text{-}\mathcal{M}od$ takes sums into products.

9.6 Let $\mathcal{B}$ be a based category. Show that the category $\mathcal{M}_\mathcal{B}$ of Mackey functors $\mathcal{B} \to \mathbb{Z}\text{-}\mathcal{M}od$ is an Abelain category satisfying $AB5$, i.e., $\mathcal{M}$ is co-complete and filtered limits of exact sequences are exact.

9.7 Let H, H' be two subgroups of G. Show that the orbit space $G/H \times G/H'$ can be identified with the set $D(H, H') = \{HgH' | g \in G\}$ of double cosets of H, H' in G.

9.8 Let G be a finite group and let $\mathcal{G}, \mathcal{G}' : GSet \to Z\text{-}\mathcal{M}od$ be Green functors such that there exists a morphism $\theta : \mathcal{G} \to \mathcal{G}'$. Show that $\mathcal{D}_{\mathcal{G}'} \subseteq \mathcal{D}_\mathcal{G}$.

If $\mathcal{G}'$ is a subfunctor of $\mathcal{G}$ such that $|G| \cdot \mathcal{G}'(*) = \mathcal{G}'(*)$, show that $\mathcal{D}_\mathcal{G} = \mathcal{D}_{\mathcal{G}'}$.

9.9 Let $(\mathcal{C}, \perp)$ be a symmetric monoidal category, S a G-set. Show that $([\underline{S}, \mathcal{C}], \perp)$ is also a symmetric monoidal category where for $\zeta, \eta \in [\underline{S}, \mathcal{C}], (\zeta \perp \eta)(s) = \zeta_s \perp \eta_s$ and $(\zeta \perp \eta)_{(g,s)} : (\zeta \perp \eta)_s \to (\zeta \perp \eta)_{gs}$.

For all $n \geq 0$, define $K_n^G(S, \mathcal{C})$ as $K_n^\perp((\underline{S}, \mathcal{C}), \perp)$. Show that $K_n^G(-, \mathcal{C}) : GSet \to \mathbb{Z}\text{-}\mathcal{M}od$ is a Mackey functor.

9.10 Let G be a finite group, S_1, S_2 G-sets. Show that

(a) If a Mackey functor M is S_1-projective and S_2-projective, then it is $S_1 \times S_2$-projective

(b) If M is S_1-projective and there exists a G-map $S_1 \to S_2$, then M is S_2-projective.

Chapter 10

Equivariant higher algebraic K-theory together with relative generalizations – for finite group actions

In this chapter, we construct and study equivariant higher K-groups whose computations will depend on those of defect basis for K_0^G functors obtained in 9.7. We shall also apply the theory in the computations of higher K-theory of grouprings. The results in this chapter are due to A. Dress and A.O. Kuku (see [52, 53]).

10.1 Equivariant higher algebraic K-theory

10.1.1 Let G be a finite group, S a G-set. Recall from (1.1.5) that we can associate to S a category $\underline{S}$ as follows: The objects of $\underline{S}$ are elements of S, while for $s, s' \epsilon S$, a morphism from s to s' is a triple (s', g, s) where $g \epsilon G$ is such that $gs = s'$. The morphism are composed by $(s'', h, s')(s', g, s) = (s'', hg, s)$. Note that any G-map $\varphi : S \to T$ gives rise to an associated covariant functor $\underline{\varphi} : \underline{S} \to \underline{T}$ where $\underline{\varphi}(s) = \varphi(s)$ and $\underline{\varphi}((s', g, s)) = (\varphi(s'), g, \varphi(s))$.

Theorem 10.1.1 *Let $\mathcal{C}$ be an exact category, S a G-set. The category $[\underline{S}, \mathcal{C}]$ of covariant functors from $\underline{S}$ to $\mathcal{C}$ is also exact.*

PROOF let ζ_1, ζ_2, and ζ_3 be functors in $[\underline{S}, \mathcal{C}]$. Define a sequence of natural transformations $\zeta_1 \to \zeta_2 \to \zeta_3$ to be exact if the sequence is exact fiberwise, i.e., for any $s \in S$, $\zeta_1(s) \to \zeta_2(s) \to \zeta_3(s)$ is exact in $\mathcal{C}$ .It can be easily checked that this notion of exactness makes $[\underline{S}, \mathcal{C}]$ an exact category. ☐

Definition 10.1.1 *We shall write for all $n \geq 0$ $K_n^G(S, \mathcal{C})$ for the nth algebraic K-group associated to the category $[\underline{S}, \mathcal{C}]$ with respect to fiberwise exact sequences.*

We now have the following.

Theorem 10.1.2

$$K_n^G(-,\mathcal{C}) : GSet \to Ab \quad \text{is a Mackey functor for all } n \geq 0.$$

PROOF   Let $\varphi : S_1 \to S_2$ be a G-map. Then φ gives rise to a restriction functor $\varphi^* : [\underline{S}_2, \mathcal{C}] \to [\underline{S}_1, \mathcal{C}]$ given by $\zeta \to \zeta \circ \varphi$ and hence a homomorphism $K_n^G(\varphi^*, \mathcal{C}) : K_n^G(S_2, \mathcal{C}) \to K_n^G(S_1, \mathcal{C})$. Also φ gives rise to an induction functor $\varphi_* : [\underline{S}_1, \mathcal{C}] \to [\underline{S}_2, \mathcal{C}]$ defined as follows: for $\zeta \in [\underline{S}_1, \mathcal{C}]$, we define $\varphi_*(\zeta) \in [\underline{S}_2, \mathcal{C}]$ by

$$\varphi_*(\zeta)(s_2) = \bigoplus_{s_1 \in \varphi^{-1}(s_2)} \zeta(s_1) \text{ and } \varphi_*(gs_2, g, s_2)$$

$$= \bigoplus_{s_1 \in \varphi^{-1}(s_2)} \zeta(gs_1, g, s_1).$$

If $\alpha : \zeta \to \zeta'$ is a natural transformation of functors in $[\underline{S}_1, \mathcal{C}]$, then we define a natural transformation of functors in $[\underline{S}_2, \mathcal{C}]$, $\varphi_*(\alpha) : \varphi_*(\zeta) \to \varphi_*(\zeta')$ by

$$\varphi_*(\alpha)(s_2) = \bigoplus_{s_1 \in \varphi^{-1}(s_2)} \alpha(s_1) : \varphi_*(\zeta)(s_2)$$

$$= \bigoplus_{s_1 \in \varphi^{-1}(s_2)} \zeta(s_1) \to \varphi_*(\zeta')(s_2)$$

$$= \bigoplus_{s_1 \in \varphi^{-1}(s_2)} \zeta'(s_1).$$

So, we have a homomorphism $K_n^G(\varphi_*, \mathcal{C}) : K_n^G(S_1, \mathcal{C}) \to K_n^G(S_2, \mathcal{C})$. It can be easily checked that

(i) $(\varphi\psi)_* = \varphi_*\psi_*$ if $\psi : S_0 \to S_1$ and $\varphi : S_1 \to S_2$ are G-maps.

(ii) $[\underline{S_1 \dot\cup S_2}_1, \mathcal{C}] \simeq [\underline{S}_1, \mathcal{C}] \times [\underline{S}_2, \mathcal{C}]$, and hence

$$K_n^G(S_1 \dot\cup S_2, \mathcal{C}) = K_n^G(S_1, \mathcal{C}) \bigoplus K_n^G(S_2, \mathcal{C}).$$

(iii) Given any pull-back diagram

$$
\begin{array}{ccc}
S_1 \times_T S_2 & \xrightarrow{\ \overline{\varphi}\ } & S_2 \\
{\scriptstyle \overline{\psi}}\downarrow & & \downarrow{\scriptstyle \psi} \\
S_1 & \xrightarrow{\ \varphi\ } & T
\end{array}
$$

in $\mathcal{C}$, we have a commutative diagram

$$
\begin{array}{ccc}
[S_1 \times_T S_2, \mathcal{C}] & \xrightarrow{\overline{\varphi_*}} & [S_2, \mathcal{C}] \\
\downarrow{\overline{\psi}^*} & & \downarrow{\psi^*} \\
[\underline{S_2}, \mathcal{C}] & \xrightarrow{\varphi_*} & [\underline{T}, \mathcal{C}]
\end{array}
$$

and hence the corresponding commutative diagram obtained by applying K_n^G. Hence $K_n^G(-, \mathcal{C})$ is a Mackey functor.

⬜

We now want to turn $K_0^G(-, \mathcal{C})$ into a Green functor. We first recall the definition of a pairing of exact categories (see [224]).

Definition 10.1.2 *Let $\mathcal{C}_1, \mathcal{C}_2, \mathcal{C}_3$ be exact categories. An exact pairing $<,>$: $\mathcal{C}_1 \times \mathcal{C}_2 \to \mathcal{C}_3$ given by $(X_1, X_2) \to < X_1, X_2 >$ is a covariant functor such that*

$$
Hom((X_1, X_2), (X_1', X_2')) =
$$
$$
= Hom(X_1, X_1') \times Hom(X_2, X_2') \to Hom(< X_1, X_2 >, < X_1', X_2' >)
$$

is biadditive and biexact (see (7.6) for more details).

Recall from (7.6) that such a pairing gives rise to a K-theoretic product $K_i(\mathcal{C}_1) \times K_j(\mathcal{C}_2) \to K_{i+j}(\mathcal{C}_3)$, and in particular a natural pairing $K_0(\mathcal{C}_1) \times K_n(\mathcal{C}_2) \to K_n(\mathcal{C}_3)$.

Now, if $\mathcal{C}_1 = \mathcal{C}_2 = \mathcal{C}_3 = \mathcal{C}$, and the pairing $\mathcal{C} \times \mathcal{C} \to \mathcal{C}$ is associative and commutative and also has a natural unit, that is, there exists an object E in $\mathcal{C}$ such that $< E, M > = < M, E > = M$ for all $M \in \mathcal{C}$, then $K_n(\mathcal{C})$ is a unitary $K_0(\mathcal{C})$-module. We shall apply this setup in the proof of 10.1.3 below.

Theorem 10.1.3 *Let $\mathcal{C}_1, \mathcal{C}_2, \mathcal{C}_3, \mathcal{C}$ be exact categories and $\mathcal{C}_1 \times \mathcal{C}_2 \to \mathcal{C}_3$ an exact pairing of categories. Then the pairing induces fiberwise a pairing $[\underline{S}, \mathcal{C}_1] \times [\underline{S}, \mathcal{C}_2] \to [\underline{S}, \mathcal{C}_3]$ and hence a pairing $K_0^G(S, \mathcal{C}_1) \times K_n^G(S, \mathcal{C}_2) \to K_n^G(S, \mathcal{C}_3)$.*
Suppose $\mathcal{C}$ is an exact category such that the pairing $\mathcal{C} \times \mathcal{C} \to \mathcal{C}$ is naturally associative and commutative and there exists $E \in \mathcal{C}$ such that $< E, M > = < M, E > = M$. Then, $K_0^G(-, \mathcal{C})$ is a Green functor and $K_n^G(-, \mathcal{C})$ is a unitary $K_0^G(-, \mathcal{C})$-module.

PROOF Let $\zeta_1 \in [\underline{S}, \mathcal{C}_1]$, $\zeta_2 \in [\underline{S}, \mathcal{C}_2]$. Define $< \zeta_1, \zeta_2 >$ by $< \zeta_1, \zeta_2 > (s) = < \zeta_1(s), \zeta_2(s) >$. This is exact with respect to fiberwise exact sequences. Now, any $\zeta_1 \in [\underline{S}, \mathcal{C}_1]$ induces an exact functor $\zeta_1^* : [\underline{S}, \mathcal{C}_2] \to [\underline{S}, \mathcal{C}_3]$ given by $\zeta_2 \to < \zeta_1, \zeta_2 >$ and hence a map $K_n^G(\zeta_1, \mathcal{C}_1) : K_n^G(S, \mathcal{C}_3) \to K_n^G(S, \mathcal{C}_3)$.

We now define a map $K_0^G(S, \mathcal{C}_1) \xrightarrow{\delta} Hom(K_n^G(S, \mathcal{C}_2), K_n^G(S, \mathcal{C}_3))$ by $\zeta_1 \to K_n^G(\zeta_1)$ and show that this is a homomorphism. This homomorphism then yields the required pairing $K_0^G(S, \mathcal{C}_1) \times K_n^G(S, \mathcal{C}_2) \to K_n^G(S, \mathcal{C}_3)$. To show that δ is a homomorphism, let $\zeta_1' \to \zeta_1 \to \zeta_1''$ be an exact sequence in $[\underline{S}, \mathcal{C}_1]$. Then we obtain an exact sequence of exact functors $\zeta_{1*}' \to \zeta_{1*} \to \zeta_{1*}'' : [\underline{S}, \mathcal{C}_2] \to [\underline{S}, \mathcal{C}_3]$ such that for each $\zeta_2 \in [\underline{S}, \mathcal{C}_2]$, the sequence $\zeta_{1*}'(\zeta_2) \to \zeta_{1*}(\zeta_2) \to \zeta_{1*}''(\zeta_2)$ is exact in $[\underline{S}, \mathcal{C}_3]$. Then, by applying Quillen's result (see theorem 6.1.1), we have $K_n^G(\zeta_{1*}') + K_n^G(\zeta_{1*}'') = K_n^G(\zeta_{1*})$.

It can be checked that given any G-map $\varphi : T \to S$, the Frobenius reciprocity law holds, i.e., for $\zeta_i \in [\underline{S}, \mathcal{C}_i], \eta_i \in [\underline{T}, \mathcal{C}_i]$, $i = 1, 2, 3$ we have a canonical isomorphism

(i) $\varphi^* < \zeta_1, \zeta_2 > \equiv < \varphi^*(\zeta_1), \varphi^*(\zeta_2) >$.

(ii) $\varphi_* < \varphi^*(\zeta_1), \eta_2 > \equiv < \zeta_1, \varphi_*(\eta_2) >$.

(iii) $\varphi_* < \eta_1, \varphi^*(\zeta_2) > \equiv < \varphi_*(\eta_1), \zeta_2 >$. It is clear that the pairing $\mathcal{C} \times \mathcal{C} \to \mathcal{C}$ induces $K_0^G(S, \mathcal{C}) \times K_0^G(S, \mathcal{C}) \to K_0^G(S, \mathcal{C})$, which turns $K_0^G(S, \mathcal{C})$ into a ring with unit such that for any G-map $\varphi : S \to T$, $K_0^G(\varphi, \mathcal{C})^*(1_{K_0^G(S, \mathcal{C})}) \equiv 1_{K_0^G(S, \mathcal{C})}$.

It is also clear that $1_{K_0^G(S, \mathcal{C})}$ acts as the identity on $K_0^G(S, \mathcal{C})$. So, $K_n^G(S, \mathcal{C})$ is a $K_0^G(S, \mathcal{C})$-module.

$\square$

10.2   Relative equivariant higher algebraic K-theory

In this section, we discuss the relative version of the theory in 10.1.

Definition 10.2.1 *Let S, T be GSets. Then the projection map $S \times T \xrightarrow{\varphi} S$ gives rise to a functor $\underline{S \times T} \xrightarrow{\varphi} \underline{S}$. Suppose that $\mathcal{C}$ is an exact category. If $\zeta \in [\underline{S}, \mathcal{C}]$, we write ζ' for $\zeta \circ \varphi : \underline{S \times T} \xrightarrow{\varphi} \underline{S} \xrightarrow{\zeta} \mathcal{C}$. Then, a sequence $\zeta_1 \to \zeta_2 \to \zeta_3$ of functors in $[\underline{S}, \mathcal{C}]$ is said to be T-exact if the sequence $\zeta_1' \to \zeta_2' \to \zeta_3'$ of restricted functors $\underline{S \times T} \to \underline{S} \to \mathcal{C}$ is split exact.*

If $\psi : S_1 \to S_2$ is a G-map, and $\zeta_1 \to \zeta_2 \to \zeta_3$ is a T-exact sequence in $[\underline{S_2}, \mathcal{C}]$, then $\zeta_1' \to \zeta_2' \to \zeta_3'$ is a T-exact sequence in $[\underline{S_1}, \mathcal{C}]$ where $\zeta_i' : \underline{S_1} \xrightarrow{\psi} \underline{S_2} \xrightarrow{\zeta_i} \mathcal{C}$. Let $K_n^G(S, \mathcal{C}, T)$ be the nth algebraic K-group associated to the exact category $[\underline{S}, \mathcal{C}]$ with respect to T-exact sequences.

Remark 10.2.1 The use of the restriction functors ζ_i' in both situations in 10.2.1 constitute a special case of the following general situation. Let $\mathcal{C}$ be an exact category and $\mathcal{B}, \mathcal{B}'$ any small categories. We can define exactness in $[\mathcal{B}, \mathcal{C}]$ relative to some covariant functor $\delta : \mathcal{B}' \to \mathcal{B}$. Thus a sequence $\zeta_1 \to \zeta_2 \to \zeta_3$ of functors in $[\mathcal{B}, \mathcal{C}]$ is said to be exact relative to $\delta : \mathcal{B}' \to \mathcal{B}$ if it is exact fiberwise and if the sequence $\zeta_1' \to \zeta_2' \to \zeta_3'$ of restricted functors $\zeta_i' := \zeta_i \circ \delta : \mathcal{B}' \xrightarrow{\delta} \mathcal{B} \xrightarrow{\zeta} \mathcal{C}$ is split exact.

Definition 10.2.2 *Let S, T be GSets. A functor $\zeta \in [\underline{S}, \mathcal{C}]$ is said to be T-projective if any T-exact sequence $\zeta_1 \to \zeta_2 \to \zeta$ is exact. Let $[\underline{S}, \mathcal{C}]_T$ be the additive category of T-projective functors in $[\underline{S}, \mathcal{C}]$ considered as an exact category with respect to split exact sequences. Note that the restriction functor associated to $S_1 \xrightarrow{\psi} S_2$ carries T-projective functors $\zeta \in [\underline{S_2}, \mathcal{C}]$ into T-projective functors $\zeta \circ \psi \in [\underline{S_2}, \mathcal{C}]$. Define $P_n^G(S, \mathcal{C}, T)$ as the nth algebraic K-group associated to the exact category $[\underline{S}, \mathcal{C}]_T$, with respect to split exact sequences.*

Remark 10.2.2 Here are some properties of the constructions above. The proofs are left as exercises.

(i) Let $\zeta : \underline{S} \to \mathcal{C}$ be an object in $[\underline{S}, \mathcal{C}]$, T an arbitrary non-empty G-set, and let $\varphi : S \times T \to S$ denote the canonical projection. Then the following conditions are equivalent:

    (a) ζ is T-projective.

    (b) The canonical map $\varphi^*(\varphi_*(\zeta)) \to \zeta$, given by the "co-diagonal": $\varphi_*(\varphi^*(\zeta))(s) = \oplus_{t \in T} \zeta(s) \to \zeta(s)$, $s \in S$, is split surjective.

    (c) ζ is isomorphic to a direct summand of $\varphi_*(\zeta)$ for some appropriate $\zeta : \underline{S} \times \underline{T} \to \mathcal{C}$.

    (d) The canonical map $\zeta \to \varphi_*(\varphi^*(\zeta))$, given by the "diagonal": $\zeta(s) \to \oplus_{t \in T} \varphi_*(\varphi^*(\zeta))(s)$, $s \in S$, is split injective.

    (e) ζ is T-injective, i.e., any T-exact sequence $\zeta \to \zeta_2 \to \zeta_3$ is exact.

(ii) Let $\varphi : S_1 \to S_2$ be a G-map. Then the induced functor $\varphi_* : [\underline{S_1}, \mathcal{C}] \to [\underline{S_2}, \mathcal{C}]$ maps fiberwise / T-exact sequences from $[\underline{S_1}, \mathcal{C}]$ into fiberwise / T-exact sequences in $[\underline{S_2}, \mathcal{C}]$ and any T-projective functors into T-projective functors.

Theorem 10.2.1 $K_n^G(-, \mathcal{C}, T)$ *and* $P_n^G(-, \mathcal{C}, T)$ *are Mackey functors from GSet to Ab for all $n \geq 0$. If the pairing $\mathcal{C} \times \mathcal{C} \to \mathcal{C}$ is naturally associative and commutative and contains a natural unit, then $K_0^G(-, \mathcal{C}, T) : GSet \to Ab$ is a Green functor, and $K_n^G(-, \mathcal{C}, T)$ and $P_n^G(-, \mathcal{C}, T)$ are $K_0^G(-, \mathcal{C}, T)$-modules.*

PROOF For any G-map $\psi : S_1 \to S_2$, the restriction functor $\psi^* : [\underline{S_2}, \mathcal{C}] \to [\underline{S_1}, \mathcal{C}]$ given by $\zeta \to \zeta \circ \psi$ carries T-exact sequences into T-exact sequences, and any T-projective functor into a T-projective functor. Hence $K_n^G(-, \mathcal{C}, T)$ and $P_n^G(-, \mathcal{C}, T)$ become contravariant functors.

Also, the induction functor $\psi_* : [\underline{S_1}, \mathcal{C}] \to [\underline{S_2}, \mathcal{C}]$ associated to $\psi : S_1 \to S_2$ p-reserves T-exact sequences and T-projective functors and hence induces homo-morphisms $K_n^G(\psi, \mathcal{C}, T)_* : K_n^G(S_1, \mathcal{C}, T) \to K_n^G(S_2, \mathcal{C}, T)$ and $P_n^G(\psi, \mathcal{C}, T)_* :$ $P_n^G(S_1, \mathcal{C}, T) \to P_n^G(S_2, \mathcal{C}, T)$, thus making $K_n^G(-, \mathcal{C}, T)$ and $P_n^G(S_1, \mathcal{C}, T)$ co-variant functors. Other properties of Mackey functors can be easily veri-fied. Observe that for any $GSet$ T, the pairing $[\underline{S}, \mathcal{C}_1] \times [\underline{S}, \mathcal{C}_2] \to [\underline{S}, \mathcal{C}_3]$ takes T-exact sequences into T-exact sequences, and so, if $[\underline{S}, \mathcal{C}_i]$, $i = 1, 2$ are considered as exact categories with respect to T-exact sequences, then we have a pairing $K_0^G(\underline{S}, \mathcal{C}_1, T) \times K_n^G(\underline{S}, \mathcal{C}_2, T) \to K_n^G(\underline{S}, \mathcal{C}_3, T)$. Also if ζ_2 is T-projective, so is $< \zeta_1, \zeta_2 >$.

Hence, if $[\underline{S}, \mathcal{C}_1]$ is considered as an exact category with respect to T-exact sequences, we have an induced pairing $K_0^G(\underline{S}, \mathcal{C}_1, T) \times P_n^G(\underline{S}, \mathcal{C}_2, T) \to$ $P_n^G(\underline{S}, \mathcal{C}_3, T)$. Now, if we put $\mathcal{C}_1 = \mathcal{C}_2 = \mathcal{C}_3 = \mathcal{C}$ such that the pairing $\mathcal{C} \times \mathcal{C} \to \mathcal{C}$ is naturally associative and commutative and $\mathcal{C}$ has a natural unit, then, as in theorem 10.1.3, $K_0^G(-, \mathcal{C}, T)$ is a Green functor and it is clear from the above that $K_n^G(-, \mathcal{C}, T)$ and $P_n^G(-, \mathcal{C}, T)$ are $K_0^G(-, \mathcal{C}, T)$-modules. $\square$

Remarks 10.2.1   (i) In the notation of theorem 10.2.1, we have the follow-ing natural transformation of functors: $P_n^G(-, \mathcal{C}, T) \to K_n^G(-, \mathcal{C}, T) \to$ $K_n^G(-, \mathcal{C})$, where T is any G-set, G a finite group, and $\mathcal{C}$ an exact cate-gory. Note that the first map is the "Cartan" map.

(ii) If there exists a G-map $T_2 \to T_1$, we also have the following natu-ral transformations $P_n^G(-, \mathcal{C}, T_2) \to P_n^G(-, \mathcal{C}, T_1)$ and $K_n^G(-, \mathcal{C}, T_1) \to$ $K_n^G(-, \mathcal{C}, T_2)$ since, in this case, any T_1-exact sequence is T_2-exact.

10.3   Interpretation in terms of group-rings

In this section, we discuss how to interpret the theories in previous sections in terms of group-rings.

10.3.1 Recall that any G-set S can be written as a finite disjoint union of transitive G-sets, each of which is isomorphic to a quotient set G/H for some subgroup H of G. Since Mackey functors, by definition, take finite disjoint unions into finite direct sums, it will be enough to consider exact categories $[G/H, \mathcal{C}]$ where $\mathcal{C}$ is an exact category.

For any ring A, let $\mathcal{M}(A)$ be the category of finitely generated A-modules and $\mathcal{P}(A)$ the category of finitely generated projective A-modules. Recall from (1.1.8)(i) that if G is a finite group, H a subgroup of G, A a commutative ring, then there exists an equivalence of exact categories $[G/H, \mathcal{M}(A)] \to \mathcal{M}(AH)$. Under this equivalence, $[G/H, \mathcal{P}(A)]$ is identified with the category of finitely generated A-projective left AH-modules.

We now observe that a sequence of functors $\zeta_1 \to \zeta_2 \to \zeta_3$ in $[G/H, \mathcal{M}(A)]$ or $[G/H, \mathcal{P}(A)]$ is exact if the corresponding sequence $\zeta_1(H) \to \zeta_2(H) \to \zeta_3(H)$ of AH-modules is exact.

Remarks 10.3.1 (i) It follows that for every $n \geq 0$, $K_n^G(G/H, \mathcal{P}(A))$ can be identified with the nth algebraic K-group of the category of finitely generated A-projective AH-modules while $K_n^G(G/H, \mathcal{M}(A)) = G_n(AH)$ if A is Noetherian. It is well known that $K_n^G(G/H, \mathcal{P}(A)) = K_n^G(G/H, \mathcal{M}(A))$ is an isomorphism when A is regular.

(ii) Let $\varphi : G/H_1 \to G/H_2$ be a G-map for $H_1 \leq H_2 \leq G$. We may restrict ourselves to the case $H_2 = G$, and so, we have $\varphi^* :$ $[G/G, \mathcal{M}(A)] \to [G/H, \mathcal{M}(A)]$ corresponding to the restriction functor $\mathcal{M}(AG) \to \mathcal{M}(AH)$, while $\varphi_* : [G/H, \mathcal{M}(A)] \to [G/G, \mathcal{M}(A)]$ corresponds to the induction functor $\mathcal{M}(AH) \to \mathcal{M}(AG)$ given by $N \to AG \otimes_{AH} N$. Similar situations hold for functor categories involving $\mathcal{P}(A)$. So, we have corresponding restriction and induction homomorphisms for the respective K-groups.

(iii) If $\mathcal{C} = \mathcal{P}(A)$ and A is commutative, then the tensor product defines a naturally associative and commutative pairing $\mathcal{P}(A) \times \mathcal{P}(A) \to \mathcal{P}(A)$ with a natural unit, and so, $K_n^G(-, \mathcal{P}(A))$ are $K_0^G(-, \mathcal{P}(A))$-modules.

10.3.2 We now interpret the relative situation. So let T be a G-set. Note that a sequence $\zeta_1 \to \zeta_2 \to \zeta_3$ of functors in $[G/H, \mathcal{M}(A)]$ or $[G/H, \mathcal{P}(A)]$ is said to be T-exact if $\zeta_1(H) \to \zeta_2(H) \to \zeta_3(H)$ is AH'-split exact for all $H' \leq H$ such that $T^{H'} \neq \emptyset$ where $T^{H'} = \{t \in T| \ gt = t \ \forall g \in H'\}$. In particular, the sequence is G/H-exact (resp. G/G-exact) if and only if the corresponding sequence of AH-modules (resp. AG-modules) is split exact. If ϵ is the trivial subgroup of G, it is G/ϵ-exact if it is split exact as a sequence of A-modules.

So, $K_n^G(G/H, \mathcal{P}(A), T)$ (resp. $K_n^G(G/H, \mathcal{M}(A), T)$) is the nth algebraic K-group of the category of finitely generated A-projective AH-modules (resp. category of finitely generated AH-modules) with respect to exact sequences that split when restricted to the various subgroups H' of H such that $T^{H'} \neq \emptyset$. Moreover, observe that $P_n^G(G/H, \mathcal{P}(A), T)$ (resp. $P_n^G(G/H, \mathcal{M}(A), T)$) is an algebraic K-group of the category of finitely generated A-projective AH-modules (resp. finitely generated AH-modules) that are relatively H'-projective for subgroups H' of H such that $T^{H'} \neq \emptyset$ with respect to split exact sequences. In particular, $P_n^G(G/H, \mathcal{P}(A), G/\epsilon) = K_n(AH)$. If A is commutative, then $K_0^G(-, \mathcal{P}(A), T)$ is a Green functor, and $K_n^G(-, \mathcal{P}(A), T)$ and $P_n^G(-, \mathcal{P}(A), T)$ are $K_0^G(-, \mathcal{P}(A), T)$-modules.

Now, let us interpret the map, associated to G-maps $S_1 \to S_2$. We may specialize to maps $\varphi : G/H_1 \to G/H_2$ for $H_1 \leq H_2 \leq G$, and for convenience we may restrict ourselves to the case $H_2 = G$, in which case we write $H_1 = H$. In this case, $\varphi^* : [\overline{G/G, \mathcal{M}(A)}] \to [\overline{G/H, \mathcal{M}(A)}]$ corresponds to the restriction of AG-modules to AH-modules, and $\varphi_* : [\overline{G/H, \mathcal{M}(A)}]$ corresponds to the induction of AH-modules to AG-modules.

We hope that this wealth of equivariant higher algebraic K-groups will satisfy a lot of future needs, and moreover, the way they have been produced systematically will help to keep them in some order and to produce new variants of them whenever desired.

Since any $GSet$ S can be written as a disjoint union of transitive G-sets isomorphic to some coset-set G/H, and since all the above K-functors satisfy the additivity condition, the above identifications extend to K-groups, defined on an arbitrary G-set S.

10.4   Some applications

10.4.1 We are now in position to draw various conclusions just by quoting well-established induction theorems concerning $K_0^G(-, \mathcal{P}(A))$ and $K_0^G(-, \mathcal{P}(A), T)$, and more generally $R \otimes_Z K_0^G(-, \mathcal{P}(A))$ and $R \otimes_Z K_0^G(-, \mathcal{P}(A), T)$ for R, a subring of $\mathcal{Q}$, or just any commutative ring (see 9.6, 9.7). Since any exact sequence in $\mathcal{P}(A)$ is split exact, we have a canonical identification $K_0^G(-, \mathcal{P}(A)) = K_0^G(-, \mathcal{P}(A), G/\epsilon)$ (ϵ the trivial subgroup of G) and thus may direct our attention to the relative case only.

So, let T be a G-set. For p a prime and q a prime or 0, let $\mathcal{D}(p, T, q)$ denote the set of subgroups $H \leq G$ such that the smallest normal subgroup H_1 of H with a q-factor group has a normal Sylow-subgroup H_2 with $T^{H_2} \neq \emptyset$ and a cyclic factor group H_1/H_2. Let $\mathcal{H}_q$ denote the set of subgroups $H \leq G$, which are q-hyperelementary, i.e., have a cyclic normal subgroup with a q-factor group (or are cyclic for $q = 0$).

For A and R being commutative rings, let $\mathcal{D}(A, T, R)$ denote the union of all $\mathcal{D}(p, T, q)$ with $pA \neq A$ and $qR \neq R$, and let $\mathcal{H}_R$ denote the set of all $\mathcal{H}_q$ with $qR \neq R$. Then, it has been proved (see [47, 48, 49]), or theorem 9.7.5 that $R \otimes_Z K_0^G(-, \mathcal{P}(A), T)$ is S-projective for some G-set S if $S^H \neq \emptyset$ for all $H \in \mathcal{D}(A, T, R) \cup \mathcal{H}_R$. Moreover (see [49]), if A is a field of characteristic $p \neq 0$, then $K_0^G(-, \mathcal{P}(A), T)$ is S-projective already if $S^H \neq \emptyset$ for all $H \in \mathcal{D}(A, T, R)$. (Also see theorems 9.7.5, 9.7.7, and remarks 9.7.3).

10.4.2 Among the many possible applications of these results, we discuss just one special case. Let $A = k$ be a field of characteristic $p \neq 0$, let $R = \mathbb{Z}(\frac{1}{p})$, and let $S = \cup_{H \in \mathcal{D}(k, T, R)} G/H$. Then, $R \otimes K_0^G(-, \mathcal{P}(k), T)$, and thus $R \otimes_Z K_0^G(-, \mathcal{P}(k), T)$ and $R \otimes_Z P_0^G(-, \mathcal{P}(k), T)$ are S-projective. Moreover, the Cartan map $P_n^G(-, \mathcal{P}(k), T) \to K_n^G(-, \mathcal{P}(k), T)$ is an isomorphism for any G-set S for which the Sylow-p-subgroups H of the stabilizers of the elements in X have a non-empty fixed point set T^H in T, since in this case T-exact sequences over X are split exact and thus all functors $\zeta : \underline{X} \to \mathcal{P}(k)$ are T-projective, i.e., $[X, \mathcal{P}(k)]_T \hookrightarrow [X, \mathcal{P}(k)]$ is an isomorphism if $[X, \mathcal{P}(k)]$ is taken to be exact with respect to T-exact and thus split exact sequences. This implies in particular that for all G-sets X, the Cartan map

$$P_n^G(X \times S, \mathcal{P}(k), T) \to K_n^G(X \times S, \mathcal{P}(k), T)$$

is an isomorphism since any stabilizer group of an element in $X \times S$ is a subgroup of a stabilizer group of an element in S, and thus, by the very definition of S and $\mathcal{D}(k, T, \mathbb{Z}(\frac{1}{p}))$, has a Sylow-$p$-subgroup H with $T^H \neq \emptyset$. This finally implies that $P_n^G(-, \mathcal{P}(k), T)_S \to K_n^G(-, \mathcal{P}(k), T)_S$ is an isomorphism. So, by the general theory of Mackey functors,

$$\mathbb{Z}(\frac{1}{p}) \otimes P_n^G(-, \mathcal{P}(k), T) \to \mathbb{Z}(\frac{1}{p}) \otimes K_n^G(-, \mathcal{P}(k), T)$$

is an isomorphism. The special case $(T = G/\epsilon)$ $P_n^G(-, \mathcal{P}(k), G/\epsilon)$, is just the K-theory of finitely generated projective kG-modules and $K_n^G(-, \mathcal{P}(k), G/\epsilon)$ the K-theory of finitely generated kG-modules with respect to exact sequences. Thus we have proved the following.

Theorem 10.4.1 *Let k be a filed of characteristic p, G a finite group. Then, for all $n \geq$), the Cartan map $K_n(kG) \to G_n(kG)$ induces isomorphisms*

$$\mathbb{Z}(\frac{1}{p}) \otimes K_n(kG) \to \mathbb{Z}(\frac{1}{p}) \otimes G_n(kG).$$

Here are some applications of 10.4.1. These applications are due to A.O. Kuku (see [108, 112, 114]).

Theorem 10.4.2 *Let p be a rational prime, k a field of characteristic p, G a finite group. Then for all $n \geq 1$,*

(i) $K_{2n}(kG)$ is a finite p-group.

(ii) The Cartan homomorphism $\varphi_{2n-1} : K_{2n-1}(kG) \to G_{2n-1}(kG)$ is surjective, and $Ker\varphi_{2n-1}$ is the Sylow-p-subgroup of $K_{2n-1}(kG)$.

PROOF

(i) In theorem 10.4.1, it was proved that for all $n \geq 0$, the Cartan map $\varphi_{2n} : K_{2n}(kG) \to G_{2n}(kG)$ induces an isomorphism $K_n(kG) \otimes \mathbb{Z}(\frac{1}{p}) = G_n(kG) \otimes \mathbb{Z}(\frac{1}{p})$ i.e., φ_n is an isomorphism mod p-torsion. Now, for all $n \geq 0$, $G_{2n}(kG) = 0$ since kG is a finite ring (see theorem 7.1.12(ii)). So, $K_{2n}(kG) = Ker\varphi_{2n}$ is a finite group. It is also a p-group since φ_{2n} is a monomorphism mod p-torsion. So (i) is proved.

(ii) $G_{2n-1}(kG)$ is a finite group of order relatively prime to p. So, $|Coker\varphi_{2n-1}|$ is power of p and divides $|G_{2n-1}(kG)|$, which is $\equiv -1$ (mod p). This is possible if and only if $Coker\varphi_{2n-1} = 0$. Hence $Coker\varphi_{2n-1} = 0$ for all $n \geq 1$, and so, each φ_{2n-1} is surjective.

Now, by theorem 7.1.12(i) $K_{2n}(kG)$ is finite, and so, $Ker\varphi_{2n-1}$ is a finite p-group since φ_{2n-1} is a monomorphism mod p-torsion. Moreover, $G_{2n-1}(kG)$ has order $\equiv -1$ (mod p). Hence $Ker\varphi_{2n-1}$ is a Sylow-p-subgroup of $K_{2n-1}(kG)$.

$\square$

Corollary 10.4.1 Let k be a field of characteristic p, C a finite EI category. Then, for all $n \geq 0$, the Cartan homomorphism $K_n(kC) \rightarrow G_n(kC)$ induces isomorphism

$$\mathbb{Z}(\frac{1}{p}) \otimes K_n(kC) \cong \mathbb{Z}(\frac{1}{p}) \otimes G_n(kC).$$

Corollary 10.4.2 Let R be the ring of integers in a number field F, m a prime ideal of R lying over a rational prime p. Then for all $n \geq 1$,

(a) the Cartan map $K_n((R/m)C) \rightarrow G_n((R/m)C)$ is surjective.

(b) $K_{2n}((R/m)C)$ is a finite p-group.

Finally, with the identification of Mackey functors : $GSet \rightarrow Ab$ with Green's G-functors $\underline{\delta}G \rightarrow Ab$ as in 9.1.1 and above interpretations of our equivariant theory in terms of grouprings, we now have, from the forgoing, the following result, which says that higher algebraic K-groups are hyperelementary computable. First, we define this concept.

Definition 10.4.1 *Let G be a finite group, $\mathcal{U}$ a collection of subgroups of G closed under subgroups and isomorphic images, A a commutative ring with identity. Then a Mackey functor $M : \delta G \rightarrow A\text{-}Mod$ is said to be $\mathcal{U}$-computable if the restriction maps $M(G) \rightarrow \prod_{H \in \mathcal{U}} M(H)$ induces an isomorphism $M(G) \simeq \varprojlim_{H \in \mathcal{U}} M(H)$ where $\varprojlim_{H \in \mathcal{U}}$ is the subgroup of all $(x) \in \prod_{H \in \mathcal{U}} M(H)$ such that for any $H, H' \in \mathcal{U}$ and $g \in G$ with $gH'g^{-1} \subseteq H$, $\varphi : H' \rightarrow H$ given by $h \rightarrow ghg^{-1}$, then $M(\varphi)(x_H) = x_{H'}$.*

Now, if A is a commutative ring with identity, $M : \delta G \rightarrow \mathbb{Z}\text{-}Mod$ a Mackey functor, then $A \otimes M : \delta G \rightarrow A\text{-}Mod$ is also a Mackey functor where $(A \otimes M)(H) = A \otimes M(H)$. Now, let $\mathcal{P}$ be a set of rational primes, $\mathbb{Z}_{\mathcal{P}} = \mathbb{Z}[\frac{1}{q} \mid q \notin \mathcal{P}]$, $C(G)$ the collection of all cyclic subgroups of G, $h_{\mathcal{P}}C(G)$ the collection of all $\mathcal{P}$-hyperelementary subgroups of G, i.e.,

$$h_{\mathcal{P}}C(G) = \{H \leq G | \exists H' \trianglelefteq H, H' \in C(G), H/H' \quad \text{a p-group for some } p \in \mathcal{P}\}.$$

Then, we have the following theorem.

Theorem 10.4.3 *Let R be a Dedekind ring, G a finite group, M any of the Green modules $K_n(R-)$, $G_n(R-)$, $SK_n(R-)$, $SG_n(R-)$, Cl_n over $G_0(R-)$; then $\mathbb{Z}_p \otimes M$ is $h_{\mathcal{P}}(C(G))$-computable.*

PROOF In view of theorem 9.3.1, it suffices to show that $\oplus_{H \in h_{\mathcal{P}}(C(G))} \mathbb{Z}_{\mathcal{P}} \otimes G_0(RH) \to \mathbb{Z}_p \otimes G_0(RG)$ is surjective. To do this, it suffices to show that for each p, $\oplus_{H \in h_p(C(G))} \mathbb{Z}_p \otimes G_0(RH) \to \mathbb{Z}_p \otimes G_0(RG)$ is surjective. Now, it is known that any torsion element in $G_0(RG)$ is nilpotent (see remarks and definition 9.7.1 and lemma 9.7.8 or [48, 209]). It is also known that $\oplus_{H \in h_p(C(G))} \mathbb{Q} \otimes G_0(RH) \to \mathbb{Q} \otimes G_0(RG)$ is surjective (see theorem 9.7.5 or [48, 207]). Hence, $\oplus_{H \in h_p(C(G))} \mathbb{Z}_p \otimes G_0(RH) \to \mathbb{Z}_p \otimes G_0(RG)$ is surjective, and the theorem is proved. ▢

Exercises

10.1 Let $\mathcal{C}$ be an exact category, S, T G-sets, $\zeta : \underline{S} \to \mathcal{C}$ an object of $[\underline{S}, \mathcal{C}]$. Show that ζ is T-projective if and only if it is T-injective.

10.2 Let $\varphi : S_1 \to S_2$ be a G-map (S_1, S_2 G-sets, G a finite group) $\mathcal{C}$ an exact category. Show that the induced functor $\varphi_* : [\underline{S}_1, \mathcal{C}] \to [\underline{S}_2, \mathcal{C}]$ maps T-exact sequences in $[\underline{S}_1, \mathcal{C}]$ to T-exact sequences in $[\underline{S}_2, \mathcal{C}]$ and T-projective functors to T-projective functors.

Chapter 11

Equivariant higher K-theory for profinite group actions

In this chapter, we extend the theory and results in chapter 10 to the case of profinite group actions. This extension is due to A.Kuku (see [109]).

11.1 Equivariant higher K-theory – (Absolute and relative)

11.1.1 Let I be a filtered index set (i.e., I is a partially ordered set such that for $i, j \in I$ there exists $k \in I$ such that $i \leq k$, $j \leq k$).

Note that I is a category whose objects are elements of I, with exactly one morphism from i to j if $i \leq j$, and no morphism otherwise.

Suppose that C is a category. A projective (resp. injectively) filtered system of C-objects is covariant (resp. contravariant) functor $F : I \to C$. A projective limit (X, μ_i) of a projectively filtered system of C-objects is a covariant (resp. contravariant) functor $F : I \to C$. A projective limit (X, μ_i) of a projectively filtered system of C-objects is a C-object X together with maps $\mu_i : X \to F_i$ such that for all $i, j \in I$ with $i < j$ and C-morphisms $\varphi_{ij} : F_i \to F_j$, we have $\mu_j = \varphi_{ij}\mu_i$, and moreover, given another C-object Y and morphisms $\nu_i : Y \to F_i$ with the same properties, then there exists exactly one map $\rho : Y \to X$ such that $\mu_i\rho = \nu_i$ for any $i \in I$.

Note that (X, μ_i) is well defined up to isomorphism by F. We shall denote the limit by $\varprojlim F$ or $\varprojlim F_i$. The injective $\varinjlim F$ is analogously defined.

Although we shall apply the above definition to some other situations, we mention a particular example just to fix notations. Let I be the set of open normal subgroups H of a profinite group G, ordered by set inclusion, $F_H = G/H$, and for $H \leq H'$, $\varphi_{H,H'} : G/H \to G/H'$ the canonical map. Then $G = \varprojlim F$.

Note that each G/H is a finite group. We shall henceforth denote G/H by G_H.

11.1.2 (a) Let G be a profinite group. A G-set is a finite set with discrete topology on which G acts continuously by permutations on the left, i.e.,

$G \times S \to S$ is continuous. The G-sets form a category $GSet$ with an initial object that we shall denote by φ, final object that shall denote by $\star = G/G$, sums (disjoint union) and products (Cartesian products), and more generally, finite projective and injective limits. If S, T are G-sets, we write $GSet(S, T)$ for the set of $GSet$-morphism from S to T.

(b) A simple G-set has the form G/H where H is some open subgroup of G, and so, any G-set S has the form $\cup G/H$ where H runs through some finite collection of open subgroups of G.

(c) If S is a G-set, we can associate with S a category $\underline{S}$ as follows – the objects of $\underline{S}$ are elements of S, while for $s, t \in S, \underline{S} = \{(g, s)|g \in G, gs = t\}$, with composition defined for $t = gs$ by $(h, t) \circ (g, s) = (hg's)$ and the identity morphism $s \to s$ given by (e, s) where e is the identity element of G.

(d) Now, let $\mathcal{C}$ be an exact category, $[\underline{S}, \mathcal{C}]$ the category of covariant functors from $\underline{S}$ to $\mathcal{C}$, which factor through some finite quotient of G, that is, such that we have a functor $\varinjlim_H [\underline{S}^H, \mathcal{C}] \to [\underline{S}, \mathcal{C}]$ where H runs through all open normal subgroups of G that act trivially on S, and $\underline{S}^H$ corresponds to S as a G/H-set. If $\zeta \in [\underline{S}, \mathcal{C}]$, we shall write ζ_s for $\zeta(s)$. It can easily be checked that $[\underline{S}, \mathcal{C}]$ is an exact category (just as in the case for G finite) where a sequence $\zeta' \to \zeta \to \zeta''$ is said to be exact if it is exact fiberwise, i.e., $\zeta'_s \to \zeta_s \to \zeta''_s$ is exact for all $s \in S$. We now define $K_n^G(S, \mathcal{C})$ as the nth algebraic K-group associated with $[\underline{S}, \mathcal{C}]$ with fiberwise exact sequences.

(e) Let S, T be G-sets. Just as for G finite (see definition 10.2.1), we have the notion of a sequence of functors in $[\underline{S}, \mathcal{C}]$ being T-exact, and hence the definition of nth algebraic K-group associated with the category $[\underline{S}, \mathcal{C}]$ with respect to exact sequences that we denote by $K_n^G(-, \mathcal{C}, T)$.

(f) Finally, the definition of T-projective functors in $[\underline{S}, \mathcal{C}]$ is as in definition 10.2.2, and so, we define $P_n^G(S, \mathcal{C}, T)$ as the nth algebraic K-group associated with the full subcategory $[\underline{S}, \mathcal{C}]_T$ of T-projective functors in $[\underline{S}, \mathcal{C}]$.

We now record the following theorem whose proof is similar to that for finite groups.

Theorem 11.1.1 Let G be a profinite group, T a G-set, $\mathcal{C}$ an exact category. Then $K_n^G(-, \mathcal{C})$, $K_n^G(-, \mathcal{C}, T)$, $P_n^G(-, \mathcal{C}, T) : GSet \to \mathbb{Z}\text{-}\mathcal{M}od$ are Mackey functors. If the pairing $\mathcal{C} \times \mathcal{C} \to \mathcal{C} : (M, N) \to M \circ N$ is naturally associative and commutative, and $\mathcal{C}$ has an object E such that $E \circ M = M \circ E = M$ for all $M \in \mathcal{C}$, then $K_0^G(-, \mathcal{C})$, $K_0^G(-, \mathcal{C}, T)$ become Green functors and $K_n^G(-, \mathcal{C})$ becomes $K_0^G(-, \mathcal{C})$-modules, while $K_n^G(-, \mathcal{C}, T)$, $P_n^G(-, \mathcal{C}, T)$ become $K_0^G(-, \mathcal{C}, T)$-modules.

Remarks 11.1.1 Since we saw in 11.1.2(b) that a G-set S has the form $\cup G/H$ where H runs through a finite collection of open subgroups of G, and the K-functors 11.1.1, being Mackey functors, take disjoint union into direct sums, then to understand how the functors behave on a G-set S, it suffices to understand their behavior on transitive G-sets G/H. The interpretation of the behavior for the functors in the latter case in terms of group-rings is similar to that for G finite, and we omit the details (see 10.3). However, we observe that if e is the identity of G, and H an open subgroup of G, $\mathcal{P}(R)$ the category of finitely generated projective modules over a commutative regular ring R, we can identify $K_n^G(G/H, \mathcal{P}(R), G/e)$ with $G_n(RH)$ while we identify $P_n^G(G/H, \mathcal{P}(R), G/e)$ with $K_n(RH)$ for any ring R with identity.

In this section, we indicate how to obtain the following result, which is an extension of a similar result in 10.4.1 for finite groups.

Theorem 11.1.2 *Let G be a profinite group, k a field of characteristic p. Then for each $n \geq 0$, the Cartan homomorphism $K_n(kG) \to G_n(kG)$ induces an isomorphism $\mathbb{Z}(\frac{1}{p}) \otimes K_n(kG) \to \mathbb{Z}(\frac{1}{p}) \otimes G_n(kG)$.*

Remarks 11.1.2 (a) In view of the identifications in 11.1.1, we only have to see that for each $n \geq 0$, the Cartan homomorphism $P_n^G(G/G, \mathcal{P}(k), G/e) \to K_n^G(-, \mathcal{P}(k), G/e)$ induces an isomorphism

$$\mathbb{Z}(\frac{1}{p}) \otimes P_n^G(-, \mathcal{P}(k), G/e) \simeq \mathbb{Z}(\frac{1}{p}) \otimes K_n^G(-, \mathcal{P}(k), G/e).$$

(b) Let $G \xrightarrow{\varphi} G_1$ be a homomorphism of profinite groups. Then φ induces a functor $\hat{\varphi} : G_1 Set \to GSet$ given by $S \to S|_G$ where $S|_G$ is obtained from S by restriction of the action of G_1 on S to G via φ, i.e., for $s \in S$, $g \in G$, $gs = \varphi(g)s$.

Also, if $\alpha : S \to T$ is a G_1-map, then α is also a G-map $\alpha|_G = \varphi(\alpha) : S|_G \to T|_G$

So, we have a canonical functor $\underline{S} \xrightarrow{\delta} \underline{S}$ given by $s \to s$; $(g, s) \to (\varphi(g), s)$.

If $\mathcal{C}$ is an exact category, then δ induces an exact functor $[\underline{S}, \mathcal{C}] \to [\underline{S}|_G, \mathcal{C}]$ (see 1.1.5 for definitions). Note that $\zeta|_G$ has the same fibers as ζ with G-action defined by restricting the G_1-action to G via φ. Hence, we have a homomorphism $\delta : K_n^{G_1}(S, \mathcal{C}) \to K_n^G(S|_G, \mathcal{C})$ of K-groups. In particular, if $S = G_1/G_1$, then $S|_G = G/G$, and we have a homomorphism $K_n^{G_1}(G_1/G_1, \mathcal{C}) \to K_n^G(G/G, \mathcal{C})$. We now apply this situation to the canonical map $G \to G_H$ from a profinite group G to $G_H = G/H$ where H is an open normal subgroup of G and $\mathcal{C} = \mathcal{P}(R)$, and get a homomorphism $K_n^{G_H}(G_H/G_H, \mathcal{P}(R)) \xrightarrow{\chi} K_n^G(G/G, \mathcal{P}(R))$ where $\mathcal{P}(R)$ is the category of finitely generated projective R-modules.

(c) Note that the functor $I \to \mathbb{Z}\text{-}\mathcal{M}od$ given by $H \to K_n^{G_H}(G_H/G_H, \mathcal{P}(R))$ is an injectively filtered system of Abelian groups since if $H, H' \in I$ and $H \leq H'$, then we have a homomorphism $G_H \to G_{H'}$, and hence a functor $G_H/G_H \to G_H/G_{H'}$, which induces an exact functor $[G_{H'}/G_{H'}, \mathcal{P}(R)] \to [G_H/G_H, \mathcal{P}(R)]$ which then induces a homomorphism $K_n^{G_{H'}}(G_{H'}/G_{H'}, \mathcal{P}(R)) \xrightarrow{\chi_{H,H'}} K_n^{G_H}(G_H/G_H, \mathcal{P}(R))$ of K-groups. Moreover, the homomorphisms $\chi_H : K_n^{G_H}(G_H/G_H, \mathcal{P}(R)) \to K_n^G(G/G, \mathcal{P}(R))$, are compatible with $\chi_{H,H'}$, and so, we have a well-defined map $\varinjlim_H K_n^{G_H}(G_H/G_H, \mathcal{P}(R)) \to K_n^G(G/G, \mathcal{P}(R))$.

By a similar argument, if T is G_H-set, we have a homomorphism $K_n^{G_H}(G_H/G_H, \mathcal{P}(R), T) \to K_n^G(G/G, \mathcal{P}(R), T|_G)$, and in particular, if e_H is the identity element of G_H and e the identity element of G, we have homomorphisms

$$K_n^{G_H}(G_H/G_H, \mathcal{P}(R), G_H/e_H) \to K_n^G(G/G, \mathcal{P}(R), G/e), \quad \text{and hence}$$

$$\varinjlim_H K_n^{G_H}(G_H/G_H, \mathcal{P}(R), G_H/e_H) \to K_n^G(G/G, \mathcal{P}(R), G/e).$$

Similarly, we obtain

$$P_n^{G_H}(G_H/G_H, \mathcal{P}(R), G_H/e_H) \to P_n^G(G/G, \mathcal{P}(R), G/e).$$

We now have the following.

Theorem 11.1.3 *Let G be a profinite group, $I = \{H\}$ the filtered index set of open normal subgroups of G, $\mathcal{P}(R)$ the exact category of finitely generated projective R-modules. Then, in the notation of remarks 11.1.2(c), the induced maps*

(i) $\varinjlim_H K_n^{G_H}(G_H/G_H, \mathcal{P}(R)) \to K_n^G(G/G, \mathcal{P}(R))$

(ii) $\varinjlim_H K_n^{G_H}(G_H/G_H, \mathcal{P}(R), G_H/e_H) \to K_n^G(G/G, \mathcal{P}(R), G/e)$

(iii) $\varinjlim_H P_n^{G_H}(G_H/G_H, \mathcal{P}(R), G_H/e_H) \to P_n^G(G/G, \mathcal{P}(R), G/e)$

are isomorphisms.

PROOF   Note that $K_n^H(G_H/G_H, \mathcal{P}(R)) = K_n$ of the exact category $[G_H/G_H, \mathcal{P}(R)]$ where $[G_H/G_H, \mathcal{P}(R)]$ can be identified with the category $\overline{\mathcal{P}(R)}_{G_H}$ of G_H-objects in $\mathcal{P}(R)$, i.e., with the category $[G_H, \mathcal{P}(R)]$ of functors from G_H (considered as a category with one object with morphisms elements of G) to $\mathcal{P}(R)$. So, $\varinjlim_H K_n(G_H/G_H, \mathcal{P}(R)) = \varinjlim_H K_n(G_H, \mathcal{P}(R))$. Now, the functor $H \to [G_H, \mathcal{P}(R)]$ from I to the category of small categories is a filtered injective system of small categories (with the homomorphisms $G_H \xrightarrow{\varphi_{H,H'}} G_{H'}$) inducing exact functors $[G_{H'}, \mathcal{P}(R)] \xrightarrow{\varphi_{H,H'}} [G_H, \mathcal{P}(R)]$, and moreover, the canonical homomorphisms $G \xrightarrow{\varphi_H} G_H$ induces exact functors

$[\underline{G}_H, \mathcal{P}(R)] \xrightarrow{\varphi_H} [\underline{G}, \mathcal{P}(R)] : \zeta \to \zeta \circ \varphi_H$, which commutes with the $\varphi_{H,H'}$ and hence induces a map $\varinjlim_H [\underline{G}_H, \mathcal{P}(R)] \xrightarrow{\varphi} [\underline{G}, \mathcal{P}(R)]$, which is easily checked to be an isomorphism of categories since $G = \varinjlim_H G_H$.

Now, $K_n^G(G, \mathcal{P}(R)) = K_n(\varinjlim_H [\underline{G}_H, \mathcal{P}(R)]) = \varinjlim_H K_n[\underline{G}_H, \mathcal{P}(R)]$ by Quillen's result (see [165] page 96.), i.e.,

$K_n^G(G/G, \mathcal{P}(R)) = \varinjlim_H K_n^{G_H}[G_H/G_H, \mathcal{P}(R)]$ as required.

The proof of (ii) and (i) are similar and are omitted. In fact, (ii) is a restatement of (i) via the identification in remarks 11.1.1. □

PROOF of 11.1.2 Now, let G be a profinite group, $I = (H)$ the filtered index set of open normal subgroups of G. Then by 11.1.3 we have

$$\mathbb{Z}(\frac{1}{p}) \otimes \varinjlim_{\overline{H}} K_n^{G_H}(G_H/G_H, \mathcal{P}(k), G_H/e_H) \simeq \mathbb{Z}(\frac{1}{p}) \otimes K_n^G(G/G, \mathcal{P}(k), G/e)$$

and

$$\mathbb{Z}(\frac{1}{p}) \otimes \varinjlim_{\overline{H}} K_n^{G_H}(G_H/G_H, \mathcal{P}(k), G_H/e_H) \simeq \mathbb{Z}(\frac{1}{p}) \otimes K_n^G(G/G, \mathcal{P}(K), G/e).$$

Now, consider the following commutative diagram

$$
\begin{array}{ccc}
\mathbb{Z}(\frac{1}{p}) \otimes P_n^G(G/G, \mathcal{P}(k), G/e) & \to & \mathbb{Z}(\frac{1}{p}) \otimes K_n^G(G/H, \mathcal{P}(K), G/e) \\
\uparrow & & \uparrow \\
\varinjlim_{\overline{H}} \mathbb{Z}(\frac{1}{p}) \otimes P_n^{G_H}(G_H/G_H, \mathcal{P}(k), G_H/e_H) & \to & \varinjlim_{\overline{H}} \mathbb{Z}(\frac{1}{p}) \otimes K_n^{G_H}(G_H/G_H, \mathcal{P}(k), G_H/e_H)
\end{array}
$$

Note that each G_H is a finite group, and so, it follows from theorem 10.4.1 that the Cartan map

$$P_n^{G_H}(G_H/G_H, \mathcal{P}(k), G_H/e_H) \to K_n^{G_H}(G_H/G_H, \mathcal{P}(k), G_H/e_H)$$

induces an isomorphism

$$\mathbb{Z}(\frac{1}{p}) \otimes P_n^{G_H}(G_H/G_H, \mathcal{P}(k), G_H/e_H) \simeq \mathbb{Z}(\frac{1}{p}) \otimes K_n^{G_H}(G_H/G_H, \mathcal{P}(k), G_H/e_H).$$

So, the bottom arrow of the square is an isomorphism. Since the vertical arrows are also isomorphisms, it means that the top arrow is also an isomorphism. □

11.2 Cohomology of Mackey functors (for profinite groups)

In this section, we call attention to the fact that if G is a profinite group, the category $GSet$ of G-sets is a category with finite sums, final object, and

finite pull-backs, and cohomology theory can be defined for Mackey functors
$GSet \to \mathbb{Z}$-$\mathcal{M}od$.
Consequently, vanishing theorems are obtained for the cohomology of the K-
functors of sections 11.1 and also for the cohomology of profinite groups.

11.2.1 (a) First note that if S is any G-set, $M : GSet \to \mathbb{Z}$-$\mathcal{M}od$ a Mackey
functor, then $M_S : GSet \to \mathbb{Z}$-$\mathcal{M}od$ defined by $M_S(T) = M(S \times T)$ is
a Mackey functor, and moreover, the projection map $pr : S \times T \to T$
defines natural transformations if $\theta_S : M_S \to M$ (resp. $\theta^S : M \to M_S$),
and M is then said to be S-projective (resp. S-injective) if θ_S (resp.
θ^S) is split surjective (resp. injective). Note that M is S-projective
if and only if M is S-injective (see proposition 9.1.1 for various other
equivalent properties of S-projectivity).

(b) Let S, T be G-sets, and $f : S \to *$, $g : T \to *$ the unique maps from S to
$*$ and T to $*$, respectively.
If $M : GSet \to \mathbb{Z}$-$\mathcal{M}od$ is a Green functor, we write $K_S(M) = Ker(M(*)$
$\xrightarrow{f^*} M(S))$ and $I_S(M) = Im(M(S) \xrightarrow{f_*} M(*))$.
Now, If $\mathcal{G} : GSet \to \mathbb{Z}$-$\mathcal{M}od$ is a Green functor such that M is a $\mathcal{G}$-
module, it follows from sections 9.2, 9.3 that the pairing $\mathcal{G} \times M \to M$ de-
fines a pairing $\mathcal{H}^i\mathcal{G} \times \mathcal{H}^j M \to \mathcal{H}^{i+j} M$ $(i, j \geq 0)$ and also $\hat{\mathcal{H}}^i\mathcal{G} \times \hat{\mathcal{H}}^j M \to$
$\hat{\mathcal{H}}^{i+j} M$ $(i, j \in \mathbb{Z})$. In particular, putting $i = j = 0$, $M = \mathcal{G}$, we see that
$\mathcal{H}^0\mathcal{G}, \hat{\mathcal{H}}^0\mathcal{G}$ are Green functors, $\mathcal{H}^i M$ are $\mathcal{H}^0\mathcal{G}$-modules, and $\hat{\mathcal{H}}_i$ are $\hat{\mathcal{H}}^0\mathcal{G}$-
modules. Hence, if one can find $n \in \mathbb{N}$ such that $n1_{\mathcal{G}(*)} \in K_S\mathcal{G} + I_T\mathcal{G}$,
then $n1_{\hat{\mathcal{H}}^0\mathcal{G}(*)} \in I_T\hat{\mathcal{H}}^0\mathcal{G}$, and so, n annihilates $K_T\hat{\mathcal{H}}^i_S$, for all $i \in \mathbb{Z}$. In
particular, when $T = S$ (i.e., $f = g$) and hence $K_T\hat{\mathcal{H}}^i\mathcal{G} = \hat{\mathcal{H}}^i\mathcal{G}$, we have
$n \cdot \hat{H}^i_S\mathcal{G} = 0$ for all $i \in \mathbb{Z}$.

11.2.2 Our next step is to indicate how to get a suitable n that annihilates
the cohomology groups in terms of the order of G. This would come from
theorem 11.2.1 that follows. First, we need some preliminary definition and
remarks.

(a) A super-natural product is a formal product $\prod p^{n_p}$ where the product is
taken over all prime numbers and n_p is an integer $0 \leq n_p < \infty$. If G is
a profinite group, H' a closed subgroup of G, then the index $(G : H')$
of H' in G is defined by $(G : H') = 1$ c.m. $(G/H, H'/(H' \cap H))$ where
H ranges over all open normal subgroups of G. The order of G written
$|G|$ is defined by $|G| = (G : 1) = 1$ c.m. $|G/H|$, H ranging over open
normal subgroups of G.

(b) Let U be a family of open subgroup of G, $S_U = \cup_{H \in U} G/H$. Suppose that
P is a set of rational primes. Define $h_P U = \{H \leq G|$ there exist $p \in$
$P, H' \leq H, g \in G, H' \in U$ with H' open in G, H'/H a p-group, and

$gH'g^{-1} \le H\}$. If for any G-set S, we write $G_S = \{g \in G | gs = s$ for all $s \in S\}$, and $N = \cap_{\substack{g \in G \\ K \in U}} (gKg^{-1})$, then $N = G_{S_U}$.

Note that any G-set has the form S_U for some U.

Theorem 11.2.1 *Let U be a finite set of open subgroups of G, P a set of rational primes, $S = S_U$, $(G : G_S)'_P$ the maximal divisor of $(G : G_S)$, which contains no prime divisor in P. Then, for any Mackey functor $M : GSet \to \mathbb{Z}\text{-}\mathcal{M}od$, we have*

$$(G : G_S)'_P M(*) \subseteq I_{S_{k_P U}}(M) + K_S(M).$$

PROOF Similar to that of theorem 9.5.1. ⧠

Theorem 11.2.2 *Let $\mathcal{C}$ be an exact category such that the pairing $\mathcal{C} \times \mathcal{C} \to \mathcal{C} : (X, Y) \to X \circ Y$ is naturally associative and commutative, and $\mathcal{C}$ has an object E such that $E \circ X = X \circ E$ for all $X \in \mathcal{C}$. Let S, T be G-sets, P a set of rational primes. Suppose that M is any of the functors $K_n^G(-, \mathcal{C})$, $K_n^G(-, \mathcal{C}, T)$, $P_n^G(-, \mathcal{C}, T)$, then $(G : G_S)'_P \hat{\mathcal{H}}^i(M) = 0$, for all $i \in \mathbb{Z}$.*

PROOF By 11.2.1(b) and theorem 11.2.1 $(G : G_S)\hat{\mathcal{H}}_S^i \mathcal{G} = 0$ for $\mathcal{G} = K_0^G(-, \mathcal{C})$ and $\mathcal{G} = K_0^G(-, \mathcal{C}, T)$. Now, $K_n^G(-, \mathcal{C})$ is a $K_0^G(-, \mathcal{C})$-module, and $K_n^G(-, \mathcal{C}, T)$, $P_n^G(-, \mathcal{C}, T)$ are $K_0^G(-, \mathcal{C}, T)$-modules. So, for $M = K_n^G(-, \mathcal{C})$, $K_n^G(-, \mathcal{C}, T)$, $P_n^G(-, \mathcal{C}, T)$, we have $(G : G_S)'_P \hat{\mathcal{H}}^i M = 0$ for all $i \in \mathbb{Z}$. ⧠

Our second conclusion concerns the vanishing of cohomology groups of G.

11.2.3 Let G be a profinite group, A any $\mathbb{Z}G$-module. For $S \in GSet$, let $\hat{A}(S) = GSet(S, A)$.

Then, $GSet(S, A)$ is an Abelian group where, for $f_1, f_2 \in GSet(S, A)$ we define $(f_1 + f_2)(s) = f_1(s) + f_2(s)$. One can prove that $\overline{A} : GSet \to \mathbb{Z}\text{-}\mathcal{M}od$ is a Mackey functor (see 9.1.1 for a similar proof for G finite). If H is an open normal subgroup of G, $\overline{A}(G/H) = GSet(G/H, A) = A^H = H^0(H, A)$, the *zero*[th] cohomology group of H with coefficients in A. Moreover, $\hat{\mathcal{H}}^i \overline{A}(*) = \hat{\mathcal{H}}^i(G, A)$, and the $\hat{\mathcal{H}}^i \overline{A}$ are $\hat{\mathcal{H}}^0 \overline{A}$-modules by 11.2.1(c). More generally, if we write $\hat{\mathcal{H}}^i \overline{A}(S) = \hat{\mathcal{H}}_i(S, A)$ for any G-sets S, then we have as below.

Theorem 11.2.3 *Let G be a profinite group, S a G-set, A a $\mathbb{Z}G$-module, P a set of rational primes, then $(G, G_S)'_P \hat{\mathcal{H}}^i(S, A) = 0$ for all $i \in \mathbb{Z}$.*

PROOF

Similar to that of 11.2.2 by applying 11.2.1(b) and theorem 11.2.1 since $\hat{\mathcal{H}}^i \overline{A}$ are $\hat{\mathcal{H}}^0 \overline{A}$-modules. ⧠

Remarks 11.2.1 The above result 11.2.3 applies to $G = Galk_s/k$, the Galois group of the separable closure k_s of a filed k, $A = k_s^*$, the units of k_s.

Exercises

11.1 Let G be a profinite group. For any G-set B, define functor $\bar{B} : GSet \to \mathbb{Z}\text{-}\mathcal{M}od$ by $\bar{B}(S) = Hom_{GSet}(S, B)$. Show that $\bar{B}$ is a Mackey functor.

11.2 Let G be a profinite group, S a G-set.

(a) Show that the category $GSet, GSet/S$ are based categories.

(b) Define the Burnside ring $\Omega(G)$ of G by $\Omega(G) := \Omega(GSet/*)$. Also for any closed subgroup H of G, define $\varphi_H : \Omega(G) \to \mathbb{Z}$ by $\varphi_H(S) = |S^H|$. Now, let R be an integral domain. Show that any homomorphism $\varphi : \Omega(G) \to R$ factors through some φ_H. Hence, show that for any prime ideal $\underline{p}$ of $\Omega(G)$, there exists some closed subgroup $H \leq G$ and some characteristic p (prime or zero) such that

$$\underline{p} = \underline{p}(H, p) := \{x \in \Omega(G) | \varphi_H(x) \equiv 0(p)\}.$$

Show also that $\underline{p}(H, p) = \underline{p}(H', p)$ iff H and H' are conjugate in G.

11.3 Let G be a profinite group and let $\mathcal{U}$ be a finite set of open subgroups of G, $\mathcal{P}$ a set of rational primes, $S = S_\mathcal{U} = \bigcup_{H \in \mathcal{U}} G/H$ $(G : G_S)'_\mathcal{P}$ the maximal division of $(G : G_S)$ that contains no prime division in $\mathcal{P}$. Show that for any Mackey functor $M : GSet \to \mathbb{Z}\text{-}\mathcal{M}od$, we have $(G : G_S)'_\mathcal{P}(M(*)) \subseteq I_{S_{h_\mathcal{P}\mathcal{U}}}(M) + K_S(M)$ (in the notation of 11.2.2(b)).

11.4 Let G be a profinite group and let $\mathcal{G} : GSet \to \mathbb{Z}\text{-}\mathcal{M}od$ be a Green functor such that multiplication $\mathcal{G}(*) \times \mathcal{G}(*) \to \mathcal{G}(*)$ is surjective. Show that there exists a unique family $\mathcal{D}_\mathcal{G}$ of closed subgroups of G such that

(i) $\mathcal{D}_\mathcal{G}$ is a closed subset of the totally disconnected space $\mathcal{C}(G)$ of all closed subgroups of G.

(ii) If H, H' are subgroups of G, $g \in G$, $H \in \mathcal{D}_\mathcal{G}$, and $gH'g^{-1} \subseteq H$, then $H' \in \mathcal{D}_\mathcal{G}$.

(iii) For any map $\varphi : S \to *$, M a G-module, the map $\varphi_* : M(S) \to M(*)$ is surjective iff for any $H \in \mathcal{D}_\mathcal{G}$ there exists $s \in S$ such that $H \leq G_S$ (i.e., iff $S^H \neq \emptyset$ for all $H \in \mathcal{D}_\mathcal{G}$).

11.5 Let K be a field and E a finite or infinite Galois extension of K with Galois group G. Let $\mathcal{A}(K,E)$ be the category of K-algebras R such that $E \otimes_K R$ is a product of a finite number of copies of E. Show that there is an equivalence of categories $\mathcal{A}(K,E) \simeq GSet$ where $R \in \mathcal{A}(K,E)$ goes to S_R and S_R is the set of K-algebra homomorphisms $R \to E$, which becomes a G-set via the action of G on E, the inverse equivalence being given by $S \in GSet$ going to $R_S := Hom_{GSet}(S,E)$.

Show that if $G = Gal(E/K)$, then the Burnside ring $\Omega(G)$ can be interpreted as the Grothendieck ring $\Omega(K,E)$ generated by objects R in $\mathcal{A}(K,E) \simeq GSets$ with sums given by direct sum and products by $\otimes_K$.

11.6 Let G be a profinite group, $S(G)$ the set of all closed subgroups of G. For any pair (H,H') of open subgroups H,H' of G with $H \leq H'$ and H' open normal in G, define $O_{H,H'} = \{K \in S(G) | K \cdot H = H'\}$. Show that $S(G)$ is a topological space with the $O_{H,H'}$ forming a sub-basis of open sets. Show also that $S(G)$ is a compact, totally disconnected Hausdorff space and that the set of open sets in dense in $S(G)$.

Prove that G acts continuously on $S(G)$ and that the orbit space $S(G)\backslash G$ is the set $S_c(G)$ of conjugacy classes of closed subgroups, and that $S_c(G)$ is also a compact totally disconnected Hausdorff space.

Chapter 12

Equivariant higher K-theory for compact Lie group actions

The aim of this chapter is to extend the theory in chapter 10 to compact Lie groups. The work reported here is due to A. Kuku [116] based on T. Tom Dieck's induction theory for compact Lie groups [219].

12.1 Mackey and Green functors on the category $\mathcal{A}(G)$ of homogeneous spaces

---

$(12.1)^A$ The Abelian group $\mathcal{U}(G, X)$, G a compact Lie group, X a G-space; the category $\mathcal{A}(G)$

12.1.1 Let G be a compact Lie group, X a G-space. The component category $\pi_0(G, X)$ is defined as follows : Objects of $\pi_0(G, X)$ are homotopy classes of maps $\alpha : G/H \to X$ where H is a closed subgroup of G. A morphism from $[\alpha] : G/H \to X$ to $[\beta] : G/K \to X$ is a G-map $\sigma : G/H \to G/K$ such that $\beta\sigma$ is G-homotopic to α.

Note that since $Hom(G/H, X) \simeq X^H$ where $\varphi \to \varphi(eH)$, we could consider objects of $\pi_0(G, X)$ as pairs (H, c) where $c \in \pi_0(X^H) =$ the set of path components of X^H.

12.1.2 A G-ENR (Euclidien Neighborhood Retract) is a G-space that is G-homeomorphic to a G-retract of some open G-subset of some G-module V. Let Z be a compact G-ENR, $f : Z \to X$ a G-map. For $\alpha : G/H \to X$ in $\pi_0(G, X)$, we identify α with the path component X_α^H into which G/H is mapped by α.

Put $Z(f, \alpha) = Z^H \cap f^{-1}(X_\alpha^H) :=$ subspace of Z^H mapped under f into X_α^H. The action of $N_\alpha H/H$ on Z^H induces an action of $N_\alpha H/H$ on $Z(f, \alpha)$, i.e., $Z(f, \alpha)$ is an $Aut(\alpha)$-space (see [219]). Note that $Aut(\alpha) = \{\sigma : G/H \to G/H \mid \alpha\sigma \simeq \alpha\}$, and $N_\alpha(H)_H$, the isotropy group of $\alpha \in \pi_0(G, X)$, is isomorphic to $Aut(\alpha)$.

12.1.3 Let $\{Z_i\}$ be a collection of G-ENR, $f_i : Z_i \to X$, G-maps. Say that $f_i : Z_i \to X$ is equivalent to $f_j : Z_j \to X$ if and only if for each $\alpha : G/H \to X$ in $\pi_0(G, X)$, the Euler characteristic $\chi(Z(f_i, \alpha)/Aut(\alpha)) = \chi(Z(f_j, \alpha))/Aut(\alpha)$.
Let $\mathcal{U}(G, X)$ be the set of equivalence classes $[f : Z \to X]$ where addition is given by $[f_0 : Z_0 \to X] + [f_1 : Z_1 \to X] = [f_0 + f_1 : Z_0 + Z_1 \to X]$; the identity element is $\varphi : X \to X$; and the additive inverse of $[f : Z \to A]$ is $[f \circ p : Z \times A \to Z \to X]$, where A is a compact G-ENR with trivial G-action and $\chi(A) = -1$. Then, $\mathcal{U}(G, X)$ is the free Abelian group generated by $[\alpha], \alpha \in \pi_0(G, X)$, i.e., $[f : Z \to X] = \Sigma n(\alpha)$ where $G/H \times E^n \subset Z$ is an open n-cell of Z, and the restriction of f to $G/H \times E^n$ defines an element $[\alpha]$ of $\mathcal{U}(G, X)$.
The cell is called an n-cell of type α. Let $n(\alpha) = \sum(-1)^i n(\alpha, i)$ = number of i-cells of type α (see [219]).
If X is a point, write $\mathcal{U}(G)$ for $\mathcal{U}(G, X)$.

12.1.4 For a compact Lie group G, the category $\mathcal{A}(G)$ is defined as follows: $ob\mathcal{A}(G) :=$ homogeneous spaces G/H. The morphisms in $\mathcal{A}(G)(G/H, G/K)$ are the elements of the Abelian group $\mathcal{U}(G, G/H \times G/K)$ and have the form $\alpha : G/L \to G/H \times G/K$, which can be represented by diagram $\{G/H \xleftarrow{\alpha} G/L \xrightarrow{\beta} G/K\}$, so that $\mathcal{U}(G, G/H \times G/K)$ = free Abelian group on the equivalence classes of diagrams $G/H \xleftarrow{\alpha} G/L \xrightarrow{\beta} G/K$ where two such diagrams are equivalent if there exists an isomorphism $\sigma : G/L \to G/L'$ such that the diagram

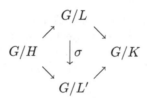

commutes up to homotopy.
Composition of morphisms is given by a bilinear map

$$\mathcal{U}(G, G/H_1 \times G/H_2) \times \mathcal{U}(G, G/H_2 \times G/H_3) \to \mathcal{U}(G, G/H_1 \times G/H_3)$$

where the composition of $(\alpha, \beta_1) : A \to G/H_1 \times G/H_2$ and $(\beta_2, \gamma) : B \to G/H_2 \times G/H_3$ yields a G-map $(\alpha\overline{\alpha}, \gamma\overline{\gamma}) : C \to G/H_1 \times G/H_3$, where $\overline{\gamma}, \overline{\alpha}$ are maps $\overline{\gamma} : C \to B$ and $\overline{a} : C \to A$, respectively.

12.1.5 (i) Each morphism $G/H \xleftarrow{\alpha} G/L \xrightarrow{\beta} G/K$ is the composition of special types of morphisms

$$G/H \xleftarrow{\alpha} G/L \xrightarrow{id} G/L \quad \text{and} \quad G/L \xleftarrow{id} G/L \xrightarrow{\beta} G/K$$

(ii) Let π_0 (or G) be the homotopy category of the orbit category or (G), that is, the objects of π_0 (or G) are the homogeneous G-spaces G/H, and morphisms are homotopy classes $[G/L \to G/K]$ of G-maps $G/L \to G/K$. We have a covariant functor π_0 (or G) $\to \mathcal{A}(G)$ given by $[G/L \xrightarrow{\beta} G/K] \to (G/L \xleftarrow{id} G/L \xrightarrow{\beta} G/K)$ and a contravariant functor π_0 (or G) $\to \mathcal{A}(G)$ given by

$$[G/H \to G/L] \to (G/H \leftarrow G/L \xrightarrow{id} G/L).$$

(iii) Addition is defined in $\mathcal{A}(G)(G/H, G/K) = \mathcal{U}(G, G/H \times G/K)$ by

$$(G/H \leftarrow G/L \to G/K) + (G/H \leftarrow G/L' \to G/K)$$

$$= (G/H \leftarrow (G/L) \dot{\cup} (G/L') \to G/K)$$

where $(G/L) \dot{\cup} (G/L')$ is the topological sum G/L and G/L'.

$(12.1)^B$ Mackey and Green functors on $\mathcal{A}(G)$

12.1.6 Let R be a commutative ring with identity. A Mackey functor M from $\mathcal{A}(G)$ to R-$\mathcal{M}$od is a contravariant additive functor. Note that M is additive if

$$M : \mathcal{A}(G)(G/H, G/K) \to R\text{-}\mathcal{M}\text{od}(M(G/K), M(G/H))$$

is an Abelian group homomorphism.

Remarks 12.1.1 M comprises of two types of induced morphisms:

(i) If $\alpha : G/H \to G/K$ is a G-map, regarded as an ordinary morphism $\alpha_! : G/H \xleftarrow{id} G/H \xrightarrow{\alpha} G/K$ of $\mathcal{A}(G)$, we have an induced morphism

$$M(\alpha_!) = M^*(\alpha) =: \alpha^* : M(G/K) \to M(G/H)$$

(ii) If α in (i) is induced from $H \subset K$, i.e., $\alpha(gH) = gK$, call α^* the restriction morphism.

(iii) If we consider α as a transfer morphism $\alpha^! : G/H \leftarrow G/K \to G/K$ in $\mathcal{A}(G)$, then we have

$$M(\alpha^!) =: M_*(\alpha) =: \alpha_* : M(G/H) \to M(G/K)$$

and call α_* the induced homomorphism associated to α.

12.1.7 Let M, N, L be Mackey functors on $\mathcal{A}(G)$. A pairing $M \times N \to L$ is a family of bilinear maps $M(S) \times N(S) \to L(S) : (x, y) \to x \cdot y$ $(S \in \mathcal{A}(G))$ such that for each G-map $f : G/H = S \to T = G/K$, we have

$$L^* f(x, y) = (M^* f(x) \cdot (N^* f(y))) \quad (x \in M(T), y \in N(T));$$
$$x \cdot (N_* f y) = L_* f((M^* f x) \cdot y) \quad (x \in M(T), y \in N(T));$$
$$(M^* f x) \cdot y = L_* f(x \cdot (N^* f y)) \quad (x \in M(T), y \in N(T)).$$

A Green functor $V : \mathcal{A}(G) \to R\text{-}\mathcal{M}od$ is a Mackey functor together with a pairing $V \times V \to V$ such that for each object S, the map $V(S) \times V(S) \to V(S)$ turns $V(S)$ into an associative R-algebra such that f^* preserves units.

If V is a Green functor, a left V-module is a Mackey functor M together with a pairing $V \times M \to M$ such that $M(S)$ is a left $V(S)$-module for every $S \in \mathcal{A}(G)$.

Remarks 12.1.2 The Mackey functor as defined in 12.1.6 is equivalent to the earlier definitions in 9.1.1 as well as in chapter 10, 11 defined for finite and profinite groups G as functors from the category $GSet$ to $R\text{-}\mathcal{M}od$. Observe that if $(M_*, M^*) = M$ is a Mackey functor (bifunctor) $GSet \to R\text{-}\mathcal{M}od$, we can get $\overline{M} : \mathcal{A}(G) \to R\text{-}\mathcal{M}od$ by putting $\overline{M}(G/H) = M_*(G/H) = M^*(G/H)$ on objects while a morphism $G/H \xleftarrow{\alpha} G/L \xrightarrow{\beta} G/K$ in $\mathcal{A}(G)$ is mapped onto $M(G/H) \xleftarrow{M_*(\alpha)} M(G/L) \xleftarrow{M^*(\beta)} M(G/K)$ in $R\text{-}\mathcal{M}od$. Then, M is compatible with composition of morphisms.

Conversely, let $\overline{M} : \mathcal{A}(G) \to R\text{-}\mathcal{M}od$ be given and for $(\alpha, \beta) \in \mathcal{A}(G)(G/H, G/K)$, let $\overline{M}^*(\alpha)$ and $\overline{M}_*(\alpha)$ be as defined in remark 12.1.1. Then, we can extend $\overline{M}$ additively to finite G-sets to obtain Mackey functors as defined in 9.1.1.

12.1.8 (Universal example of a Green functor) Define $\overline{V}(G/H) := \mathcal{U}(G, G/H)$, $\mathcal{U}(H, G/G) := \overline{V}(H)$. Now, consider $\mathcal{U}(G, G/H) = \mathcal{U}(G, G/G \times G/H)$ as a morphism set in $\mathcal{A}(G)$. Then, the composition of morphisms

$$\mathcal{U}(G, G/G \times G/K) \times \mathcal{U}(G, G/H \times G/K) \to \mathcal{U}(G, G/G \times G/H)$$

defines an action of $\overline{V}$ on morphisms.

Theorem 12.1.1 $\overline{V} : \mathcal{A}(G) \to \mathbb{Z}\text{-}\mathcal{M}od$ *is a Green functor, and any Mackey functor* $M : \mathcal{A}(G) \to \mathbb{Z}\text{-}\mathcal{M}od$ *is a* $\overline{V}$-*module.*

12.2   An equivariant higher K-theory for G-actions

12.2.1 Let G be a compact Lie group, X a G-space. We can regard X as a category $\underline{X}$ as follows. The objects of $\underline{X}$ are elements of X, and for $x, x' \in X$, $\underline{X}(x, x') = \{g \in G \mid gx = x'\}$ (see [135]).

12.2.2 Let X be a G-space, $\mathcal{C}$ an exact category.
Let $[\underline{X}, \mathcal{C}]$ be the category of functors $\underline{X} \to \mathcal{C}$. Then $[\underline{X}, \mathcal{C}]$ is an exact category where a sequence $0 \to \zeta' \to \zeta \to \zeta'' \to 0$ is exact in $[\underline{X}, \mathcal{C}]$ if and only if

$$0 \to \zeta(x)' \to \zeta(x) \to \zeta''(x) \to 0$$

is exact in $\mathcal{C}$, for all $x \in X$. In particular, for $X = G/H$ in $\mathcal{A}(G)$, $[\underline{G/H}, \mathcal{C}]$ is an exact category.

Example 12.2.1 The most important example of $[\underline{G/H}, \mathcal{C}]$ is when $\mathcal{C}$ is the category $\mathcal{M}(\mathcal{C})$ of finite-dimensional vector spaces over the field $\mathbb{C}$ of complex numbers. Here, the category $[\underline{G/H}, \mathcal{M}(\mathbb{C})]$ can be identified with the category of G-vector bundles on the compact G-space G/H, where for any $\zeta \in [\underline{G/H}, \mathcal{M}(\mathbb{C})]$, $x \in G/H$, $\zeta(x) \in \mathcal{M}(\mathbb{C})$ is the fiber $\hat{\zeta}_x$ of the vector bundle $\hat{\zeta}$ associated with ζ. Indeed, $\hat{\zeta}$ is completely determined by $\zeta_{\bar{e}}$ where $\bar{e} = eH$ (see [184, 186]).

Definition 12.2.1 *For $X = G/H$ and all $n \geq 0$, define $K_n^G(X, \mathcal{C})$ as the nth algebraic K-group of the exact category $[\underline{X}, \mathcal{C}]$ with respect to fiberwise exact sequence introduced in 12.2.2.*

Theorem 12.2.1   *(i) for all $n \geq 0$, $K_n^G(-, \mathcal{C}) : \mathcal{A}(G) \to \mathbb{Z}\text{-}\mathcal{M}od$ is a Mackey functor.*

(ii) If there is a pairing $\mathcal{C} \times \mathcal{C} \to \mathcal{C}$, $(A, B) \to A \circ B$ which is naturally associative and commutative, and there exists $E \in \mathcal{C}$ such that $E \circ M = M \circ E = M$ for all $M \in \mathcal{C}$, then $K_0^G(-, \mathcal{C})$ is a unitary $K_0^G(-, \mathcal{C})$-modules for each $n \geq 0$.

Before proving 12.2.1, we first briefly discuss pairings and modules structures on higher K-theory of exact categories.

12.2.3 Let $\mathcal{C}$, $\mathcal{C}_1$, and $\mathcal{C}_2$ be three exact categories and $\mathcal{C}_1 \times \mathcal{C}_2$ the product category. An exact pairing $\mathcal{C}_1 \times \mathcal{C}_2 \to \mathcal{C}$: $(M_1, M_2) \to M_1 \circ M_2$ is a covariant functor from $\mathcal{C}_1 \times \mathcal{C}_2 \to \mathcal{C}$ such that $\mathcal{C}_1 \times \mathcal{C}_2((M_1, M_2), (M_1', M_2')) = \mathcal{C}_1(M_1, M_1') \times \mathcal{C}_2(M_2, M_2') \to \mathcal{C}(M_1 \circ M_2, M_1' \circ M_2')$ is bi-additive and bi-exact, that is, for a fixed M_2, the functor $\mathcal{C}_1 \to \mathcal{C}$ given by $M_1 \to M_1 \circ M_2$ is additive and exact, and for fixed M_1, the functor $\mathcal{C}_2 \to \mathcal{C} : M_2 \to M_1 \circ M_2$ is additive and exact. It follows from [   ] that such a pairing gives rise to a K-theoretic product $K_i(\mathcal{C}_1) \times K_j(\mathcal{C}_2) \to K_{i+j}(\mathcal{C})$, and in particular to natural pairing $K_0(\mathcal{C}_1) \circ K_n(\mathcal{C}_2) \to K_n(\mathcal{C})$ which could be defined as follows.

Any object $M_1 \in \mathcal{C}_1$ induces an exact functor $M_1 : \mathcal{C}_2 \to \mathcal{C} : M_2 \to M_1 \circ M_2$ and hence a map $K_n(M_1) : K_n(\mathcal{C}_2) \to K_n(\mathcal{C})$. If $M_1' \to M_1 \to M_1''$ is an exact sequence in $\mathcal{C}_1$, then we have an exact sequence of exact functors $M_1^{*'} \to M_1^{*} \to M_1^{*''}$ from $\mathcal{C}_2$ to $\mathcal{C}$ such that for each object $M_2 \in \mathcal{C}_2$ the sequence $M_1^{*'}(M_2) \to M_1^{*}(M_2) \to M_1^{*''}(M_2)$ is exact in $\mathcal{C}$, and hence, by a result of Quillen (see 6.1.4), induces a relation $K_n(M_1^{*'}) + K_n(M_1^{*''}) = K_n(M_1^{*})$. So, the map $M_1 \to K_n(M_1) \in Hom(K_n(\mathcal{C}_2), K_n(\mathcal{C}))$ induces a homomorphism $K_0(\mathcal{C}_1) \to Hom(K_n(\mathcal{C}_2), K_n(\mathcal{C}))$ and hence a pairing $K_0(\mathcal{C}_1) \times K_n(\mathcal{C}_2) \to K_n(\mathcal{C})$.

If $\mathcal{C}_1 = \mathcal{C}_2 = \mathcal{C}$ and the pairing $\mathcal{C} \times \mathcal{C} \to \mathcal{C}$ is naturally associative (and commutative), then the associated pairing $K_0(\mathcal{C}) \times K_0(\mathcal{C}) \to K_0(\mathcal{C})$ turns $K_0(\mathcal{C})$ into an associative (and commutative) ring that may not contain the identity. Now, suppose that there is a pairing $\mathcal{C} \circ \mathcal{C}_1 \to \mathcal{C}_1$ that is naturally associative with respect to the pairing $\mathcal{C} \circ \mathcal{C} \to \mathcal{C}$; then, the pairing $K_0(\mathcal{C}) \times K_0(\mathcal{C}_1) \to K_0(\mathcal{C}_1)$ turns $K_0(\mathcal{C}_1)$ into a $K_0(\mathcal{C})$-module, which may or may not be unitary. However, if $\mathcal{C}$ contains a unit, i.e., an object E such that $E \circ M = M \circ \mathcal{C}$ are naturally isomorphic to M for each $\mathcal{C}$-object M, then the pairing $K_0(\mathcal{C}) \times K_n(\mathcal{C}_1) \to K_n(\mathcal{C}_1)$ turns $K_n(\mathcal{C}_1)$ into a unitary $K_0(\mathcal{C})$-module.

Proof of 12.2.1

(i) It is clear from the definition of $K_n^G(G/H, \mathcal{C})$ that for any $G/H \in \mathcal{A}(G)$, $K_n^G(G/H, \mathcal{C}) \in \mathbb{Z}\text{-}\mathcal{M}od$. Now suppose that $(G/H \overset{\alpha}{\leftarrow} G/L \overset{\beta}{\to} G/K) \in \mathcal{A}(G)(G/H, G/K)$, then $G/H \overset{\alpha}{\leftarrow} G/L \overset{\beta}{\to} G/K \in \mathcal{A}(G)(G/H, G/K)$ goes to

$$K_n^G(G/H, \mathcal{C}) \leftarrow K_n^G(G/L, \mathcal{C}) \leftarrow K_n^G(G/K, \mathcal{C})$$

in $\mathbb{R}\text{-}\mathcal{M}od$ $(K_n^G(G/K, \mathcal{C}))$, $K_n^G(G/H, \mathcal{C})$. If we write K_n^G for $K_n^G(-, \mathcal{C})$, then

$$K_n^G(G/H \leftarrow (G/L) \dot{\cup} (G/L') \to G/K)$$

$$= K_n^G(G/H \leftarrow G/L \to G/K) + K_n^G(G/H \leftarrow G/L' \to G/K).$$

Hence $K_n^G(-, \mathcal{C})$ is a Mackey functor.

(ii) From the discussion in 12.1.2, it is clear that if we put $\mathcal{C}_1 = \mathcal{C}_2 = \mathcal{C} = [G/H, \mathcal{C}]$, then

$$K_0^G(G/H, \mathcal{C}) \times K_n^G(G/H, \mathcal{C}) \to K_n^G(G/H, \mathcal{C})$$

turns

$$K_n^G(G/H, \mathcal{C}) \quad \text{into} \quad K_0^G(G/H, \mathcal{C})\text{-modules}.$$

Hence the result.

Examples 12.2.1 (i) In general $[G/H, \mathcal{C}]$ = category of H-representations in $\mathcal{C}$. Hence $[G/G, \mathcal{C}]$ = category of G representations in $\mathcal{C} = \mathcal{M}(\mathbb{C})$, the category of finite dimensional vector space over the complex numbers $\mathbb{C}$, $K_0^G(G/G, \mathcal{M}(\mathbb{C}))$ is the complex representation ring denoted by $R_{\mathbb{C}}(G)$ or simply $R(G)$ in the literature.

(ii) If $\mathcal{C} = \mathcal{M}(R)$:=category of finitely generated R-modules, where R is a Noetherian ring compatible with the topological structure of G, then $K_n^G(G/H, \mathcal{M}(R)) \simeq G_n(RH)$.

(iii) If $\mathcal{C} = \mathcal{P}(R)$ = category of finitely generated projective R-modules, we have

$$K_n^G(G/H, \mathcal{P}(R)) = G_n(H, R), \quad \text{and}$$

when R is regular, $G_n(R, H) \simeq G_n(RH)$.

12.3 Induction theory for equivariant higher K-functors

In this section, we discuss the induction properties of the equivariant K-functors constructed in section 12.2 leading to the proof of Theorem 12.3.3 below.

Definition 12.3.1 *Let G be a compact Lie group and $M : \mathcal{A}(G) \to \mathbb{Z}\text{-}\mathcal{M}od$ a Mackey functor. A finite family $\Sigma = (G/H_j)_{j \in J}$ is called an inductive system. Such a system yields two homomorphisms, $p(\Sigma)$ (induction map) and $i(\Sigma)$ (restriction maps) defined by*

$$p(\Sigma) : \oplus_{j \in J} M(G/H_j) \to M(G/G)$$
$$(x_j | j \in J) \mapsto \sum_{j \in J} p(H_j)_*(x_j)$$
$$i(\Sigma) : M(G/G) \to \oplus_{j \in J} M(G/H_j)$$
$$x \mapsto (p(H_j)^* x | j \in J)$$

Note that $p(H)$ denotes the unique morphism $G/H \to G/G$. Σ is said to be projective if $p(\Sigma)$ is surjective and Σ is said to be injective if $i(\Sigma)$ is injective. Note that the identity $[id]$ of $\mathcal{U}(G, G/K \times G/H)$ has the form

(I) $\qquad [id] = \Sigma_\alpha n_\alpha [\alpha : G/L_\alpha \to G/K \times G/H]. \qquad$ *see 12.1.4.*

12.3.1 Let $S(K, H)$ be the set of α over which the summation (I) is taken, and let $\alpha = (\alpha(1), \alpha(2))$ be the component of α, where $\alpha(1) : G/L_\alpha \to G/K$, $\alpha(2) : G/L_\alpha \to G/H$. Define induction map

$$p(\Sigma, G/H) : \oplus_{j \in J} (\oplus_{\alpha \in S(H_j, H)} M(G/L_\alpha)) \to M(G/H)$$

by

$$(x(j,\alpha)) \mapsto \sum_{j \in J}(\sum_{\alpha \in S(H_j, H)} n_\alpha \alpha(2)_* x(j, \alpha))$$

and restriction maps

$$i(\Sigma, G/H) : M(G/H) \to \oplus_{j \in J}(\oplus_{\alpha \in S(H_j, H)} M(G/L_\alpha))$$

by

$$x \to (\alpha(2)^* x | (\alpha \in (S(H_J, H))j \in J)).$$

Theorem 12.3.1 *Let $M = K_n^G(-, \mathcal{C})$, $V = K_0^G(-, \mathcal{C})$ be, respectively, Mackey and Green functors $\mathcal{A}(G) \to \mathbb{Z}\text{-}\mathcal{M}od$ as defined in 12.1.6. If Σ is projective for V, then for each homogeneous space G/H, the induction map $p(\Sigma, G/H)$ is split surjective, and the restriction map $i(\Sigma, G/H)$ is split injective.*

PROOF   Since $p(\Sigma)$ is surjective for V, there exist elements $x_j \in V(G/H_i)$ such that $\Sigma(H_j)_* x_j = 1$. Define a homomorphism

$$q(\Sigma, G/H) : M(G/H) \to \oplus_j(\oplus_\alpha M(G/L_\alpha))$$

by $q(\Sigma, G/H)x = \alpha(1)^* x_j \cdot \alpha(2)^* x$ such that $\alpha \in S(H_j, h)$, $j \in J$. Then, $p(\Sigma, G/H)q(\Sigma, G/H)$

$$= \sum_j(\sum_\alpha n_\alpha \alpha(2)_*(\alpha(1)^* \cdot \alpha(2)^* x))$$

$$= \sum_j(\sum_\alpha n_\alpha \alpha(2)_* \alpha(1)^*) x$$

$$= \sum p(H)^* p(H_j)_* x_j \cdot x$$

$$= \sum p(H)^* (\sum p(H_j)_* x_j) x$$

$$= p(H)^*(1)x = 1 \cdot x = x.$$

So, $p(\Sigma, G/H)q(\Sigma, G/H)$ is the identity. Hence, $q(\Sigma, G/H)$ is a splitting for $p(\Sigma, G/H)$. We can also define a splitting $j(\Sigma, G/H)$ for $i(\Sigma, G/H)$ by $j(\Sigma, G/H) : \oplus_j(\oplus_\alpha(MG/L_\alpha)) \to M(G/H)$ where

$$x(j, \alpha) \mapsto \sum_j(\sum_\alpha n_\alpha \alpha(2)_* \alpha(1)^* x_j \circ x(j, \alpha)).$$

☐

Remarks 12.3.1 As will be seen in lemma 12.3.1, $V = K_0^G(-\mathcal{C})$ has "defect sets" $D(V)$, and so, $\Sigma = \{G/H \mid H \in D(V)\}$ is projective for V. It would then mean that an induction theorem for $K_0^G(-, \mathcal{C})$ implies a similar theorem for $K_n^G(-, \mathcal{C})$.

Definition 12.3.2 *A finite set E of conjugacy classes (H) is an induction set for a Green functor V if $\oplus V(G/H) \to V(G/G)$ given by*

$$(x(H)) \to \sum p(H)_* x(H) \quad \text{is surjective.}$$

Define $E \le F$ if and only if for each $(H) \in E$ there exists $(K) \in E$ such that $(H) \le (K)$, i.e., H is a subconjugate to K. Then, $\le$ is a partial ordering on induction sets.

Lemma 12.3.1 *Every Green functor V possesses a minimal induction set $D(V)$, called the defect set of V.*

For proof see [219]. Hence $K_0^G(-, \mathcal{C})$ has defect sets.

12.3.2 Let V be a Green functor and M be a V-module. Define homomorphisms

$$p_1, p_2 : \oplus_{i,j \in J} \oplus_{\alpha \in S(i,j)} M(G/L_\alpha) \to \oplus_{k \in J} M(G/H_k)$$

by

$$p_2(x(i,j,\alpha)) = \sum_{i,j \in J} \sum_{\alpha \in S(i,j)} \eta_\alpha \alpha(2)_* x(i,j,\alpha)$$

and

$$p_1(x(i,j,\alpha)) = \sum_{i,j \in J} \sum_{\alpha \in S(i,j)} \eta_\alpha \alpha(1)_* x(i,j,\alpha)$$

where $S(i,j) = S(H_i, H_j)$ and $\alpha \in S(i,j)$ is in the decomposition of $[id] \in \mathcal{U}(G, G/H_i \times G/H_j)$.

Theorem 12.3.2 *Let $M = K_n^G(-, \mathcal{C})$. Then there exists an exact sequence*

$$\oplus_{i,j \in J}(\oplus_{\alpha \in S(i,j)} M(G/L_\alpha)) \overset{p_2 - p_1}{\to} \oplus_{k \in j} M(G/H_k) \overset{p}{\to} M(G/G) \to 0.$$

PROOF   We have seen in 12.3.1 that p is surjective through the construction of a splitting homomorphism q such that $pq = $ identity. We now construct a homomorphism q_1 such that $(p_2 - p_1)q_1 + qp = id$ from which exactness follows.
Since p_2 is defined as $\oplus_{k \in J} p(\Sigma G/H_k)$ we define $q_1 = \oplus_{k \in J} q(G/H_k)$ and obtain, as in the proof of lemma 12.3.1, that $p_2 q_1 = $ identity. One can also show that $p_1 q_1 = qp$. Hence the result. $\quad\quad\Box$

Definition 12.3.3 *A subgroup C of G is said to be cyclic if powers of a generator of C are dense in G. If p is a rational prime, then a subgroup K of G is called p-hyperelementary if there exists an exact sequence $1 \mapsto C \to K \to P \to 1$ where P is a finite p-group and C a cyclic group such that the order of C/C_0 is prime to p. Here, C_0 is the component of the identity in C. It is called hyperelementary if it is p-hyperelementary for some p.*

Let $\mathcal{H}$ be the set of hyperelementary subgroups of G. We now have the following result, which is the goal of this section and typifies results that can be obtained.

Theorem 12.3.3 *Let* $M = K_n^G(-, M(\mathbb{C}))$. *Then* $\oplus_{\mathcal{H}} M(G/H) \to M(G/G)$ *is surjective (i.e.,* M *satisfies hyperelementary induction, i.e.,* $M(G/G)$ *can be computed in terms of p-hyperelementary subgroups of* G.)

PROOF   It suffices to show that if $V = K_0^G(-, M(\mathbb{C}))$, then $\oplus_{\mathcal{H}} V(G/H) \to V(G/G)$ (I) is surjective since $K_n^G(-, \underline{M}(\mathbb{C}))$ is a V-module. Now, it is clear from example 12.2.1 that $K_0^G(G/G, \underline{M}(\mathbb{C}))$ is the complex representation ring $R(G)$. Moreover, it is known (see [184, 186]) that $R(G)$ is generated as an Abelian group by modules induced from hyperelementary subgroups of G. This is equivalent to the surjectivity of $\oplus_{H \in \mathcal{H}} V(G/H) \to V(G/G)$.   ⬛

$(12.3)^A$ Remarks on possible generalizations

12.3.3 Let $\mathcal{B}$ be a category with finite sums, a final object, and finite pullback (and hence finite coproducts).

A Mackey functor $M : \mathcal{B} \to \mathbb{Z}$-$\mathcal{M}$od is a bifunctor $M(M_*, M^*)$, M_* covariant, M^* contravariant such that $M(X) = M_*(X) = M^*(X)$ for all $X \in \mathcal{B}$ and

(i) For any pull-back diagram

$$
\begin{array}{ccc}
A' & \xrightarrow{\,p_2\,} & A_2 \\
{\scriptstyle p_1}\downarrow & & \downarrow{\scriptstyle f_1} \\
A_1 & \xrightarrow{\,f_2\,} & A
\end{array}
\quad \text{in } \mathcal{B}
$$

the diagram

$$
\begin{array}{ccc}
M(A') & \xrightarrow{\,p_{2*}\,} & M(A_i) \\
{\scriptstyle p_2^*}\uparrow & & \uparrow{\scriptstyle f_1^*} \\
M(A_i) & \xrightarrow{\,f_2^*\,} & M(A)
\end{array}
$$

(ii) M^* transforms finite coproducts in $\mathcal{B}$ over finite products in $\mathbb{Z}$-$\mathcal{M}$od.

Example 12.3.1 Now suppose that G is a compact Lie group. Let $\mathcal{B}$ be the category of G-spaces of the G-homotopy type of G-CW-complex (e.g., G-ENR, see [137, 219]). Then $\mathcal{B}$ is a category with finite coproducts (topological sums), final object, and finite pull-backs (fibered products) (see [2]). Hence a Mackey functor is defined on $\mathcal{B}$ along the lines of 9.1.

Hence, in a way analogous to what was done in chapter 10, we could define for $X, Y \in \mathcal{B}$ the notion of Y-exact sequences in the exact category $[\underline{X}, \mathcal{C}]$ (where $\mathcal{C}$ is an exact category) and obtain $K_n^G(X, \mathcal{C}, Y)$ as the nth algebraic K-group of $[\underline{X}, \mathcal{C}]$ with respect to Y-exact sequences.

We could also have the notion of an element $\zeta \in [\underline{X}, \mathcal{C}]$ being Y-projective and obtain a full subcategory $[\underline{X}, \mathcal{C}]_Y$ of Y-projective functors in $[\underline{X}, \mathcal{C}]$ so that we could obtain $P_n^G(X, \mathcal{C}, Y)$ as the nth algebraic K-group of $[X, \mathcal{C}]_Y$ with respect to split exact sequences and then show that $K_n^G(-, \mathcal{C}, Y), P_n^G(-, \mathcal{C}, Y)$ are $K_0^G(-, \mathcal{C}, Y)$-modules in a way analogous to what was done in chapter 10.

Exercise

Let G be a compact Lie group. Construct an equivariant higher K-theory for G-actions along the lines suggested in example 12.3.1.

Chapter 13

Equivariant higher K-theory for Waldhausen categories

The aim of this chapter is to extend the constructions in chapter 10 from exact categories to Waldhausen categories. In a way analogous to what was done in chapter 10, we shall construct, for any Waldhausen category W and any finite group G, functors $K_n^G(-W), \mathbb{K}_n^G(-, W, Y)$ and $\mathbb{P}_n^G(-, W, Y) : G\mathcal{S}ets \to \mathbb{Z}\text{-}\mathcal{M}od$ as Mackey functors for all $n \geq 0$, and under suitable hypothesis on W show that $\mathbb{K}_n^G(-, W), \mathbb{K}_n^G(-W, Y)$ are Green functors. We then highlight some consequences of these facts and also obtain equivariant versions of Waldhausen K-theory additivity and fibration theorems. Our main applications are to Thomason's "complicial" bi-Waldhausen categories of the form $Ch_b(\mathcal{C})$ (see example 5.4.1(ii) and example 5.4.2) where $\mathcal{C}$ is an exact category and hence to chain complexes of modules over grouprings. In particular, we prove (see theorem 13.4.1) that the Waldhausen's K-groups of the category $(Ch_b(\mathcal{M}(RG), w)$ of bounded chain complexes of finitely generated RG-modules with suitable quasi-isomorphism as weak equivalences are finite Abelian groups (see theorem 13.4.1(2)). We also present an equivariant approximation theorem for complicial bi-Waldhausen categories.

We have worked out in some detail the situation for finite group actions. However, one could work out the situation for discrete group action via the procedure outline in examples and remarks 9.3.1(ii). Also, it should be possible to construct equivariant K-theory for Waldhausen categories for profinite and compact Lie groups as was done for exact categories in chapters 11 and 12.

We already introduced Waldhausen's higher K-groups with relevant examples in 5.4. So, unexplained notations and terminologies are those used in 5.4, $(6.1)^B$ and 6.4. We recall that if W is a Waldhausen category, we shall write $\mathbb{K}(\mathcal{E})$ for the Waldhausen K-theory space or spectrum, and $\mathbb{K}_n(W)$ for Waldhausen's n^{th} K-group $\pi_n(\mathbb{K}(W))$ (see definition 5.4.5). The results in this chapter are due to A. Kuku (see [122]).

13.1   Equivariant Waldhausen categories

Definition 13.1.1 *Let G be a finite group, X a G-set. The translation category of X is a category $\underline{X}$ whose objects are elements of X and whose morphisms $Hom_{\underline{X}}(x, x')$ are triples (g, x, x') where $g \in G$ and $gx = x'$.*

Theorem 13.1.1 *Let W be a Waldhausen category, G a finite group, $\underline{X}$ the translation category of a G-set X, $[\underline{X}, W]$ the category of covariant functors from $\underline{X}$ to W. Then $[\underline{X}, W]$ is a Waldhausen category.*

PROOF   Say that a morphism $\zeta \to \eta$ in $[\underline{X}, W]$ is a cofibration if $\zeta(x) \mapsto \eta(x)$ is a cofibration in W. So, isomorphisms are cofibrations in $[\underline{X}, W]$. Also, if $\zeta \mapsto \eta$ is a cofibation and $\eta \to \delta$ is a morphism in $[\underline{X}, W]$, then the push-out $\zeta \underset{\eta}{\cup} \delta$ defined by $(\zeta \underset{\eta}{\cup} \delta)(x) = \zeta(x) \underset{\eta(x)}{\cup} \delta(x)$ exists since $\zeta(x) \underset{\eta(x)}{\cup} \delta(x)$ is a push-out in W for all $x \in X$. Hence, coproducts also exist in $[\underline{X}, W]$.

Also, define a morphism $\zeta \to \eta$ in $[\underline{X}, W]$ as a weak equivalence if $\zeta(x) \to \eta(x)$ is a weak equivalence in W for all $x \in X$. It can be easily checked that the weak equivalences contain all isomorphisms and also satisfy the gluing axiom, i.e., if $\begin{smallmatrix} \delta & \leftarrow & \zeta & \mapsto & \eta \\ \downarrow\sim & & \downarrow\sim & & \downarrow\sim \\ \delta' & \leftarrow & \zeta' & \mapsto & \eta' \end{smallmatrix}$ is a commutative diagram where the vertical maps are weak equivalences and the two right horizontal maps are cofibrations, then the induced maps $\eta \underset{\zeta}{\cup} \delta \to \eta' \underset{\zeta'}{\cup} \delta'$ are also a weak equivalence.   ☐

Remarks and definitions 13.1.1   (i) Note that each cofibration $\zeta \mapsto \eta$ has a cokernel $\eta/\zeta = \delta$ defined by $\delta(x) = \eta(x)/\zeta(x)$ for all $x \in X$. We shall call $\zeta \mapsto \eta \twoheadrightarrow \delta$ a cofibration sequence in $[\underline{X}, W]$.

(ii) One can easily check that the cofibration sequences $\zeta \mapsto \eta \twoheadrightarrow \delta$ in $[\underline{X}, W]$ also form a Waldhausen category, which we shall denote by $\mathcal{E}[\underline{X}, W]$ and call an extension category of $[\underline{X}, W]$.

(iii) If W is saturated, then so is $[\underline{X}, W]$. For if $f : \zeta \to \zeta', g : \zeta' \to \eta$ are composable arrows in $[\underline{X}, W]$, and gf is a weak equivalence, then for any $x \in X$, $(gf)(x) = g(x)f(x)$ is a weak equivalence in W. But then, $f(x)$ is a weak equivalence iff $g(x)$ is for all $x \in X$. Hence f is a weak equivalence iff g is.

Example 13.1.1   (i) Let $W = Ch_b(\mathcal{C})$, ($\mathcal{C}$ an exact category) be a complicial bi-Waldhausen category. Then, for any small category ℓ, $[\ell, W]$ is also a complicial bi-Waldhausen category (see [57]). Hence, for any G-set X, $[\underline{X}, Ch_b(\mathcal{C})]$ is a complicial bi-Waldhausen category. We shall be interested in the cases $[\underline{X}, Ch_b(\underline{P}(R))]$ $[\underline{X}, Ch_b(\mathcal{M}'(R)]$ and $[\underline{X}, Ch_b(\mathcal{M}(R))]$, R a ring with identity.

(ii) Here is another way to see that $[\underline{X}, Ch_b(\mathcal{C})]$ is a complicial bi-Waldhausen category.

One can show that there is an equivalence of categories $[\underline{X}, Ch_b(\mathcal{C})] \xrightarrow{F} Ch_b([\underline{X}, \mathcal{C}])$ where F is defined as follows:

For $\zeta_* \in [\underline{X}, Ch_b(\mathcal{C})]$, $\zeta_*(x) = \{\zeta_r(x)\}$, $\zeta_r(x) \in \mathcal{C}$ where $a \leq r \leq b$ for some $a, b \in \mathbb{Z}$, and where each $\zeta_r \in [\underline{X}, \mathcal{C}]$. Put $F(\zeta_*) = \zeta'_* \in Ch_b[\underline{X}, \mathcal{C}]$ where $\zeta'_* = \{\zeta'_r\}$, $\zeta'_r(x) = \zeta_r(x)$.

Definition 13.1.2 *Let $X < Y$ be G-sets, and $\underline{X \times Y} \xrightarrow{\varphi} \underline{X}$ the functor induced by the projection $X \times Y \xrightarrow{\phi} X$. Let W be a Waldhausen category. If $\zeta \in ob[\underline{X}, W]$, we shall write ζ' for $\zeta \circ \varphi : \underline{X \times Y} \to \underline{X} \to W$. Call a cofibration $\zeta \mapsto \eta$ in $[\underline{X}, W]$ a Y-cofibration if $\zeta' \to \eta'$ is a split cofibration in $[\underline{X \times Y}, W]$. Call a cofibration sequence $\zeta \mapsto \eta \twoheadrightarrow \delta$ in $[\underline{X}, W]$ a Y-cofibration sequence if $\zeta' \to \eta' \to \delta'$ is a split cofibration sequence in $[\underline{X \times Y}, W]$.*

We now define a new Waldhausen category $^Y[\underline{X}, W]$ as follows: $ob(^Y[\underline{X}, W]) = ob[\underline{X}, W]$. Cofibrations are Y-cofibrations and weak equivalences are the weak equivalence in $[\underline{X}, W]$.

Definition 13.1.3 *With the notations as in definition 13.1.2, an object $\zeta \in [\underline{X}, W]$ is said to be Y-projective if every Y-cofibration sequence $\zeta \mapsto \eta \twoheadrightarrow \delta$ in $[\underline{X}, W]$ is a split cofibration sequence. Let $[\underline{X}, W]_Y$ be the full subcategory of $[\underline{X}, W]$ consisting of Y-projective functors. Then, $[\underline{X}, W]_Y$ becomes a Waldhausen category with respect to split cofibrations and weak equivalences in $[\underline{X}, W]$.*

13.2   Equivariant higher K-theory constructions for Waldhausen categories

---

$(13.2)^A$ Absolute and relative equivariant theory

Definition 13.2.1 *Let G be a finite group, X a G-set, W a Waldhausen category, $[\underline{X}, W]$ the Waldhausen category defined in 13.1. We shall write $\mathbb{K}^G(X, W)$ for the Waldehausen K-theory space (or spectrum) $\mathbb{K}([\underline{X}, W])$ and $\mathbb{K}_n^G(X, W)$ for the Waldhausen K-theory group $\pi_n(\mathbb{K}([\underline{X}, W]))$.*

For the Waldhausen category $^Y[\underline{X}, W]$, we shall write $\mathbb{K}^G(X, W, Y)$ for the Waldhausen K-theory space (or spectrum) $\mathbb{K}(^Y[\underline{X}, W])$ with corresponding n^{th} K-theory groups $\mathbb{K}_n^G(X, W, Y) := \pi_n(\mathbb{K}^Y[\underline{X}, W])$.

Finally, we denote by $\mathbb{P}^G(X, W, Y)$ *the Waldhausen K-theory space (or spectrum)* $\mathbb{K}([X, W]_Y)$ *with corresponding n^{th} K-theory group* $\pi_n(\mathbb{K}([X, W]_Y))$, *which we denote by* $\mathbb{P}_n^G(X, W, Y)$.

Theorem 13.2.1 *Let W be a Waldhausen category, G a finite group, X any G-set. Then, in the notation of definition 13.2.1, we have $\mathbb{K}_n^G(-, W)$, $\mathbb{K}_n^G(-, W, Y)$, and $\mathbb{P}_n^G(-, W, Y)$ are Mackey functors $GSet \to Ab$.*

PROOF Any G-map $f : X_1 \to X_2$ defines a covariant functor $f : \underline{X_1} \to \underline{X_2}$ given by $x \to f(x), (g, x, x') \mapsto (g, f(x), f(x'))$, and an exact restriction functor $f^* : [\underline{X_2}, W] \to [\underline{X_1}, W]$ given by $\zeta \to \zeta \circ f$. Also f^* maps cofibration sequence to cofibration sequences and weak equivalence to weak equivalences. So, we have an induced map $\mathbb{K}_n^G(f, W)^* : \mathbb{K}_n^G(X_2, W) \to K_n^G(X_1, W)$ making $\mathbb{K}_n^G(-, W)$ contravariant functor $GSet \to Ab$. The restriction functor $[\underline{X_2}, W] \to [\underline{X_1}, W]$ carries Y-cofibrations over $\underline{X_2}$ to Y-cofibrations over X_1 and also Y-projective functors in $[\underline{X_2}, W]$ to Y-projective functors in $[\underline{X_1}, W]$. Moreover, it preserves weak equivalences in both cases. Hence we have induced maps

$$\mathbb{K}_n^G(f, W, Y)^* : \mathbb{K}_n^G(X_2, W, Y) \to \mathbb{K}_n^G(X_1, W, Y)$$

$$\mathbb{P}_n^G(f, W, Y)^* : \mathbb{P}_n^G(X_2, W, Y) \to \mathbb{P}_n^G(X_1, W, Y)$$

making $K_n^G(-, W, Y)^*, \mathbb{P}_n^G(-, W, Y,)^*$ contravariant functors $GSet \to Ab$.

Now, any G-map $f : X_1 \to X_2$ also induces an "induction functor" $f_* : [\underline{X_1}, W] \to [\underline{X_2}, W]$ as follows. For any functor $\zeta \in ob[\underline{X_1}, W]$, define $f_*(\zeta) \in [\underline{X_2}, W]$ by $f_*(\zeta)(x_2) = \bigoplus_{x_1 \in f^{-1}(x_2)} \zeta(x_1) : f_*(\zeta)(g, x_2, x_2') = \bigoplus_{x_1 \in f^{-1}(x_2)} \zeta(g, x_1, gx_1)$. Also, for any morphism $\zeta \to \zeta'$ in $[\underline{X_1}, W]$, define $(f_*(\alpha)(x_2) = \bigoplus_{x_1 \in f^{-1}(x_2)} \alpha(x_1); f_*(\zeta)(x_2) = \bigoplus_{x_1 \in f^{-1}(x_2)} \zeta x_1) \to f_*(\zeta')(x_2) = \bigoplus_{x_1 \in f^{-1}(x_2)} \zeta'(x_1)$. Also, f_* preserves cofibrations and weak equivalences. Hence, we have induced homomorphisms $\mathbb{K}_n^G(f, W) : K_n^G(X_1, W) \to K_n^G(X_2, W)$, and $K_n^G(-, W)$ is a covariant functor $GSet \to Ab$. Also, the induction functor preserves Y-cofibrations and Y-projective functors as well as weak equivalences. Hence, we also have induced homomorphisms

$$\mathbb{K}_n^G(f, W, Y)_* : \mathbb{K}_n^G(X_1, W, Y) \to \mathbb{K}_n^G(X_2, W, Y)$$

$$\text{and} \quad \mathbb{P}_n^G(f, W, Y)_* : \mathbb{P}_n^G(X_1, W, Y) \to \mathbb{P}_n^G(X_2, W, Y).$$

making $\mathbb{K}_n^G(-, W, Y)$, and $\mathbb{P}_n^G(-, W, Y)$ covariant functions $GSet \to Ab$.

Also, for morphisms $f_1 : X_1 \to X, f_2 : X_2 \to X$ in $GSet$ and any pull-back diagram

$$
\begin{array}{ccc}
X_1 \underset{X}{\times} X_2 & \xrightarrow{\ f_2\ } & X_2 \\
\Big\downarrow{f_1} & & \Big\downarrow{f_2} \\
X_1 & \xrightarrow{\ f_1\ } & X
\end{array}
\qquad (I)
$$

we have a commutative diagram

$$
\begin{array}{ccc}
[X_1 \underset{X}{\times} X_2, W] & \longrightarrow & [\underline{X}_2, W] \\
\Big\downarrow & & \Big\downarrow \\
[X_1, W] & \longrightarrow & [X, W]
\end{array}
\qquad (II)
$$

and hence, the commutative diagrams obtained by applying $\mathbb{K}_n^G(-, W)$, $\mathbb{K}_n^G(-, W, Y)$ to diagram (II) above and applying $\mathbb{P}_n^G(-, W, Y)$ to diagram (III) below

$$
\begin{array}{ccc}
[X_1 \underset{X}{\times} X_2, W]_Y & \longrightarrow & [\underline{X}_2, W]_Y \\
\Big\downarrow & & \Big\downarrow \\
[X_2, W]_Y & \longrightarrow & [X, W]_Y
\end{array}
\qquad (III)
$$

show that Mackey properties are satisfied. Hence $\mathbb{K}_n^G(-, W), \mathbb{K}_n^G(-, W, Y)$, and $\mathbb{P}_n^G(-, W, Y)$ are Mackey functors. □

Theorem 13.2.2 *Let W_1, W_2, W_3 be Waldhausen categories and $W_1 \times W_2 \to W_3$, $(A_1, A_2) \to A_1 \circ A_2$ an exact pairing of Waldhausen categories. Then, the pairing induces, for a G-set X, a pairing $[\underline{X}, W_1] \times [\underline{X}, W_2] \to [\underline{X}, W_3]$ and hence a pairing*

$$
\mathbb{K}_0^G(X, W_1) \times K_n^G(X, W_2) \to K_n^G(X, W_3).
$$

Suppose that W is a Waldhausen category such that the pairing is naturally associative and commutative and there exists $E \in W$ such that $E \circ X = X \circ E = X$; then $K_0^G(-, W)$ is a Green functor and $K_n^G(-, W)$ is a unitary $K_0^G(-, W)$-module.

PROOF The pairing $W_1 \times W_2 \to W_3, (X_1, X_2) \to X_1 \circ X_2$ induces a pairing $[\underline{X}, W_1] \times [\underline{X}, W_2] \to [\underline{X}, W_3]$ given by $(\zeta_1, \zeta_2) \to \zeta_1 \circ \zeta_2$ where $(\zeta_1 \circ \zeta_2)(x) = \zeta_1(x) \circ \zeta_2(x)$. Now, any $\zeta_1 \in [\underline{X}, W_1]$ induces a functor $\zeta_1^* : [\underline{X}, W_2] \to [\underline{X}, W_3]$ given by $\zeta_2 \to \zeta_1 \circ \zeta_2$, which preserves cofibrations and weak equivalences and hence the map

$$
\mathbb{K}_n^G(\zeta_1^*) : \mathbb{K}_n^G(X, W_2) \to \mathbb{K}_n^G(X, W_3).
$$

Now, define a map:

$$\mathbb{K}_0^G(X, W_1) \xrightarrow{\delta} Hom(\mathbb{K}_n^G(X, W_2), \mathbb{K}_n^G(X, W_3)) \qquad (I)$$

by $[\zeta_1] \to \mathbb{K}_n^G(\zeta_1^*)$. We now show that this map is a homomorphism. Let $\zeta_1' \mapsto \zeta_1 \to \zeta_1''$ be a cofibration sequence in $[\underline{X}, W_1]$. Then, we obtain a cofibration sequence of functors $\zeta_1'^* \mapsto \zeta_1^* \to \zeta_1''^* : [\underline{X}, W_2] \to [\underline{X}, W_3]$ such that for each $\zeta_2 \in [\underline{X}, W_2]$, the sequence $\zeta_1'^*(\zeta_2) \to \zeta_1^*(\zeta_2) \to \zeta_1''^*(\zeta_2)$ is a cofibration sequence in $[\underline{S}, W_3]$. Then, by applying the additivity theorem for Waldhausen categories (see $(6.1)^B$), we have $\mathbb{K}_n^G(\zeta_1'^*) + \mathbb{K}_n^G(\zeta_1''^*) = \mathbb{K}_n^G(\zeta_1^*)$. So, δ is a homomorphism and hence we have a pairing $\mathbb{K}_0^G(S, W_1) \times \mathbb{K}_n^G(S, W_2) \to K_n^G(S, W_3)$.

One can easily check that for any G-map $\varphi : X' \to X$ the Frobenius reciprocity law holds, i.e.,
for $\xi_i \in [\underline{X}, W_i], \eta_i \in [\underline{X}', W_i]$, $i = 1, 2$, we have canonical isomorphisms

$$f_*(f^*(\zeta_1) \circ \zeta_2) \cong \zeta_1 \circ f_*(\zeta_2),$$
$$f_*(\zeta_1 \circ f^*(\zeta_2)) \cong f_*(\zeta_1) \circ \zeta_2, \quad \text{and}$$
$$f^*(\zeta_1 \circ \zeta_2) \cong f^*(\zeta_1) \circ f^*(\zeta_2).$$

Now, the pairing $W \times W \to W$ induces $\mathbb{K}_0^G(X, W) \times \mathbb{K}_0^G(X, W) \to K_0^G(X, W)$, which turns $K_0^G(X, W)$ into a ring with unit such that for any G-map $f : X \to Y$, we have $\mathbb{K}_0^G(f, W)_*(1_{K_0^G(X,W)}) \equiv 1_{K_0^G(Y,W)}$. Then, $1_{K_0^G(X,W)}$ acts as the identity on $K_0^G(X, W)$. So, $K_0^G(X, W)$ is a $K_0^G(X, W)$-module. ∎

Theorem 13.2.3 *Let Y be a G-set, W a Waldhausen category. If the pairing $W \times W \to W$ is naturally associative and commutative, and W contains a natural unit, then $K_0^G(-, W, Y) : GSet \to Ab$ is a Green functor, and $\mathbb{K}_n^G(-, W, Y)$ and $\mathbb{P}_n^G(-, W, Y,)$ are $\mathbb{K}_0^G(-, W, Y)$-modules.*

PROOF   Note that for any G-set Y, the pairing $[\underline{X}, W] \times [\underline{X}, W] \to [\underline{X}, W]$ takes Y-cofibration sequence to Y-cofibration sequences, and Y-projective functors, to Y-projective functors, and so, we have induced pairing $^Y[\underline{X}, W] \times^Y [\underline{X}, W] \to^Y [\underline{X}, W]$ inducing a pairing $\mathbb{K}_0^G(X, W, Y) \times \mathbb{K}_0^G(X, W, Y) \to$
$\mathbb{K}_0^G(X, W, Y)$, as well as induced pairing $^Y[\underline{X}, W] \times [\underline{X}, W]_Y \to [\underline{X}, W]_Y$ yielding K-theoretic pairing $\mathbb{K}_0^G(X, W, Y) \times \mathbb{P}_0^G(X, W, Y) \to \mathbb{P}_0^G(X, W, Y)$. If $W \times W$ is naturally associative and commutative and W has a natural unit, then $K_0^G(-, W)$ is a Green functor, and $P_n^G(-, W, Y)$ and $K_n^G(-, W, Y)$ are $K_0^G(-, W, Y)$-modules. ∎

Remarks 13.2.1   (1) It is well known that the Burnside functor $\Omega : GSet \to Ab$ is a Green functor, that any Mackey functor $M :$

$GSet \rightarrow Ab$ is an Ω-module, and that any Green functor is an Ω-algebra (see theorems 9.4.1 and 9.4.2). Hence, the above K-functors $\mathbb{K}_n^G(-, W, Y), \mathbb{P}_n^G(-, W, Y)$ and $\mathbb{K}_n^G(-, W)$ are Ω-modules, and under suitable hypothesis on W, $K_0^G(-, W.Y)$ and $K_0^G(-, W)$ are Ω-algebras.

(2) Let M be any Mackey functor $GSet \rightarrow Ab$, X a G-set. Define $K_M(X)$ as the kernel of $M(G/G) \rightarrow M(X)$ and $I_M(X)$ as the image of $M(X) \rightarrow M(G/G)$. An important induction result is that $|G|M(G/G) \subseteq K_M(X) + I_M(X)$ for any Mackey functor M and G-set X. This result also applies to all the K-theoretic functors defined above.

(3) If M is any Mackey functor $GSet \rightarrow Ab$, X a G-set, define a Mackey functor $M_X : GSet \rightarrow Ab$ by $M_X(Y) = M(X \times Y)$. The projection map pr: $X \times Y \rightarrow Y$ defines a natural transformation $\Theta_X : M_X \rightarrow M$ where $\Theta_X(Y) = \text{pr} : M(X \times Y) \rightarrow M(Y)$. M is said to be X-projective if Θ_X is split surjective (see definition 9.1.4).

Now, define the defect base D_M of M by $D_M = \{H \leq G | X^H \neq \emptyset\}$ where X is a G-set (called the defect set of M) such that M is Y-projective iff there exists a G-map $f : X \rightarrow Y$ (see proposition 9.1.1(c)). If M is a module over a Green function $\mathcal{G}$, then M is X-projective iff $\mathcal{G}$ is X-projective iff the induction map $\mathcal{G}(X) \rightarrow \mathcal{G}(G/G)$ is surjective (see theorem 9.3.1). In general, proving induction results reduce to determining $GSets$ X for which $\mathcal{G}(X) \rightarrow \mathcal{G}(G/G)$ is surjective, and this in turn reduces to computing $D_{\mathcal{G}}$ (see 9.6.1).

Hence, one could apply induction techniques to obtain results on higher K-groups, which are modules over the Green functors $\mathbb{K}_0^G(-, W)$ and $K_0^G(-, W, Y)$ for suitable W (e.g., $W = Ch_b(\mathcal{C})$, $\mathcal{C}$ a suitable exact category (see example 5.4.2).

(4) One can show via general induction theory principles that for suitably chosen W, all the higher K-functors $\mathbb{K}_n^G(-, W), \mathbb{K}_n^G(-, W, Y)$ and $\mathbb{P}_n^G(-, W, Y)$ are "hyperelementary computable" (see chapters 10, 12).

$(13.2)^B$ Equivariant additivity theorem

In this subsection, we present an equivariant version of additivity a theorem for Waldhausen Categories. First wse review the non-equivariant situation.

Definition 13.2.2 *Let W, W' be Waldhausen categories. Say that a sequence $F' \mapsto F \twoheadrightarrow F''$ of exact functors $F', F, F'' : W \rightarrow W'$ is a cofibration sequence of exact functors if each $F'(A) \mapsto F(A) \twoheadrightarrow F''(A)$ is a cofibration in W',* and if for every cofibration $A \mapsto B$ in W, $F(A) \underset{F'(A)}{\cup} F'(B) \to F(B)$ is a cofibration in W'.

Theorem 13.2.4 (Additivity theorem) [165, 224] Let W, W' be Waldhausen categories, and $F' \mapsto F \to F''$ a cofibration sequence of exact functors from W to W'. Then, $F_* \simeq F'_* + F''_* : \mathbb{K}_n(W) \to \mathbb{K}_n(W')$.

Theorem 13.2.5 (Equivariant Additivity Theorem) Let W, W' be Waldhausen categories, X, Y, G-sets, and $F' \mapsto F \to F''$ cofibration sequence of exact functors from W to W'. Then, $F' \mapsto F \to F''$ induces a cofibration sequence $\widehat{F}' \mapsto \widehat{F} \to \widehat{F}''$ of exact functors from $[\underline{X}, W]$ to $[\underline{X}, W']$, from $^Y[\underline{X}, W]$ to $^Y[\underline{X}, W']$, and from $[\underline{X}, W]_Y$ to $[\underline{X}, W']_Y$, and so, we have induced homomorphisms

$$\widehat{F}_* \cong \widehat{F}'_* + \widehat{F}''_* : \mathbb{K}_n^G(\underline{X}, W) \to \mathbb{K}_n^G(\underline{X}, W'),$$
$$\mathbb{K}_n^G(\underline{X}, W, Y) \to \mathbb{K}_n^G(\underline{X}, W', Y),$$
$$and \quad \mathbb{P}_n^G(\underline{X}, W, Y) \to \mathbb{P}_n^G(\underline{X}, W', Y).$$

PROOF First note that $[\underline{X}, W], [\underline{X}, W']$; $^Y[\underline{X}, W]$, $^Y[\underline{X}, W']$; and $[\underline{X}, W]_Y$, $[\underline{X}, W']_Y$ are all Waldhausen categories. Now, define $\widehat{F}', \widehat{F}$ and $\widehat{F}''$: $[\underline{X}, W]$, $[\underline{X}, W']$ by $\widehat{F}'(\zeta)(x) = \widehat{F}'(\zeta(x)), \widehat{F}'(\zeta)(x) = \widehat{F}(\zeta(x))$, and $\widehat{F}''(\zeta)(x) = \widehat{F}''(\zeta(x))$. Then one can check that $\widehat{F}' \to \widehat{F} \to \widehat{F}''$ is a cofibration sequence of exact functors $[X, W] \to [\underline{X}, W']$, $^Y[\underline{X}, W] \to ^Y[X, W']$, and $[\underline{X}, W]_Y \to [\underline{X}, W']_Y$. Result then follows by applying theorem 13.2.4. □

$(13.2)^C$ Equivariant Waldhausen fibration sequence

In this subsection, we present an equivariant version of Waldhausen fibration sequence. First, we define the necessary notion and state the non-equivariant version.

Definition 13.2.3 (Cylinder functors) A Waldhausen category has a cylinder functor if there exists a functor $T : ArW \to W$ together with three natural transformations p, j_1, j_2 such that to each morphism $f : A \to B$, T assigns an object Tf of W and $j_1 : A \to Tf$, $j_2 : B \to Tf$, $p : Tf \to B$, satisfying certain properties (see [224, 225]).
Cylinder Axiom. For all $f, p : Tf \to B$ is in $\omega(W)$.

13.2.1 Let W be a Waldhausen category. Suppose that W has two classes of weak equivalences $\nu(W), \omega(W)$ such that $\nu(W) \subset \omega(W)$. Assume that $\omega(W)$

satisfies the saturation and extension axioms and has a cylinder functor T, which satisfies the cylinder axiom. Let W^ω be the full subcategory of W whose objects are those $A \in W$ such that $0 \to A$ as in $\omega(W)$. Then, W^ω becomes a Waldhausen category with $co(W^\omega) = co(W) \cap W^\omega$ and $\nu(W^\omega) = \nu(W) \cap (W^\omega)$.

Theorem 13.2.6 (Waldhausen fibration sequence) [225]. *With the notations and hypothesis of 13.2.1 suppose that W has a cylinder functor, T, which is a cylinder functor for both $\nu(W)$ and $\omega(W)$. Then, the exact inclusion functors $(W^\omega, \nu) \to (W, \nu)$ and $(W, \nu) \to (W, \omega)$ induce a homotopy fiber sequence of spectra*

$$\mathbb{K}(W^w \omega \nu) \to \mathbb{K}(W, \nu) \to \mathbb{K}(W, \omega)$$

and hence a long exact sequence

$$\mathbb{K}_{n+1}(W, \omega) \to \mathbb{K}_n(W^\omega, \nu) \to \mathbb{K}_n(W, \nu) \to \mathbb{K}_n(W, \omega) \to .$$

13.2.2 Now, let W be a Waldhausen category with two classes of weak equivalences $\nu(W)$ and $\omega(W)$ such that $\nu(W) \subset \omega(W)$. Then, for any G-set X, $[\underline{X}, W]$ is a Waldhausen category with two choices of weak equivalence $\hat{\nu}[\underline{X}, W]$ and $\hat{\omega}[\underline{X}, W]$, and $\hat{\nu}[\underline{X}, W] \subseteq \hat{\omega}[\underline{X}, W]$ where a morphism $\zeta \xrightarrow{f} \zeta'$ is in $\hat{\nu}[\underline{X}, W]$ (resp. $\hat{\omega}[\underline{X}, W]$ if $f(x) : \zeta(x) \to \zeta'(x)$ is in νW (resp. $\omega(W)$). One can easily check that if $\omega(W)$ satisfies the saturation axiom, so does $\hat{\omega}[\underline{X}, W]$.

Suppose that $\omega(W)$ has a cylinder functor $T : ArW \to W$, which also satisfies the cylinder axiom such that for all $f : A \to B$, in W, the map $p : Tf \to B$ is in $\omega(W)$; then T induces a functor $\widehat{T} : Ar([\underline{X}, W]) \to [\underline{X}, W]$ defined by $\widehat{T}(\zeta \to \zeta')(x) = T(\zeta(x) \to \zeta'(x))$ for any $x \in X$. Also, for a map $f : \zeta \to \zeta'$ in $[\underline{X}, W]$ the map $\hat{p} : \widehat{T}(f) \to \zeta' \in \hat{\omega}([\underline{X}, W])$. Let $[\underline{X}, W]^{\hat{\omega}}$ be the full subcategory of $[\underline{X}, W]$ such that $\zeta_0 \to \zeta \in \hat{\omega}[\underline{X}, W]$ where $\zeta_0(x) = 0 \in W$ for all $x \in X$. Then $[\underline{X}, W]^{\hat{\omega}}$ is a Waldhausen category with $co([\underline{X}, W]^{\hat{\omega}}) = co([\underline{X}, W) \cap [\underline{X}, W)^{\hat{\omega}})$ and $\hat{\nu}[\underline{X}, W) \cap [\underline{X}, W]^{\hat{\omega}}$. We now have the following

Theorem 13.2.7 (Equivariant Waldhausen fibration sequence) *Let W be a Waldhausen category with a cylinder functor T and which also has a cylinder functor for $\nu(W)$ and $\omega(W)$. Then, in the notation of 13.2.2, we have exact inclusions $([\underline{X}, W]^{\hat{\omega}}, \hat{\nu}) \to ([\underline{X}, W], \hat{\nu})$ and $([\underline{X}, W], \hat{\nu}) \to ([\underline{X}, W], \hat{\omega})$ which induce a homotopy fibre sequence of spectra*

$$\mathbb{K}([\underline{X}, W]^{\hat{\omega}}, \hat{\nu}) \to \mathbb{K}([\underline{X}, W], \hat{\nu}) \to \mathbb{K}([\underline{X}, W], \hat{\omega})$$

and hence a long exact sequence

$$\cdots \mathbb{K}_{n+1}([\underline{X}, W], \hat{\omega}) \to \mathbb{K}_n([\underline{X}, W]^{\hat{\omega}}, \hat{\nu}) \to \mathbb{K}_n([\underline{X}, W], \hat{\nu}) \to \mathbb{K}_n([\underline{X}, W], \hat{\omega})$$

PROOF Similar to that of theorem 13.2.6. $\qquad\qquad$ ▯

13.3 Applications to complicial bi-Waldhausen categories

In this section, we shall focus attention on Waldhausen categories of the form $Ch_b(\mathcal{C})$ where $\mathcal{C}$ is an exact category.

13.3.1 Recall from definition 10.1.1 that if $\mathcal{C}$ is an exact category, and X, Y, G-sets, $K_n^G(X, \mathcal{C})$ is the nth (Quillen) algebraic K-group of the exact category $[X, \mathcal{C}]$ with respect to fiberwise exact sequences; $K_n^G(X, \mathcal{C}, Y)$ is the nth (Quillen) algebraic K-group of the exact category $[\underline{X}, \mathcal{C}]$ with respect to Y-exact sequences; while $P_n^G(X, \mathcal{C}, Y)$ is the nth (Quillen) algebraic K-group of the category $[\underline{X}, \mathcal{C}]$ of Y-projective functors in $[\underline{X}, \mathcal{C}]$ with respect to split exact sequences.

We now have the following result.

Theorem 13.3.1 *Let G be a finite group, X, Y, G-sets, $\mathcal{C}$ an exact category. Then,*

(1) $K_n^G(X, \mathcal{C}) \cong \mathbb{K}_n^G(X, Ch_b(\mathcal{C}))$.

(2) $K_n^G(X, \mathcal{C}, Y) \cong \mathbb{K}_n^G(X, Ch_b(\mathcal{C}), Y)$.

(3) $P_n^G(X, \mathcal{C}, Y) \cong \mathbb{P}_n^G(X, Ch_b(\mathcal{C}), Y)$.

PROOF

(1) Note that $[\underline{X}, \mathcal{C}]$ is an exact category and $[\underline{X}, Ch_b(\mathcal{C})] \simeq Ch_b([\underline{X}, \mathcal{C}])$ is a complicial bi-Waldhausen category. Now, identify $\zeta \in [\underline{X}, \mathcal{C}]$ with the object ζ_* in $Ch_b[\underline{X}, \mathcal{C}]$ defined by $\zeta_*(x)$ = chain complex consisting of a single object $\zeta(x)$ in degree zero and zero elsewhere. The result follows by applying the Gillet - Waldhausen theorem.

(2) Recall that $\mathbb{K}_n^G(X, Ch_b(\mathcal{C}), Y)$ is the Waldhausen K-theory of the Waldhausen category $^Y[\underline{X}, Ch_b(\mathcal{C})]$ where $ob^Y[\underline{X}, Ch_b(\mathcal{C})] = ob[\underline{X}, Ch_b(\mathcal{C})]$, cofibrations are Y-cofibrations in $[\underline{X}, Ch_b(\mathcal{C})]$, and weak equivalences are the weak equivalences in $[\underline{X}, Ch_b(\mathcal{C})]$. Also, $K_*^G(X, \mathcal{C}, Y)$ is the Quillen K-theory of the exact category $[\underline{X}, \mathcal{C}]$ with respect to Y-exact sequences. Denote this exact category by $^Y[\underline{X}, \mathcal{C}]$. We can define an inclusion functor $[\underline{X}, \mathcal{C}] \subseteq Ch_b(^Y[\underline{X}, \mathcal{C}]) \cong {}^Y[\underline{X}, Ch_b(\mathcal{C})]$ as in (1) and apply the Gillet - Waldhausen theorem.

(3) Just as in the last two cases, we can define an inclusion functor from the exact category $[\underline{X}, \mathcal{C}]_Y$ to the Waldhausen category $Ch_b([\underline{X}, \mathcal{C}]_Y \simeq [\underline{X}, Ch_b(\mathcal{C})]_Y$ and apply the Gillet - Waldhausen theorem.

〇

Theorem 13.3.2 (Equivariant Approximation Theorem) *Let*
$W = Ch_b(\mathcal{C})$ *and* $W' = Ch_b(\mathcal{C}')$ *be two complicial bi-Waldhausen categories*
where C, C' *are exact categories.* $F : W \to W'$ *an exact functor. Suppose*
that the induced map of derived categories $D(W) \to D(W')$ *is an equivalence*
of categories. Then, for any G-set X, *the induced map of spectra* $\mathbb{K}(F)$:
$\mathbb{K}([\underline{X}, W]) \to \mathbb{K}([\underline{X}, W'])$ *is a homotopy equivalence.*

PROOF   An exact functor $F : Ch_b(\mathcal{C}) \to Ch_b(\mathcal{C}')$ induces a functor

$$\widehat{F} : [\underline{X}, Ch_b(\mathcal{C})] \to [X, Ch_b(\mathcal{C}')], \quad \zeta \to \widehat{F}(\zeta),$$

where $\widehat{F}(\zeta)(x) = F(\zeta(x))$. Now, suppose that the induced map $D(Ch_b(\mathcal{C}) \to D(Ch_b(\mathcal{C}'))$ is an equivalence of categories. Note that $D(Ch_b(\mathcal{C}))$ (resp. $D(Ch_b(\mathcal{C}'))$ is obtained from $Ch_b(\mathcal{C})$ (resp. $Ch_b(\mathcal{C}')$) by formally inverting quasi-isomorphisms. Now, a map $\zeta \to \eta$ in $[\underline{X}, Ch_b(\mathcal{C})$ is a quasi-isomorphism iff $\zeta(x) \to \eta(x)$ is a quasi-isomorphism in $Ch_b(\mathcal{C})$. The proof is now similar to [57] 5.2.   〇

13.4   Applications to higher K-theory of grouprings

13.4.1   (1) Recall from examples 1.1.5(i) that if $X = G/H$ where H is a subgroup of G and R is a commutative ring with identity, we can identify $[G/H, \mathcal{M}'(R)]$ with $\mathcal{M}'(RH)$ and $[G/H, \mathcal{P}(R)]$ with $\mathcal{P}_R(RH)$. Hence, we can identify $[G/H, Ch_b(\mathcal{M}'(R))]$ with $Ch_b(\mathcal{M}'(RH))$ and $[G/H, Ch_b(\mathcal{P})]$ with $Ch_b(\mathcal{P}_R(RH))$. So, we can identify $K_n^G(G/H, \mathcal{M}'(R))$ with $K_n(\mathcal{M}'(R)) = G_n(RH)$ when R is Noetherian. By theorem 13.3.1, we can identify $\mathbb{K}_n^G(G/H, Ch_b(\mathcal{M}'(R)))$ with $\mathbb{K}_n(Ch_b(\mathcal{M}'(RH))) \simeq G_n(RH)$ by the Gillet - Waldhausen theorem. Also, $K_n^G(G/H, \mathcal{P}(R)) \simeq K_n(\mathcal{P}_R(RH)) \simeq \mathbb{K}_n(Ch_b\mathcal{P}_R(RH)) \simeq G_n(R, H)$ by the Gillet - Waldhausen theorem.

(2) With the notations above, we can identify $K_n^G(G/H, \mathcal{M}'(R), Y)$ (resp. $K_n^G(G/H, \mathcal{P}(R), Y)$ with Quillen K-theory of the exact category $\mathcal{M}'(RH)$ (resp. $\mathcal{P}_R(RH)$) with respect to exact sequences that split when restricted to the various subgroups H' of H with a non-empty fixed point set $T^{H'}$ (see 10.3.2). In particular,

$$K_n^G(G/H, \mathcal{M}'(R), G/e) \simeq K_n^G(G/H, \mathcal{M}'(R)) \simeq K_n(\mathcal{M}'(RH) \simeq G_n'(RH).$$

$$K_n^G(G/H, \mathcal{M}(R), G/e) \simeq K_n^G(G/H, \mathcal{P})(R)) \simeq K_n(\mathcal{P}(RH) \simeq G_n(R, H).$$

Hence, we also have

$$\mathbb{K}_n^G(G/H, Ch_b(\mathcal{M}'(R)), G/e) \simeq \mathbb{K}_n^G(G/H, Ch_b(\mathcal{M}'(R)))$$

$$\mathbb{K}_n(Ch_b(\mathcal{M}'(RG))) \simeq K_n(\mathcal{M}'(RG)) \simeq G_n'(RG).$$

by the Gillet - Waldhausen theorem.

(3) Recall from 10.3.2 that $P_n^G(G/H, \mathcal{M}'(R), Y)$ (resp. $P_n^G(G/H), \mathcal{P}(R), Y)$) are the Quillen K-groups of the exact category $\mathcal{M}'(RH)$ (resp. $\mathbb{P}_R(RH)$) that are relatively projective with respect to $D(Y, H) = \{H' \le H | Y^{H'} \neq \emptyset\}$. In particular, $P_n^G(G/H, \mathcal{P}(R), G/e) \equiv K_n(\mathcal{P}(RH) \simeq K_n(RH)$. Hence, we can identify $\mathbb{P}_n^G(G/H, Ch_b(\mathcal{P}(R)), G/e)$ with $\mathbb{K}_n(Ch_b(\mathcal{P}(RH)) \simeq K_n(RH)$ by the Gillet - Waldhausen theorem.

(4) In view of the identifications in (1),(2),(3) above, we recover the relevant results and computations in chapter 10.

13.4.2 We now record below (theorem 13.4.1) as an application of Waldhausen fibration sequence theorem 13.2.7 and Garkusha's result [56] 3.1.

Theorem 13.4.1   *(i) In the notations of 13.4.1, let R be a commutative ring with identity, G a finite group, $\mathcal{M}'(RG)$ the category of finitely presented RG-modules, $Ch_b(\mathcal{M}'(RG))$ the Waldhausen category of bounded complexes over $\mathcal{M}'(RG)$ with weak equivalences being stable quasi-isomorphism (see examples 5.4.2(iv),(v)). Then, we have a long exact sequence for all $n \ge 0$:*

$$\to \mathbb{K}_{n+1}(Ch_b(\mathcal{M}'(RG), \omega) \to \mathbb{P}_n^G(G/G, Ch_b(\mathcal{P}(R)), G/e) \ldots$$

$$\to \mathbb{K}_n^G(G/G, Ch_b(\mathcal{M}'(R), G/e) \to \mathbb{K}_n(Ch_b(\mathcal{M}'(RG), \omega) \to \ldots$$

(ii) If in (i), R is the ring of integers in a number field, then for all $n \ge 1$, $\mathbb{K}_{n+1}(Ch_b(\mathcal{M}'(RG), \omega)$ is a finite Abelian group.

PROOF   From theorem 13.3.1 we have

$$\mathbb{P}_n^G(G/G, Ch_b(\mathcal{P}(R)), G/e) \cong P_n^G(G/G, \mathcal{P}(R), G/e) \simeq K_n(RG).$$

$$\mathbb{K}_n^G(G/G.Ch_b(\mathcal{M}'(R), G/e) \simeq K_n^G(G/G, \mathcal{M}'(R), G/e) \cong G_n'(RG).$$

Hence, the long exact sequence follows from [56] 3.1.

Now, if R is the ring of integers in a number field F, then RG is an R-order in a semi-simple F-algebra FG, and so, by theorems 7.1.11 and 7.1.13, $K_n(RG), G_n(RG)$ are finitely generated Abelian groups for all $n \ge 1$. Hence, for all $n \ge 1$, $K_{n+1}(Ch_b(\mathcal{M}(RG), \omega)$ is finitely generated. So, to show that $K_{n+1}(Ch_b(\mathcal{M}(RG), \omega)$ is finite, we only have to show that it is torsion.

Now, let $\alpha_n : K_n(RG) \to G_n(RG)$ be the Cartan map that is part of the exact sequence

$$\cdots \to K_{n+1}(Ch_b(\mathcal{M}(RG)), \omega) \to K_n(RG) \xrightarrow{\alpha_n}$$

$$G_n(RG) \to K_n(Ch_b(\mathcal{M}(RG)), \omega) \to \cdots \qquad \text{(I)}$$

From this sequence we have a short exact sequence

$$0 \to \text{Coker } \alpha_{n+1} \to K_{n+1}(Ch_b(\mathcal{M}(RG)), \omega) \to \text{Ker } \alpha_n \to 0 \qquad \text{(II)}$$

for all $n \geq 1$. So, it suffices to prove that $\text{Ker} \alpha_n$ is finite and $\text{Coker} \alpha_{n+1}$ is torsion.

Now, from the commutative diagram

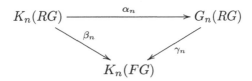

we have an exact sequence $0 \to \text{Ker } \alpha_n \to SK_n(RG) \to SG_n(RG) \to \text{Coker } \alpha_b \to \text{Coker } \beta_n \to \text{Coker } \gamma_n \to 0$. Now, for all $n \geq 1$, $SK_n(RG)$ is finite (see theorem 7.1.11). Hence $\text{Ker } \alpha_n$ is finite for all $n \geq 1$.

Also, $SG_n(RG)$ is finite for all $n \geq 1$ (see theorem 7.1.13) and $\text{Coker } \beta_n$ is torsion (see theorem 7.2.6, 1.7). Hence $\text{Coker } \alpha_n$ is torsion. So, from (II), $K_{n+1}(Ch_b(\mathcal{M}(RG)), \omega)$ is torsion. Since it is also finitely generated, it is finite. $\quad\square$

Exercise

Let k be a finite field of characteristic p, G a finite group not divisible by p, $(\mathcal{A}, \omega)$ the Waldhausen category of bounded chain complexes over $\mathcal{M}(kG)$ with weak equivalences being stable quasi-isomorphisms. Show that p does not divide $|K_{2n+1}(\mathcal{A}, \omega)|$ for all $n \geq 1$.

Chapter 14

Equivariant homology theories and higher K-theory of grouprings

One of the major goals of this book is to highlight techniques for the computations of higher K-theory of grouprings of finite and infinite groups. In particular, chapters 9 - 13 have been devoted to induction techniques for such computations in the spirit of Mackey functors and equivariant higher K-theory. Chapters 5 to 8 highlighted other techniques from higher K-theory for such computations.

In this chapter, we also aim at computing higher algebraic K-groups of discrete groups via induction techniques under the umbrella of equivariant homology theories. Indeed, an equivariant homology theory is the result of linking G-homology theories for different G with induction structure. In fact, an important criteria for G-homology theory $\mathcal{H}_n^G : GSet \to \mathbb{Z}\text{-}\mathcal{M}od$ is that it is isomorphic to some Mackey functor $GSet \to \mathbb{Z}\text{-}\mathcal{M}od$ (see theorem 14.2.4 or [14]).

We shall be particularly interested in Farrell - Jones conjecture (for algebraic K-theory) and Baum - Connes conjecture (for topological K-theory) as induction results under a unified approach through a metaconjecture due to Davis - Lück (see [40] and [138]) as well as specific induction results due to W. Lück, A. Bartels, and H. Reich (see [16]).

We only cover a rather limited aspect of this topic in this book. One should see the excellent article [138] where W. Lück and H. Reich discuss this subject in its various ramification including L-theory and many topological and geometric applications.

The significance of also discussing Baum - Connes conjecture in this unified treatment is inspired by the fact that it is well known by now that algebraic K-theory and topological K-theory of stable C^*-algebras coincide (see [205]). Hence Baum - Connes conjecture also fits into the theme of this book.

14.1   Classifying space for families and equivariant homology theory

$(14.1)^A$ Classifying spaces for families and G-homology theory

Definition 14.1.1 *Let G be a discrete group. By a family $\mathcal{F}$ of subgroups of G, we shall mean a set of subgroups closed under conjugation and taking subgroups. We shall be interested in the following examples:*

(i) *All:= all subgroups of G.*

(ii) *Fin:= all finite subgroups of G.*

(iii) *VCy:= all virtually cyclic subgroups of G (see 4.5.3 for definition).*

(iv) *Triv:= trivial family consisting of only one element, i.e., the identity element of G.*

(v) *FCy:= all finite cyclic subgroups of G.*

Note: *As we shall see later, computations of some K-groups of G can be reduced to computations of equivariant homology groups for "Fin" or "VCy".*

14.1.1 Remarks on G-spaces (G-discrete). Before proceeding further, we recall the following properties of G-spaces. Readers are referred to [29].

(i) A G-space is a topological space X equipped with a left G-action (i.e., continuous map $G \times X \to X$ $(g,x) \to gx$ such that $g(hx) = (gh)x$.

   For $x \in X, G_x = \{g \in G | gx = x\}$ is the stabilizer of x in G and $Gx = \{gx | x \in X\}$ is the orbit of x in G. Note that every discrete G-space X has the form

   $$X = \cup G/H, \ H = G_x \quad \text{for some} \quad X \in X.$$

(ii) If $H \leq G$, $N(H) = N_G(H) = \{g \in G | gH = Hg\}$ is the normalizer if H in G. Note that X^H is an $N(H)/H$-space. Call $W(H) := N(H)/H$ the Weyl group. If $K \leq G, (G/H)^K = \{gH | g^{-1}Kg \subseteq H\}$.

(iii) If X, Y are G-spaces, the mapping space $\text{Map}(X,Y)$ is the set of continuous maps with compact open topology. $\text{Map}(X,Y)$ is a G-space given by $(g,\varphi) \to g\varphi(g^{-1}), g \in G, \varphi \in \text{Map}(X,Y)$. If $H \leq G$, $\text{Map}(G/H, X) \simeq X^H$.

(iv) Let $\alpha : H \to G$ be a homomorphism of discrete groups, X an H-space. The induced G-space $\operatorname{Ind}_\alpha X := G \underset{\alpha}{\times} X := (G \times X)/H$ where H acts on $G \times X$ by $h(g, x) = (g\alpha(h^{-1}), hx)$ and where G-action on $\operatorname{Ind}_\alpha X$ is given by left G-action on $(G \times X)$, which passes on to G-action on $G \underset{\alpha}{\times} X$ by $(g, (\overline{k, x}) \mapsto (\overline{gk, x})$. If α is a monomorphism, H can thus be regarded as a subgroup of G, and we write $G \underset{H}{\times} X$ for $G \underset{\alpha}{\times} X$ and Ind_H^G for $\operatorname{Ind}_\alpha$. One defines a coinduced G-space analogously. We have $\operatorname{Map}_H(G, X)$, G considered as an H-space via left H-action, and a map $G \times \operatorname{map}_H(G, X) \to \operatorname{map}_H(G, X)$ $(g, \phi) \to g\phi : G \to X$ given by $y \to \phi(yg)$. Then, $g\varphi$ is also an H-map since $(g\varphi)(hy) = \varphi(hy\ g) = h\varphi(yg) = h(g\varphi)(y)$. So, if Z is any G-space, the H-map $Z \to X$ corresponds to G-map $Z \to \operatorname{map}_H(G, X)$.

14.1.2 Remarks on G-CW complexes

(i) Recall that a G-CW-complex consists of a G-space X together with a filtration $X^0 \subseteq X^1 \subseteq X^2 \subseteq \cdots \subseteq X$ by G-subspace X^i such that the following are satisfied:

(1) Each X^n is closed in X.

(2) $X = \cup X^n$.

(3) X^0 is a discrete subspace of X.

(4) For each $n \geq 1$, there exists a discrete G-space Δ_n together with G-maps $f : S^{n-1} \times \Delta_n \to X^{n-1}$, $\hat{f} : B^n \times \Delta_n \to X^n$ such that the following diagram

$$
\begin{array}{ccc}
S^{n-1} \times \Delta_n & \hookrightarrow & B^n \times \Delta_n \\
\downarrow & & \downarrow \hat{f} \\
X^{n-1} & \hookrightarrow & X^n
\end{array}
\qquad \text{is a push-out}
$$

where $B^n =$ unit ball in Euclidean n-space (note that each $\Delta_n = \cup G/K_\alpha$ where K_α is an isotropy group).

(5) A subspace $Y \subset X$ is closed iff $Y \cap X^n$ is closed for each $n \geq 0$. Note that if $G = (e)$, then a G-CW-complex is just an ordinary CW-complex.

(ii) A G-CW-complex is said to be finite dimensional iff $X^n = X$ for some n : dim $X :=$ least $n \geq -1$ for which $X^n = X$.

(iii) If X, Y are CW-complexes, $X \times Y$ is in general not a CW-complex. However, if we endow $X \times Y$ with weak topology ($A \subset X \times Y$ closed iff $A \cap (X_\alpha \times Y_\alpha)$ is closed for all finite subsets $X_\alpha \subset X, Y_\beta \subset Y$), then $X \times Y$ becomes a CW-complex.

(iv) For G-CW-complexes, weak G-equivalences are actually G-homotopy equivalences. This is because of Whitehead theorem, which says that "A G-map $f : X \to Y$ between G-CW-complexes is a G-homotopy equivalence if for all $H \leq G$ and all $x_0 \in X^H$, the induced map $\pi_*(X^H, x_0) \to \pi_*(Y^H, x_0)$ is bijective, i.e., f is a weak G-equivalence".

(v) The geometric realization functor $|\ | :$ Spaces $\to$ CW-complexes extends to a functor $\Gamma : G$-Spaces $\to G$-CW-complexes together with a natural transformation $\Gamma X \to X$ which is a weak equivalence for every G-space X. E.g., if X, Y are G-spaces, $\Gamma(\mathrm{Map}(X, Y))$ is a G-CW-complex (see [138]).

(vi) A G-CW-complex is a proper G-space iff all its point stabilizers (isotropy groups) are finite (see [24, 138]).

Definition 14.1.2 *Let $\mathcal{F}$ be a family of subgroups of G. A classifying space for $\mathcal{F}$ or a universal G-space with stabilizers in $\mathcal{F}$ is a contractible space $E_{\mathcal{F}}(G)$ endowed with a G-action such that $E_{\mathcal{F}}(G)^H$ is non-empty and $E_{\mathcal{F}}(G)^H$ is contractible iff $H \in \mathcal{F}$.*

Definition 14.1.3 *Let G be a discrete group, $\mathcal{F}$ a family of subgroups also closed under taking finite intersections. A model for $E_{\mathcal{F}}(G)$ is a G-CW-complex E such that all cell stabilizers lie in $\mathcal{F}$, and for any $H \in \mathcal{F}$, the fixed point set E^H is contractible.*

Remarks 14.1.1 (i) We shall by abuse of notation also write $E_{\mathcal{F}}(G)$ for the model for $E_{\mathcal{F}}(G)$ when the context is clear.

(ii) $E_{\mathcal{F}}(G)$ is unique up to G-homotopy, i.e., for any G-CW-complex X all of whose isotropy groups belong to $\mathcal{F}$, there exists up to G-homotopy precisely one G-map $X \to E_{\mathcal{F}}(G)$.

(iii) $E_{Triv}(G) = EG$ where EG is the universal covering of BG, i.e., EG is the total space of the G-principal bundle $G \to EG \to BG$.

(iv) $E_{\mathcal{F}in}(G)$ is the classifying space for proper G-actions and is denoted in the literature by $\underline{EG}$.
Note that a G-CW-complex is proper iff for all $x \in G$, G_x (stabilizer of x) is finite.

(v) A model for $E_{All}(G)$ is the one-point space G/G.

(vi) If G is a discrete subgroup of a Lie group L with finitely many path components, then for any maximal compact subgroup $K \subseteq L$, the space L/K with its left G-action is a model for $E_{\mathcal{F}in}(G)$.

14.1.3 Let G be a discrete group and R a commutative ring with identity. A G-homology theory $\mathcal{H}_*^G$ with values in $\underline{R\text{-}\mathcal{M}od}$ is a collection of covariant functors $\mathcal{H}_n^G : (G\text{-CW-pairs}) \to (R\text{-}\mathcal{M}od)$ indexed by $n \in \mathbb{Z}$ together with natural transformations $\partial_n^G(X, A) : \mathcal{H}_n^G(X, A) \to \mathcal{H}_{n-1}^G(A) := \mathcal{H}_{n-1}^G(A, \phi)$ for $n \in \mathbb{Z}$ such that the following axioms hold:

(1) **G-homotopy invariance**

For G-homotopic maps $f_0, f_1 : (X, A) \to (Y, B)$ of G-CW-pairs, we have $\mathcal{H}_n^G(f_0) = \mathcal{H}_n^G(f_1)$ for all $n \in \mathbb{Z}$.

(2) **Long exact sequence of a pair**

If (X, A) is a pair of G-CW-complexes and $i : A \to X$, $j : X \to (X, A)$ the inclusion maps, then there exists a long exact sequence

$$\cdots \to \mathcal{H}_{n+1}^G(X, A) \xrightarrow{\partial_{n+1}^G} \mathcal{H}_n^G(A) \xrightarrow{\mathcal{H}_n^G(i)} \mathcal{H}_n^G(X) \xrightarrow{\mathcal{H}_n^G(j)} \mathcal{H}_n^G(X, A) \xrightarrow{\partial_n^G} \mathcal{H}_{n-1}^G(A).$$

(3) **Excision**

Given a G-CW-pair (X, A) and a cellular G-map $f : A \to B$ of G-CW-complexes, we can endow $(X \underset{f}{\cup} B, B)$ with the induced structure of a G-CW-pair. Then, the canonical map $(F, f) : (X, A) \to (X \underset{f}{\cup} B, B)$ induces for each $n \in \mathbb{Z}$ an isomorphism

$$\mathcal{H}_n^G(F, f) : \mathcal{H}_n^G(X, A) \xrightarrow{\sim} \mathcal{H}_n^G(X \underset{f}{\cup} B, B).$$

(4) **Disjoint union axiom**

Let $\{X_i | i \in I\}$ be a collection of G-CW-complexes, and $j_i : X_i \to \coprod_{i \in I} X_i$ the canonical inclusion. Then the map

$$\bigoplus_{i \in I} \mathcal{H}_n^G(j_i) : \bigoplus_{i \in I} \mathcal{H}_n^G(X_i) \xrightarrow{\sim} \mathcal{H}_n^G(\coprod_{i \in I} X_i)$$

is an isomorphism.

14.1.4 Next, we indicate how to construct homology theories relevant to our context. We are interested in equivariant homology theories with coefficients in spectra. We now set out towards defining this concept.

14.1.5 Let $\mathcal{C}$ be a small category (e.g., $Or_{\mathcal{F}}(G) = \{G/H | H \in \mathcal{F}\}$. A contravariant $\mathcal{C}$-space (resp. $\mathcal{C}$-spectrum) X is a contravariant functor $X : \mathcal{C} \to$ Space (resp. $X : \mathcal{C} \to$ Spectra). Define a covariant $\mathcal{C}$-space or $\mathcal{C}$-spectra analogously.

Now, let $\mathcal{G}$poids be the category of small groupoids with covariant functors as morphisms. We want to describe a way of getting $O_{\mathcal{F}}(G)$-spectrum from $\mathcal{G}$poids-spectrum in order to obtain the desired induction results.

14.1.6 In $(5.2)^C$, we discussed spectra in general and their basic properties. Recall from $(5.2)^C$ that given a spectrum $\underline{E}$ and a pointed space Y_+, we can define a smash product spectrum $Y \wedge \underline{E}$ by $(Y \wedge \underline{E})(n) = Y \wedge E(n)$. Cone (Y_+) was also defined in $(5.2)^C$. We also observed that if X is a pointed space and A is a subspace of X, then any spectrum $\underline{E}$ defines a homology theory $H_n(X, A, \underline{E}) = \pi_n(X_+ \underset{A_+}{\cup} cone(A_+) \wedge \underline{E})$. We want to extend this to G-homology theories.

14.1.7 Let $\mathcal{G}poids\,^{inj}$ be the subcategory of $\mathcal{G}poids$ such that $ob(\mathcal{G}poids\,^{inj}) = ob(\mathcal{G}poids)$ and morphisms are injective (or faithful) functors. Recall that a functor $F : \mathcal{G}_0 \to \mathcal{G}_1$ from a groupoid $\mathcal{G}_0$ to a groupoid $\mathcal{G}_1$ is said to be faithful (or injective) if for any two objects $x, y \in \mathcal{G}_0$, the induced map $mor_{\mathcal{G}_0}(x, y) \to mor_{\mathcal{G}_1}(F(x), F(y))$ is injective.

Recall from 1.1.6 that if S is a G-set, then the translation category $\underline{S}$ of S is a groupoid. We shall sometimes denote this groupoid by $\underline{S}_G$ (to indicate that S is a G-set). Hence, there is a covariant functor "$-_G$", $Or_\mathcal{F}(G) \to \mathcal{G}poids\,^{inj}$ given by $G/H \to \overline{G/H}$.

Now, let $\underline{E} : \mathcal{G}poids\,^{inj} \to$ Spectra be a $\mathcal{G}poids\,^{inj}$-spectrum which sends equivalences of groupoids to weak equivalences of spectra.

(Recall that a weak equivalence of spectra is a map $f : \underline{E} \to \underline{F}$ of spectra inducing an isomorphism on all homology groups.) Then, for any discrete group G, $\underline{E}$ defines a G-homology theory associated to $Or_\mathcal{F}(G)$-spectrum

$$\underline{E} \circ \text{``} -_G \text{''}, Or_\mathcal{F}(G) \to \mathcal{G}poids\,^{inj} \xrightarrow{\underline{E}} \text{Spectra}.$$

Definition 14.1.4 *(i) To a left G-space Y, we can associate a contravariant $Or_\mathcal{F}(G)$-space $map_G(-, Y)$ defined by $map_G(-, Y) : Or_\mathcal{F}(G) \to$ Spaces : $G/H \to map_G(G/H, Y) = Y^H$. If Y is pointed, then $map_G(-, Y)$ takes values in pointed spaces.*

(ii) Let X be a contravariant and Y a covariant $\mathcal{C}$-space ($\mathcal{C}$ any small category). The balanced product of X and Y is defined as $X \times Y := \underset{\mathcal{C}}{}$

$$\underset{c \in ob(\mathcal{C})}{\coprod} X(c) \times Y(c)/ \sim \text{ where "}\sim\text{" is the equivalence relation gener-}$$

ated by $(x\varphi, y) \sim (x, \varphi y)$ for all morphisms $\varphi, c \to d$ in $\mathcal{C}$ and points $x \in X(d); y \in Y(c)$. Note that $x\phi$ is written for $X(\phi)(x)$, and φy for $Y(\phi)(y)$.

The next result, due to J.F. Davis and W. Lück, is the equivariant analogue of the result in 14.1.6.

Theorem 14.1.1 [40]. *Let $\underline{E}$ be a contravariant $Or_\mathcal{F}(G)$-spectrum. Then, $\underline{E}$ defines a G-homology theory $\mathcal{H}_n^G(-; E)$ by $\mathcal{H}_n^G(X; A; \underline{E}) = \pi_n(map_G(-, X_+ \cup_{A_+} cone(A_+)) \wedge_{Or_\mathcal{F}(G)} \underline{E})$. In particular, $H_n^G(G/H; \underline{E}) = \pi_n(\underline{E}(G/H))$.*

PROOF   See [40].   ☐

Next, we define equivariant homology theory and provide an analogue of theorem 14.4.1 for the theory.

Definition 14.1.5 *Let A be a commutative ring with identity. An equivariant homology theory $\mathcal{H}_*^?$ with values in A-$\mathcal{M}od$ consists of a G-homology theory $\mathcal{H}_*^G$ with values in A-$\mathcal{M}od$ for each G together with the following induction structure:*

(i) *If $\alpha : H \to G$ is a group homomorphism, and (X, A) a proper CW-pair such that $\operatorname{Ker} \alpha$ acts freely on X, then for each $n \in \mathbb{Z}$, there are natural isomorphims $\operatorname{ind}_\alpha : H_n^H(X, A) \xrightarrow{\simeq} H_n^G(\operatorname{ind}_\alpha X, A)$ such that we have compatibility with boundary homomorphism, i.e., $\partial_n^G \circ \operatorname{ind}_\alpha = \operatorname{ind}_\alpha \circ \partial_n^H$.*

(ii) **Functoriality** *Suppose that $\beta : G \to K$ is another group homomorphism such that $\operatorname{Ker}(\beta\alpha)$ acts freely on X. Then, for $n \in \mathbb{Z}$, $\operatorname{ind}_{\beta\alpha} = H_n^K(f_1) \circ \operatorname{ind}_\beta \circ \operatorname{ind}_\alpha : \mathcal{H}_n^H(X, A) \to \mathcal{H}_n^K(\operatorname{ind}_{\beta\alpha}(X, A))$ where*

$$f_1 : \operatorname{ind}_\beta \operatorname{ind}_\alpha(X, A) \xrightarrow{\simeq} \operatorname{ind}_{\beta\alpha}(X, A) \quad (k, g, x) \to (k_\beta(g), x)$$

is a natural K-homeomorphism.

(iii) **Compatibility with conjugation** *If (X, A) is a proper G-CW-pair, and $g \in G$, $n \in \mathbb{Z}$, then the homomorphism $\operatorname{ind}_{c(g):G \to G} : \mathcal{H}_n^G(X, A) \to \mathcal{H}_n^G(\operatorname{ind}_{c(g):G \to G}(X, A))$ agrees with $\mathcal{H}_n^G(f_2)$ where f_2 is a G-homeomorphism $f_2 : (X, A) \to \operatorname{ind}_{c(g):G \to G}(X, A)$ sending x to $(1, g^{-1}c)$ and $c(s) : G \to G$ takes g^1 to gg^1g^{-1}.*

Example 14.1.1 Bredon homology Let X be a G-CW-complex, $\mathcal{F}(X)$ a family of subgroup of G that occur as stabilizers of G-actions on X, $\mathcal{F}$ a family of subgroups of G containing $\mathcal{F}(X)$. Let Δ_i be discrete G-set occurring in the definition of the i-skeleton of X (see 14.1.2(i)). Then, $\Delta_i = \dot{\cup} G/K_\alpha$, $K_\alpha \in \mathcal{F}(X)$. Now, X defines a contravariant functor $Or_{\mathcal{F}(X)}(G) \to (CW - \mathcal{C}exes)$ given by $G/H \to \operatorname{map}_G(G/H, X) = X^H$. The cellular chain complex $\underline{C}_*(X) = \{\underline{C}_i(X)\}$ where $C_i(X) = \mathbb{Z}(\Delta_i)$, and $C_*(X^H) = \{C_i(X^H) = \mathbb{Z}(\Delta_i^H)\}$. So we have a functor

$$\underline{C}_*(X) : Or_{\mathcal{F}(X)}(G) \to (CW - \mathcal{C}exes) \to \text{(cellular chain complexes)}$$
$$G/H \to X^H \to C_*(X^H).$$

Let A be a commutative ring with identity and $M : Or_{\mathcal{F}(X)}(G) \to A$-$\mathcal{M}od$ a covariant functor. Then Bredon homology of $X : H_n^G(X, M) := H_*(\underline{C}_*(X) \underset{\mathbb{Z}or_{\mathcal{F}(X)}(G)}{\otimes} M)$. Note that $H_n^G(X, M)$ is the same for any $\mathcal{F}$ containing $\mathcal{F}(X)$. Note that $\mathcal{H}_n^G(-, M)$ defines an equivariant homology theory $\mathcal{H}_n^G(-, M)$.

Remarks 14.1.2 Recall that the category $\mathcal{M}od_{\mathcal{F}}^R(G)$ of contravariant functors $Or_{\mathcal{F}}(G) \rightarrow R\text{-}\mathcal{M}od$ is an Abelian category where R is a commutative ring with 1, and so, the usual homological algebra can be done inside $\mathcal{M}od_{\mathcal{F}}^R(G)$. An object $P \in \mathcal{M}od_{\mathcal{F}}^R(G)$ is projective if $\operatorname{Hom}_{\mathcal{M}od_{\mathcal{F}}^R(G)}(P, \cdot)$: $\mathcal{M}od_{\mathcal{F}}^R(G) \rightarrow R\text{-}\mathcal{M}od$ is exact. One can construct projective objects in $\mathcal{M}od_{\mathcal{F}}^R(G)$, and such objects have the form $P_K : Or_{\mathcal{F}}(G) \rightarrow R\text{-}\mathcal{M}od$: $G/H \rightarrow R\operatorname{Hom}_{Or_{\mathcal{F}}(G)}(G/H, G/K)$ where $K \in \mathcal{F}$ (see [156]). In particular, $\mathcal{M}od_{\mathcal{F}}^R(G)$ has enough projectives, i.e., for every $M \in \mathcal{M}od_{\mathcal{F}}(G)$, there exists an epimorphism $P \rightarrow M$ with P projective. If we write $P_*(M) \twoheadrightarrow M$ for a projective resolution of M, then for each $N \in \mathcal{M}od_{\mathcal{F}}^R(G)$, we have a cochain complex $\operatorname{Hom}_{\mathcal{M}od_{\mathcal{F}}^R(G)}(P_*(M), N)$ and hence

$\operatorname{Ext}^i(M, N) = H^i(\operatorname{Hom}_{\mathcal{M}od_{\mathcal{F}}^R(G)}(P_*(M), N)$ $i \geq 0$.

Similarly, the covariant functors $N : Or_{\mathcal{F}}(G) \rightarrow R\text{-}\mathcal{M}od$ form an Abelian category.

Let $G\text{-}\mathcal{M}od_{\mathcal{F}}^R$ be the category of covariant functors $Or_{\mathcal{F}}(G) \rightarrow R\text{-}\mathcal{M}od$. Note that for $M \in \mathcal{M}od_{\mathcal{F}}^R(G), N \in G - \mathcal{M}od_{\mathcal{F}}^R, M \otimes_{\mathcal{F}} N$ is defined, and so, is $(?) \otimes_{\mathcal{F}} N : \mathcal{M}od_{\mathcal{F}}^R(G) \rightarrow R\text{-}\mathcal{M}od$. Also, for any projective object $P \in \mathcal{M}od_{\mathcal{F}}^R(G), P \otimes_{\mathcal{F}} (-)$ is exact. So, if $P_*(M)$ is a projective resolution of M, we can compute $Tor_i(M, N)$ as the homology groups of the complex $P_*(M) \otimes_{\mathcal{F}} N$. See [138, 156] for details.

14.2 Assembly maps and isomorphism conjectures

14.2.1 Let $\mathcal{F}$ be a subfamily of a family $\mathcal{F}'$ of subgroups of a discrete group G. Since all the isotropy groups of $E_{\mathcal{F}}(G)$ also lie in $\mathcal{F}'$, then by the universal property of the construction, there is a G-map $E_{\mathcal{F}}(G) \rightarrow E_{\mathcal{F}'}(G)$ unique up to G-homotopy.

Now, let $\mathcal{H}_*^G$ be a G-homology theory. The relative assembly map $A_{\mathcal{F} \rightarrow \mathcal{F}'}$ is the homomorphism $A_{\mathcal{F} \rightarrow \mathcal{F}'} : \mathcal{H}_n^G(E_{\mathcal{F}}(G)) \rightarrow \mathcal{H}_n^G(E_{\mathcal{F}'}(G))$. If $\mathcal{F}' = $ All, then $E_{\mathcal{F}'}(G) = pt$, and $A_{\mathcal{F} \rightarrow \mathcal{F}'}$ is just called the assembly map, and we just write this map as $A_{\mathcal{F}}$.
Note that $\{1\} \subset FCy \subset \text{Fin} \subset \text{VCy} \subset \text{All}$, and so, we can factorize $A_{(1) \rightarrow \text{All}}$ into relative assembly maps.

Example 14.2.1 The relative assembly map $A_{\text{Fin} \rightarrow \text{VCy}} : \mathcal{H}_n^G(E_{\mathcal{F}in}(G) \rightarrow \mathcal{H}_n^G(E_{\text{VCy}}(G))$ is an isomorphism iff for all virtually cyclic subgroups V of G, the assembly map $A_{\mathcal{F}in} = A_{\mathcal{F}in \rightarrow \text{All}} : H_n^V(E_{\mathcal{F}in}(V)) \rightarrow \mathcal{H}_n^V(pt)$ is an isomorphism.

14.2.2 (i) **Metaconjecture** Let $E_{\mathcal{F}}(G) \to pt$ be the natural projection. The metaconjecture says that the assembly map $A_{\mathcal{F}} : \mathcal{H}_n^G(E_{\mathcal{F}}(G)) \to \mathcal{H}_n^G(pt)$ is an isomorphism for all $n \in \mathbb{Z}$.

(ii) **Farrell - Jones conjecture** Let R be a ring with identity and $\underline{K}R$ the K-theory spectrum of R (see 5.2.2). Then, it is well known that $\mathcal{H}_n^G(pt, \underline{K}R) \simeq K_n(RG)$ (see 5.2.2). Farrell - Jones conjecture says that for all $n \in \mathbb{Z}$, $\mathcal{H}_n^G(E_{\mathrm{VCy}}(G), \underline{K}R) \to \mathcal{H}_n^G(pt, \underline{K}R) \simeq K_n(RG)$ is an isomorphism.

An analogous conjecture of Farrell - Jones can be stated for L-theory, but we do not intend to go into this.

(iii) **Baum - Connes conjecture** Let C^*-$\mathcal{A}lg$ be the category of C^*-algebras. Then there exists a functor $\underline{K}^{top} : C^*$-$\mathcal{A}lg \to$ Spectra such that
$\pi_n(\underline{K}^{top}(C_r^*(G))) = K_n(C_r^*(G))$ for any discrete group and any $n \in \mathbb{Z}$. Baum - Connes conjecture says that $A_{\mathcal{F}in} : \mathcal{H}_n^G(E_{\mathcal{F}in}(G), \underline{K}^{ton}) \to H_n^G(pt, \underline{K}^{ton}) \simeq K_n(C_r^*(G))$ induced by the projection $E_{\mathcal{F}in}(G) \to pt$ is an isomorphism. We shall explain the constructions above in 14.4.

(iv) **Fibered isomorphism conjecture** A group G together with a family of subgroups $\mathcal{F}$ is said to satisfy the isomorphism conjecture (in the range $\leq N$) if the projection map $pr : E_{\mathcal{F}}G \to \{pt\}$ induces an isomorphism $\mathcal{H}_n^G(pr) : \mathcal{H}_n^G(E_{\mathcal{F}}G) \xrightarrow{\simeq} \mathcal{H}_n^G(pt)$ for any $n \in \mathbb{Z}$ with $n \leq N$.

The pair $(G, \mathcal{F})$ satisfies the fibered isomorphism conjecture (in the range $\leq N$) if for each group homomorphism $\varphi : H \to G$ the pair $(H, \varphi^*\mathcal{F})$ satisfies the isomorphism conjecture (in the range $\leq N$).

Note that $\varphi^*\mathcal{F} := \{H' \leq H/\varphi(H') \in \mathcal{F}\}$ and that $\varphi : H \to G$ is an inclusion of subgroups; then, $\varphi^*\mathcal{F} = H \cap \mathcal{F} = \{H' \subseteq H | H' \in \mathcal{F}\} = \{K \cap H/K \in \mathcal{F}\}$.

Remarks 14.2.1 Note that if $\varphi : H \to G$ is a group homomorphism and $(G, \mathcal{F})$ satisfies the fibered isomorphism conjecture, then $(H, \varphi^*\mathcal{F})$ also satisfies the fibered isomorphism conjecture.

Next, we state the transitivity principle for equivariant homology as follows.

Theorem 14.2.1 *Let* $\mathcal{F} \subset \mathcal{F}'$ *be two families of subgroups of a discrete group* G, N *an integer. If for every* $H \in \mathcal{F}'$ *and every* $n \leq N$, *the map induced by the projection* $\mathcal{H}_n^H(E_{\mathcal{F} \cap H}(H)) \to \mathcal{H}_n^H(\{pt\})$ *is an isomorphism, then for every* $n \leq N$, *the map* $\mathcal{H}_n^G(E_{\mathcal{F}}(G)) \to \mathcal{H}_n^G(E_{\mathcal{F}'}(G))$ *induced up to* G-*homotopy by the* G-*map* $E_{\mathcal{F}}(G) \to E_{\mathcal{F}'}(G)$ *is an isomorphism.*

PROOF See [138].   ⬚

The above result implies the following transitivity principle for the fibered isomorphism conjecture.

Theorem 14.2.2 [138] (Transitivity principle for fibered isomorphism conjecture) *Let $\mathcal{F} \subset \mathcal{F}'$ be two families of subgroups of G. Suppose that every group $H \in \mathcal{F}'$ satisfies the fibered isomorphism conjecture for $\mathcal{F} \cap H$ in the range $\leq N$. Then $(G, \mathcal{F}')$ satisfies the fibered isomorphism conjecture (in the range $\leq N$), iff $(G, \mathcal{F})$ satisfies the (fibered) isomorphism conjecture in the range $\leq N$.*

PROOF   See [138].   □

Next we have the following result due to A. Bartels and W. Lück (see [14]).

Theorem 14.2.3 [14] *Let $\mathcal{F}$ be a class of finite groups closed under isomorphisms and taking finite subgroups. Assume that every finite group L satisfies the fibered isomorphism conjecture (in the range $\leq N$) with respect to the family $\mathcal{F}(L) = \{K \leq L | K \in \mathcal{F}\}$. Suppose that G is a group. Then, G satisfies the fibered isomorphism conjecture in the range $\leq N$ with respect to the family $\mathcal{F}(G)$ iff G satisfies the fibered isomorphism conjecture in the range $\leq N$ with respect to the family $VCy(G) = \{H \leq G | H \in VCy\}$.*

PROOF   Recall that $V \in VCy$ if either $V \in \mathcal{F}$ or there exists an extension $1 \to \mathbb{Z} \to V \to F \to 1$ with $F \in \mathcal{F}$. Let $\mathcal{F}'$ be the class of V's for which we have such an extension. Now, let $V \in VCy$. In view of theorem 14.2.1 we have to show that V satisfies the fibered isomorphism conjecture in the range $\leq N$ for the family $\mathcal{F}'(V) = \mathcal{F}'(G) \cap V$. If $V \in \mathcal{F}$, then the result follows from the hypothesis. Now, suppose we can write $1 \to \mathbb{Z} \to V \xrightarrow{p} F \to 1$ where $F \in \mathcal{F}$. Since F satisfies the fibered isomorphism conjecture in the range $\leq N$, it follows that V satisfies the fibered isomorphism conjecture (in the range $\leq N$) for $p^*\mathcal{F}$ by remarks 14.2.1. Now, $p^*\mathcal{F}(F) \subset \mathcal{F}'(V)$. By applying lemma 14.2.8 below, we see that V satisfies fibered isomorphism conjecture (in the range $\leq N$) for $\mathcal{F}'(V)$.   □

Lemma 14.2.1 *Let $\mathcal{F} \subset \mathcal{F}'$ be families of subgroups of a discrete group G. Suppose that G satisfies fibered isomorphism conjecture (in the range $n \leq N$) for the family $\mathcal{F}$. Then, G satisfies the fibered isomorphism conjecture in the range $n \leq N$ for the family $\mathcal{F}'$.*

The next result applies induction theory, earlier extensively discussed in chapter 9, for finite groups in the context of Mackey and Green functors, to G-homology theories. This application is due to A. Bartels and W. Lück, see [14].

Theorem 14.2.4 Induction criteria for G-homology theories

Let G be a finite group, $\mathcal{F}$ a family of subgroups of G, A a commutative ring with identity, and $\mathcal{H}_n^G$ a G-homology theory with values in A-$\mathcal{M}$od. Suppose that the following conditions are satisfied:

(1) *There exists a Green functor $\mathcal{G}$: GSet $\to$ A-$\mathcal{M}$od such that $\bigoplus\limits_{H \in \mathcal{F}} \mathcal{G}(pr_H)$: $\bigoplus\limits_{H \in \mathcal{F}} \mathcal{G}(G/H) \to \mathcal{G}(G/G)$ is surjective where pr_H : $G/H \to G/G$ is the projectiion.*

(2) *For every $n \in \mathbb{Z}$, there is a (left) $\mathcal{G}$-module M such that the covariant functor M_* : GSets $\to$ A-$\mathcal{M}$od is naturally equivalent to the covariant functor $\mathcal{H}_n^G$: GSets $\to$ A-$\mathcal{M}$od : $S \to \mathcal{H}_n^G(S)$.*

   Then, the projection $pr : E_\mathcal{F}(G) \to G/G$ induces for all $n \in \mathbb{Z}$ an A-isomorphism $\mathcal{H}_n^G(pr) : \mathcal{H}_n^G(E_\mathcal{F}(G/?)) \xrightarrow{\sim} \mathcal{H}_n^G(G/G)$, and the canonical map $\operatorname*{colim}\limits_{Or_\mathcal{F}(G)} \mathcal{H}_n^G(G/?) \cong \mathcal{H}_n^G(G/G)$ is bijective.

The proof of theorem 14.2.4 makes use of the following lemma.

Lemma 14.2.2 [14] *Let G be a finite group and M a Mackey functor GSet $\to$ A-$\mathcal{M}$od, and S a finite non-empty G-set. Assume that M is S-projective and let $\mathcal{F}(S)$ be the family of subgroups $H \subseteq G$ such that $S^H \neq \emptyset$. Let $\underline{A}_{\mathcal{F}(S)}$ be the contravariant functor $Or(G) \to$ A-$\mathcal{M}$od given by*

$$G/H \to \begin{cases} A \text{ if } H \in \mathcal{F}(S) \\ 0 \text{ if } H \notin \mathcal{F}(S) \end{cases}, \text{ and which takes a morphism to either id} : A \to A$$

or the zero map.

Then there are natural A-isomorphisms

$$Tor_p^{AOr_\mathcal{F}(G)}(\underline{A}_{\mathcal{F}(S)}, M) \cong \begin{cases} M(G/G) & p = 0 \\ 0 & p \geq 1 \end{cases}$$

$$Ext^p_{AOr_\mathcal{F}(G)}(\underline{A}_{\mathcal{F}(S)}, M) \xrightarrow{\sim} \begin{cases} M(G/G) & p = 0 \\ 0 & p \geq 1 \end{cases}$$

PROOF See [14]. ⬜

Proof of theorem 14.2.4. Let $S = \bigcup\limits_{H \in \mathcal{F}} G/H$. Then, it follows from condition (1), proposition 9.1.1 and theorem 9.3.1 that M is S-projective. We deduce from the second condition and lemma 14.2.2 that there is a canonical A-isomorphism

$$Tor_p^{AOr_\mathcal{F}(G)}(\underline{A}_{\mathcal{F}(S)}, \mathcal{H}_q^G(G/?)) \xrightarrow{\sim} \begin{cases} \mathcal{H}_q^G(G/G) & p = 0 \\ 0 & p \geq 1 \end{cases}$$

for all $q \in \mathbb{Z}$. The cellular $AOr(G)$-chain complex of $E_\mathcal{F}(G)$ is a projective $AOr(G)$-resolution of $\underline{A}_{\mathcal{F}(S)}$. Hence, $Tor_p^{AOr_\mathcal{F}(G)}(\underline{A}_{\mathcal{F}(S)}, \mathcal{H}_q^G(G/?))$ agrees with the Bredon homology $H_p^{AOr(G)}(E_{\mathcal{F}}G; \mathcal{H}_q^G(G/?))$. But this is exactly the E^2-term in the equivariant Atiyah - Hirzebruch spectral sequence, which converges to $\mathcal{H}_{p+q}^G(E_{\mathcal{F}}G)$. Hence, the spectral sequence yields an isomorphism $\mathcal{H}_n^G(E_{\mathcal{F}}(G)) = \mathcal{H}_n^G(G/G)$ for all $n \in \mathbb{Z}$, and this isomorphism can easily be identified with the A-map $\mathcal{H}_n^G(pr)$. There is also a natural identification

$$\mathrm{Tor}_*^{AOr_{\mathcal{F}}(G)}(\underline{A}_{\mathcal{F}(S)}, \mathcal{H}_q^G(G/?)) \xrightarrow{\sim} \operatorname*{colim}_{Or_{\mathcal{F}}(G)} \mathcal{H}_q^G(G/?).$$

Hence the result.

14.3   Farrell - Jones conjecture for algebraic K-theory

14.3.1 In $(4.5)^A$ and 14.2.3(ii) we already introduced the Farrell - Jones (F/J) conjecture, namely that if R is a ring with identity and $\underline{K}R$ the K-theory spectrum of R, then for all $n \in \mathbb{Z}$,

$$\mathcal{H}_n^G(E_{V_{C_y}}(G), \underline{K}R) \to \mathcal{H}_n^G(pt, \underline{K}R) \simeq K_n(RG)$$

is an isomorphism. Note that $\underline{K}R$ is the non-connective K-theory spectrum such that $\pi_n(\underline{K}R) = \mathrm{Quillen}\ K_n(R), n \geq 0$, and $\pi_n(\underline{K}R) = \mathrm{Bass}\ K_n R$, for $n \leq 0$.

First, we state the following result, which says that for some regular rings R, F/J conjecture can be stated with Fin replacing V_{C_y}.

Theorem 14.3.1 *Let R be a regular ring in which the orders of all finite subgroups of G are invertible. Then, for every $n \in \mathbb{Z}$, the relative assembly map $A_{\mathrm{Fin} \to V_{C_y}}$ for algebraic K-theory, namely*

$$A_{\mathrm{Fin} \to V_{C_y}} : \mathcal{H}_n^G(E_{\mathrm{Fin}}(G), \underline{K}R) \to \mathcal{H}_n^G(E_{V_{C_y}}(G), \underline{K}R)$$

is an isomorphism.

In particular, if R is a regular ring that is a $\mathbb{Q}$-algebra (e.g., a field of characteristic zero), then $A_{\mathrm{Fin} \to V_{C_y}}$ is an isomorphism for all groups G.

PROOF   See [138], Proposition 2.14.   ☐

14.3.2 Next, we indicate how to see that, for every $n \in \mathbb{Z}$ and any group G, if the Farrell - Jones conjecture holds, then $H_n^G(E_{\mathrm{Fin}}(G), \underline{K}R)$ is a direct summand of $K_n(RG)$.

First, we have the following

Theorem 14.3.2 [138] *For any group G and any ring R as well as any $n \in \mathbb{Z}$, the relative assembly map*

$$A_{\text{Fin} \to V_{C_y}} : \mathcal{H}_n^G(E_{\text{Fin}}(G), \underline{K}R) \to \mathcal{H}_n^G(E_{V_{C_y}}(G), \underline{K}R)$$

is split surjective.

Hence, if the Farrell - Jones conjecture $\mathcal{H}_n^G(E_{\text{Fin}}(G), \underline{K}R) \xrightarrow{\sim} \mathcal{H}_n^G(pt, \underline{K}R) \simeq K_n(RG)$ holds, then $\mathcal{H}_n^G(E_{\text{Fin}}(G), \underline{K}R)$ is a direct summand of $K_n(RG)$, i.e., $K_n(RG) \simeq \mathcal{H}_n^G(E_{\text{Fin}}(G), \underline{K}R) \oplus ?$ where $? = \mathcal{H}_n^G(E_{V_{C_y}}(G), E_{Fin}(G), \underline{K}R)$.

PROOF   See [138].   ∎

The next results provide more information on "?".

Theorem 14.3.3   (i) $\mathcal{H}_n^G(E_{V_{C_y}}(G), E_{\text{Fin}}(G), \underline{K}\mathbb{Z}) = 0$ *for $n < 0$.*

(ii) $\mathcal{H}_n^G(E_{V_{C_y}}(G), E_{\text{Fin}}(G), \underline{K}\mathbb{Z}) \underset{\mathbb{Z}}{\otimes} Q = 0$ *for $n \in \mathbb{Z}$.*

Hence, the Farrell - Jones conjecture for algebraic K-theory predicts that $A_{\text{Fin}} : \mathcal{H}_n^G(E_{\text{Fin}}(G), \underline{K}\mathbb{Z}) \otimes \mathbb{Q} \simeq K_n(\mathbb{Z}G) \underset{\mathbb{Z}}{\otimes} Q$ is always an isomorphism.

Remark 14.3.1   (i) Note that $\mathcal{H}_n^G(E_{V_{C_y}}(G), E_{\text{Fin}}(G), \underline{K}R)$ plays the role of the Nil groups for a general group.

The higher Nil groups have been shown to vanish rationally in 7.5.

(ii) Our next aim is to provide a brief discussion of induction theory that helps to reduce the family V_{C_y} to smaller families in the statements of the Farrell - Jones conjecture for Algebraic K-theory. These results are due to A. Bartels and W. Lück (see [14]).

14.3.3 Recall from chapter 9 that for a prime p, a finite group is said to be p-elementary if $G \simeq C \times P$ where C is a cyclic group and P is a p-group such that $(|C|, p) = 1$.

A finite group G is said to be p-hyperelementary if G can be written as an extension $1 \to C \to G \to P \to 1$ where C is a cyclic group and P is a p-group such that $(|C|, p) = 1$. Say that G is elementary or hyperelementary if it is p-elementary or p-hyperelementary for some p. Let $\mathcal{E}_p$ (resp. $\mathcal{H}_p$) be the class of p-elementary (resp. p-hyperelementary) subgroups of G and $\mathcal{E}$ (resp. $\mathcal{H}$) for the set of elementary (resp. hyperelementary) subgroups of G.

If $\mathcal{F} \subset \text{Fin}$, we denote by $\mathcal{F}' \subseteq V_{C_y}$ the class of subgroups V for which there exists an extension $1 \to \mathbb{Z} \to V \to H \to 1$ for a group $H \in \mathcal{F}$, or for which $V \in \mathcal{F}$ holds.

We now state the Induction theorem for algebraic K-theory.

Theorem 14.3.4 *Let G be a group and N an integer. Then,*

(i) G satisfies the (fibered) isomorphic conjecture (in the range $\leq N$) for algebraic K-theory with coefficients in R for the family V_{C_y} iff G satisfies the fibered isomorphic conjecture (in the range $\leq N$) for algebraic K-theory with coefficients in R for the family $\mathcal{H}'$.

(ii) Let p be a prime; then G satisfies the (fibered) isomorphic conjecture (in the range $\leq N$) for algebraic K-theory with coefficients in R for the family V_{C_y} after applying $\mathbb{Z}_{(p)}\otimes_{\mathbb{Z}}$ iff G satisfies the fibered isomorphic conjecture (in the range $\leq N$) for algebraic K-theory with coefficients in R for the family $\mathcal{H}'_p$ after applying $\mathbb{Z}_{(p)}\otimes_{\mathbb{Z}}$.

(iii) Suppose that R is regular, and $\mathbb{Q} \subseteq R$. Then, the group G satisfies the isomorphism conjecture (in the range $\leq N$) for algebraic K-theory with coefficients in R for the family V_{C_y} iff G satisfies the isomorphism conjecture (in the range $\leq N$) for algebraic K-theory with coefficients in R for the family $\mathcal{H}$.

If we assume that R is regular, and $\mathbb{C} \subseteq R$, then we can replace $\mathcal{H}$ by $\mathcal{E}$.

(iv) Suppose that R is regular and $\mathbb{Q} \subset R$. Let p be a prime. Then, G satisfies the isomorphism conjecture (in the range $\leq N$) for algebraic K-theory with coefficients in R for the family V_{C_y} after applying $\mathbb{Z}_{(p)}\otimes_{\mathbb{Z}}$-iff G satisfies the isomorphism conjecture (in the range $\leq N$) for algebraic K-theory with coefficients in R for the family $\mathcal{H}_p$ after applying $\mathbb{Z}_{(p)}\otimes_{\mathbb{Z}}$.

If we assume that R is regular, and $\mathbb{C} \subset R$, then we can replace $\mathcal{H}_p$ by $\mathcal{E}_p$.

Remark 14.3.2 The proof of the above result uses the following lemma 14.3.1, which we shall state to show the connections to G-homology for discrete groups. Before stating the lemma, we make the following observation. Let A be a commutative ring with identity such that A is as flat as $\mathbb{Z}$-module, or equivalently, A is torsion free as an Abelian group. Let $\varphi : G \to G_1$ be a group homomorphism. Recall from 9.3.4 (ii) that if X is a G_1-CW-complex, then $X|_G$ is a G-CW-complex obtained from X by restricting the action of G_1 on X to G via φ. Then, $X \to A \otimes_{\mathbb{Z}} \mathcal{H}^G_\alpha(X|_G, \underline{K}R)$ defines a G_1-homology theory. We shall write $\varphi^* X$ for $X|_G$.

Lemma 14.3.1 Let $\mathcal{F}$ be a class of finite groups closed under isomorphism and taking subgroups. Let $\varphi : G \to G_1$ be a group homomorphism with G_1 finite. Then the G_1-homology theory $A \otimes_{\mathbb{Z}} \mathcal{H}^G_\alpha(\varphi^*_-; \underline{K}R)$ satisfies the assumptions in theorem 14.2.4 for the familiy $\mathcal{F}(G_1) = \{h \geq G_1 | H \in \mathcal{F}\}$ in the following cases:

(i) $\mathcal{F} = $ class $\mathcal{H}$ of hyperelementary groups and $A = \mathbb{Z}$.

(ii) $\mathcal{F} = $ is the class of elementary groups and $A = \mathbb{Z}$ provided $\mathbb{C} \subset R$.

(iii) *For a given prime p, the family $\mathcal{F}$ is the class $\mathcal{H}_p$ of p-hyperelementary groups and $\Lambda = \mathbb{Z}_{(p)}$.*

(iv) *For a given prime p, the family $\mathcal{F}$ is the class $\mathcal{E}_p$ of p-elementary groups and $A = \mathbb{Z}_{(p)}$ provided $\mathbb{C} \subseteq R$.*

(v) *$\mathcal{F}$ is the class FC_y of finite cyclic groups and $A = Q$.*

Proof of lemma 14.3.1 Recall from 1.4.1(i) that for a groupoid $\mathcal{G}$, $Sw^f(\mathcal{G})$ has the structure of a commutative ring with identity. Also, since $\mathcal{G}$ is an EI category, it follows from the discussion of pairings and module structures in $(7.5)^E$ that $K_n(R\mathcal{G})$ is a $Sw^f(\mathcal{G})$-module for all $n \geq 0$.

Now, if A is a commutative ring with identity as in lemma 14.3.1, $\varphi : G \to G_1$ a group homomorphism with G_1 finite (G discrete), then $A \otimes Sw^f$ ("-"$_{G_1}$) : $G_1\mathrm{Sets} \to A\text{-}\mathcal{M}od$ is a Green functor and $\Lambda \otimes K_n^{G_1}(\varphi^*-, \mathcal{P}(R))$ a $A \otimes Sw^f$("-")-module.

There is a natural equivalence of covariant functors $G_1\mathrm{Sets} \to \mathbb{Z}\text{-Mod}$ from $K_n^G(\varphi^*(\_), \mathcal{P}(R))$ to $\mathcal{H}_n^G(\varphi^*-, \underline{K}R)$. So, one needs only check that

(i) The map $\underset{H \in \mathcal{H}}{\oplus} Sw^f(H) \to Sw^f(G)$ is surjective.

(ii) The map $\underset{H \in \mathcal{E}}{\oplus} Sw(\mathbb{C}H) \to Sw(\mathbb{C}G)$ is surjective.

(iii) For any given prime p, the map $\underset{H \in \mathcal{H}_p}{\oplus} Sw^f(H)_{(p)} \to Sw^f(G)_{(p)}$ is surjective.

(iv) For a given prime p, the map $\oplus Sw(\mathbb{C}H)_{(p)} \to Sw(\mathbb{C}G)_{(p)}$ is surjective.

(v) The map $\underset{H \in FC_y}{\oplus} \mathbb{Q} \underset{\mathbb{Z}}{\otimes} Sw^f(H) \to \mathbb{Q} \underset{\mathbb{Z}}{\otimes} Sw^f(G)$ is surjective.

Now, (i) and (v) follow from [207] since $Sw^f("-"^{''}_G)$ is isomorphic to $Sw("-"^{''}_G)$. It follows from [207] and [209] that any torsion element in $Sw(G)$ is nilpotent, and so, by (v) and theorem 9.3.1, we have that (iii) follows from [218], 6.3.3. Moreover, (ii) and (iv) follow from [187], theorems 27 and 28.

Remarks 14.3.1 (i) Theorem 14.3.4 is a consequence of lemma 14.3.1 and theorem 14.2.4. See [14] for details.

(ii) We close this section with a brief discussion of Farrell - Jones conjecture for torsion-free groups. The Farrell-Jones conjecture for algebraic K-theory in this case has a simpler form and other interesting consequences. For higher algebraic theory, the conjecture has been verified for torsion-free groups of geometric type (see [16]).

14.3.4 Farrell - Jones conjecture for torsion-free groups Let G be a torsion-free group, R a regular ring. Then, the assembly map $H_n(BG, \underline{K}R) \to K_n(RG)$ is an isomorphism for all $n \in \mathbb{Z}$.

Remarks 14.3.2   (i) Note that for a torsion-free G, the family VC_y of all virtually cyclic subgroups reduces to the family C_{yc} of all cyclic subgroups.

(ii) The next result provides an example of torsion-free groups for which Farrell - Jones conjecture for higher K-theory has been verified.

Theorem 14.3.5 [16] *Let R be an associative ring with identity, G the fundamental group of a closed Riemannian manifold with strictly negative curvature. Then, for all $n \in \mathbb{Z}$, the assembly map $A_{C_{yc}} : \mathcal{H}_n^G\left(E_{C_{yc}} ; \underline{K}R\right) \to K_n(RG)$ induced by the projection $E_{C_{yc}} \to pt$ is an isomorphism.*

PROOF   See [16] theorem 1.4.   ▯

Corollary 14.3.1 [16] Let R be an associative ring with identity, G the fundamental group of a closed Riemannian manifold with strictly negative sectional curvature. Then, for all $n \in \mathbb{Z}$, the assembly map for NK-groups

$$A_{C_{yc}} : \mathcal{H}_n^G\left(E_{C_{yc}}, \underline{NK}R\right) \to \mathcal{H}_n^G\left(pt, \underline{NK}R\right) \simeq NK_n(RG)$$

is an isomorphism.

PROOF   Follows from the splitting $\underline{K}R[t] \simeq \underline{K}R \vee \underline{NK}R$ and the fact that the isomorphism result for any two of the assembly maps associated to $\underline{K}R$, $\underline{K}R[t]$ and $\underline{NK}R$ implies the isomorphism result for the third.   ▯

14.4   Baum - Connes conjecture

There has been a lot of vigorous research on the Baum - Connes conjecture in the last two decades, and several areas of mathematics, notably K-theory, topology, geometry, and theory of operator algebras have been further enriched through these research efforts.

There are several formulations of the conjecture, and we do intend to review some of them in this section. However, we shall also focus on the formulation of the conjecture through Davis - Lück assembly map and equivariant homology theories that will blend into our earlier themes of a unified treatment of this and Farrell - Jones conjecture to obtain induction-type results.

$(14.4)^A$ Generalities on Baum - Connes conjecture

Definition 14.4.1 *Let G be a discrete group, $\ell^2(G)$ the Hilbert space of a square summable complex valued functions on G, i.e., any element $f \in \ell^2(G)$ can be written in the form $f = \sum\limits_{g \in G} \lambda_g g$, $\lambda_g \in \mathbb{C}$, $\sum\limits_{g \in G} |\lambda_g|^2 < \infty$.*

Let $\mathbb{C}G$ be the complex groupring (a group algebra). We could think of elements of $\mathbb{C}G$ as formal sums $\sum\limits_{g \in G} \lambda_g g$ where all except a finite number of λ_g are zero or complex-valued functions on G with finite support. Hence $\mathbb{C}G$ is a subspace of $\ell^2(G)$.

There is a left regular representation λ_G of G on the space $\ell^2(G)$ given by

$$\lambda_G(g) \cdot \left(\sum_{h \in G} \lambda_h h \right) := \sum_{h \in G} \lambda_h gh, \ g \in G, \ \sum_{h \in G} \lambda_h h \in \ell^2(G).$$

This unitary representation extends linearly to $\mathbb{C}G$.

We now define the reduced C^*-algebra $C^*_r G$ of G as the norm closure of the image $\lambda_G(\mathbb{C}G)$ in the C^*-algebra of bounded operators on $\ell^2(G)$.

(Recall that a C^*-algebra A is a Banach$*$-algebra satisfying $|a^*a| = |a|^2$ for all $a \in A$.)

14.4.1 Let A be a unital C^*-algebra. Then $GL_n(A)$ is a topological group and we have continuous embeddings given by $GL_n(A) \hookrightarrow GL_{n+1}(A) : X \to \left(\begin{smallmatrix} X & 0 \\ 0 & 1 \end{smallmatrix} \right)$. Let $GL(A) = \varinjlim\limits_{n} GL_n(A)$ and give $GL(A)$ the direct-limit topology. We now define the higher topological K-theory of A as follows. For all $n \geq 1$, define $K_n(A) = \pi_{n-1}(GL(A)) = \pi_{n-1}(BGL(A))$.

Remarks/Examples 14.4.1   (i) $K_0(\mathbb{C}) \simeq \mathbb{Z}$ and $K_1(\mathbb{C}) = \pi_0(GL(\mathbb{C})) = 0$ since $GL_n(\mathbb{C})$ is connected for $n \in \mathbb{N}$.

(ii) $K_n(A)$ satisfies Bott periodicity, i.e., there is a natural isomorphism $K_n(A) \simeq K_{n+2}(A)$ for all $n \geq 0$.

Hence, K-theory of C^*-algebras is a $\mathbb{Z}/2$-graded theory.

(iii) If G is a finite group, then $C^*_r G \simeq \mathbb{C}G$. Hence, $K_0(C^*_r G) \simeq R(G)$, the additive group of the complex representation ring of G.

14.4.2 For a CW-complex X, let $K_*(X)$ be the K-homology of X, i.e., the homology theory associated to the topological complex K-theory spectrum $\underline{K}^{\text{top}}$ (usually denoted by BU). This homology theory is dual to topological K-theory and $K_*(X) = H_*(X, \underline{K}^{\text{top}})$. Note that complex K-homology theory is 2-periodic, i.e., $K_n(X) \cong K_{n+2}(X)$.

If G is a discrete group, and X_+ is X with a disjoint base point, then $K_n(X) =$

$\pi_n(X_+ \wedge \mathrm{BU}$. There exists an assembly map $K_*(BG) \to K_*(C_r^*G)$ (see [24, 138]).

14.4.3 Baum - Connes conjecture for torsion-free groups If G is a torsion-free group, then the assembly map $\bar{\mu}^* : K_*(BG) \to K_*(C_r^*G)$ is an isomorphism.

Remarks/Examples 14.4.2 (i) The requirement that G be torsion free in the conjecture 14.4.3 is essential. Indeed, $\bar{\mu}^*$ is not an isomorphism if G is not torsion free. For, suppose $G = \mathbb{Z}/2$, $C_r^* \simeq \mathbb{C}G \simeq \mathbb{C} \oplus \mathbb{C}$ to be a $\mathbb{C}$-algebra.

Hence

$$K_0(C_r^*G) \cong K_0(\mathbb{C}) \oplus K_0(\mathbb{C}) \simeq \mathbb{Z} \oplus \mathbb{Z}. \tag{I}$$

Also, since rational homology of finite groups are zero in positive degrees, we have, by using homological algebra, that

$$K_0(BG) \underset{\mathbb{Z}}{\otimes} Q \simeq \oplus limits_{n=0}^{\infty} H_{2n}(BG; Q) \simeq Q. \tag{II}$$

It follows from (I) and (II) that $\bar{\mu}^*$ cannot be an isomorphism.

(ii) If G is not torsion free, we shall write $K_n^G(X)$ for G-equivariant K-homology of a topological space X. It is also well known (see [24, 138]) that there is an assembly map $\mu_* : K_*^G(E_{\mathrm{Fin}}(G)) \to K_*(C_r^*G)$. We now state the conjecture for a general discrete group.

14.4.4 Baum - Connes conjecture for a discrete group Let G be a discrete group. Then, the assembly map $\mu_* : K_*^G(E_{\mathrm{Fin}}(G)) \to K_*(C_r^*G)$ is an isomorphism.

Remarks 14.4.1 (i) Note that if G is torsion free, then $K_*(BG) = K_*^G(E_{\mathrm{Fin}}(G))$ and conjectures 14.4.3 and 14.4.4 coincide.

(ii) We also have a real version of the Baum - Connes conjecture where the K-homology is replaced by KO-homology, and on the right-hand side, C_r^* is replaced by the real reduced C^*-algebra $C_{r,\mathbb{R}}^*G$.

(iii) There is a connection between Baum - Connes conjecture and idempotents in C_r^*G leading to the so-called trace conjecture and Kadison conjecture, etc.

Let M be a closed manifold and $D : C^\infty(E) \to C^\infty(F)$ an elliptic differential operator between two bundles on M. Let $\widetilde{M} \to M$ be a normal covering of M with deck transformation group G (see [138, 156]). Then we can lift D to $\widetilde{M}$ and obtain an elliptic G-equivariant differential operator

$\tilde{D} : C^\infty(\tilde{E}) \to C^\infty(\tilde{F})$. Since the action is free, we can define an analytic index $\mathrm{ind}_G(\tilde{D}) \in K_{\dim M}(C_r^*G)$.

If G is torsion free, one can show (see [138, 156]) that the image of μ_0 (in 14.4.4) coincides with the subset of $K_0(C_r^*G)$ consisting of $\mathrm{ind}_G(\tilde{D})$. So, if μ_0 is surjective, then, for each $x \in K_0(C_r^*G)$ there exists a differential operator D such that $x = \mathrm{ind}_G(\tilde{D})$. Hence, $\dim_G(x) \in \mathbb{Z}$. Hence, we have

14.4.5 Trace conjecture Let G be a torsion-free discrete group. Then, $\dim_G(K_0(C_r^*(G)) \subset \mathbb{Z}$.

We also have the Kadison conjecture as follows.

14.4.6 Kadison conjecture If G is a torsion-free discrete group, then the only idempotent elements in C_r^*G are 0 and 1.

The next result provides a connection between the trace conjecture and the Kadison conjecture.

Lemma 14.4.1 [138] *The trace conjecture 14.4.5 implies the Kadison conjecture.*

PROOF   See [138]. ⬚

Next we state the idempotent conjecture.

Conjecture 14.4.1 — idempotent conjecture Let G be a torsion-free group and R an integral domain. Then, the only idempotents in RG are 0 and 1.

Note that if $R \subseteq \mathbb{C}$, then the Kadison conjecture 14.4.6 implies the idempotent conjecture.

14.4.7 The functor KK^G and the category $\underline{KK}^G$

Before discussing analytic and other formulations of the Baum - Connes conjecture, we briefly introduce the functor KK^G and the category $\underline{KK}^G$. We shall focus on discrete groups G but note that similar constructions hold for a locally compact group G.

(i) Let B be a C^*-algebra, $\mathcal{E}$ a right B-module. A B-valued scalar product on $\mathcal{E}$ is a B-bilinear map $<,>_B \colon \mathcal{E} \times \mathcal{E} \to B$ such that $< x, y >_B = < y, x >^*$ and $< x, x >_B \geq 0$ ($< x, x >_B = 0$ iff $x = 0$).

If $\mathcal{E}$ is complete with respect to the norm $\|x\|^2 = |< x, x >_B|$, call $\mathcal{E}$ a Hilbert C^*-module over B or just Hilbert B-module.

(ii) Let $\mathcal{E}$ be a Hilbert B-module, $L(\mathcal{E})$ the set of operators $T : \mathcal{E} \to \mathcal{E}$ having an adjoint operator $T^* : \mathcal{E} \to \mathcal{E}$ such that $< Tx, y >_B = < x, T^*y >_B$ for every $x, y \in \mathcal{E}$). An operator in $L(\mathcal{E})$ of the form $\sum \theta_{y_i, z_i} : x \to \sum\limits_{i=1}^{n} < x, y_i > z_i$, $y_i, z_i \in \mathcal{E}$ is called a finite-rank operator. An operator in $L(\mathcal{E})$ is said to be compact if it is a norm limit of a sequence of finite-rank operators.

Let $B_0(\mathcal{E})$ denote the set of compact operators in $L(\mathcal{E})$.

(iii) Let G be a discrete group. A G-C^*-algebra is a C^*-algebra A endowed with an action of G by $*$-automorphisms.

Let A, B be G-C^*-algebras. A cycle over (A, B) is a triple (U, π, F) where

(a) U is a representation of G on some Hilbert B-module $\mathcal{E}$ such that U is unitary, i.e., $< U(g)\xi|U(g)\eta >_B = g < \xi|\eta >_B$ for all $g \in G, \xi, \eta \in \mathcal{E}$.

(b) π is a representation of A on $\mathcal{E}$ such that $< \pi(a)\xi|\eta >_B = < \xi|\pi(a^*)\eta >_B$ for all $a \in G$ and π is covariant, i.e., $U(g)\pi(a)U(g^{-1}) = \pi(ga)$ for all $g \in G, a \in A$.

(c) F is an operator on $\mathcal{E}$, self adjoint with B-valued scalar product $< F(\xi)|\eta >_B = < \xi|F(\eta) >_B$. Moreover, $\pi(F^2 - 1), [\pi(a), F], [U(g), F]$ $a \in A, g \in G$ are compact.

A cycle (U, π, F) is even if $\mathcal{E}$ is $\mathbb{Z}_2$-graded, U, π preserve grading, and F reverses it, i.e., if $\mathcal{E} = \mathcal{E}_0 \oplus \mathcal{E}_1$, $U = \left(\begin{smallmatrix} u_0 & 0 \\ 0 & u_1 \end{smallmatrix} \right) \pi = \left(\begin{smallmatrix} \pi_0 & 0 \\ 0 & \pi_1 \end{smallmatrix} \right)$, $F = \left(\begin{smallmatrix} 0 & p^* \\ p & 0 \end{smallmatrix} \right)$. A cycle is odd otherwise.

A cycle $\alpha = (U, \pi, F)$ is degenerate if for all $a \in A$ $[\pi(a), F], \pi(a)[F^2 - 1], [U(g), F] = 0$.

Two cycles $\alpha_0 = (U_0, \pi_0, F_0)$ and $\alpha_1 = (U_1, \pi_1, F_1)$ are said to be homotopic if $U_0 = U_1$, $\pi_0 = \pi_1$, and there exists a norm continuous path $(F_t)_{t \in [0,1]}$ connecting F_0 and F_1.

Two cycles α_0, α_1 are said to be equivalent (written $\alpha_0 \sim \alpha_1$) if there exist two generate cycles β_0, β_1 such that $\alpha_0 \oplus \beta_0$ is homotopic to $\alpha_1 \oplus \beta_1$.

Now, define $KK_0^G(A, B) :=$ set of equivalence classes of even cycles $KK_1^G(A, B) :=$ set of equivalence classes of odd cycles. Note that $KK_0^G(A, B)$ and $KK_1^G(A, B)$ are Abelian groups. $KK_i^G(A, \mathbb{C}) = K_i^G(A)$ is equivariant K-homology of A; $KK_i^G(\mathbb{C}, B) = K_i(B) := K$-theory of B.

If G is the trivial group, we write $KK_i(A, B)$ $i = 0, 1$.

(iv) $KK_i^G(-, -)$ is a bivariant functor contravariant in A and covariant in B.

Proof. If $\alpha = (U, \pi, F) \in KK_i^G(A, B)$, $\theta : C \to A$, then $\theta^* \alpha = (U, \pi\theta, F) \in KK_i^G(C, B)$.

If $\theta : B \to C$, let $\mathcal{E} \otimes_B C$ be a Hilbert C-module, $\alpha \in KK_i^G(A, B)$; then, $\theta_* \alpha = (U \otimes 1, \pi \otimes 1, F \otimes 1) \in KK_i^G(A, C)$.

(v) Let $\underline{KK}^G$ be the category obtained as follows: $\mathrm{ob}\underline{KK}^G$ consists of separable G-C^*-algebras. Morphisms set $\underline{KK}^G(A, B)$ from A to B consists of elements of $KK_0^G(A, B)$. In [151], R. Meyer and R. Nest show that $\underline{KK}^G$ is a triangulated category.

An object $A \in \underline{KK}^G$ is said to be compactly induced if it is $\underline{KK}^G$-equivalent to $\mathrm{Ind}_H^G A'$ for some finite subgroup $H \leq G$ and some $H - C^*$-algebra A'. Let $\mathcal{CI} \subseteq \underline{KK}^G$ be the full subcategory of compactly induced objects, and $(\mathcal{CI})$ is the localizing subcategory generated by $\mathcal{CI}$. Define $\mathcal{CI}$-approximation $\overline{A}$ of A in $\underline{KK}^G$ as a morphism in $\overline{A} \to A$ where $\overline{A} \in (\mathcal{CI})$ such that $\underline{KK}^G(P, \overline{A}) \simeq \underline{KK}^G(P, A)$ for all $P \in (\mathcal{CI})$. In [151], Meyer and Nest used this setup to define Baum - Connes conjecture via localization of functors from $\underline{KK}^G$ to an Abelian category.

14.4.8 Before closing this subsection, we discuss briefly the analytic (classical) formulation of the Baum - Connes conjecture. So, let G be a locally compact group or discrete group G, X is a locally compact Haussdorff G-space X such that X is proper and G-compact.

We shall write KK^G for the Kasparov's equivariant KK-functor (see [97]). Note that $KK^G(C_0(X), \mathbb{C})$ can be identified with the G-equivariant K-homology of X and that there are canonical maps $\mu_i^r : KK_i^G(C_0(X), \mathbb{C}) \to K_i(C_r^*(G))$ where $C_0(X)$ is the C^*-algebra of continuous complex-valued functions on X vanishing at infinity.

Let $\underline{EG}$ be the universal space for proper actions of G. Note that we had earlier identified $\underline{EG}$ with $E_{\mathrm{Fin}}(G)$. The equivariant K-homology of $\underline{EG}$, $RK_i^G(\underline{EG})$, $i = 0, 1$, is defined as the inductive limit of $KK_i^G(C_0(X), \mathbb{C})$ where X varies over all possible locally compact G-proper and G-compact subsets of $\underline{EG}$.

Since KK_i^G and K_i commute with the process of taking inductive limits, one could now define

$$\mu_i^r : RK_i^G(\underline{EG}) \to K_i(C_r^*G), \qquad i = 0, 1 \tag{I}$$

and the Baum - Connes conjecture says that this μ_r is an isomorphism.

Remarks 14.4.2 (i) Note that when G is discrete, the formulation of the BC-conjecture in 14.4.4 is equivalent to that in 14.4.8.

(ii) **Groups for which the conjecture is known to be true**
The conjecture is known to be true for the following classes of groups

(a) All amenable groups. These include all finite groups, all Abelian, nilpotent, and solvable groups. Note that the class of all amenable groups is closed under taking subgroups, quotients, extensions, and direct unions.

Recall that a finitely generated discrete group G is called amenable if for any given finite set S of generators (such that $1 \in S$ and $s \in S \Rightarrow s^{-1} \in S$), there exists a sequence of finite subsets X_i of G such that

$$\frac{|SX_j = \{sx|s \in S, x \in X\}|}{|X_j|} \xrightarrow{j \to \infty} 1.$$

An arbitrary discrete group is called amenable if each finitely generated subgroup is amenable.

(b) One relator groups, i.e., groups with a presentation $G =< g_1, \ldots g_r | r >$ with only one relation r.

(c) Groups with the Haagerup property.

These are groups which admit an isometric action on some affine Hilbert H-space that is proper, i.e., such that $g_n V \xrightarrow{n \to \infty} \infty$ for every $v \in H$ whenever $g_n \xrightarrow{n \to \infty} \infty$ in G. Such groups include amenable groups, Coxeter groups, groups acting on trees, etc. (see [138, 156]).

(d) Discrete subgroups of $\mathrm{Sp}(n, 1)$, $\mathrm{SO}(a, 1)$, and $\mathrm{SU}(n, 1)$.

(e) G a subgroup of a word hyperbolic group.

(f) Artin full braid group.

(iii) **Baum - Connes conjecture with coefficients**

There is also Baum - Connes conjecture with coefficients. Let A be a C^*-algebra on which a discrete group G acts by automorphisms. Let $C_c(G, A)$ be the space of finitely supported functions $f : G \to A$. For $f, g \in C_c(G, A)$, say $f = \sum_{s \in G} f(s)s, g = \sum_{t \in G} g(t)t$ define twisted convolution by $f *_\alpha g = \sum_{s,t \in G} f(s)\alpha_s(gt)st$ where $\alpha : G \to Aut(A)$. For each $t \in G$, we have $(f \times_\alpha g)(t) = \sum_{s \in G} f(s)\alpha_s(g(s^{-1}t))$. $C_c(G, A)$ is a $*$-algebra whose involution is given by $f^*(s) = \alpha_s(f(s^{-1}))$ for all $f \in C_c(G, A)$, $s \in G$.

Define $\ell^2(G, A) = \{\xi : G \to A | \sum_{s \in G} \xi(s)^*(s)$ converges in $A\}$.

The norm $||\xi|| = || \sum_{s \in G} \xi(s)^*(\xi(s))||_A$ turns $\ell^2(G, A)$ into a Banach space.

The left regular representation $\lambda_{G,A}$ of $C_c(G, A)$ on $\ell^2(G, A)$ is given by $(\lambda_{G,A}(f)\xi)(g) = \sum_{s \in G} \alpha_{g^{-1}}(f(s)\xi(s^{-1}g))$ for each $f \in C_c(G, A)$, $\xi \in$

$\ell^2(G, A)$ $g \in G$, and so, $C_c(G, A)$ acts on $\ell^2(G, A)$ by bounded operators.

Definition *The reduced crossed product* $A \rtimes_r G$ *is the operator norm closure of* $\lambda_{GA}(C_c(G, A))$ *in* $B(\ell^2(G, A))$.

Let A be a countable G-C^*-algebra. Then there is an assembly map

$$KK_n^G(E_{\text{Fin}}(G), A) \to K_n(A \rtimes_r G) \qquad (\text{I})$$

(see [85, 138]). The Baum - Connes conjecture with coefficients says that the assembly map (I) is an isomorphism.

(iv) **Groups for which Baum - Connes conjecture with coefficients are known to be true**

   (a) Groups with the Haagerup property (or equivalently a-T-menable groups). See (ii)(c).

   (b) **Groups belonging to the class** $LH\mathcal{E}\mathcal{J}\mathcal{H}$

   Note. Let $H\mathcal{T}\mathcal{H}$ be the smallest class of groups that contain all a-T-menable groups, and it contains a group G if there exists a one-dimensional contractable G-CW-complex whose stabilizes belong to $H\mathcal{T}\mathcal{H}$. Let $H\mathcal{E}\mathcal{T}\mathcal{H}$ be the smallest class of groups containing $H\mathcal{T}\mathcal{H}$ and containing a group G if either G is countable and admits a surjective map $p : G \to Q$ such that Q and $p^{-1}(F) \in H\mathcal{E}\mathcal{T}\mathcal{H}$ for every finite subgroup $F \subseteq Q$, or if G admit a 1-dimensional contractable G-CW-complex whose stabilizes belong to $H\mathcal{E}\mathcal{T}\mathcal{H}$. Let $LH\mathcal{E}\mathcal{T}\mathcal{H}$ be the class of groups G whose finitely generated subgroups belong to $H\mathcal{E}\mathcal{T}\mathcal{H}$.

   Note that $LH\mathcal{E}\mathcal{T}\mathcal{H}$ is closed under finite products, passing to subgroups and under extensions with torsion-free quotients. This class includes one-relator groups and all knot groups.

(v) **Counterexample to Baum - Connes conjecture with coefficients**
Some counterexamples to Baum - Connes conjecture with coefficients have been provided by Gromov (see [85, 138]). These examples involve cases of finitely generated groups containing arbitrary large expanders in their Cayley graph.

(vi) **Baum - Connes conjecture for the action of discrete quantum groups**
In [61], D. Goswami and A. Kuku formulated the Baum - Connes conjecture for the action of discrete quantum groups as a generalization of the classical formulation for discrete groups.

More precisely, given an action of a discrete quantum group $\mathcal{A}$ on a C^*-algebra B satisfying certain regularity assumptions (resembling the

action of proper G-compact action of classical discrete groups on some s-
paces), they at first constructed a canonical map $\mu_i, \mu_i^r (i = 0, 1)$ from the
$\mathcal{A}$-equivariant K-homology groups $KK_i^{\mathcal{A}}(B, \mathbb{C})$ to the K-groups $K_i(\widehat{\mathcal{A}})$
and $K_i(\widehat{\mathcal{A}}_r)$, respectively, where $\widehat{\mathcal{A}}, \widehat{\mathcal{A}}_r$ denote, respectively, the quantum analogue of the full and reduced C^*-algebras. They then construct
a direct family $\{\mathcal{E}_F\}$ of C^*-algebras (F varying over some index set) and
show that the natural action of $\mathcal{A}$ on $\mathcal{E}_F$ satisfy certain hypothesis that
makes it possible to define Baum - Connes maps

$$\mu_i : \varinjlim KK_i^{\mathcal{A}}(\mathcal{E}_F, \mathbb{C}) \to KK_i(\mathbb{C}, \widehat{\mathcal{A}})$$

$$\mu_i^r : \varinjlim KK_i^{\mathcal{A}}(\mathcal{E}_F, \mathbb{C}) \to KK_i(\mathbb{C}, \widehat{\mathcal{A}}_r)$$

so that in the classical case when $\mathcal{A} = C_0(G)$, the isomorphism μ_i^r is
equivalent to Baum - Connes conjecture.

They verified the conjecture for finite dimensional quantum groups and
showed that μ_i^r is surjective for the dual of $SU_q(2)$. For details, see [61].

(vii) **Baum - Connes conjecture via localization of categories**
In [151], R. Meyer and R. Nest formulated the Baum - Connes conjecture
via localization of categories as follows, where G is a countable locally
compact group.

Recall from 14.4.14 (v) the category $\underline{KK}^G$ whose objects are separable
G-C^*-algebras and whose morphism set is $KK_0^G(A, B)$ for any two objects $A, B \in KK_0^G$.

As in 14.4.14(v), let $\mathcal{CI} \subseteq \underline{KK}^G$ be the full subcategory of compactly
induced objects and $(\mathcal{CI})$ the localization subcategory generated by $\mathcal{CI}$.
Let $\bar{A} \to A$ be a $\mathcal{CI}$-approximation of $A \in \mathcal{CI}$. Then, $\bar{A} \in (\mathcal{CI})$. If
$F : \underline{KK}^G \to \mathcal{C}$ be any homological functor from $\underline{KK}^G$ to an Abelian
category $\mathcal{C}$, then, the localization LF of F defined by $LF(A) := F(\bar{A})$
is also a homological functor $\underline{KK}^G \to \mathcal{C}$, and LF is equipped with
a natural transformation $LF(A) \to F(A)$. If $F(A) = K_*(A \rtimes_r G)$,
then the map $LF(A) \to F(A)$ is isomorphic to the BC-map. Hence
$K_*^{\text{top}}(G, A) \simeq LF(A)$ (see [151]).

14.5   Davis - Lück assembly map for BC conjecture and its identification with analytic assembly map

14.5.1 The C_r^*-category $C_r^*\mathcal{G}$ ($\mathcal{G}$ a groupoid) Let $\mathcal{C}$ be a category, R
a commutative ring with identity. The R-category associated to $\mathcal{C}$ denoted
by $R\mathcal{C}$ is defined as follows: $\text{ob}(R\mathcal{C}) = \text{ob}(\mathcal{C})$, $\text{Hom}_{R\mathcal{C}}(x, y) = $ free R-module
generated by $\text{Hom}_{\mathcal{C}}(x, y)$.

Let $\mathcal{G}$ be a groupoid, R a commutative ring with identity and with involution. Then $R\mathcal{G}$ has the structure of an R-category with involution by defining $(\sum_{i=1}^{n} \lambda_i f_i)^* := \sum_{i=1}^{n} \bar{\lambda}_i f_i^{-1}$, $\lambda_i \in R$, $f_i \in \text{Mor}\mathcal{G}$.

One can complete the category with involution, $\mathbb{C}\mathcal{G}$, to a C^*-category $C_r^*\mathcal{G}$ defined as follows: $\text{ob}C_r^*\mathcal{G} = \text{ob}\mathcal{G}$. Let x, y be two objects of $\mathcal{G}$. If $\text{Hom}_{\mathcal{G}}(x, y) = \emptyset$, put $\text{Hom}_{C_r^*\mathcal{G}}(x, y) = 0$. If $\text{Hom}_{\mathcal{G}}(x, y) \neq \emptyset$, choose $z \in \text{ob}\mathcal{G}$ such that $\text{Hom}_{\mathcal{G}}(z, x) \neq \emptyset$.

For any set S, let $\ell^2(S)$ be a Hilbert space with S as a basis, and for S_1, S_2, let $B(\ell^2(S_1), \ell^2(S_2))$ be the space of bounded linear operators from $\ell^2(S_1)$ to $\ell^2(S_2)$. Now, define a $\mathbb{C}$-linear map $i_{x,y,z} : \mathbb{C}\text{Hom}_{\mathcal{G}}(x, y) \to B(\ell^2(\text{Hom}_{\mathcal{G}}(z, x)), \ell^2(\text{Hom}_{\mathcal{G}}(z, y)))$, which takes $f \in \text{Hom}_{\mathcal{G}}(x, y)$ to the bounded linear operator from $\ell^2(\text{Hom}_{\mathcal{G}}(z, x))$ to $\text{Hom}_{\mathcal{G}}(z, y))$ given by composition with f.

For $u \in \text{Hom}_{\mathbb{C}\mathcal{G}}(x, y) = \mathbb{C}\text{Hom}_{\mathcal{G}}(x, y)$, define the norm $||u||_{x,y} := ||i_{x,y,z}(u)||$. Then $||u||_{x,y}$ is independent of the choice of z. The Banach space of morphisms in $C_r^*\mathcal{G}$ from x to y is the completion of $\text{Hom}_{\mathbb{C}\mathcal{G}}(x, y)$ with respect to the norm $|| \ ||_{x,y}$. We shall denote the induced norm on the completion $\text{Hom}_{C_r^*\mathcal{G}}(x, y)$ of $\text{Hom}_{\mathbb{C}\mathcal{G}}(x, y)$ also by $|| \ ||_{x,y}$.

Composition defines a $\mathbb{C}$-bilinear map $\text{Hom}_{\mathbb{C}\mathcal{G}}(x, y) \times \text{Hom}_{\mathbb{C}\mathcal{G}}(y, z) \to \text{Hom}_{\mathbb{C}\mathcal{G}}(x, z)$ satisfying $||g \cdot f||_{x,z} \leq ||g||_{y,z}||f||_{x,y}$. Hence it induces a map on the completions

$$\text{Hom}_{C_r^*\mathcal{G}}(x, y) \times \text{Hom}_{C_r^*(\mathcal{G})}(y, z) \to \text{Hom}_{C_r^*\mathcal{G}}(x, z).$$

If G is a group, then G defines a groupoid $\mathcal{G}$ with one object, and $C_r^*\mathcal{G}$ is the reduced group C^*-algebra C_r^*G.

Let C^*-$\mathcal{C}at$ be the category of small C^*-categories. We now have a functor $C_r^* : \mathcal{G}poids^{inj} \to C^*$-$\mathcal{C}at : \mathcal{G} \to C_r^*\mathcal{G}$.

If $F : \mathcal{G}_0 \to \mathcal{G}_1$ is faithful functor of groupoids, then F guarantees that the map $\text{Hom}_{\mathbb{C}\mathcal{G}_0}(x, y) \to \text{Hom}_{\mathbb{C}\mathcal{G}_1}(F(x), F(y))$ extends to $\text{Hom}_{C_r^*\mathcal{G}_0}(x, y) \to \text{Hom}_{C_r^*\mathcal{G}_1}(F(x), F(y))$ for all $x, y \in \text{ob}(\mathcal{G}_0)$.

Note that, in general, the assignment of the C^*-algebra C_r^*H to a group H cannot be extended to a functor from the category of groups to the category of C^*-algebras. For example, the reduced C^*-algebra $C_r^*(\mathbb{Z} \times \mathbb{Z})$ of the free group on two letters is simple and hence has no C^*-homomorphism to the reduced C^*-algebra $\mathbb{C}$ of the trivial group.

14.5.2 The Ω-spectrum $\mathbb{K}^{\text{top}}(\mathcal{C})$, $\mathcal{C}$ a C_r^*-category

(i) Let $\mathcal{C}$ be an R-category. Define a new R-category $\mathcal{C}_\oplus$, called the symmetric monoidal R-category associated to $\mathcal{C}$, with an associative and commutative sum "$\oplus$" as follows: $\text{ob}(\mathcal{C}_\oplus)$ are n-tuples $\underline{x} = (x_1, \ldots, x_n)$, $x_i \in \text{ob } \mathcal{C}$ $n = 0, 1, 2, \ldots$. For $\underline{x} = (x_1, x_2, \ldots, x_n)$ and $\underline{y} = (y_1, \ldots, y_n)$, $\text{Hom}_{\mathcal{C}_\oplus}(\underline{x}, \underline{y}) = \bigoplus_{\substack{1 \leq i \leq n \\ 1 \leq j \leq n}} \text{Hom}_{\mathcal{C}}(x_i, y_i)$. Note that $\text{Hom}_{\mathcal{C}_\oplus}(\underline{x}, \underline{y})$ is an R-module. If $f : \underline{x} \to \underline{y}$ is a morphism in $\mathcal{C}_\oplus$, denote by $f_{ij} : x_i \to y_j$ the

component which belongs to $i \in \{1, \ldots, m\}$ and $j \in \{1, \ldots, n\}$.
The composite of $f : \underline{x} \to \underline{y}$ and $g : \underline{y} \to \underline{z}$ is defined by $(g \cdot f)_{i,k} = \sum_{j=1}^{n} g_{ik} \cdot f_{ij}$. Sum in $\mathcal{C}_{\oplus}$ is defined by $\underline{x} \oplus \underline{y} = (x_1, \ldots, x_m, y_1, \ldots, y_n)$ and satisfies $(\underline{x} \oplus \underline{y}) \oplus \underline{z} = (\underline{x} \oplus (\underline{y} \oplus \underline{z})); \underline{x} \oplus \underline{y} \simeq \underline{y} \oplus \underline{x}$.

(ii) Recall from 4.4 that the idempotent completion $\mathbb{P}(\mathcal{C})$ of a category $\mathcal{C}$ is defined as follows: $\mathrm{ob}(\mathbb{P}(\mathcal{C}))$ consist of pairs (x, p) where $p : x \to x$ is a morphism such that $p^2 = p$. A morphism $f : (x, p) \to (y, q)$ is a morphism $f : x \to y$ such that $qfp = f$. Note that if $\mathcal{C}$ is an R-category (resp. symmetric monoidal R-category), then $\mathbb{P}(\mathcal{C})$ is also an R-category (resp. symmetric monoidal R-category).

Also recall from example 5.3.1(i) that for any category $\mathcal{C}$, ob $(\mathrm{Iso}(\mathcal{C})) = \mathrm{ob}\mathcal{C}$ and morphisms in $\mathrm{Iso}(\mathcal{C})$ are isomorphisms in $\mathcal{C}$.

(iii) Now, let $\mathcal{C}$ be a symmetric monoidal R-category all of whose morphisms are isomorphisms. The group completion of $\mathcal{C}$ is a symmetric monoidal category $\mathcal{C}^{\wedge}$ defined as follows: $\mathrm{ob}(\mathcal{C}^{\wedge}) = \{(x, y) | x, y \in \mathrm{ob}(\mathcal{C})\}$.
A morphism from (x, y) to $(\overline{x}, \overline{y})$ in $\mathcal{C}^{\wedge}$ is given by the equivalence class of triples (z, f, g) where $z \in \mathrm{ob}(\mathcal{C})$ are $f : x \oplus z \to \overline{x}$, $g : y \oplus \overline{z} \to \overline{y}$ are isomorphisms. Say that two of such triples (z, f, g) and (z', f', g') are equivalent if there exists an isomorphism $h : z \to z'$ such that $f' \cdot (id_x \oplus h) = f$, $g' \cdot (id_y \oplus h) = g$. Sum in $\mathcal{C}^{\wedge}$ is given by $(x, y) \oplus (\overline{x}, \overline{y}) = (x \oplus \overline{x}, y \oplus \overline{y})$.

Note that if $\mathcal{C}$ is a C^*-category, then $\mathcal{C}_{\oplus}, \mathbb{P}(\mathcal{C})$ inherit the structure of C^*-categories where each object $p : x \to x$ in $\mathbb{P}(\mathcal{C})$ is self-adjoint and idempotent, i.e., $p^* = p$ and $p^2 = p$. Also, $\mathcal{C}_{\oplus}$, $\mathbb{P}(\mathcal{C}_{\oplus})$ and $\mathrm{Iso}(\mathbb{P}(\mathcal{C}_{\oplus}))^{\wedge}$ inherit the structure of a topological category.

(iv) Now, given topological categories $\mathcal{D}, \mathcal{D}'$, we have a homeomorphism $B(\mathcal{D}, \mathcal{D}') \to B\mathcal{D} \times B\mathcal{D}'$ induced by the projections where for any category $\mathcal{C}$, $B\mathcal{C}$ is the classifying space of $\mathcal{C}$ (see 1.1.8 of [120]).
Hence, for any C^*-category $\mathcal{C}$, we have a map
$B(\mathrm{Iso}(\mathcal{C}_{\oplus}))^{\wedge} \times B(\mathrm{Iso}(\mathbb{P}(\mathcal{C}_{\oplus}))^{\wedge} \to B(\mathrm{Iso}(\mathbb{P}(\mathcal{C}_{\oplus})))^{\wedge}$, which sends $B(\mathrm{Iso}(\mathcal{C}_{\oplus}))^{\wedge} \vee B(\mathrm{Iso}(\mathbb{P}(\mathcal{C}_{\oplus}))^{\wedge}$ to the base point $B\{0\} \subset B(\mathrm{Iso}(\mathbb{P}(\mathcal{C}_{\oplus}))^{\wedge}$.
So, we have a map $\mu : B(\mathrm{Iso}(\mathcal{C}_{\oplus}))^{\wedge} B(\mathrm{Iso}(\mathbb{P}(\mathcal{C}_{\oplus}^{\wedge}) \to B(\mathrm{Iso}(\mathbb{P}(\mathcal{C}_{\oplus}))$.

Remarks. Note that $\mathbb{C}$ can be regarded as a C^*-category with precisely one object denoted by $\underline{1}$, and so, if we let $\underline{n}$ be the n-fold sum of the object $\underline{1}$, then $\mathrm{ob}\mathbb{C}_{\oplus} = \{\underline{n} | n = 0, 1, 2 \ldots\}$ while the Banach space of morphisms $\underline{m} \to \underline{n}$ is given by $n \times m$ matrices with entries in $\mathbb{C}$. Hence, we can identify $\mathrm{Iso}(\mathbb{C}_{\oplus})$ with $\coprod_{n \geq 0} GL_n$ (where $\coprod$ is disjoint union). If we write $GL(\mathbb{C})$ for $\varinjlim GL_n(\mathbb{C})$, then $\mathbb{Z} \times GL(\mathbb{C})$ is a symmetric monoidal category whose objects are given by integers and

$$\text{Hom}_{\mathbb{Z} \times GL(\mathbb{C})}(m, n) = \begin{cases} \emptyset & \text{if } m \neq n \\ GL(\mathbb{C}) & \text{if } m = n \end{cases}$$

(see 5.3.2 to 5.3.7). There exists a functor $\text{Iso}(\mathbb{C}_\oplus) \to \mathbb{Z} \times GL(\mathbb{C})$, and so, $B\text{Iso}(\mathbb{C}_\oplus)^\wedge$ has the homotopy type of $\mathbb{Z} \times BGL(\mathbb{C})$.

(v) Now, let $b : S^2 \to B\text{Iso}(\mathbb{C}_\oplus)^\wedge$ be a fixed representative of the Bott element in $\pi_2(B\text{Iso}(\mathbb{C}_\oplus))^\wedge = K^{-2}(\text{pt})$. Then b, μ yield a map $S^2 \wedge B(\text{Iso}(\mathbb{P}(C_\oplus)))^\wedge \to B(\text{Iso}(\mathbb{P}C_\oplus))^\wedge$, natural in $\mathcal{C}$, with an adjoint

$$\beta : B(\text{Iso}(\mathbb{P}(C_\oplus))^\wedge \to \Omega^2 B(\text{Iso}(\mathbb{P}(C_\oplus)))^\wedge.$$

We now define a non-connective topological K-theory spectrum $\mathbb{K}^{\text{top}}(\mathcal{C})$ for the C^*-category $\mathcal{C}$ by the spaces

$$\begin{cases} B(\text{Iso}(\mathbb{P}(C_\oplus))^\wedge & \text{in even dimension} \\ \Omega B(\text{Iso}(\mathbb{P}(C_\oplus))^\wedge & \text{in odd dimension} \end{cases}$$

with structural maps that are identity in even dimensions and β in odd dimension. Then $\mathbb{K}^{\text{top}}(\mathcal{C})$ is an Ω-spectrum.

14.5.3 The functor $\mathbb{K}^{\text{top}} : \mathcal{G}poids^{inj} \to Spectra$

In view of 14.5.1 and 14.5.2, we now have a functor $\mathbb{K}^{\text{top}} : \mathcal{G}poids^{inj} \to Spectra$ $\mathcal{G} \to \mathbb{K}^{\text{top}}(C_r^*\mathcal{G})$ given by composing the functor $\mathcal{G}poids^{inj} \to C_r^* - Categories : \mathcal{G} \to C_r^*\mathcal{G}$ into the functor $\mathbb{K}^{\text{top}} : C_r^* Categories \to Spectra : C_r^*\mathcal{C} \to \mathbb{K}^{\text{top}}(C_r^*\mathcal{C})$.

14.5.4 Recall from 1.1.3 and 1.1.4 that to any G-set S, we can associate a groupoid $\underline{S}$ (the translation category of S). If $S = G/H$, then in the notation of 14.1.7 we have $\pi_n(\mathbb{K}^{\text{top}}(G/H)) \cong K_n(C_r^*H)$ where H is any subgroup of G. So Baum - Connes conjecture now says that $A_{fin} : \mathcal{H}_n^G(E_{fin}(G), K^{\text{top}}) \to \mathcal{H}_n^G(pt, \underline{K}^{\text{top}}) \cong K_n(C_r^*G)$ induced by the projection $E_{fin}(G) \to pt$ is an isomorphism.

14.5.5 Assembly for BC conjecture via controlled topology

(i) Let G be a discrete group, Z a G-CW-complex, $X \subset Z$ a closed G-invariant subspace, $Y = Z - X$. A subset C of Z is said to be relatively G-compact if C/G is relatively compact in Z/G. Let R be a ring with identity, also with involution. We define a category $B_G(Z, X, R)$ as follows: ob $B_G(Z, X, R)$ consists of pairs (A, f) where A is a free RG-module, $f : A \to \text{Fin}(Y)$ a G-equivariant map from A to finite subsets of Y, satisfying

(1) $f(a + b) \subseteq f(a) \cup f(b)$.

(2) $A_y = \{a \in A | f(a) \subseteq \{y\}\}$, a finitely generated RG_y-module for each $y \in Y$

(3) $A = \bigoplus_{y \in Y} A_y$ is an R-module.

(4) $\{y \in Y | A_y \neq 0\}$ is locally finite and relatively G-compact in Z.

(5) If Y has more than one point, then $f(a) = \emptyset$ iff $a = 0$.

A morphism $\varphi : (A, f) \rightarrow (B, y)$ consists of an RG-homomorphism $\varphi : A \rightarrow B$ commuting with f, g such that φ is continuously controlled at $X \subset Z$, i.e., in terms of coordinates, given $\varphi_y^z : A_y \rightarrow B_z (y, z \in Y)$ for every $x \in X$, and for every neighborhood U of x in Z there exists a smaller neighborhood $V \subset U$ of x in Z such that $\varphi_y^z = 0, \varphi_z^y = 0$ whenever $y \in Y - U, z \in V \cap Y$.

(ii) Now, let X be a G-CW-complex, $Z = X \times [0, 1]$ with subspace $X = X \times 1 \subset Z$. Let $\mathcal{B}_G(X \times [0, 1]; R) := \mathcal{B}_G(X \times [0, 1], X \times 1; R)$ and denote by $\mathcal{B}_G(X \times [0, 1]; R)_\emptyset$ the full subcategory of $\mathcal{U} = \mathcal{B}_G(X \times [0, 1]; R)$ with objects (A, f) such that the closure of the intersection

$$supp(A, f) = \overline{\{(x, t) \in (X \times [0, 1]) | A_{x,t} \neq \emptyset\}} \cap (X \times 1) = \emptyset$$

Then, $\mathcal{B}_G(X \times [0, 1]; R)_\emptyset$ is equivalent to $\mathcal{F}(RG)$, the category of the finitely free RG-modules.

(iii) Let $C_r^* \mathcal{B}_G(X \times [0, 1], \mathbb{C})$ be a category defined as follows: ob $C_r^* \mathcal{B}_G(X \times [0, 1], \mathbb{C})$ consists of objects (A, f) of $\mathcal{B}_G(X \times [0, 1], \mathbb{C})$ satisfying the extra condition that A has a G-invariant Hilbert space structure. Morphisms in $C_r^* \mathcal{B}_G(X \times [0, 1], \mathbb{C})$ are obtained by completing the subgroup of morphisms in $\mathcal{B}_G(X \times [0, 1], \mathbb{C})$, which are bounded linear operators on Hilbert spaces. This gives $C_r^* \mathcal{B}_G(X \times [0, 1], \mathbb{C})$ the stucture of a C_r^*-category.

The subcategory $C_r^* \mathcal{B}_G(X \times [0, 1], \mathbb{C})_\emptyset$ is a full subcategory consisting of objects (A, f) such that $supp(A, f) = \emptyset$.

The quotient category $C_r^* \mathcal{B}_G(X \times [0, 1], \mathbb{C})^{>0}$ has same objects as $\mathcal{B}_G(X \times [0, 1], \mathbb{C})$ with two morphisms identified in their quotient category if the difference can be approximated closely by morphisms factoring through objects of the subcategory.

(iv) **Definition** *For any G-space X, define $\mathbb{K}^{top}(X) := \mathbb{K}^{top}(C_r^* \mathcal{B}_G(X \times [0, 1], \mathbb{C})^{>0}$.*

This defines a functor $\mathbb{K}^{top} : G\text{-Spaces} \rightarrow Spectra$.

(v) Recall that a functor $F : Spaces \rightarrow Spectra$ is said to be homotopy invariant if it takes homotopy equivalences to homotopy equivalences.

A homotopy invariant functor is strongly excisive if $F(\emptyset)$ is contractible and F preserves arbitrary coproducts up to homotopy equivalence.

The following result is due to I. Hambleton and E.K. Pederson (see [77]).

Theorem 14.5.1 [77] *The functor* $\mathbb{K}^{top}$: *G-Spaces* $\to$ *Spectra is G-homotopy invariant and G-excisive with the following properties:*

(1) *If* X *is a co-compact G-space with finite isotropy, then* $\pi_*(\mathbb{K}^{top}(X)) = KK^G_{*-1}(C_0(X), \mathbb{C})$ *where* $C_0(X)$ *is the* C^**-algebra of complex-valued functions vanishing at* ∞.

(2) *For any subgroup* $H \leq G, \Omega\mathbb{K}^{top}(G/H)$ *is weakly equivalent to* $\mathbb{K}(C^*_r(H))$.

(3) *For any G-space* $X, \mathbb{K}^{top}(X)$ *is the homotopy colimit of* $\mathbb{K}^{top}$ *applied to G-compact subspace.*

The following results identifying the Davis - Lück assembly map with Baum - Connes assembly map as well as continuously controlled assembly maps are also due to D. Hamilton and E.K. Pederson.

Theorem 14.5.2 [77] *Let* G *be a discrete group and* $X = E_{\mathcal{F}in}(G)$. *Then, the continuously controlled assembly map*

$$\alpha_X : \mathbb{K}^{top}(C^*_r \mathcal{B}_G(X \times [0,1]; \mathbb{C})^{>0} \to \mathbb{K}^{top}(C^*_r \mathcal{B}_G(X \times [0,1], \mathbb{C})$$

induces the Baum - Connes assembly map

$$KK^G_i(C_0(X); \mathbb{C}) \to K_i(C^*_r(G))$$

on homotopy groups.

Theorem 14.5.3 [77] *The Davis - Lück assembly map arising from* $\mathbb{K}^{top}$ *applied to* $E_{\mathcal{F}in}(G)$ *are naturally isomorphic to the continuously controlled assembly map.*

Theorem 14.5.4 [77] *The Davis - Lück assembly map arising from* $\mathbb{K}^{top}$ *applied to* $E_{\mathcal{F}in}(G)$ *is naturally isomorphic to the Baum - Connes assembly map.*

Before we close this section, we discuss briefly the Baum - Connes conjecture vis-a-vis the family VC_y of virtually cyclic subgroups and the family FC_y of finite cyclic subgroups. The results below, 14.5.5 and 14.5.6, indicate that the relative assembly $A_{Fin \to VC_y}$ and $A_{FC_y \to Fin}$ are isomorphisms.

Theorem 14.5.5 [138] *For any discrete group* G *and any* $n \in \mathbb{Z}$, *the relative assembly map*

$$A_{\mathcal{F}in} \to VC_y : \mathcal{H}^G_n(E_{\mathcal{F}in}(G), \mathbb{K}^{top}) \to \mathcal{H}^G_n(E_{VCy}(G), K^{top})$$

is an isomorphism.

Theorem 14.5.6 [138] *For every discrete group G and any $n \in \mathbb{Z}$, the relative assembly map*

$$A_{VC_y \to Fin} : \mathcal{H}_n^G(E_{FC_y}(G), \mathbb{K}^{\text{top}}) \to H_n^G(E_{Fin}(G), \mathbb{K}^{\text{top}})$$

is an isomorphism.

Exercise

In the notation of 14.2.1, show that the relative assembly map $A_{Fin \to VCy}$: $H_n^G(E_{Fin}(G)) \to H_n^G(E_{VCy}(G))$ is an isomorphism iff for all virtually cyclic subgroups V of G, the assembly map

$$A_{Fin} = A_{Fin \to All} : H_n^V(E_{Fin}(V)) \to H_n^V(pt)$$

is an isomorphism.

Appendices

A   Some computations

I: K_0

(1) If G is a finite group of square-free order, then $NK_0(\mathbb{Z}G) = 0$.

(This result is due to D. Harmon; see [78].)

(2) Let R be integers in a number field F, Λ an R-order in a semi-simple F-algebra Σ.

Then,

   (i) $G_0(\Lambda), K_0(\Lambda)$ are finitely generated Abelian groups.

   (ii) $SK_0(\Lambda)$, $SG_0(\Lambda)$ are finite groups, if Λ satisfies the "Cartan condition".

(These results are due to H. Bass; see [20] or theorems 2.2.1 and 2.2.2.)

(3) Let R be a Noetherian ring with 1, G a finite group. Then, in the notation of 2.4,

   (i) $G_0(RG) \cong \oplus_{C \in X(G)} G_0(R\langle C \rangle)$.

   (ii) If C is a cyclic group of order n, then $G_0(\mathbb{Z}\langle C \rangle) = \oplus_{d/n}(\mathbb{Z} \oplus Cl(\mathbb{Z}[\zeta_d, \frac{1}{d}]))$.

   (iii) If $H = G \rtimes G_1$, G Abelian, G_1 any finite group such that the action of G_1 on G stabilizes every cocyclic subgroup of G. Then, $G_0(\mathbb{Z}H) \cong \oplus_{C \in X(G)} G_0(\mathbb{Z}\langle C \rangle \# G_1)$.

(These results are due to D. Webb; see [230] or 2.4.)

(4) Let D_{2n} be a dihedral group of order $2n$. Then, $G_0(\mathbb{Z}D_{2n}) \simeq \mathbb{Z}^\varepsilon \oplus \oplus_{q|n}(\mathbb{Z} \oplus Cl(\zeta_d, \frac{1}{d})_+)$. (See remarks 2.4.5 and [230].)

$$\text{where } \varepsilon = \begin{cases} 2 & \text{if } n \text{ is odd} \\ d & \text{if } n \text{ is even.} \end{cases}$$

(The result is due to D. Webb; see 2.4 or [230].)

(5) Let G be the generalized quaternion group Q_{4n} where $n = s^2 n', n'$ odd. (See remarks 2.4.6 (iv).) Then,

$$G_0(\mathbb{Z}G) = \mathbb{Z}^\varepsilon \oplus_{d|n} , (\mathbb{Z}^{s+2} \oplus_{i=0}^{s} Cl(\mathbb{Z}[\zeta_{2^i d}, \frac{1}{2^i d}]) \oplus Cl_{H_{2^{s+1}d}}(\mathbb{Z}[\zeta_{2^{s+1}d}, \frac{1}{2^{s+1}d}])$$

$$\text{where } \varepsilon = \begin{cases} 2 & \text{if } s > 0 \\ 1 & \text{if } s = 0. \end{cases}$$

Note The result above is due to D. Webb. (See [230], or 2.6.)

(6) Let R be a left Noetherian ring of finite global dimension. Then $K_0(R) \simeq K_0(R[X_1, X_2 \cdots , X_n])$.

(This result is due to A. Grothendieck; see [188].)

(7) Let R be a Dedekind domain. Then each $P \in \mathcal{P}(R[X_1, X_2, \ldots, X_n])$ has the form $P = R[X_1, X_2, \ldots, X_n] \otimes_R Q$ where $Q \in \mathcal{P}(R)$.

If R is a principal ideal domain, then each $P \in \mathcal{P}(R[X_1, X_2, \ldots, X_n])$ is free over $R[X_1, X_2, \ldots, X_n]$.

(The result above is due independently to D. Quillen and A. Suslin; see [203] or [167].)

(8) Let G be a cyclic group of order p. Then $K_0(\mathbb{Z}G) \simeq K_0(\mathbb{Z}[\zeta]) \simeq \mathbb{Z} \oplus Cl(\mathbb{Z}[\zeta]))$ where $\mathbb{Z}[\zeta]$ is the ring of integers in the cyclotomic field $Q(\zeta)$ and ζ is a primitive pth root of unity.

(This result is originally due to D.S. Rim; see [176, 177] or theorem 4.1.2.)

(9) Let R be the ring of integers in a number field F, Λ any R-order in a semi-simple F-algebra. Then,

    (a) $Cl(\Lambda)$ is a finite group.

    (b) If $\Lambda = \mathbb{Z}C_p$, (p a prime) then $Cl(\Lambda) \simeq Cl(\mathbb{Z}[\zeta])$ where ζ is a primitive pth root of 1 (Result (b) is due to D. S. Rim; see [177].)

(10) Let G be a finite p-group of order p^s, R the ring of integers in a p-adic field F, A a maximal R-order in a central division algebra D. Then, $SK_0(G, \mathcal{P}(A)) \simeq G_0(AG)$ is a finite cyclic p-group of order $\leq p^s$.

This is due to A. Kuku (see theorem 7.7.4 or [110].)

II: K₁

(1) Let G be a finite group. Then $K_1(\mathbb{Z}G), Wh(G)$ are finitely generated Abelian groups, and rank $K_1(\mathbb{Z}G) =$ rank $Wh(G) = r - q$

$$\text{where } r = \text{number of irreducible } \mathbb{R}G\text{-modules}$$
$$q = \text{number of irreducible } \mathbb{Q}G\text{-modules.}$$

(This result is due to H. Bass; see [20].)

(2) Let G be a finite group. Then $SK_1(\mathbb{Z}G)$ is isomorphic to the full torsion subgroup of $Wh(G)$. (See [159].)

(This result is due to C.T.C. Wall; see [229].)

(3) Let G be a finite group. Then $SK_1(\mathbb{Z}G) = 0$ if

   (a) G is a cyclic group;

   (b) G is a dihedral group;

   (c) G is a quaternion or semi-dihedral 2-group;

   (d) $G \simeq C_{p^n} \times C_p$ for any prime p and all n;

   (e) $G \simeq (C_2)^n$ for some n;

   (f) G is any iterated products or wreath products of symmetric and dihedral groups;

   (g) $G = S_n$, all n.

   Note

   (i) (b) above is due to B.A. Magurn see [143], (a) and (e) are due to H. Bass (see [20]). (d) is due to T.Y. Lam (see [126]). (c) and (f) are due to R. Oliver see [159].

   (ii) The next results (4) to (10) are all due to R. Oliver; see [159].

(4) If G is Abelian, then $SK_1(\mathbb{Z}G) = 0$ iff $G \simeq \mathbb{Z}_2^s$ (some positive integer s) or each Sylow subgroup of G has the form $\mathbb{Z}_{p^n}$ or $\mathbb{Z}_p \times \mathbb{Z}_{p^n}$.

(5)

$$SK_1(\mathbb{Z}A_n) = \begin{cases} \mathbb{Z}_3 & \text{if } n = \Sigma_{i=1}^r 3^{m_i} \geq 27 \text{ and } m_1 > \cdots > m_r \geq 0 \\ & \qquad\qquad\qquad\qquad \text{and } \Sigma m_i \text{ is odd} \\ 0 & \text{otherwise.} \end{cases}$$

(6) $SK_1(\mathbb{Z}G) = \mathbb{Z}_p^{p-1}$ if G is non-Abelian of order p^3.

(7) If $|G| = 16$, then

$$SK_1(\mathbb{Z}G) = \begin{cases} 1 & \text{if } G^{ab} = \mathbb{Z}_2 \times \mathbb{Z}_2 \text{ or } \mathbb{Z}_2 \times \mathbb{Z}_2 \times \mathbb{Z}_2 \\ \mathbb{Z}_2 & \text{if } G^{ab} = \mathbb{Z}_4 \times \mathbb{Z}_2 \end{cases}$$

(8)

$$\left. \begin{array}{l} SK_1(\mathbb{Z}[PSL(2,q)]) = \mathbb{Z}_3 \\ SK_1(\mathbb{Z}[SL(2,q)]) \quad = \mathbb{Z}_3 \times \mathbb{Z}_3 \end{array} \right\} if\ q = 3$$

$$SK_1(\mathbb{Z}[PSL(2,q)]) = SK_1(\mathbb{Z}[SL(2,q)]) = 0 \text{ otherwise.}$$

(9) $SK_1(\mathbb{Z}G) = \mathbb{Z}_p^{p-1}$ if G is non-Abelian of order p^3.

(10) $Wh(G) = 0$ if $|G| = 4$ or if $G \cong \mathbb{Z}$.

(11) Let R be a Dedekind domain with quotient field K a global field, Γ a maximal R-order in a central simple K-algebra A. Then, in the notation of (Remarks 3.2.1) $nrK_1(\Lambda) = R^* \cap K^+$ where $K^+ = \{a \in K | a_{\underline{p}} > 0$ at each prime $\underline{p}$ of K ramified in $A\}$.

(This result is due to R. Swan; see [39, 213].)

(12) Let R be a discrete valuation ring with quotient field K and finite residue class field. Let Γ be a maximal order in a central simple K-algebra A. Then, $nrK_1(\Gamma) = nr\Gamma^* = R^*$.

(This result is also due to R. Swan; see [39, 213].)

(13) If a ring Λ satisfies SR_n, then

   (i) $GL_m(A)/E_m(A) \to GL(A)/E(A)$ is onto for $m \geq n$ and injective for all $m > n$.

   (ii) $E_m(A) \lhd GL_m(A)$ if $m \geq n + 1$.

   (iii) $GL_m(A)/E_m(A)$ is Abelian for $m > n$.

   Note If R is a Dedekind domain and Λ is an R-order, then Λ satisfies SR_2.

   (The above results are due to H. Bass and L. Vaserstein; see [17, 18, 20, 221].)

(14) Let R be a left Noetherian ring of finite global dimension; then,

$$K_1(R) \approx K_1(R[X_1 \cdots X_n]).$$

(This result is due to Bass, Heller, and Swan; see [20].)

(15) Let R be the ring of integers in a number field F, Λ an R-order in a semi-simple F-algebra.

Then,

  (i) $K_1(\Lambda)$ is a finitely generated Abelian group.

  (ii) $SK_1(\Lambda)$ is a finite group.

(This result is due to H. Bass; see [20].)

(16) Let R be a discrete valuation ring with finite residue field $\overline{R} = R/p$ and with quotient field F. Let D be central division algebra over F, Γ a maximal R-order in D. Then, $SK_1(\Gamma)$ is a cyclic group of order $(q^n - 1)/q - 1$ where $q = (\overline{R})$.

(This result is due to A. Kuku and M.E. Keating; see [105, 99].)

(17) Let F be a global field with ring of integers R, D a skewfield with center F. For each prime ideal $\underline{p}$ of F, let $\hat{D}_{\underline{p}}$ be the p-adic completion of D. Then, $\hat{D}_{\underline{p}} \approx M_{m_{\underline{p}}}(D(\underline{p}))$ where $D(\underline{p})$ is a central $\hat{F}_{\underline{p}}$-algebra with index $m_{\underline{p}}$ (i.e., $\dim_{\hat{F}_{\underline{p}}}(D(\underline{p}) = m_{\underline{p}}^2)$. Let Γ be a maximal R-order in a central division algebra over F. Put $q_{\underline{p}} = |R/\underline{p}|$.

Then,

  (a) $SK_1(\Gamma) \approx \coprod_{\underline{p}} SK_1(\hat{\Gamma}_{\underline{p}})$, and

  (b) There exists exact sequences

$$0 \to SK_1(\Gamma) \to K_1(\Gamma) \to R^* \cap K^* \to 0.$$

(This result is due to M. Keating; see [99].)

(18) Let G be a non-Abelian p-group (p a prime)

Then,

  (a) $Cl_1(\mathbb{Z}G) \neq 0$ unless $p = 2$ and G^{ab} has exponent 2.

  (b) $SK_1(\mathbb{Z}G) = Cl_1(\mathbb{Z}G) \simeq (\mathbb{Z}/p)^{p-1}$ if p is odd and $|G| = p^3$.

(This result is due to R. Oliver; see [159].)

(19) If G is any quaternion or semi-dihedral 2-group, then $Cl_1(\mathbb{Z}G \times (C_2)^k) \simeq (\mathbb{Z}/2)^{2^k - k - 1}$.

(This result is due to R. Oliver; see [159].)

(20) Let R be the ring of integers in a number field. Then, $SK_1(RC_n) = 0$ for any finite cyclic group C_n of order n.

(This result is due to Alperin, Dennis, Oliver, and Stein; see [5].)

(21) $SK_1(\mathbb{Z}G) = 0$ if $G \simeq C_{p^n}$ or $C_{p^n} \times C_p$ for any prime p and any n; if $G = (C_2)^n$, any n or if G is any dihedral, quaternion, or semi-dihedral 2-group. Conversely, if G is a p-group and $Cl_1(\mathbb{Z}G) = 0$, then either G is one of the groups above or $p = 2$ and $G^{ab} = (C_2)^n$ for some n.

(This result is due to R. Oliver; see [159].)

(22) Let G be a finite group, p a rational prime. Then $SK_1(\mathbb{Z}G)(p) = 0$ if the Sylow p-subgroup $S_p(G)$ of G is isomorphic to C_{p^n} or $C_{p^n} \times C_p$(any n).

(This result is due to R. Oliver; see [159].)

(23) Let G be a finite Abelian group and $r(G)$ the product of distinct primes p dividing $|G|$, for which $S_p(G)$ is not cyclic. Then $exp(SK_1(\mathbb{Z}G)) = \varepsilon \cdot gcd(\exp(G), \frac{|G|}{r(G) \cdot exp(G)})$ where $\varepsilon = \frac{1}{2}$ if

   (a) $G \cong (C_2)^n$ for some $n \geq 3$ or

   (b) $S_2(G) \simeq C_{2^n} \times C_{2^n}$ for some $n \geq 3$ or

   (c) $S_2(G) \simeq C_{2^n} \times C_{2^n} \times C_2$ for some $n \geq 2$,

and $\varepsilon = 1$ otherwise.

(This result is due to Alperin, Dennis, Oliver, and Stein; see [5].)

(24) Let F be a p-adic field, R the ring of integers of F, Γ a maximal R-order in a central division algebra D over $F, \underline{m} = rad\ R$. Then $SK_1(\Gamma, \underline{m}) = 0$.

(This result is due to A. Kuku; see [107].)

(25) Let R be the ring of integers in a number field or p-adic field F. Let Λ be any R-order in a semi-simple algebra, $\underline{a}$ a two-sided ideal of Λ. Then, $SK_1(\Lambda, \underline{a})$ is a finite group.

(For number fields, the result is due to H. Bass; see [20]. For p-adic fields, the result is due to A. Kuku; see [105].)

(26) Let p be a rational prime and G a quaternion or dihedral group of order 8. Then, $SK_1(\hat{\mathbb{Z}}_pG) = 0$.

(This result is due to A. Kuku; see [104].)

(27) Let G be any finite group, p a rational prime. Then, $SK_1(\hat{\mathbb{Z}}_pG)$ is a finite p-group.

(This result is due to C.T.C Wall; see [229, 159].)

III: K_2

(1) Let R be the ring of integers in a number field, Λ any R-order in a semi-simple F-algebra Σ. Then, $K_2(\Lambda)$ is a finite group. Hence $K_2(RG)$ is finite for any finite group G. $G_2(\Lambda)$, $G_2(RG)$ are also finite groups.

(This result is due to A. Kuku and is in fact part of a general result that says that for all $n \geq 1, K_{2n}(\Lambda), G_{2n}(A)$ are finite; see theorem 7.2.7 or [121].)

(2) Let F be a number field, R the ring of integers in F, and D a finite-dimensional central division F-algebra with square-free index, Λ a maximal R-order in D. Then, in the notation of $(3 \cdot 3)^B, K_2\Lambda \simeq K_2^+R = Ker(K_2R \to \prod_{\text{real ramified}}\{\pm 1\})$.

(This result is due to X. Guo, A. Kuku, and H. Qin; see [73].)

(3) Let F be a number field and D a central division algebra over F. Then, $div(K_2(D)) \subseteq WK_2(D)$ and $div(K_2(D))(l) \simeq WK_2(D)(l)$ for all odd primes l. If the index of D is square free, then

(a) $div(K_2(D)) \simeq div(K_2(F))$.

(b) $WK_2(D) \simeq WK_2(F)$ and $|WK_2(D)/divK_2(D)| \leq 2$.

(This result is due to Guo and Kuku; see [72].)

(4) Let F be a number field and D a central division algebra F with square-free index. Then, every element of $K_2(D)$ is a symbol of the form $\{a, b\}$ for $a \in F^*, b \in D^*$.

(This result is due to X. Guo, A. Kuku, and H. Qin; see [73].)

Note The next three results (5), (6), (7) are due to B. Magurn (see [142, 144]).

(5) Let F be a finite field of characteristics p, and G a finite group whose Sylow-p-subgroup is a cyclic direct factor. Then $K_2(FG) = 0$. For any Abelian group G, $K_2(FG) = 0$ iff the Sylow-p-subgroup of G is cyclic.

(6) Suppose that p is a prime and F a finite field of characteristic p. Let A be a p'-group, C_i cyclic p'-groups, and G_i semidirect products $C_i \rtimes Z_p$, for $i = 1, 2$. Then,

(a) $K_2(FA) = 0$.

(b) $K_2(F[G_1 \times A]) = 0$.

(c) $K_2(F[G_1 \times G_2 \times A]) \simeq K_2(F[G_1^{ab} \times G_2^{ab} \times A])$.

Hence, if D_n is the dihedral group of order $2n$, and n and m odd, then

$$K_2(F_2[D_n \times D_m]) \simeq K_2(F_2[\mathbb{Z}_2 \times \mathbb{Z}_2]) \simeq \mathbb{Z}_2^3.$$

(7) Let F be a finite field with 2^f elements and G a finite Abelian group of order n, 2-rank t and 4-rank ≤ 1. Then,

$$K_2(\mathbb{F}G) \simeq \mathbb{Z}_2^{f(n/2^t)(t-1)(2^t-1)}.$$

(8) Let $\mathbb{F}_q$ be a finite field of order $q = p^f$. Then, $K_2(\mathbb{F}_q C_p^r) \simeq C_p^{f(r-1)(p^r-1)}$.

(This result is due to Dennis, Keating, and Stein; ee [41].)

IV: Negative K-theory

(1) Let G be a finite group of order s, R the ring of integers in a number field F. Then, $K_{-n}(RG) = 0$ for all $n > 1$. For any prime ideal $\underline{p}$ of R, let $f, f_{\underline{p}}, r_{\underline{p}}$, respectively, be the number of isomorphism classes of irreducible $F, \hat{F}_{\underline{p}}$ and $R/\underline{p}$ representations of G. Then, $K_{-1}(RG)$ is a finitely generated Abelian group and rank $K_{-1}(RG) = f + \Sigma_{\underline{p}|sR}(f_{\underline{p}} - r_{\underline{p}})$.

(This result is due to D. Carter; see [32] or Corollary 4.4.2.)

(2) Let $V = G \rtimes_\alpha T$ be a virtually infinite cyclic group, i.e., V is the semi-direct product of a finite group G of order r with an infinite cyclic group $T = \langle t \rangle$ with respect to the automorphism $\alpha : G \to G : g \to tgt^{-1}$. Let R be the ring of integers in a number field. Then

   (a) $K_n(RV) = 0$ for all $n < -1$.

   (b) $K_{-1}(RV)$ is finitely generated.

(The result above for RV is due to Kuku and Tang; see [123]. An earlier version for $\mathbb{Z}V$ is due to Farrell and Jones; see [55].)

(3) Let R be a regular ring. Then, for any triple $\underline{R} = (R, B_0, B_1)$ (see definition 4.5.2), we have $Nil_n^W(\underline{R}) = 0$ for all $n \in \mathbb{Z}$.

(This result is due to F. Waldhausen; see [224].)

(4) Let R be a quasi-regular ring. Then, for any triple $\underline{R} = (R, B_0, B_1)$ we have
$$Nil_n^W(\underline{R}) = 0 \text{ for all } n \leq -1.$$

(This result is due to F. Conolly and M. Da Silva; see [36].)

(5) For the triple $\underline{R} = (R; R^\alpha, R^\beta)$, let $R_\rho := \rho(\underline{R})$ (in the notation of 4.5.5). Let γ be the ring automorphism of $\begin{pmatrix} R & 0 \\ 0 & R \end{pmatrix}$ given by $\gamma : \begin{pmatrix} a & 0 \\ 0 & b \end{pmatrix} \to \begin{pmatrix} \beta^{(b)} & 0 \\ 0 & \alpha(b) \end{pmatrix}$.

Then, there is a ring isomorphism
$$\mu : R_\rho \simeq \begin{pmatrix} R & 0 \\ 0 & R \end{pmatrix}_\gamma [x].$$

(This result is due to A. Kuku and G. Tang; see theorem 7.5.6 or [123]. Note that this result, which expresses R_ρ as a twisted polynomial ring, greatly facilitates computations of $K_n(\underline{R}), NK_n(\underline{R})$ for all $n \in \mathbb{Z}$.)

(6) (a) Let R be a regular ring. Then, $NK_n(R; R^\alpha, R^\beta) = 0$ for all $n \in \mathbb{Z}$.

(b) If R is quasi-regular, then $NK_n(R, R^\alpha, R^\beta) = 0$ for all $n \leq 0$.

(This result is due to A. Kuku and G. Tang; see theorem 7.5.7 or [123]) and is a consequence of (5) on the previous page.)

(7) Let $V = G_0 *_H G_1$, $[G_0 : H] = 2 = [G_1, H]$. Then, $NK_n(\mathbb{Z}H, \mathbb{Z}[G_0 - H], \mathbb{Z}[G_1 - H]) = 0$ for $n \leq -1$.

(This result is due to A. Kuku and G. Tang; see theorem 7.5.8 or [123].)

V: Higher K-theory

(1) Let A be any finite ring. Then, for all $n \geq 1$,

    (i) $K_n(A)$ is a finite group.

    (ii) $G_n(A)$ is a finite group.

    (iii) $G_{2n}(A) = 0$.

Note Above results apply to finite groupings, e.g., $A = RG$, G any finite group, R any finite ring.

(The result above is due to A. Kuku; see theorem 7.1.2 or [106, 112].)

(2) Let F be a p-adic field with ring of integers R, Γ a maximal R-order is a semi-simple F-algebra Σ. Then, for all $n \geq 1$,

    (i) $SK_{2n}(\Gamma) = 0$.

    (ii) $SK_{2n-1}(\Gamma) = 0$ iff Σ is unramified over its center.

Note The result above applies to grouprings $\Gamma = RG$ where the order of G is relatively prime to p.

(This result is due to A. Kuku; see theorem 7.1.3 or [107].)

(3) Let K be a p-adic field with ring of integers R, and let k be the residue field of K. Assume that $(|k| = q)$. Let Γ be a maximal R-order in a central division algebra D of dimension t^2 over K. Then, for all $n \geq 1, |SK_{2n-1}(\Gamma)| = (q^{nt} - 1)/(q - 1)$.

(This result is due to M. E. Keating; see [99] or theorem 7.1.2.)

(4) Let F be a complete discretely valued field with finite residue field of characteristic p (e.g., a p-adic field). Let R be the ring of integers of F, Γ a maximal order in a central division algebra, $\overline{\Gamma}$ the residue class field of Γ. Then, for all $n \geq 1$, $K_n(\Gamma) \otimes \hat{\mathbb{Z}}_q \simeq K_n(\overline{\Gamma}) \otimes \hat{\mathbb{Z}}_q$ for $q \neq p$.

(This result is due to A. Suslin and A.V. Yufryakov; see [204].)

(5) Let R be the ring of integers in a number field F, Λ any R-order in a semi-simple F-algebra Σ. Then, for all $n \geq 1$.

(a) $K_n(\Lambda)$ is a finitely generated Abelian group.

(b) $SK_n(\Lambda)$ is a finite group.

(c) $SK_n(\hat{\Lambda}_{\underline{p}})$ is finite (or zero) for any prime ideal $\underline{p}$ of R.

(This result is due to A.O. Kuku; see theorem 7.1.11 or [112, 113].)

Note The result holds for $\Lambda = RG, \hat{\Lambda}_{\underline{p}} = \hat{R}_{\underline{p}}G$).

(6) Let R be the ring of integers in a number field F, Λ any R-order in a semi-simple F-algebra Σ. Then, for all $n \geq 1$,

(a) $G_n(\Lambda)$ is a finitely generated Abelian group.

(b) $SG_{2n-1}(\Lambda)$ is finite and $SG_{2n}(\Lambda) = 0$.

(c) $SG_{2n-1}(\hat{\Lambda}_{\underline{p}})$, $SG_{2n-1}(\hat{\Lambda}_{\underline{p}})$ are finite of order relatively prime to the rational prime p lying below $\underline{p}$, and $SG_{2n}(\Lambda_{\underline{p}}) = SG_{2n}(\hat{\Lambda}_{\underline{p}}) = 0$.

(This result is due to A. Kuku; see theorem 7.1.15 or [109, 110].)

Note also that the results hold for $\Lambda = RG, \hat{\Lambda}_{\underline{p}} = \hat{R}_{\underline{p}}G, \Lambda_{\underline{p}} = R_{\underline{p}}G$.

(7) Let R be a Dedekind domain with quotient field F, Λ an R-order in a semi-simple F-algebra. Assume that

(i) $SG_1(\Lambda) = 0$.

(ii) $G_n(\Lambda)$ is a finitely generated Abelian group.

(iii) $R/\underline{p}$ is finite for all primes $\underline{p}$ of R.

(iv) If ζ is an l^s-th root of unity for any rational prime l and positive integer s, $\overline{R}$ the integral closure of R in $F(\zeta)$, then $SG_1(\overline{R} \otimes_R \Lambda) = 0$. Then $SG_n(\Lambda) = 0$ for all $n \geq 1$.

Hence, if R is the ring of integers in a number field F, and G a finite group, then $SG_n(RG) = 0$ for all $n \geq 1$.

(This result is due to R. Laubenbacher and D. Webb; see theorem 7.1.15 or [131].)

(8) Let R be the ring of integers in a number field F, A any R-algebra finitely generated as an R-module. Then, $G_n(A)$ is a finitely generated Abelian group.

(This result is due to A. Kuku; see theorem 7.1.14 or [112].)

(9) Let G be a finite p-group. Then, for all $n \geq 1$,

(a) $SK_{2n-1}(\mathbb{Z}G)$ is a finite p-group.

(b) For any rational prime l, $SK_{2n-1}(\hat{\mathbb{Z}}G)$ is a finite p-group (or zero).

(This result is due to A. Kuku; see theorem 7.1.17 or [121].)

(10) Let R be the ring of integers in a number field F, Λ any R-order in a semi-simple F-algebra Σ, Γ a maximal R-order containing Λ. Then, for all $n \geq 2$, rank $K_n(\Lambda) = $ rank $K_n(\Gamma) = $ rank $G_n(\Lambda) = $ rank $K_n(\Sigma)$. Hence, if G is a finite group, then,

$$\text{rank } K_n(RG) = \text{rank } K_n(\Gamma) = \text{rank } G_n(RG) = \text{rank } K_n(FG).$$

(This result is due to A. O. Kuku; see theorem 7.2.1 or [115].)

(11) Let R be the ring of integers in a number field F, Λ any R-order in a semi-simple F-algebra. Then, for all $n \geq 1$, $K_{2n}(\Lambda)$, $G_{2n}(\Lambda)$ are finite groups.

Hence, $K_{2n}(RG), G_{2n}(RG)$ are finite groups for any finite group G and all $n \geq 1$.

(This result is due to A. Kuku; see theorem 7.2.7 or [12].)

(12) Let C_p be a cyclic group of order p (p a prime). Assume that $n \geq 2$ and $p \equiv 3(4)$ or $p \equiv 5(8)$. Then we have isomorphisms

$$K_{2n-1}(\mathbb{Z}C_p)(2) \cong K_{2n-1}(\mathbb{Z})(2)$$

and

$$K_{2n-2}(\mathbb{Z}C_p)(2) \cong K_{2n-2}(\mathbb{Z}(\zeta_p))(2) \oplus K_{2n-2}(\mathbb{Z})(2)$$

(This result is due to P. Ostvaer; see [160].)

(13) Let $p \equiv 5(8)$ and let n be an odd integer. Then we have isomorphism

$$K_{2n-1}(\mathbb{Z}C_p)(2) \simeq K_{2n-1}(\mathbb{Z})(2).$$

Moreover, for $n \equiv 3(4)$, we have an isomorphism $K_{2n-2}(\mathbb{Z}C_p)(2) \simeq K_{2n-2}(\mathbb{Z}[\zeta_p])(2) \oplus K_{2n-2}(\mathbb{Z})(2)(I)$. If $n \equiv 1(4)$, then (I) is true provided p is a two-regular prime.

(Recall that a prime number p is two-regular if the 2-rank of the Picard group of the two-integers in $Q(\zeta_p)$ equals zero, and 2 is a primitive root $mod(p)$.)

(The result above is due to P. Ostvaer; see [160].)

(14) Let G be a finite Abelian group, R a Noetherian ring. Then, in the notation of 7.3.1, $G_n(RG) \simeq \oplus_{C \in X(G)} G_n(R\langle C \rangle)$ where C runs through the cyclic quotient of G.

(This result is due to D. Webb; see theorem 7.3.1 or [232].)

(15) Let H be a non-abelian group of order pq, $p|q-1$. Let G_1 denote the unique subgroup of order p of $Gal(Q(\zeta_q)/Q)$. Then,

$$G_n(\mathbb{Z}H) \simeq G_n(\mathbb{Z}) \oplus G_n(\mathbb{Z}[\zeta_p, \frac{1}{p}] \oplus G_n(\mathbb{Z}[\zeta_q, \frac{1}{q}]^{G_1}))\text{for all } n \geq 0.$$

(This result is due to D. Webb; see proposition 7.3.1 or [231].)

(16) Let D_{2s} be the Dihedral group of order $2s$. Then,

$$G_n(\mathbb{Z}D_{2s}) \simeq G_n(\mathbb{Z}) \oplus G_n(\mathbb{Z}[\frac{1}{2}])^\epsilon \oplus \bigoplus_{\substack{d|s \\ d>2}} G_n(\mathbb{Z}[\zeta_d, \frac{1}{d}]+).$$

$$\text{where } \varepsilon = \begin{cases} 1 & \text{if } n \text{ is odd} \\ 2 & \text{if } n \text{ is even,} \end{cases}$$

and $\mathbb{Z}[\zeta_d, \frac{1}{d}]_+$ is the complex conjugation-invariant subring of $\mathbb{Z}[\zeta_d, \frac{1}{d}]$.

(This result is due to D. Webb; see proposition 7.3.2 or [231].)

(17) For the symmetric group S_3, $G_3(\mathbb{Z}S_3) \simeq \mathbb{Z}/48 \oplus \mathbb{Z}/48$.

(This result is due to D. Webb; see example 7.3.1 or [231].)

(18) Let H be the generalized quaternion group of order 4.2^s. Then, in the notation of 7.3.3,

$$G_*(\mathbb{Z}H) \simeq \oplus G_*(\mathbb{Z}[\zeta_{2^j}, \frac{1}{2^j}]) \oplus G_*(\Gamma[\frac{1}{2^{s+1}}]) \oplus G_*(\mathbb{Z}[\frac{1}{2}])^2.$$

(This result is due to D. Webb; see proposition 7.3.3 or [231].)

(19) Let R be a Noetherian ring and H a finite p-group (p a prime), Γ a maximal $\mathbb{Z}$-order in QH containing $\mathbb{Z}H$. Then, in the notation of 7.3.11, we have

$$G_n(RH) \simeq \oplus G_n(R \otimes_\mathbb{Z} \Gamma_\rho)$$

where $X(H)$ is the set of irreducible rational representation of H.

(This result is due to Hambleton, Taylor, and Williams; see theorem 7.3.5 or [76]. The result is also true for nilpotent groups; see [76].)

(20) Let R be the ring of integers in a number field, Λ an R-order is a semi-simple F-algebra. Then,

(a) $Cl_n(\Lambda)$ is a finite group for all $n \geq 1$.

(b) For all $n \geq 1$, p-torsion in $Cl_{2n-1}(\Lambda)$ can occur only for primes p lying below prime ideals $\underline{p}$ at which $\hat{\Lambda}_{\underline{p}}$ is not maximal.

(c) For any finite group G, and all $n \geq 1$, the only possible p-torsion in $Cl_{2n-1}(RG)$ is for those primes p dividing the order of G.

(Results (b) (c) are due to M. Kolster and R. Laubenbacher; see theorem 7.4.4 and Corollary 7.4.1 or [102]. Result (a) is due to A. Kuku; see theorem 7.1.11(ii).)

(21) Let S_r be the symmetric group on r letters, and let $n \geq 0$. Then, $Cl_{4n+1}(\mathbb{Z}S_r)$ is a finite 2-torsion group, and the only possible odd torsion in $Cl_{4n-1}(\mathbb{Z}S_r)$ are for odd primes p such that $\frac{p-1}{2}$ divides n.

(This result is due to Kolster and Laubenbacher; see theorem 7.4.6 or [102].)

(22) Let D_{2r} be the dihedral group of order $2r$. If the local Quillen - Lichtenbaum conjecture is true, then $Cl_{4n+1}(\mathbb{Z}D_{2r})$ is a finite 2-torsion group.

(This result is due to Kolster and Laubenbacher; see theorem 7.4.7 or [102].)

(23) Let F be a number field with ring of integers R, Λ a hereditary R-order in a semi-simple F-algebra or an Eichler order in a quaternion algebra. Then the only p-torsion possible in $Cl_{2n}(\Lambda)$ is for those primes p lying below the prime ideals $\underline{p}$ of R at which $\hat{\Lambda}_{\underline{p}}$ is not maximal.

(This result is due to Guo and Kuku and is proved for "generalized Eichler orders", which combine the properties of hereditary and Eichler orders; see theorem 7.4.10 or [74].)

(24) Let R be the ring of integers in a number field F, Λ any R-order in a semi-simple F-algebra Σ, α an autopmorphism of Λ. Then, for all $n \geq 0$,

   (a) $NK_n(\Lambda, \alpha)$ is s-torsion for some positive integer s. Hence, rank $K_n(\Lambda_\alpha(t)) = $ rank $K_n(\Lambda)$ and is finite.

   (b) If G is a finite group of order r, then $NK_n(RG, \alpha)$ is r-torsion where α is the automorphism of RG induced by that of G.

(This result is due to A.O. Kuku and G. Tang; see theorem 7.5.4 or [123].)

(25) Let R be the ring of integers in a number field F, $V = G \rtimes_\alpha T$ the semi-direct product of a finite group G of order r with an infinite cyclic group $T = \langle t \rangle$ with respect to the automorphism $\alpha : G \to G : g \to tgt^{-1}$. Then, for all $n \geq 0$, we have

   (a) $G_n(RV)$ is a finitely generated Abelian group.
   (b) $NK_n(RV)$ is r-torsion.

(This result is due to A.O. Kuku and G. Tang; see theorem 7.5.5 or [123].)

(26) Let V be a virtually infinite cyclic group of the form $V = G_0 *_H G_1$ when the groups $G_i, i = 0, 1$ and H are finite, and $[G_i : H] = 2$. Then, in the notation of 7.5.3, the Nil groups $NK_n(\mathbb{Z}H; \mathbb{Z}[G_0 - H], \mathbb{Z}[G_1 - H])$ are $|H|$-torsion for all $n \geq 0$.

(This result is due to A. Kuku and G. Tang; see theorem 7.5.8 or [123].)

(27) Let F be a number field and Σ a semi-simple F-algebra, $WK_n(\Sigma)$ the Wild kernel of Σ (see definition 7.1.1). Then, $WK_n(\Sigma)$ is a finite group for all $n \geq 0$.

(This result is due to X. Guo and A. Kuku; see theorem 7.1.8 or [72].)

(28) Let k be a field of characteristic p, and G a finite or profinite group. Then, for all $n \geq 0$, the Cartan map $K_n(kG) \rightarrow G_n(kG)$ induces isomorphisms $\mathbb{Z}(\frac{1}{p}) \otimes K_n(kG) \rightarrow \mathbb{Z}(\frac{1}{p}) \otimes G_n(kG)$.

(For finite groups the result is due to A. Dress and A. Kuku; see theorem 10.4.1 or [53]. The extension to profinite groups is due to A. Kuku; see theorem 11.1.2 or [109].)

(29) Let p be a rational prime, k a field of characteristic p, G a finite group. Then, for all $n \geq 1$,

  (a) $K_{2n}(kG)$ is a finite p-group.

  (b) The Cartan map $\varphi_{2n-1} : K_{2n-1}(kG) \rightarrow G_{2n-1}(kG)$ is surjective, and $Ker\varphi_{2n-1}$ is Sylow-p-subgroup of $K_{2n-1}(kG)$.

(This result is due to A. Kuku; see theorem 10.4.2 or [112].)

(30) Let k be a field of characteristic p, $\mathcal{C}$ a finite EI category. Then, the Cartan homomorphism $K_n(k\mathcal{C}) \rightarrow G_n(k\mathcal{C})$ induces isomorphism $\mathbb{Z}(\frac{1}{p}) \otimes K_n(k\mathcal{C}) \simeq \mathbb{Z}(\frac{1}{p}) \otimes G_n(k\mathcal{C})$.

(This result is due to A. Kuku; see theorem 7.6.5 or [114].)

(31) Let R be a Dedekind ring, G a finite group, M any of the Green modules $K_n(R-), G_n(R-), SK_n(R-), SG_n(R-), Cl_n(R-)$ over Green ring $G_0(R-)$, $\mathcal{P}$ a set of rational primes, $\mathbb{Z}_\mathcal{P} = \mathbb{Z}[\frac{1}{q}|q \notin \mathcal{P}]$. Then, $\mathbb{Z}_\mathcal{P} \otimes M$ is hyperelementary computable.

(This result is due to A. O. Kuku; see theorem 10.4.3 or [108].)

(32) Let $\mathcal{C}$ be an exact category such that there exists a pairing $\mathcal{C} \times \mathcal{C} \rightarrow \mathcal{C} :$ $(X, Y) \rightarrow X \cdot Y$, which is naturally associative and commutative and such that $\mathcal{C}$ has an object E such that $E \circ X = X \circ E$ for all $X \in \mathcal{C}$. Let G be a profinite group and S, T, G-sets, $\mathcal{P}$ a set of natural primes.

Let M be any of the functors $K_n^G(-,\mathcal{C},T), P_n^G(-,\mathcal{C},T)$. Then, in the notation of 11.2.1 and 11.2.2 we have $(G:G_S)_{\not{p}}\mathcal{H}(M) = 0$ for all $i \in \mathbb{Z}$.

(This result is due to A. Kuku; see theorem 11.2.2 or [109].)

(33) Let G be a compact Lie group, $\mathcal{M}(\mathbb{C})$ the category of finite-dimensional complex vector spaces. Then, in the notation and terminology of 12.3, $K_n^G(-\mathcal{M}(C))$ is hyperelementary computable (see theorem 12.3.3 or [116]).

(This result is due to A. Kuku.)

(34) Let $\mathcal{C}$ be a finite EI category, R the ring of integers in a number field of F. Then, for all $n \geq 0$,

    (a) $K_n(R\mathcal{C})$ is a finitely generated Abelian group.

    (b) $G_n(R\mathcal{C})$ is a finitely generated Abelian group.

    (c) $SG_n(R\mathcal{C}) = 0$.

    (d) $SK_n(R\mathcal{C})$ is a finite group.

(This result is due to A. Kuku; see corollary 7.6.12 and 7.6.15.)

(35) Let R be the ring of integers in a number field, G a finite group. Then, the Waldhausen's K-groups of the category $(Ch_b(\mathcal{M}(RG)),\omega)$ of bounded complexes of finitely generated RG-modules with stable quasi-isomorphisms as weak equivalences are finite Abelian groups.

(This result is due to A.O. Kuku; see chapter 13 or [122].)

(36) Let $\mathcal{C}$ be an exact category such that $K_n(\mathcal{C})$ is finitely generated. Then, $K_n(\mathcal{C}) \otimes \hat{\mathbb{Z}}_\ell \simeq K_n^{pr}(\mathcal{C},\hat{\mathbb{Z}}_\ell) \simeq K_n(\mathcal{C},\hat{\mathbb{Z}}_\ell)$ are an ℓ-complete profinite Abelian group.
E.g., $\mathcal{C} = \mathcal{P}(\Lambda), \mathcal{M}(\Lambda)$ where Λ is an order in a semi-simple algebra over a number field.

(This result is due to A.O. Kuku; see 8.2.3 or [117].)

(37) Let p be a rational prime, F a p-adic field, R the ring of integers of F, Λ any R-order in a semi-simple F-algebra Σ, ℓ a rational prime such that $\ell \neq p$. Then, for all $n \geq 2$,

    (i) $K_n^{pr}(\Sigma,\hat{\mathbb{Z}}_\ell) \simeq K_n(\Sigma,\hat{\mathbb{Z}}_\ell)$ is an ℓ-complete profinite Abelian group.

    (ii) $G_n^{pr}(\Lambda,\hat{\mathbb{Z}}_\ell) \simeq G_n(\Lambda,\hat{\mathbb{Z}}_\ell)$ is an ℓ-complete profinite Abelian group.

(This result is due to A.O. Kuku; see theorems 8.3.1 and 8.3.3 or [117].)

(38) Let R be the ring of integers in a number field F, Λ any R-order in a semi-simple F-algebra Σ, $\underline{p}$ a prime ideal of R, ℓ a rational prime such that $\ell \neq char(R/\underline{p})$. Then, for all $n \geq 1$,

(i) $G_n(\Lambda)_\ell$ are finite groups.

(ii) $K_n(\hat{\Sigma}_p)_\ell$ are finite groups.

(iii) If Λ satisfies the hypothesis of theorem 8.1.2, we also have that $K_n(\hat{\Lambda}_p)_\ell$ are finite groups.

(The result above is due to A.O. Kuku; see theorems 8.3.4 and 8.3.5 or [117].)

(39) Let R be the ring of integers in a p-adic field F, Λ any R-order in a semi-simple F-algebra, $K_n^c(\Lambda)$ continuous K-theory of Λ. Then, for all $n \geq 1$,

(i) $K_n^c(\Lambda)$ is a profinite group.

(ii) $K_{2n}^c(\Lambda)$ is a pro-p-group.

(The result above is due to A.O. Kuku; see theorem 8.4.1 or [117].)

B   Some open problems

1. Let G be a finite group, p a rational prime.

   Question Is $SK_{2n-1}(\widehat{\mathbb{Z}}_p G)$ a finite p-group for all $n \geq 1$?

   For $n = 1$, a positive answer is due to C.T.C. Wall (see [229]). For G a finite p-group, and for all $n \geq 1$, an affirmative answer is due to A. Kuku (see [121] or theorem 7.1.17).

2. Let T be a category whose objects have the form $\mathbf{R} = (R, B_0, B_1)$ where R is a ring and B_0, B_1 are R-bimodules (see 4.5.4). Then, as in 4.5.5, there exists a functor $\rho : T \to \text{Rings}$ defined by $\rho(\mathbf{R}) := R_\rho$ as in 4.5.5. Recall that $NK_n(R_\rho) := \text{Ker}(K_n(R_\rho) \to K_n\left(\begin{smallmatrix} R & 0 \\ 0 & R \end{smallmatrix}\right))$ for all $n \in \mathbb{Z}$. For $\mathbf{R} \in T$, the Waldhausen Nilgroups $Nil_{n-1}^W(\mathbf{R})$ are also defined (see definition 4.5.2). In [36], F. Conolly and M. Da Silva prove that, for $n \leq 1$, $NK_n(R_\rho) \simeq Nil_{n-1}^W(\mathbf{R})$.

   Question Is $NK_n(R_\rho) \simeq Nil_{n-1}^W(\mathbf{R})$ for all $n > 1$?

3. Torsion in even-dimensional higher class groups

   Let F be a number field with ring of integers R, Λ any R-order in a semi-simple F-algebra. In [102], (also see 7.4) M. Kolster and R. Laubenbacher proved that for all $n \geq 1$, the only p-torsion possible in $Cl_{2n-1}(\Lambda)$ is for those rational primes p that lie under the prime ideals of R at which Λ is not maximal. The question arises whether a similar result is true for $Cl_{2n}(\Lambda)$ for all $n \geq 1$. Guo and Kuku proved in [74] (also see 7.4) that the result holds for "generalized Eichler orders", which include Eichler orders in quaternion algebras and hereditary orders. It is still open to prove this result for arbitrary orders. Note that when $\Lambda = RG$, G a finite group, the anticipated result implies that p-torsion occurs in $Cl_{2n}(RG)$ only for primes p dividing the order of G.

4. Let F be a number field, D a central division algebra over F, $WK_n(D) n \geq 0$ the wild kernel of D (see definition 7.1.1).

   Question Is $WK_n(D)(l) \simeq divK_n(D)(l)$ for all $n \geq 1$, and any rational prime l?

   In connection with this question, X. Guo and A. Kuku proved in [72] that if $(D : F) = m^2$, then,

   (i) $div(K_n(D))(l) = WK_n(D)(l)$ for all odd primes l and all $n \leq 2$;

   (ii) If l does not divide m, then $divK_3(D)(l) = WK_3(D)(l) = 0$;

   (iii) If $F = Q$ and l does not divide m, then $div\ K_n(D)(l) \subset WK_n(D)(l)$ for all n. (see [72] theorem 3.4).

420 *A.O. Kuku*

Note that the question above is a generalization of a conjecture in [12] by Banaszak et al. that, for any number field F, $WK_n(F)(l) = divK_n(F)(l)$. This conjecture was shown in [12] to be equivalent to the Quillen - Lichtenbaum conjecture under certain hypothesis.

5. Let C be an arbitrary EI category, R a commutative ring with identity.

   Problem Analyze the group structure of $K_n(RC)$.

   Note that A. Kuku showed in [114] (see 7.6) that if R is the ring of integers in a number field, and C a finite EI category, then $K_n(RC)$ is a finitely generated Abelian group.

6. Compute $K_n(R), n \geq 0, R$ an Artinian ring with maximal ideal $\underline{m}$, such that $\underline{m}^r = 0$ for some integer r. Note that $G_n(R)$ is well understood via Devissage. (see Example 6.1.1).

7. Generalize the Devissage theorem to Waldhausen categories.

8. Formulate a profinite K-theory for Waldhausen categories. Note that this has been done by Kuku for exact categories (see [117] or Chapter 8) with applications to orders and groupings.

9. Let X be an H-space, m a positive integer. Then the mod-m Hurentz map $h_n : \pi_n(X, \mathbb{Z}/m) \to H_n(X, \mathbb{Z}/m)$ is defined by $\alpha \in [M_m^n, X] \to \alpha(\varepsilon_n)$ where ε_n is the canonical generator of $H_n(M_m^n, \mathbb{Z}/m)$ corresponding to $1 \in H_n(S^n, \mathbb{Z}/m)$ under the isomorphism $\alpha : H_m(M_m^n, \mathbb{Z}/m) \simeq H_n(S^n, \mathbb{Z}/m) \simeq \mathbb{Z}/m$. Compute the kernel and cokernel of h_n especially for $X = BQC, C = \mathcal{P}(\Lambda), \mathcal{M}(\Lambda)$ where Λ are R-orders, R being the ring of integers in number fields and p-adic fields.

10. Generalize Soule's construction of $H_{\acute{e}t}^k(A, \mu_\varepsilon^{\otimes i})$ for commutative rings to non-commutative rings, e.g., A = maximal order in a central division algebra over number fields and hence to some accessible non-commutative groupings. Then, compute the resulting etále non-commutative Chern characters $K_n(A) \to H_{er}^k(A, \mu_\varepsilon^{\otimes i})$.

11. Generalize the profinite K-theory for exact categories discussed in chapter 8 to Waldhausen categories W and apply this to the category $W = Ch_b(C)$ of bounded chain complexes in an exact category C where $C = \mathcal{P}(\Lambda)$, and $\mathcal{M}(\Lambda)$ and Λ are orders in semi-simple algebras over number fields and p-adic fields.

12. Compute $K_n(RG)$ for various finite and/or discrete groups G. Obtain decompositions for $K_n(RG)$ analogous to those discussed in 7.3.

13. Let $V = G \rtimes_\alpha T$ be a virtually infinite cyclic group, R the ring of integers of a number field F. It is known (see 7.5.5(ii)) that $K_{-1}(RV)$ is a finitely generated Abelian group.

    Question What is the rank of $K_{-1}(RV)$?

14. Prove (or disprove) the Hsiang vanishing conjecture, which says that for any discrete group G, $K_n(\mathbb{Z}G) = 0$ for all $n \le -2$.

    Note This conjecture has been proved by Farrell and Jones for all subgroups of co-compact discrete subgroups of Lie groups (see [55]); also for finite groups by Carter (see theorem 4.4.7 or [32]) and for virtually infinite cyclic groups $V = G \rtimes_\alpha T$. (See theorem 7.5.5 (1).)

15. Let F be a totally real field with ring of integers R, Λ any R-order in a semi-simple F-algebra.

    Question Is $K_{4n+3}(\Lambda)$ a finite group for all $n \ge 1$?

    Note The above result is true for Λ commutative.

16. Let R be a discrete valuation ring with unique maximal ideal $\underline{m} = sR$. Let F be the quotient field of R. Then, $F = R(1/s)$. Put $k = R/\underline{m}$. Gerstein's conjecture says that for all $n \ge 0$, we have a short exact sequence

$$0 \to K_n(R) \xrightarrow{\alpha_n} K_n(F) \xrightarrow{\beta_n} K_{n-1}(k) \to 0.$$

    This conjecture is known to be true in several cases outlined in Examples 6.2.1(iii).

    Problem Prove (or disprove) this conjecture for the case $char(R) = 0$ or $char(k) = p$.

17. Let R be the ring of integers in a number field F, Λ any R-order in a semi-simple F-algebra Σ, $\underline{p}$ a prime ideal of R, ℓ a rational prime.

    Question Is $K_n^{pr}(\hat{\Lambda}_{\underline{p}}, \mathbb{Z}_\ell)$ ℓ-complete? Note that it is weakly ℓ-complete (see remarks 8.3.2).

18. Let A be a ring and S a central multiplicative system in A. Let $\mathcal{H}_S(A)$ be the category of finitely generated S-torsion A-modules of finite homological dimension.

    Problem Obtain a good understanding of $K_n(\mathcal{H}_S(A))$ for various rings A and, in particular, for $A = RG$, G a finite group, R the ring of integers in a number field or p-adic field.

    Let $\mathcal{B}$ be the category of modules $M \in \mathcal{H}_S(A)$ with resolution $0 \to R^n \to R^n \to M \to 0$. Is $0 \to K_n(\mathcal{B}) \to K_n(\mathcal{H}_S(A)) \to K_n(A) \to K_n(A_S)$ exact for all $n \ge 1$? This is true for $n = 0$.

References

[1] J.F. Adams, *Vector fields on spheres*, Ann. Math. **75** (1962) 603 - 632.

[2] G. Allday and V. Puppe, *Cohomological Methods in Transformation Groups*, Cambridge Univ. Press (1993).

[3] R.C. Alperin, R.K. Dennis, and M. Stein, The Non-triviality of $SK_1(\mathbb{Z}G)$, Lect. Notes in Math. **353** (1973), Springer-Verlag, 1 - 7.

[4] R.C. Alperin, R.K. Dennis, and M. Stein, SK_1 *of finite Abelian groups*, Invent. Math. **86** (1986).

[5] R.C. Alperin, R.K. Dennis, R. Oliver and M. Stein, SK_1 *of finite Abelian groups II*, Invent. Math. **87** (1987) 253 - 302.

[6] S. Araki and H. Toda, *Multiplicative sructures on mod-q cohomology theories*, Osaka J. Math. (1965) 71 - 115.

[7] D. Arlettaz, *Algebraic K-theory of rings from a topological view point*, Publ. Mat. **44** (2000) 3 - 84.

[8] D. Arlettaz and G. Banaszak, *On the non-torsion elements in the algebraic K-theory of rings of integers*, J. Reine Angrew. Math. **461** (1995) 63 - 79.

[9] E. Artin and J. Tate, *Class Field Theory*, W.A. Benjamin, New York, 1967.

[10] M.F. Atiyah, *K-theory*, W.A. Benjamin, New York.

[11] A. Bak and U. Rehmann, *The congruence subgroup of metaplectic problems for $SL_{n \geq 2}$ of division algebras*, J. Algebra **78** (1982) 475 - 547.

[12] G. Banaszak, *Generalisations of the Moore exact sequences of the wild kernel for higher K-groups*, Composito Math. **86** (1993) 281 - 305.

[13] G. Banaszak and P. Zelewski, Continuous K-theory, K-theory J. **9** (1995) 379 - 393.

[14] A. Bartels and W. Lück, *Induction theorems and isomorphism conjectures for K- and L-theory* (preprint).

[15] A. Bartels, F.T. Farrell, L.E. Jones, and H. Reich, *On the isomorphism conjecture in Algebraic K-theory*, Topology **43**(1) (2004) 157 - 231.

424   *References*

[16] A. Bartels and H. Reich, *On the Farrell-Jones conjecture for higher Algebraic K-theory*, Preprint SFB 478, Munsten, 2003.

[17] H. Bass, *K-theory and stable algebra*, IHES **22** (1964) 1 - 60.

[18] H. Bass, *Dirichlet unit theorem, induced character and Whitehead group of finite groups*, Topology **4** (1966) 391 - 410.

[19] H. Bass, *Lenstra's calculation of $G_0(R\pi)$ and applications to Morse-Smale diffeomorphisms*, SLN **88** (1981) 287 - 291.

[20] H. Bass, *Algebraic K-theory*, W.A. Benjamin, New York, 1968.

[21] H. Bass, J. Milnor and J.P. Serre, *Solution of the congruence subgroup problem for $SL_n(n \geq 3)$ and $SP_{2n}(n \geq 2)$*, IHES **33** (1967) 59 - 137.

[22] H. Bass and M.P. Murthy, *Gröthendieck groups and Picard groups of Abelian group-rings*, Ann. Math. **86** (1967) 16 - 73.

[23] H. Bass, A.O. Kuku, and C. Pedrini (Eds.), *Algebraic K-theory and its applications*, ICTP K-theory Proceedings, World Scientific, 1999.

[24] P. Baum, A. Connes, and N. Higson, *Classifying space for proper actions and K-theory of group C^*-algebras*, Contemp. Math. AMS **167** (1994) 240 - 291.

[25] J. Berrick, *An Approach to Algebraic K-theory*, Pitman, 1982.

[26] A. Borel, *Stable real cohomology of arithmetic groups*, Ann. Sci. Ecole Norm. Sup. (4) **7** (1994) 235 - 272.

[27] A. Borel and J.P. Serre, *Corners and arithmetic groups*, Comm. Math. Helv. **48** (1974) 436 - 491.

[28] N. Boubaki, *Algebre Commutative*, Herman, Paris, 1965.

[29] G. Bredon, *Introduction to Compact Transformation Groups*, Academic Press, New York, 1972.

[30] W. Browder, *Algebraic K-theory with coefficients $\mathbb{Z}/p$*, Lect. Notes in Math. **657** Springer-Verlag, New York, 40 - 84.

[31] A. Cartan and S. Eilenberg, *Homological Algebra*, Princeton Univ. Press, Princeton, 1956.

[32] W. Carter, *Localization in lower Algebraic K-theory*, Comm. Algebra **8**(7) 603 - 622.

[33] R. Charney, *A note on excision in K-theory*, Lect. Notes in Math. **1046** (1984) Springer-Verlag, New York, 47 - 54.

[34] S. Chase, D.K. Harrison and A. Rosenberg, *Galois theory and Galois cohomology of commutative rings*, Mem. Am. Math. Soc. **52**, 15 - 33.

[35] A. Connes, *Non-commutative Geometry*, Academic Press, New York, London, 1994.

[36] F. Connolly and M. Da Silva, *The groups $NK_0(\mathbb{Z}\pi)$ are finite generated $\mathbb{Z}[N]$ modules if π is a finite group*, Lett. Math. Phys. **9** (1995) 1 - 11.

[37] F.X. Connelly and S. Prassidis, *On the exponent of NK_0-groups of virtually infinite cyclic groups* (preprint).

[38] C.W. Curtis and I. Reiner, *Methods of Representation Theory I*, John Wiley & Sons, New York, 1981.

[39] C.W. Curtis and I. Reiner, *Methods of Representation Theory II*, John Wiley & Sons, New York, 1987.

[40] J.F. Davis and W. Lück, *Spaces over a category and assembly maps in isomorphism conjectures in K- and L-theory*, K-theory **15** (1998) 201 - 252.

[41] R.K. Dennis, M.E. Keating and M.R. Stein, *Lower bounds for the order of $K_2(\mathbb{Z}G)$ and $Wh_2(G)$*, Math. Ann. **233** (1976) 97 - 103.

[42] R.K. Dennis and R.C. Laubenbacher, *Homogeneous functions and Algebraic K-theory* (preprint).

[43] D.N. Diep, A.O. Kuku and N.Q. Tho, *Non-commutative Chern characters of compact Lie group C^*-algebras*, K-theory **17** (1999) 195 - 208.

[44] D.N. Diep, A.O. Kuku and N.Q. Tho, *Non-commutative Chern Characters of compact quantum groups*, J. Algebra **226** (2000) 311 - 331.

[45] A.W.M. Dress, *A characterization of solvable groups*, Math. Z. **110** (1969) 213 - 217.

[46] A.W.M. Dress, *The Witt ring as a Mackey functor*, Appendix A to Belefeld notes on Representation theory of finite groups (1970).

[47] A.W.M. Dress, *Vertices of integral representation*, Math. Z. **114** (1977) 159 - 169.

[48] A.W.M. Dress, *Contributions to the theory of induced representations*, Algebraic K-theory II, Springer Lect. Notes **342** (1973) 183 - 240.

[49] A.W.M. Dress, *On relative Gröthendieck rings*, Springer Lect. Notes **448** (1975) 79 - 131.

[50] A.W.M. Dress, *Induction and structure theorems for orthogonal representations of finite groups*, Ann. Math. **102** (1975) 291 - 325.

[51] A.W.M. Dress, *The weak local-global principle in algebraic K-theory*, Comm. Algebra **3**(7) (1975) 615 - 661.

[52] A.W.M. Dress and A.O. Kuku, *A convenient setting for equivariant higher algebraic K-theory*, Proc. of the Oberwolfach Conference (1980), Springer Lect. Notes **966** (1982) 59 - 68.

[53] A.W.M. Dress and A.O. Kuku, *The Cartan map for equivariant higher algebraic K-groups*, Comm. Algebra **9**(7) (1981) 727 - 746.

[54] F.T. Farrell and L.E. Jones, *Isomorphic conjectures in algebraic K-theory*, J. Am. Math. Soc. **6** (1993) 249 - 297.

[55] F.T. Farrell and L.E. Jones, *The lower algebraic K-theory of virtually infinite cyclic groups*, K-theory **9** (1995) 13 - 30.

[56] G. Garkusha, *On the homotopy cofibre spectrum of $K(R) \to G(R)$*, J. Algebra Appl. (1), **3** (2002) 327 - 334.

[57] G. Garkusha, *Systems of diagram categories and K-theory*, Math. Z. (to appear).

[58] S.M. Gersten, *Higher K-theory of rings*, Lect. Notes in Math., Springer-Verlag **341**, 43 - 56.

[59] S.M. Gersten, *Some exact sequences in the higher K-theory of rings*, SLN **341** (1973) 211 - 244.

[60] S.M. Gersten, *The localization theorem for projective modules*, Comm. Algebra **2** (1974) 307 - 350.

[61] D. Goswami and A.O. Kuku, *A complete formulation of the Baum - Connes conjecture for the action of discrete quantum groups*, K-theory **30** (2003) 341 - 363.

[62] D. Grayson, *Higher Algebraic K-theory II* (after Daniel Quillen), Lect. Notes in Math. **551** (1976) 217 - 240.

[63] J.A. Green, *On the indecomposable representations of a finite group*, Math. Z. **70** (1959) 430 - 445.

[64] J.A. Green, *Axiomatic representation theory for finite groups*, J. Pure Appl. Algebra **1**(1) (1971) 41 - 77.

[65] S. Green, D. Handelman and P. Roberts, *K-theory of finite dimensional division algebras*, J. Pure Appl. Algebra **12** (1978) 153 - 158.

[66] J.P.C. Greenlees, *Some remarks on projective Mackey functors*, J. Pure Appl. Algebra **81** (1992) 17 - 38.

[67] J.P.C. Greenlees and J.P. May, *Generalized Tate cohomology*, Mem. AMS **543** (1995).

[68] J.P.C. Greenlees, *Rational Mackey functors for compact Lie groups*, Proc. Lond. Math. Soc. **76**(3) (1998) 549 - 578.

[69] J.P.C. Greenlees, J.P. May, and J.M. James (Eds.), Equivariant stable homotopy theorem, *Handbook of Algebraic Topology* (1999), Elsevier, Amsterdam.

[70] J.P.C. Greenlees and J.P. May, *Some remarks on the structure of Mackey functors*, Proc. Am. Math. Soc. **115**(1), 237 - 243.

[71] M. Gromov, *Spaces and Questions*, Geom. Funct. Anal. (special volume Part I) GAFA (2000) 118 - 161.

[72] X. Guo and A.O. Kuku, *Wild kernels for higher K-theory of division and semi-simple algebras*, Beitrage zür Algebra und Geometrie (to appear).

[73] X. Guo, A.O. Kuku, and H. Qin, K_2 *of division algebras*, Comm. Algebra **33**(4) (2005) 1073 - 1081.

[74] X. Guo and A.O. Kuku, *Higher class groups of generalised Eichler orders*, Comm. Algebra **33** (2005) 709 - 718.

[75] X. Guo and A.O. Kuku, *Higher class groups of locally triangular orders over number fields* (preprint).

[76] I. Hambleton, L. Taylor and B. Williams, *On $G_n(RG)$ for G a finite nilpotent group*, J. Algebra **116** (1988) 466 - 470.

[77] I. Hambleton and E.K. Pedersen, *Identifying assembly maps in K- and L-theory*, Math. Ann. **328**(1) (2004) 27 - 58.

[78] D. Harmon, NK_1 *of finite groups*, Proc. Am. Math. Soc. **100**(2) (1987) 229 - 232.

[79] R. Hartshore, *Algebraic Geometry*, Springer-Verlag, New York, 1977.

[80] A. Hatcher and J. Wagoner, *Pseudo-isotopies on comapct manifolds*, Asterisq **6** (1973).

[81] A. Heller, *Some exact sequences in algberaic K-theory*, Topology **4** (1965) 389 - 408.

[82] A. Heller and I. Reiner, *Gröthendieck groups of orders in semi-simple algebras*, Trans. AMS **112** (1964) 344 - 355.

[83] D.G. Higman, *The units of group rings*, Proc. Lond. Math. Soc. **46** (1940) 231 - 248.

[84] N. Higson and G. Kasparov, *E-theory and KK-theory for groups which act properly and isometrically on Hilbert spaces*, Invent. Math. **144**(1) (2001) 23 - 74.

[85] N. Higson, V. Lafforgue and G. Skandalis, *Counter-examples to the Baum-Connes conjecture*, Geom. Funct. Anal. **12**(2) (2002) 330 - 354.

[86] O.J. Hilton and U. Stammback, *A Course in Homological Algebra*, Springer-Verlag, Heidelberg, 1971.

[87] W.C. Hsiang, *Geometric applications of Algebraic K-theory*, Proc. Int. Congress Math. Warsaw **1** (1983) 99 - 118.

[88] H. Innasaridze, *Algebraic K-theory*, Kluwer Publisher, 1995.

[89] N. Jacobson, *Basic Algebra II*, Freeman & Co., 1980.

[90] G.D. James, *The representation theory of the symmetric groups*, Lect. Notes in Math. **682**, Springer-Verlag, 1976.

[91] U. Jannsen, *Continuous etale cohomology*, Math. Ann. **280** No. 988, 207 - 245.

[92] G. Janus, *Algebraic Number Fields*, Academic Press, London, 1973.

[93] D. Juan-Pineda and S. Prassidis, *On the lower Nil groups of Waldhausen*, Forum Math. **13**(2) (2001) 261 - 285.

[94] D. Juan-Pineda and S. Prassidis, *On the Nil groups of Waldhausen Nils'*, Topology Appl. **146/147** (2005) 489 - 499.

[95] M. Karoubi, *K-theory: An Introduction*, Springer-Verlag, Heidelberg, 1978.

[96] M. Karoubi, A.O. Kuku, and C. Pedrini (Eds.), *Contempory developments in algebraic K-theory*, ICTP Lect. Notes Series (2003), No. 15, viii-536.

[97] G.G. Kasparov, *Equivariant KK-theory and the Nivikov conjecture*, Invent. Math. **91** (1988) 147 - 201.

[98] B. Keller, Derived categories and their uses, *Handbook of Algebra*, Vol. 1, 671 - 701.

[99] M.E. Keating, *Values of tame symbols on division algebra*, J. Lond. Math. Soc. **2**(14) (1976) 25 - 30.

[100] M.E. Keating, G_1 *of integral group-rings*, J. Lond. Math. Soc. **14** (1976) 148 - 152.

[101] M.E. Keating, *K-theory of triangular rings and orders*, Lect. Notes in Math., Springer-Verlag, New York, Berlin (1984) 178 - 192.

[102] M. Kolster and R.C. Laubenbacher, *Higher class groups of orders*, Math. Z. **228** (1998) 229 - 246.

[103] A.O. Kuku, *Some algebraic K-theoretic applications of LF and NF functors*, Proc. Am. Math. Soc. **37**(2) (1973) 363 - 365.

[104] A.O. Kuku, *Whitehead group of orders in p-adic semi-simple algebras*, J. Alg. **25** (1973) 415 - 418.

[105] A.O. Kuku, *Some finiteness theorems in the K-theory of orders in p-adic algebras.* J. Lond. Math. Soc. **13**(1) (1976) 122 - 128.

[106] A.O. Kuku, SK_n *of orders and* G_n *of finite rings*, Lect. Notes in Math. **551**, Springer-Verlag (1976) 60 - 68.

[107] A.O. Kuku, SG_n *of orders and group rings*, Math. Z. **165** (1979) 291 - 295.

[108] A.O. Kuku, *Higher algebraic K-theory of group rings and orders in algebras over number fields*, Comm. Algebra **10**(8) (1982) 805 - 816.

[109] A.O. Kuku, *Equivariant K-theory and the cohomology of profinite groups*, Lect. Notes in Math. **1046** (1984), Springer-Verlag, 234 - 244.

[110] A.O. Kuku, *K-theory of group rings of finite groups over maximal orders in division algebras*, J. Algebra **91**(1) (1984) 18 - 31.

[111] A.O. Kuku, *Axiomatic theory of induced representations of finite groups*, Les cours du CIMPA, No. 5, Nice, France, 1985.

[112] A.O. Kuku, K_n, SK_n *of integral group rings and orders*, Contemp. Math. AMS **55** (1986) 333 - 338.

[113] A.O. Kuku, *Some finiteness results in the higher K-theory of orders and group-rings*, Topology Appl. **25** (1987) 185 - 191.

[114] A.O. Kuku, *Higher K-theory of modules over EI categories*, Afrika Mat. **3** (1996) 15 - 27.

[115] A.O. Kuku, *Ranks of K_n and G_n of orders and group rings of finite groups over integers in number fields*, J. Pure Appl. Algebra **138** (1999) 39 - 44.

[116] A.O. Kuku, *Equivariant higher K-theory for compact Lie group actions*, Bietrage zür Algebra und Geometrie **41**(1) (2000) 141 -150.

[117] A.O. Kuku, *Profinite and continuous higher K-theory of exact categories, orders and group-rings*, K-theory **22** (2001) 367 - 392.

[118] A.O. Kuku, Classical algebraic *K*-theory: the functors K_0, K_1, K_2, *Handbook of Algebra* **3** (2003) Elsevier, 157 - 196.

[119] A.O. Kuku, *K-theory and representation theory – contempory developments in algebraic K-theory I*, ICTP Lect. Notes Series, No. 15 (2003) 259 - 356.

[120] A.O. Kuku, Higher Algebraic *K*-theory, *Handbook of Algebra*, **4** (2006) Elsevier, 3 - 74.

[121] A.O. Kuku, *Finiteness of higher K-groups of orders and group-rings*, K-theory (2006), paper has appeared online.

[122] A.O. Kuku, *Equivariant higher algebraic K-theory for Waldhausen categories*, Bietrage zür Algebra und Geometrie – Contributions to Algebra and Geometry (to appear).

[123] A.O. Kuku and G. Tang, *Higher K-theory of group-rings of virtually infinite cyclic groups*, Math. Ann. **325** (2003) 711 - 725.

[124] A.O. Kuku and M. Mahdavi-Hezavehi, *Subgroups of $GL_1(R)$ for local rings R*, Commun. in Algebra **32**(15) (2004), 1895 - 1902.

[125] A.O. Kuku and G. Tang, *An explicit computation of "bar" homology groups of a non-unital ring*, Beitrage zür Algebra und Geometrie **44**(2) (2003), 375 - 382.

[126] T.Y. Lam, *Induction techniques for Gröthendieck groups and Whitehead groups of finite groups*, Ann. Sci. Ecole Norm. Sup. Paris **1** (1968) 91 - 148.

[127] T.Y. Lam, *Artin exponent for finite groups*, J. Algebra **9** (1968) 94 - 119.

[128] T.Y. Lam and I. Reiner, *Relative Gröthendieck rings*, Bull. Am. Math. Soc. (1969) 496 - 498.

[129] T.Y. Lam and I. Reiner, *Relative Gröthendieck groups*, J. Algebra **11** (1969) 213 - 242.

[130] T.Y. Lam and I. Reiner, *Restriction maps on relative Gröthendieck groups*, J. Algebra **14** (1970) 260 - 298.

[131] R.C. Laubenbacher and D. Webb, *On SG_n of orders*, J. Algebra **133** (1990) 125 - 131.

[132] R. Lee and R.H. Szcarba, *The group $K_3(\mathbb{Z})$ is cyclic of order 48*, Ann. Math. **104** (1976) 31 - 60.

[133] H.W. Lenstra, *Gröthendieck groups of Abelian group rings*, J. Pure Appl. Algebra **20** (1981) 173 - 193.

[134] H. Lindner, *A remark on Mackey functors*, Manuscripta Math. **18** (1976) 273 - 278.

[135] J.L. Loday, *Cyclic Homology*, Springer-Verlag, Berlin, 1992.

[136] J.L. Loday, *K-theorie algebrique et representation de groupes*, Ann. Sci. Ecole Norm. Sup. 4eme series **9** (1976) 309 - 377.

[137] W. Lück, *Transformation groups and algebraic K-theory*, Springer-Verlag, Berlin, Heidelberg, New York, LN408, 1989.

[138] W. Lück and H. Reich, The Baum - Connes and the Farrell - Jones conjectures in *K*- and *L*-theory, *Handbook of K-theory II*, Vol. 2, 703 - 842 (2005).

[139] S. Maclane, *Natural associativity and commutativity*, Rice University Studies (1963).

[140] S. Maclane, *Homology*, Springer-Verlag, Berlin, 1963.

[141] S. Maclane, *Categories for the Working Mathematician*, Springer-Verlag, Berlin, 1971.

[142] B.A. Magurn, *An Algebraic Introduction to K-theory*, Cambridge University Press, 2002.

[143] B.A. Magurn, SK_1 *of dihedral groups*, J. Algebra **51** (1978) 399 - 415.

[144] B.A. Magurn, *Explicit K_2 of some finite group-rings* (preprint).

[145] M. Mandell and J.P. May, *Equivariant orthogonal spectra and S-modules*, Mem. AMS **159** (2002) 755.

[146] M. Mathey and G. Mislin, *Equivariant K-homology and restriction to finite cyclic subgroups* (preprint).

[147] J.P. May et al., *Equivariant homotopy and cohomology theories*, NSE-CBNS Regional Conference series in Math. **91** (1996).

[148] A.S. Merkurjev, *K-theory of simple algebras*, Proc. Symp. Pure Math. **58**(1), 65 - 83.

[149] A.S. Merkurjev, K_2 *of fields and the Brauer group*, Contemp. Math. AMS **55** (1983) 529 - 546.

[150] A.S. Merkurjev and A.A. Suskin, *K-cohomology of Severi-Brauer varieties and norm residue homomorphisms*, Izv. Akad. Nauk USSR **46**, 1011 - 1096.

[151] R. Meyer and R. Nest, *The Baum-Connes conjecture via localisation of categories* (preprint).

[152] J. Milnor, *On axiomatic homology theory*, Pacific J. Math. **12** (1962) 337 - 345.

[153] J. Milnor, *Whitehead torsion*, Bull. Am. Math. Soc. **72** (1966) 358 - 426.

[154] J. Milnor, *Algebraic K-theory and quadratic forms*, Invent. Math. **9** (1970) 318 - 344.

[155] J. Milnor, *Introduction to algebraic K-theory*, Princeton University Press, Princeton, 1971.

[156] G. Mislin and A. Vallette, *Proper Group Actions and the Baum - Connes Conjecture*, Advanced Courses in Mathematics, CRM Barcelona Birkhäuser, 2003.

[157] C. Moore, *Group extensions of p-adic and adelic linear groups*, Publ. Math. IHES **353**.

[158] J. Neisendorfer, *Primary homotopy theory*, Mem. Am. Math. Soc. **232** (1980) AMS Providence.

[159] R. Oliver, *Whitehead groups of finite groups*, Cambridge University Press, Cambridge, 1988.

[160] P.A. Ostvaer, *Calculation of two-primary Algebraic K-theory of some group-rings*, K-theory **16** (1999) 391 - 397.

[161] F. Peterson, *Generalized cohomology groups*, Am. J. Math. **78** (1956) 259 - 282.

[162] D. Quillen, *On the cohomolgy and K-theory of the general linear groups over a finite field*, Ann. Math. **96** (1972) 552 - 586.

[163] D. Quillen, *Cohomology of groups*, Proc. ICM Nice, 1970 **2** (1971), Gauthier-Villars, Paris, 47 - 52.

[164] D. Quillen, *Higher K-theory for categories with exact sequences. New developments in topology*, London Math. Soc. Lect. Notes **11** (1974), Cambridge University Press, 95 - 110.

[165] D. Quillen, *Higher algebraic K-theory I*, K-theory I, Lect. Notes in Math. **341** (1973), Springer-Verlag, 85 - 147.

[166] D. Quillen, *Finite generation of the groups K_i of rings of algebraic integers*, Algebraic K-theory I Lect. Notes in Math. **341** (1973), Springer-Verlag, 195 - 214.

[167] D. Quillen, *Projective modules over polynomial rings*, Invent. Math. **36** (1976) 167 - 171.

[168] A. Ranicki, *Lower K- and L-theory*, Cambridge University Press, Cambridge, 1992.

[169] U. Rehmann and U. Stuhler, *On K_2 of finite dimensional division algebras over arithmetic fields*, Invent. Math. **50** (1978) 75 - 90.

[170] I. Reiner, *Integral representations of cyclic groups of prime order*, Proc. Am. Math. Soc. **8** (1957) 142 - 146.

[171] I. Reiner, *Maximal Orders*, Academic Press, London, 1975.

[172] I. Reiner, *Integral representation ring of finite groups*, Mich. Math. J. **12** (1965) 11 - 22.

[173] I. Reiner, *A survey of integral representation theory*, Bull. Am. Math. Soc.

[174] I. Reiner and K.W. Roggenkamp, *Integral representations*, Springer-Verlag Lect. Notes **744**, Berlin (1979).

[175] I. Reiner and S. Ullom, *Class groups of integral group-rings*, Trans. AMS **170** (1972) 1 - 30.

[176] S. Rim, *Modules over finite groups*, Ann. Math. **69** (1959) 700 - 712.

[177] S. Rim, *Projective class groups*, Trans. AMS **98** (1961) 459 - 467.

[178] K.W. Roggerkamp, *Lattices over orders II*, Springer Lect. Notes **142** (1970).

[179] J. Rognes, $K_4(\mathbb{Z})$ *is the trivial group*, Topology **39** (2000) 267 - 281.

[180] J. Rognes and C. Weibel, *Two primary Algebraic K-theory of rings of integers in number fields*, J. Am. Math. Soc. **13** (2000) 1 - 54.

[181] J. Rosenberg, *Algebraic K-theory and its Applications*, Springer-Verlag, Berlin, 1994.

[182] W. Scharlan, *Quadratic and Hermitian Forms*, Springer-Verlag, Berlin, 1985.

[183] P. Scott and C.T.C. Wall, *Topological methods in group theory*, LMS Lect. Notes series **36** (1979).

[184] G. Segal, *Equivariant K-theory*, Publ. Math. J. IHES **34** (1968).

[185] G. Segal, *Categories and cohomology theories*, Topology **13** (1974) 293 - 312.

[186] G. Segal, *Representation ring of compact Lie group*, IHES **34** (1968) 113 - 128.

[187] J.P. Serre, *Linear Representation of Finite Groups*, Springer-Verlag, New York, 1977.

[188] J.P. Serre, *Modules projectifs et espaces fibres vectorielle*, Expose 23 Seminaire P. Dubreil 1957/1958, Secretariat Mathematique, Paris (1958).

[189] J.P. Serre, *Arithmetic groups in homological group theory*, LMS Lect. Notes **36**, 105 - 136.

[190] J.P. Serre, *Local Fields*, Springer-Verlag, Berlin, 1979.

[191] C. Sherman, *Group representations and Algebraic K-theory*, Lect. Notes in Maths. **966**, Springer-Verlag, New York, Berlin (1982) 208 - 243.

[192] K. Shimakawa, *Multiple categories and algebraic K-theory*. J. Pure Appl. Algebras **41** (1986) 285 - 300.

[193] K. Shimakawa, *Mackey structures on equivariant algebraic K-theory*, K-theory **5** (1992) 355 - 371.

[194] K. Shimakawa, *Infinite loop G-spaces asscociated to monordial G-graded categories*, **25** (1989), RIMS Kyoto University, 239 - 262.

[195] L. Siebenmann, *Obstruction to finding a boundary for an open manifold of dimension greater than two*, Princeton University Thesis (1965); *The structure of tame ends*, Notices AMS **13** (1968) 862.

[196] C. Soulé, *Groupes de Chon et K-theorie des varietes sur un corps fini,* Math. Ann. **268** (1984) 317 - 345.

[197] E.A. Spanier, *Algebraic Topology,* McGraw-Hill, 1966.

[198] V. Srinivas, *Algebraic K-theory,* Prog. Math. **90** (1991) Birkhauser.

[199] D. Sullivan, *Genetics of homotopy theory and the Adam's conjecture,* Ann. Math. (1966).

[200] A.A. Suslin, *Excision in integral algebraic K-theory,* Proc. Steklov Inst. Math. **208** (1995) 255 - 279.

[201] A.A. Suslin, *Homology of GL_n, characteristic classes and Milnor K-theory,* Springer Lect. Notes **1046** (1984) 357 - 375.

[202] A.A. Suslin, *Stability in Algebraic K-theory,* Lect. Notes in Math. **966** Springer-Verlag (1982) 304 - 333.

[203] A.A. Suslin, *Projective modules over polynomial rings,* Dokl Akad. Nauk **219** (1976) 221 - 238.

[204] A.A. Suslin and A.V. Yufryakov, *K-theory of local division algebras,* Soviet Math Docklady **33** (1986) 794 - 798.

[205] A.A. Suslin and M. Wodzicki, *Excision in algebraic K-theory,* Ann. Math. (2) **136**(1), (1992) 51 - 122.

[206] R. Swan, *Projective modules over finite groups,* Bull. Am. Math. Soc. **65** (1959) 365 - 367.

[207] R. Swan, *Induced representation and projective modules,* Ann. Math. **71** (1960) 552 - 578.

[208] R. Swan, *Vector bundles and projective modules,* Trans. AMS **105** (1962) 264 - 277.

[209] R. Swan, *Gröthendieck ring of a finite group,* Topology **2** (1963) 85 - 110.

[210] R. Swan, *Nonabelian homological algebras and K-theory,* Proc. Symp. Pure Math. **XVII** (1970) 88 - 123.

[211] R. Swan, *Algebraic K-theory,* Proc. ICM Nice, 1970 **I** (1971), Cauthier-Villar, Paris, 191 - 199.

[212] R. Swan, *Excision in algebraic K-theory,* J. Pure Appl. Algebra **1** (1971) 221 - 252.

[213] R. Swan, *K-theory of finite groups and orders,* Springer-Verlag Lect. Notes **149** (1970).

[214] R. Swan, *Topological examples of projective modules,* Trans. AMS **230** (1977) 201 - 234.

[215] R. Swan, *Algebraic K-theory*, Lect. Notes in Math. **76** (1968), Springer-Verlag.

[216] R.W. Thomason and T. Trobaugh, *Higher Algebraic K-theory of schemes and derived categories*, The Gröthendieck Feschrift II. Progress in Math. Vol. 88. Birkhauser (1990) 247 - 435.

[217] T. Tom-Dieck, *Equivariant homology and Mackey functors*, Math. Ann. (1973).

[218] T. Tom-Dieck, *Transformation groups and representation theory*, Springer-Verlag Lect. Notes **766** (1979).

[219] T. Tom-Dieck, *Transformation Groups*, Walter de Grugster, Berlin, 1987.

[220] W. van der-Kallen, *Homology stability for linear groups*, Invent. Math. **60** (1980) 269 - 295.

[221] L.N. Vaserstein, *K-theory and the congruence subgroup problems*, Math. Notes **5** (1969) 141 - 148. Translated from Mat. Zamethi **5** (1969) 233 - 244.

[222] M.F. Vigneras, *Arithmetique des algebras de quaternions*, Lect. Notes in Math. **800** Springer-Verlag (1980).

[223] J.B. Wagoner, *Continuous cohomology and p-adic K-theory*, Lect. Notes in Math. **551**, Springer-Verlag, New York, 241 - 248.

[224] F. Waldhausen, *Algebraic K-theory of generalized free products*, Ann. Math. **108** (1978) 135 - 256.

[225] F. Waldhausen, *Algebraic K-theory of spaces*, Lect. Notes in Math. **1126** Springer-Verlag, 1985.

[226] C.T.C. Wall, *Finiteness conditions for CW-complexes*, Ann. Math. **81** (1965) 56 - 69.

[227] C.T.C. Wall, *Survey of non-simply-connected manifolds*, Ann. Math. **84** (1966) 217 - 276.

[228] C.T.C. Wall, *Foundations of algebraic L-theory*, Algebraic K-theory III, Lect. Notes in Math. **343** (1973), Springer-Verlag, 266 - 300.

[229] C.T.C. Wall, *Norms of units in group rings*, Proc. Lond. Math. Soc. **29** (1974) 593 - 632.

[230] D. Webb, *Gröthendieck groups of dihedral and quaternion group rings*, J. Pure Appl. Algebra **35** (1985) 197 - 223.

[231] D. Webb, *The Lenstra map on classifying spaces and G-theory of group rings*, Invent. Math. **84** (1986) 73 - 89.

[232] D. Webb, *Quillen G-theory of Abelian group rings*, J. Pure Appl. Algebra **39** (1986) 177 - 195.

[233] D. Webb, *Higher G-theory of nilpotent group rings*, J. Algebra (1988) 457 - 465.

[234] D. Webb, *G-theory of group rings for groups of square-free order*, K-theory **1** (1987) 417 - 422.

[235] D. Webb and D. Yao, *A simple counter-example to the Hambleton-Taylor-Williams conjecture K- theory*, **7**(6) (1993) 575 - 578.

[236] C. Weibel, *Mayer-Vietorics sequence and mod-p K-theory*, Lect. Notes in Math. **966** (1982), Springer-Verlag, New York, 390 - 407.

[237] C. Weibel, *Mayer-Vietorics sequences and module structures on NK_2*, Lect. Notes in Math., Springer-Verlag, New York, 466 - 493.

[238] C. Weibel, *Introduction to Homological Algebra*, Cambridge University Press, 1998.

[239] C. Weibel, *Etale Chern characters at prime 2 Algebraic K-theory and Algebraic Topology*, NATO ASI series C. **407**, Kluwer (1993) 249 - 286.

[240] C. Weibel, *Virtual K-book*, http://math.rutgers.edu/weibel/kt.

[241] E. Weiss, *Algebraic Number Theory*, McGraw-Hill, New York, 1963.

[242] M. Weiss and B. Williams, *Automorphisms of manifolds and Algebraic K-theory*, K-theory **1**(6) (1988) 575 - 626.

[243] G.W. Whitehead, *Elements of Homotopy Theory*, Springer-Verlag, New York, 1978.

[244] J.H.C. Whitehead, *Simplicial spaces, nucleii and M-groups*, Proc. Lond. Math. Soc. **45** (1939) 243 - 327.

[245] J.H.C. Whitehead, *Simple homotopy types*, Am. J. Math. **72** (1950) 1 - 57.

Index

Milton Keynes UK
Ingram Content Group UK Ltd.
UKHW021902071024
449327UK00021B/1603